A TEXTBOOK OF SOUND

MAXWELL PRESS

A TEXTBOOK OF SOUND

A.B. WOOD D.Sc. (Manchester), F.Inst.P.

Senior Scientific Officer, Admiralty Research Laboratory, Teddington, Middlesex; Member of Council of the Physical Society of London; late Oliver Lodge Fellow and Lecturer in Physics, Liverpool University

MAXWELL PRESS

Chennai Trichy New Delhi

MAXWELL PRESS

This edition has been published in india by arrangement with Carson Books, UK

ISBN 978-93-90063-59-8 Maxwell Press

No. 44, Nallathambi Street,
Triplicane, Chennai 600 005

MJP 905

Publisher's Note

The legacy of a country is in its varied cultural heritage, historical literature, developments in the field of economy and science. The top nations in the world are competing in the field of science, economy and literature. This vast legacy has to be conserved and documented so that it can be bestowed to the future generation. The knowledge of this legacy is slowly getting perished in the present generation due to lack of documentation.

Keeping this in mind, the concern with retrospective acquiring of rare books has been accented recently by the burgeoning reprint industry. Maxwell Press is gratified to retrieve the rare collections with a view to bring back those books that were landmarks in their time.

In this effort, a series of rare books would be republished under the banner, "Maxwell Press". The books in the reprint series have been carefully selected for their contemporary usefulness as well as their historical importance within the intellectual. We reconstruct the book with slight enhancements made for better presentation, without affecting the contents of the original edition.

Most of the works selected for republishing covers a huge range of subjects, from history to anthropology. We believe this reprint edition will be a service to the numerous researchers and practitioners active in this fascinating field. We allow readers to experience the wonder of peering into a scholarly work of the highest order and seminal significance.

MAXWELL PRESS

PREFACE

THE change which has taken place during the past ten or twenty years in the study of mechanical vibrations and sound is my excuse for writing this book when such treatises as those of Rayleigh and Lamb already exist. These treatises, founded so firmly on sound mechanical and mathematical principles, must for a long time to come be regarded as the basis of further developments. Since Rayleigh's *Sound* first appeared many treatises have been written, but, with the exception of a few which have been published since I commenced to write the present volume, the majority have dealt with the subject in the time-honoured way with insufficient reference to the change in the methods of treatment which has been so conspicuous in recent years. I have consequently followed, whenever possible, the theoretical treatment of Rayleigh and Lamb, but have also endeavoured to deal with the ever-increasing bulk of new data and methods of investigation which is now available. Whilst writing the book my doubts have increased regarding a suitable choice of title. Murray's *New English Dictionary* defines 'Sound' as "(1) The sensation produced in the organs of hearing when the air is set in vibration in such a way as to effect these; also that which is or may be heard; the external object of audition, or the properties of bodies by which this is produced; (2) the particular auditory effect produced by a special cause"; whilst 'Acoustics' is defined as "The science of sound, and the phenomenon of hearing." Now the subject as I have treated it, in the light of recent developments, embraces much more than these definitions imply. I have regarded it as including not only the range of audible frequencies with which the subject was formerly concerned, but also the regions of inaudible vibrations of very low and very high frequency. A new concise description is required which embraces vibratory systems of *all* frequencies. The indiscriminate use of the titles 'Sound' and 'Acoustics' is not surprising in view of the dictionary definitions to which I have referred. I have, however, in the present volume endeavoured to reserve the use of the word 'acoustics' (Greek ακουω, hear) for that aspect of the subject more immediately connected with hearing. For want of a more appropriate word I have regarded 'sound' as referring to vibrations of all frequencies, audible or otherwise.

In addition to the large extension of frequency-range, new apparatus and new methods of investigation have been introduced which have completely transformed all branches of the subject. The very important analogy existing between electrical and mechanical vibrating systems is now in general use. Not only are electrical methods extensively applied in the production and reception of mechanical vibrations, but the theory of one is closely interlinked with that of the other. The familiar conceptions of electrical potential, current, and impedance are now applied with almost equal facility to mechanical systems; we have mechanical filters analogous to electrical filters, and complex mechanical vibrating systems (*e.g.* such as a gramophone sound box or a 'loud speaker') are now designed in strict accordance with electrical principles. Electrical methods are extensively used also as a means of amplifying *mechanical* vibrations, the feeble effects at a sound-receiving surface being converted into electrical currents, amplified electrically and reproduced as mechanical vibrations of greatly increased intensity. It seems desirable that these and the many other important developments should be recorded side by side with the earlier work of Rayleigh and his followers.

I have endeavoured to write the book in a manner suited to the requirements of university students, but it is hoped that it may also prove of some value to those interested in the more technical, or applied, aspects of the subject. For those requiring further information a plentiful supply of references to original papers is provided.

In spite of the care I have taken in reading through the MS. I am still conscious of mistakes, for which I ask the indulgence of my readers. I hope, however, that none of them are of a serious nature.

In conclusion, I should like to express my thanks to numerous authors for their kindness in supplying me with reprints of their original papers, and to publishers for permission to use their blocks for illustrations. Specific acknowledgments will be found in the text. My thanks are also due to the Admiralty for permission to write and publish this book.

A. B. W.

Admiralty Research Laboratory
Teddington

CONTENTS

SECTION III

TRANSMISSION OF SOUND

SECTION IV

RECEPTION, TRANSFORMATION, AND MEASUREMENT OF SOUND ENERGY

SECTION V

TECHNICAL APPLICATIONS

INTRODUCTION

We may regard sound either as a sensation or the stimulus which produces the sensation. The physicist is usually concerned with the stimulus—that is, in the phenomena which occur external to the ear. Just as it is customary to regard 'light' as including 'invisible' radiations in the ultra-violet and the infra-red, so we shall regard sound as including mechanical vibrations of all frequencies, audible or otherwise. In this respect sound may be considered as including that branch of mechanics which deals with alternating or vibratory motion. The study of mechanical vibrations is closely analogous to that of alternating currents, the mathematical theory being almost identical when the members of certain pairs of fundamental quantities, *e.g.* inertia-inductance, stiffness-capacity, friction-resistance, are interchanged. To those familiar with alternating-current theory, this analogy will prove of considerable assistance in dealing with the theoretical aspect of sound. Recent developments in the use of high-frequency alternating currents, in radio-telephony, have their counterpart in the production of mechanical vibrations of the same order of frequency. Alternating currents, of whatever frequency, provide also a very convenient means of exciting the corresponding mechanical vibrations and *vice versa.*

It requires little effort to demonstrate that all sounding bodies are in a state of vibration. In some cases the vibrations can be seen directly as a blurred outline, or optical magnification may be necessary to reveal them, whilst in other cases more delicate mechanical or electrical tests are necessary. The simple experiment with the pith-ball pendulum is a very convincing way of demonstrating the vibration of tuning-forks, bells, or similar sources of sound. The experiment becomes more convincing when applied to the case of a short steel bar in which the longitudinal vibrations are above the upper audible limit.

In order to convey the sound from the vibrating body to the ear, or other form of receiver, at a distant point it is necessary that the intervening medium should be capable of transmitting the vibrations. In this connection it is customary to refer to the classical experiments of Hawksbee (*Phil. Trans.*, 1705), in which a bell is caused to ring under an evacuated glass jar. When the jar is full of air at atmospheric pressure the sound is clearly heard; after the air is withdrawn the sound is hardly perceptible. Tyndall (*Sound*, p. 7, 1867), with an improved pump and the use of hydrogen gas to replace the air in the jar, rendered the bell quite silent. It is important of course, in this experiment, to ensure that the vibrations of the bell are not communicated to the air outside the jar *via* the supports, *i.e.* the bell should be suspended by thin strings which are incapable of transmitting an appreciable amount of the vibration.

We are naturally inclined to regard the atmosphere as the universal medium for the transmission of sound. Speech, music, and all the familiar noises of everyday life are conveyed to the ear through the surrounding air. It is therefore not very surprising that until comparatively recent times the practical utilisation of other media as vehicles for the transmission of sound has been somewhat neglected. The exigencies of war made a study of sound propagation in the sea a necessity, and we were thus brought to realise that this relatively homogeneous medium is almost ideal for the purpose. As a result of this study many important applications of under-water sound transmission have been developed. On account of the heterogeneous nature of the earth's crust little practical development has so far been possible in its use as a medium for sound transmission. The passage of sound through solid wires and bars is of course familiar to all, but we have only limited experience yet of sound propagation through an extended solid mass, comparable in bulk with the atmosphere or the sea.*

To complete the natural sequence of events a suitable receiver is required to collect the sound energy after transmission through the intervening elastic medium. In the case of sound transmission through the atmosphere the ear forms a natural, though not always the most convenient receiver. When liquids or solids transmit the sound other forms of receiver are generally necessary or preferable.

The production, transmission, and reception of sound has now developed into a science with many important technical applications. Before dealing with the subject in detail, however, we shall recall briefly a few of the more familiar points, with the object of defining certain terms which will occur frequently in the text.

Musical Sounds and Noises – There can be no hard-and-fast

* This disregards the case of propagation of earthquake waves, natural or artificial, through the various layers forming the earth's crust and interior.

line of division between a musical sound and a noise, for some noises have a more or less musical character, whilst some musical sounds are not free from noise. The extreme cases, of course, raise no difficulty. Musical sounds are characterised principally by smoothness and a pleasing effect on the auditory nerves, whereas noises are of an irregular or explosive nature. A musical sound is regular or periodic, *i.e.* the vibrations repeat themselves at regular intervals. The converse is, of course, not necessarily true, for the periodic repetition of a noise—like a clock tick—would not constitute a musical sound unless the noise were repeated at a frequency above the 'lower audible limit.' If the edge of a card is struck by the teeth of a rotating wheel the sounds heard might therefore be regarded as a succession of noises up to a speed of about 20 taps per second, merging into a musical sound, of a harsh character, at higher speeds.

We shall be principally concerned in the following pages with musical or periodic sounds of frequency within, or above, the audible range.

Frequency. Pitch - The *frequency* of a regular or periodic vibration is the number of vibrations performed per second. The *period* of a vibration is the reciprocal of the frequency, *i.e.* the time taken to complete one vibration. Musical sounds arrange themselves in a natural order according to *pitch*. The latter depends solely on the predominant frequency of the vibrations—the greater this frequency the higher the pitch. This is simply verified by means of the toothed wheel and card mentioned above or by means of a siren which produces a sound whose fundamental frequency is proportional to the product of speed (revs. per sec.) and the number of holes in the revolving disc.

Musical Intervals. Diatonic Scale - In the case of musical or periodic sounds within the audible range we recognise certain corresponding differences of pitch called *intervals* which are always regarded as the same, for the same relationship, wherever they occur in the scale. Physically, the intervals are distinguished by the property that the frequencies corresponding to the respective pitches are in a simple numerical ratio. The more important consonant intervals, with their frequency ratios, are:

Unison	1 : 1	Fifth	3 : 2
Minor Third	6 : 5	Major Sixth	8 : 5
Major Third	5 : 4	Minor Sixth	5 : 3
Fourth	4 : 3	Octave	2 : 1

The notes whose frequencies are multiples of that of a given one, the *fundamental*, are called *harmonics*. A slight mistuning of either note comprising a consonant interval produces the phenomenon of *beats*. If the mistuning is more serious, the beats become more rapid, a sense of *discord* is experienced, and ultimately two independent notes of the diatonic scale are recognised. The frequency-ratios which define each note of the *Diatonic Musical Scale* (the basis of European music) are as follow :—

C	D	E	F	G	A	B	C
1	$1\frac{1}{8}$	$1\frac{1}{4}$	$1\frac{1}{3}$	$1\frac{1}{2}$	$1\frac{2}{3}$	$1\frac{7}{8}$	2.

The same series of ratios applies to any other octave which may be chosen.

Intensity. Loudness – These terms refer to the external, or physical, and the internal, or subjective, aspects respectively. The intensity of a sound refers to a definite physical quantity which determines the rate of supply of vibrational energy. This energy is proportional to the square of the *amplitude* of vibration (this being measured by the maximum excursion of the vibrator from its central position). Loudness corresponds to the degree of sensation, depending on the intensity of the sound and the sensitiveness of the ear under the particular conditions. Near the limits of audibility the loudness may be very feeble, although the intensity be very great. In this connection also it is important to note that the sensation of loudness is not simply proportional to intensity (or to amplitude). The range of intensity to which the ear can accommodate itself is indeed remarkable. Thus extremely feeble vibrations having an amplitude of the order 10^{-8} cm. can be detected by the ear, the same ear remaining undamaged by the sound of a violent explosion at a short range. The ear can thus easily accommodate itself to deal with sounds varying in amplitude in the ratio of 1 to 10^6, or in intensity in the ratio 1 to 10^{12}. The relation between sensation and stimulus is generally expressed by Weber's law: "The increase of stimulus necessary to produce the minimum perceptible increase of sensation is proportional to the pre-existing stimulus." From this law Fechner derived the relation

$$S = k \log I \qquad \text{or} \qquad dS/dI = k/I,$$

S being the magnitude of the sensation, I the intensity of the stimulus, and k a constant. Such a law is difficult to verify experimentally on account of the uncertainty in the measure of

S the sensation. The fact remains, however, that the sensitiveness dS/dI of the ear diminishes rapidly with increase of total intensity of the sound. Many feeble sounds which are easily audible in the 'stilly night' could not be heard in the 'noisy day.'

Quality. Wave-form – Sounds of the same pitch and loudness but produced by different means are distinguished by their 'quality.' Thus the same note produced by a voice, a piano, and a violin has distinct characteristics which are at once recognisable by the ear. The statement that notes of the same pitch have the same frequency is therefore somewhat indefinite; the same '*predominant* frequency' is implied. Very few sounds can be regarded as 'pure,' *i.e.* free from overtones. A note sounded on the piano is complex since it may contain, in addition to the fundamental note, a large number of overtones. The presence or otherwise of these overtones decides the *quality* of the sound. A tuning-fork emits a note which is almost pure, whereas the same note on a violin may contain many important overtones. These overtones reveal to the ear the nature of the instrument producing the sound. Analysis of the wave-form of a particular sound reveals which harmonics are present and their relative intensities. Wave-form is therefore an indication of 'quality.'

Velocity – The passage of sound from the vibrating source through the intervening medium is not instantaneous. The flash and the report of a distant gun are separated by an appreciable time-interval, indicating a finite velocity for the sound as it travels through the atmosphere. The time taken for the flash to reach the observer being quite negligible, it is possible to determine the velocity of the sound wave in a very simple and direct manner provided the distance between the gun and the observer is known. Of course there are various other factors involved in a reliable determination of the velocity, but we need not consider them at present. Sound travels with different velocities in different media. For example, sound travels faster in a long iron pipe or bar than in the surrounding air. This is easily verified by an observer listening at the end of a long iron rail whilst a confederate strikes the rail at the opposite end. Two sounds will be heard, the first to arrive *via* the iron, the second *via* the air. The velocity of sound in air is about 330 metres/sec., in water about 1500 metres/sec., and in steel about 5100 metres/sec.—covering a range of approximately 16 : 1

Within wide limits the velocity of sound is approximately independent of intensity or frequency. Otherwise distant music would be confused and discordant. When the amplitude of the vibration is very great, however, as in the vicinity of a violent explosion, this simple law no longer holds, for the variations of density of the medium are then comparable with the normal density.

Propagation of Sound. Longitudinal Wave-Motion–Sound, in transmission through an elastic medium, has all the characteristics of wave-motion. Just as ripples spread outwards in two dimensions from a stone thrown into a pond, so waves of sound spread, in three dimensions, in the elastic medium in contact with a vibrating body or 'source' of sound. The movement of any particle of the medium is, however, purely local, each particle making small to-and-fro excursions in a manner similar to that of the vibrating body itself. This local motion of the particles must be clearly distinguished from the motion of the disturbance which travels forward from layer to layer of the medium with an ever-increasing radius. It is the state of to-and-fro motion which advances, the medium as a whole, after the disturbance has passed, remaining in its initial position. As we shall see, the velocity with which the wave travels is dependent on the elasticity and density of the medium. This *wave*-velocity must not be confused with the variable *particle* velocity involved in the to-and-fro vibration. Sound waves differ from water waves, or ripples, in one important respect, viz. the vibratory motion takes place in the *same* line as the direction of advance of the wave, whereas in the case of ripples the to-and-fro motion is at right angles to the direction of propagation of the wave. The former type of motion, as in the sound wave, is called *longitudinal*; the latter, transverse. The general character of longitudinal wave-motion may be demonstrated by means of a long helical spring (*e.g.* 6 feet long, 3 or 4 inches diameter, of wire 0·06 inch thick) supported horizontally at frequent intervals by threads. If one end be moved to and fro longitudinally, a corresponding motion will travel, like a wave, to the other end. This simple apparatus may be used to demonstrate effects of reflection, etc., with which we are not at present concerned.*

* A useful wave-model has recently been demonstrated by Mr. T. B. Vinycomb (Phys. Soc. Exhibition, 1927). This will be found valuable as a means of observing in 'slow motion' the various wave phenomena encountered in the study of sound.

Order of Treatment – The logical order of treatment of the theory of sound involves three fundamental considerations. First, a vibrating body or 'sound source' is necessary; second, an elastic medium in contact with the body is required to transmit the vibrations to a distant point; and third, some form of receiver is necessary to absorb the energy from the medium and to reconvert it into a form which is convenient for observation. We shall follow this order of treatment as far as practicable.

SECTION I

THEORY OF VIBRATIONS

THE simplest and most fundamental type of sound sensation is that which corresponds to a simple harmonic motion—*i.e.* to the simplest mathematical form of periodic function. Simple harmonic motions may vary in period and amplitude but in no other manner; they are consequently ideal for the production of 'simple' or 'pure' tones. Another important feature of this form of motion is the possibility of transmission from one medium to another without change of form. Again, it has been proved by Fourier that the most complex form of periodic motion can be analysed into a series of simple harmonic motions having frequencies which are multiples of that of the complex motion.

The vibrations of a tuning-fork may approximate closely to a simple harmonic motion, the sensation resulting being described as a 'pure tone.'

VIBRATIONS OF A PARTICLE

Simple Harmonic Motion – The study of the motion of vibrating bodies, sources of sound, is complicated by the presence of the surrounding medium. This medium withdraws energy from the sounding body and consequently exerts forces of reaction on it. We shall for the present, however, ignore such forces and consider only the vibrations of a simple isolated system.

The typical case of simple harmonic motion is that of a particle attracted towards a fixed point O in the line of motion with a force varying as the distance from that point. If s denote the force at unit distance, the force at a distance x will be $-sx$, the sign being always opposite to that of x. Thus, if m is the mass of the particle,

$$m\frac{d^2x}{dt^2}=-sx \quad . \quad . \quad . \quad . \quad (1)$$

If we write

$$n^2=s/m \quad . \quad . \quad . \quad . \quad (2)$$

we have

$$\frac{d^2x}{dt^2}+n^2x=0 \quad . \quad . \quad . \quad . \quad (3)$$

of which the solution is

$$x = A \cos nt + B \sin nt \quad . \quad . \quad . \quad (4)$$

or substituting $A = a \cos \epsilon$ and $B = -a \sin \epsilon$,

$$x = a \cos (nt + \epsilon) \quad . \quad . \quad . \quad (5)$$

where the constants A, B, or a, ϵ, are arbitrary. The motion is therefore periodic, the values of x and dx/dt recurring whenever nt increases by 2π. The maximum displacement a is called the *amplitude* of the oscillation. The *period*

$$T = 2\pi/n = 2\pi\sqrt{m/s} \quad . \quad . \quad (6)$$

is always the same whatever the initial conditions of amplitude a or phase ϵ, hence the oscillations are said to be '*isochronous.*'

The type of vibration represented by (5) is of fundamental importance. The equation shows that the particle oscillates continuously between two points which are at a distance a from the centre O. By simple differentiation, *the velocity of the particle*, which moves as in (5), is given by

$$dx/dt = -an \sin (nt + \epsilon) \quad . \quad . \quad . \quad (7)$$

and *the acceleration*

$$d^2x/dt^2 = -an^2 \cos (nt + \epsilon) \quad . \quad . \quad . \quad (8)$$

$= -n^2x$, as postulated in (1). The meaning of equation (5) is best exemplified in the case of a point P moving uniformly with constant angular velocity n in a circular path, of radius 'a,' see fig. 1. The orthogonal projection M of P on the fixed diameter AOB will perform simple harmonic motion in accordance with equation (5). The angle $\theta = (nt + \epsilon)$ indicating the position of P, and hence of M, after a time t, is called the *phase* at that particular instant; whilst the angle ϵ, the starting position P_0 when $t = 0$, is known as the *initial phase* or '*epoch.*'

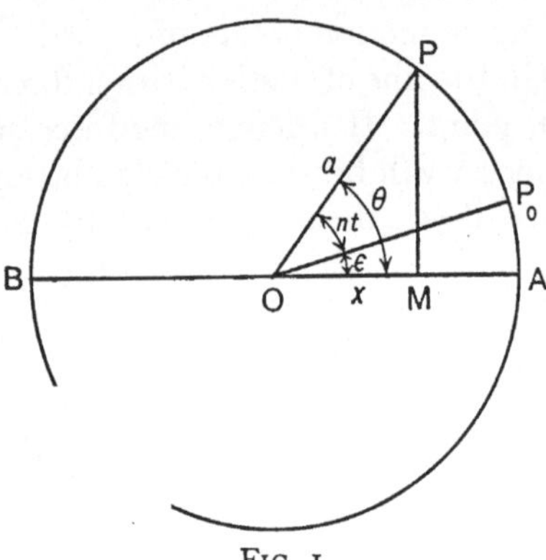

FIG. 1

If we denote OM by x it will be seen that *the displacement of M from* O is

$$x = a \cos \theta = a \cos (nt + \epsilon).$$

A simple harmonic motion may consequently be described as the orthogonal projection of a uniform circular motion. The *periodic time* T is the time of revolution $2\pi/n$ of the point P, and the *amplitude* is the radius 'a' of the circle.

The *frequency* * of the motion, *i.e.* number of oscillations of the point M per second, is $n/2\pi$, or the number of revolutions of P per second. The expressions for the velocity and acceleration of the point M are obtained by considering that these are equal to the projections on AB of the corresponding quantities for P. Since the linear velocity of P is an, then *the velocity of M* is equal to $-an \sin(nt+\epsilon)$, the minus sign being a consequence of the negative direction of the velocity of M when θ is positive. The acceleration of the point P is directed radially inwards towards the centre O of the circle and is equal to $a^2n^2/a = an^2$. The resolved component of this along AB gives *the acceleration of M*, viz. $-an^2 \cos(nt+\epsilon)$, or $-n^2x$.

A useful means of studying periodic motions is the graphical method in which displacements, velocities, and accelerations are plotted as ordinates with the time 't' or the phase angle θ (that is $(nt+\epsilon)$) as abscissæ. Such curves are shown in fig. 2. Curve

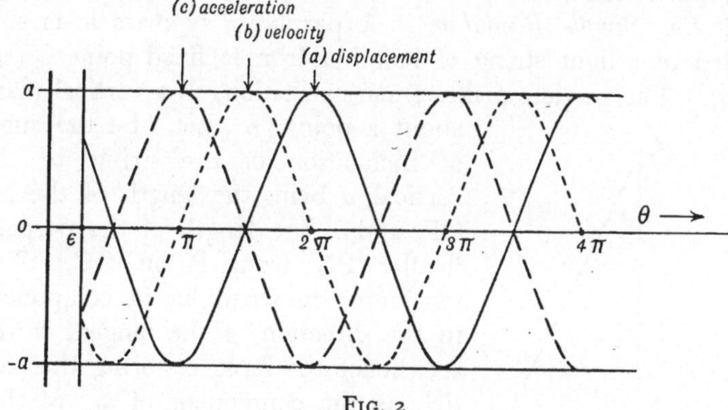

FIG. 2

(*a*) represents the displacement x as a function of θ, curve (*b*) the velocity dx/dt, and curve (*c*) the acceleration d^2x/dt^2. The ordinates for these three curves are not plotted to the same scale in the figure—the values shown must be multiplied by n and n^2 respectively to give the true values of velocity and acceleration. It will be seen that the velocity is always 90° (or $\pi/2$), and the

* The quantity 'n' which characterises frequency in equations of periodic motion has no well-recognised descriptive name. The term 'pulsatance' is sometimes used. Thus pulsatance $n = 2\pi \times$ frequency.

acceleration 180° (or π) out of phase with the displacement. In the light of equations (5), (7), and (8) above it is evident that the *slope* and *curvature* of the displacement curve in fig. 2 represent particle *velocity* and *acceleration* respectively. The velocity is a maximum (equal to na) when the displacement is zero, and the acceleration is a minimum (equal to $-n^2a$) when the displacement is a maximum (equal to a).

Examples of Simple Harmonic Motion – In the form of motion we have just considered the force is always directed to a central, or equilibrium, position and varies as the distance from that position. As equation (6) indicates, the periodic time of such a simple oscillation is $T=2\pi\sqrt{m/s}$, i.e. *the period increases with the mass m and decreases with the 'stiffness' or 'elastic' constant s* (s being the force required to produce unit displacement of the mass from its zero position). This relation is of very wide application in the study of sound vibrations; in fact by means of this principle it is sometimes possible to obtain limits for a periodic time of a vibrating system, which cannot easily be calculated exactly. The following examples are given as typical of simple harmonic motion :—

(1) *The Simple Pendulum* – A particle P of mass m is suspended by a light string of length l from a fixed point C (see fig. 3). The particle makes small oscillations in a vertical plane about a point O. Let θ be the angle of inclination of the string to the vertical, a being the length of the arc OP, and x the length of the perpendicular PN from P on OC. The tension of the string has no component in the direction of the tangent to the arc, hence the force restoring the particle is the component of mg in that direction, viz. $-mg \sin \theta$. To the first order of approximation we may regard $\sin \theta$ as equal to θ and the chord x equal to the arc a, *when the angle of swing is small.* Consequently we may write: Restoring force $=-mgx/l$, therefore the motion is simple harmonic, and s (the restoring force per unit displacement) $=-mg/l$. Consequently the periodic time

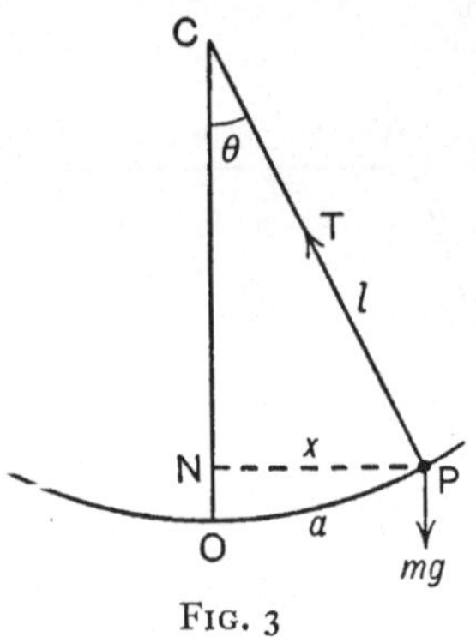

FIG. 3

$$T=2\pi\sqrt{m/s}=2\pi\sqrt{ml/mg}=2\pi\sqrt{l/g}.$$

This case of the simple pendulum may be used to illustrate many important principles in the theory of sound. In spite of the fact that the periodic time and amplitudes involved are so widely different in magnitude, the treatment is fundamentally the same.

(2) *The vertical Vibrations of a Mass suspended on a Helical Spring* – To a first order of approximation this case also provides a good example of simple harmonic motion. Assuming the spring obeys Hooke's law, the restoring force is proportional to the displacement x, and neglecting the mass of the spring,* we have: Restoring force $=-sx$, Mass moved $=m$, Periodic time $T=2\pi\sqrt{m/s}$. This result is conveniently expressed in terms of the statical increase of length which is produced by the weight when hanging in equilibrium. Denoting this by l, we have $sl=mg$, so that the period is $T=2\pi\sqrt{l/g}$, the same as that of a simple pendulum of length l.

It requires little effort to suggest further examples of the same nature. In all such cases when we are dealing with a central force proportional, but of opposite sign, to the displacement the motion is simple harmonic with a period $2\pi\sqrt{m/s}$— the greater the 'stiffness' the smaller the period, the greater the 'inertia' the greater the period. Expressed in terms of frequency of vibration the rule becomes—*the smaller the 'inertia' or the greater the 'stiffness,' the higher the frequency; frequency* $=\frac{1}{2\pi}\sqrt{s/m}$.

Energy of a Particle in Simple Harmonic Motion – We have seen that the velocity v of the particle in S.H.M. is given by $\frac{dx}{dt}=-an\sin(nt+\epsilon)$. Consequently the kinetic energy at this instant is

$$E=\tfrac{1}{2}mv^2=\tfrac{1}{2}ma^2n^2\sin^2(nt+\epsilon).$$

Now the sum of the kinetic E and the potential P energies is a constant (since no energy enters or leaves the particle), and when one form is at a maximum value the other is zero. Thus the maximum value of E gives us the total energy of the vibrating particle, *i.e* $\frac{1}{2}ma^2n^2$. Hence $E+P=\frac{1}{2}ma^2n^2$, whence

$$P=\tfrac{1}{2}ma^2n^2\{1-\sin^2(nt+\epsilon)\}=\tfrac{1}{2}ma^2n^2\cos^2(nt+\epsilon).$$

Now the displacement at any instant t is $x=a\cos(nt+\epsilon)$. Consequently the *potential energy of the particle* at that instant is

* The inertia of the spring can be allowed for by adding one third of its mass to that of the suspended mass (see p. 14).

$P=\frac{1}{2}mn^2x^2$, *i.e.* proportional to the mass m of the particle, the square of the frequency $(n/2\pi)$, and the square of the displacement x.

Since the mean values of $\sin^2(nt+\epsilon)$ and $\cos^2(nt+\epsilon)$ are each equal $\frac{1}{2}$, we have

$$\text{Mean Kinetic Energy}=\text{Mean Potential Energy}=\tfrac{1}{4}ma^2n^2$$

and

$$\text{Maximum Energy}=\tfrac{1}{2}ma^2n^2.$$

Consequently the energy is on the average half kinetic and half potential. The energy is a maximum at the mid-point and the two turning points of the vibration, in the former case all kinetic and in the latter cases all potential.

It is often convenient to study the motion of a particle from the standpoint of potential and kinetic energy. Thus if the particle of mass m vibrate under the action of a restoring force which is $-s$ times the displacement, the kinetic and potential energies are given by

$$E=\tfrac{1}{2}m\dot{x}^2 \quad \text{and} \quad P=\int_0^x sx\,dx=\tfrac{1}{2}sx^2.$$

Since $E+P=$constant, $\frac{1}{2}m\dot{x}^2+\frac{1}{2}sx^2=$constant. On differentiating we obtain $m\ddot{x}+sx=0$, the solution of which is

$$x=a\cos(nt+\epsilon) \quad \text{where} \quad n=\sqrt{s/m}=2\pi/T,$$

and a and ϵ are constants representing the amplitude and initial phase of the motion. That is, we obtain the ordinary equation for the simple harmonic motion of the particle having a periodic time $T=2\pi\sqrt{m/s}$.

Example – As an example of the application of this method consider the case of a mass M suspended by a coiled spring, the mass of the latter being too great to be negligible. Assuming the spring of mass m to be uniformly stretched at every instant of the motion, we have $P=\frac{1}{2}sx^2$ where s is the force required to produce unit extension of the spring.

Kinetic energy of the suspended mass $=\frac{1}{2}M\dot{x}^2$,

$$\text{Kinetic energy of the spring}=\int_0^l \tfrac{1}{2}\frac{m}{l}\,.\,dy\,.\left(\frac{y}{l}\dot{x}\right)^2=\tfrac{1}{6}m\dot{x}^2,$$

where l is the length of the spring, y is the distance of the element δy from point of support. Therefore, total kinetic energy of system $=\frac{1}{2}(M+\frac{1}{3}m)\dot{x}^2$ and the periodic time $T=2\pi\sqrt{(M+\frac{1}{3}m)/s}$.

Thus, in calculating the period, one-third the mass of the spring must be added to the mass of the suspended weight.

Composition of Simple Harmonic Motions – If a single particle is acted on by a number of distinct forces, each of which acting separately would cause the particle to perform simple harmonic motion, the question arises as to the resultant motion of the particle, assuming that each force produces its own effect.

Case I – *Vibrations of the same Frequency and in the same Straight Line.* (a) *Two forces only.* Let the two vibrations be represented by

$$x_1=a_1 \cos (nt+\epsilon_1) \quad \text{and} \quad x_2=a_2 \cos (nt+\epsilon_2) \quad . \quad (1)$$

Since the two displacements are in the same straight line and each produces its own effect, the resultant displacement is $x=x_1+x_2$, that is

$$x=\mathrm{A} \cos nt-\mathrm{B} \sin nt \quad . \quad . \quad . \quad (2)$$

where

$$\left.\begin{aligned} \mathrm{A}&=a_1 \cos \epsilon_1+a_2 \cos \epsilon_2 \\ \text{and} \qquad \mathrm{B}&=a_1 \sin \epsilon_1+a_2 \sin \epsilon_2 \end{aligned}\right\} \quad . \quad . \quad . \quad (3)$$

Writing

$$\mathrm{A}=r \cos \theta \quad \text{and} \quad \mathrm{B}=r \sin \theta$$

we have

$$r^2=\mathrm{A}^2+\mathrm{B}^2 \quad \text{and} \quad \tan \theta=\mathrm{B}/\mathrm{A} \quad . \quad . \quad (4)$$

Thus

$$x=r(\cos nt \cos \theta- \sin nt \sin \theta),$$

i.e.

$$x=r \cos (nt+\theta) \quad . \quad . \quad . \quad (5)$$

This indicates that the two component simple harmonic vibrations result in a vibration of the same frequency but with amplitude and phase given by r and θ respectively. The values of r and θ may easily be deduced in terms of a_1a_2 and $\epsilon_1\epsilon_2$ from equations (3) and (4) above, whence it will be found that the resultant amplitude

$$r=\{a_1{}^2+a_2{}^2+2a_1a_2 \cos (\epsilon_1-\epsilon_2)\}^{\frac{1}{2}} \quad . \quad . \quad (6)$$

and

$$\tan \theta=\frac{a_1 \sin \epsilon_1+a_2 \sin \epsilon_2}{a_1 \cos \epsilon_1+a_2 \cos \epsilon_2} \quad . \quad . \quad (7)$$

It is a simple matter to obtain the same results graphically. It will be found that two simple harmonic oscillations in the same straight line may be combined like two forces in the usual parallelogram construction, *i.e.* according to the law of vector addition. *The diagonal of the parallelogram represents in length the amplitude, and in direction the phase of the resultant vibration.*

(b) *Any Number of Vibrations* – In a similar manner it may be shown that any number of simple harmonic vibrations of the same frequency result in a vibration of the same type.

Thus

$$\left.\begin{aligned} x &= \Sigma a \cos(nt+\epsilon) \\ &= \cos nt \Sigma a \cos\epsilon - \sin nt \Sigma a \sin\epsilon \\ &= r\cos(nt+\theta) \end{aligned}\right\} \quad . \quad . \quad (8)$$

where

$$r = \{(\Sigma a \cos\epsilon)^2 + (\Sigma a \sin\epsilon)^2\}^{\frac{1}{2}} \quad . \quad . \quad (9)$$

and

$$\tan\theta = \frac{\Sigma a \sin\epsilon}{\Sigma a \cos\epsilon} \quad . \quad . \quad . \quad (10)$$

The particular case of two forces only, (a) above, may be at once deduced from these expressions.

Regarded from the graphical standpoint, since two S.H.M.'s of the same frequency and in the same straight line may be combined, it will be seen that *any number* of such vibrations can be combined by taking the resultant of any two of them and combining it with a third vibration and so on, until the final resultant is obtained. *Such a system of vibrations is reduced to a single resultant by a vector polygon in exactly the same manner as a system of forces acting at a point.*

(c) *A large Number of Vibrations of the same Amplitude and Frequency but differing progressively in Phase* – This case is of particular importance when we come to deal with the subject of diffraction of sound waves (see p. 300). The effect of superposing m component vibrations of the same amplitude and frequency, but progressively differing in phase by ϵ, may be obtained from equations (9) and (10). We must now write

$$\Sigma a \sin\epsilon = a[\sin\epsilon + \sin 2\epsilon + \ . \ . \ . + \sin m\epsilon] = \frac{\sin(m\epsilon/2)}{\sin(\epsilon/2)} \cdot \sin\frac{m\epsilon}{2},$$

and

$$\Sigma a \cos\epsilon = a[\cos\epsilon + \cos 2\epsilon + \ . \ . \ . + \cos m\epsilon] = \frac{\sin(m\epsilon/2)}{\sin(\epsilon/2)} \cdot \cos\frac{m\epsilon}{2},$$

whence

$$R=\frac{ma\sin\theta}{\theta} \quad\text{and}\quad \theta=\frac{m\epsilon}{2} \quad . \quad . \quad (10a)$$

where 2θ is the phase difference between the first and the last of the component vibrations. The same result may be obtained, if required, from the vector polygon (see fig. 4). Commencing at O in the line OX the polygon OABP is constructed with each side of length a and making an angle ϵ with its neighbour. The resultant vibration R is given in amplitude and phase by OP the closing side of the polygon.

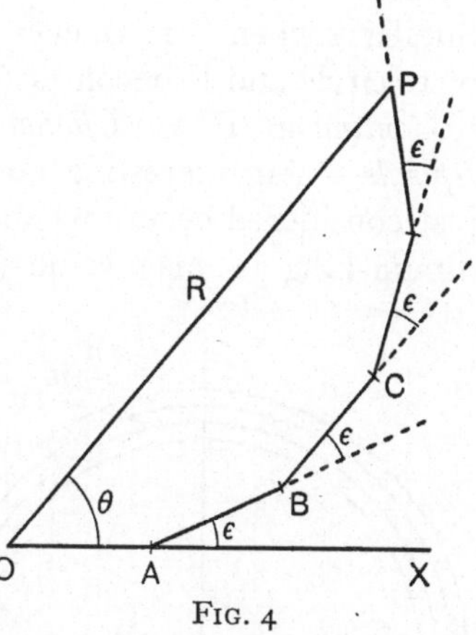

FIG. 4

Phase Change Continuous – A case of special importance arises when the component amplitudes a and phase difference ϵ may be regarded as infinitesimal. The polygon then becomes a continuous curve close by a chord. If a is represented by a length δs of the curve and ϵ by a small change of angle $\delta\theta'$, we have as before $ds/d\theta'$=(constant), or $s=\theta'\times$(constant), which defines the arc of a circle tangential to OX at O (see fig. 5). The constant in this relation is the radius r of the circle. Writing as before $\angle XOP=\theta$, then $\theta'=2\theta$ and $s=2\theta r$. The resultant vibration is represented by the closing side of the polygon, *i.e.* the chord OP joining the extremities of the arc. We see from the construction that

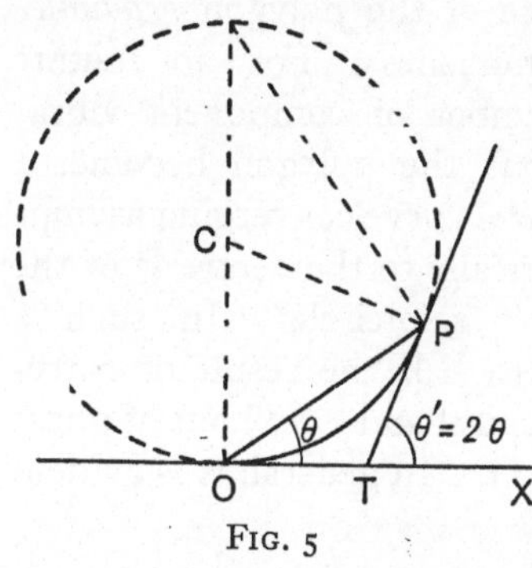

FIG. 5

$$R=\text{chord } OP=2r\sin\theta,$$

and the arc of length $2r\theta$ must also be $m\delta s$, whence $r=m\delta s/2\theta$. Consequently

$$R=\frac{m\delta s\,.\,\sin\theta}{\theta} \quad\text{or}\quad \frac{s\sin\frac{\theta'}{2}}{\frac{\theta'}{2}},$$

which is in agreement with equation (10a).

Whenever θ' is an odd multiple of π, the last component vibration being in opposite phase to the first, and θ an odd multiple of $\pi/2$, the resultant $R(=2s/\pi)$ is equal to the diameter of the semicircle formed by the vector addition of the components. Similarly when θ' is an even multiple of π, the polygon is a complete circle and R becomes zero.

Continuous Phase Change. Amplitudes gradually diminishing. Spirals – An interesting and important modification of the case just considered occurs when the successive amplitudes are slowly diminishing in magnitude (see fig. 6). From θ', 0 to π, the polygon is represented approximately by a semicircle; from π to 2π a second semicircle is added but the diameter is slightly less on account of the gradual decrease in amplitude; a third semicircle of still smaller diameter is added between 2π and 3π, and so on. Actually, of course, the radius of curvature of the polygon *gradually* diminishes. For an infinite number of component vibrations the polygon becomes a *spiral* curve converging asymptotically to the centre P of the first semicircle. In such a case the resultant $R(=OP)$ is equal to half the resultant corresponding to the first half-turn of the spiral and is $\pi/2$ out of phase with the first component vibration at O. In general, it is evident that

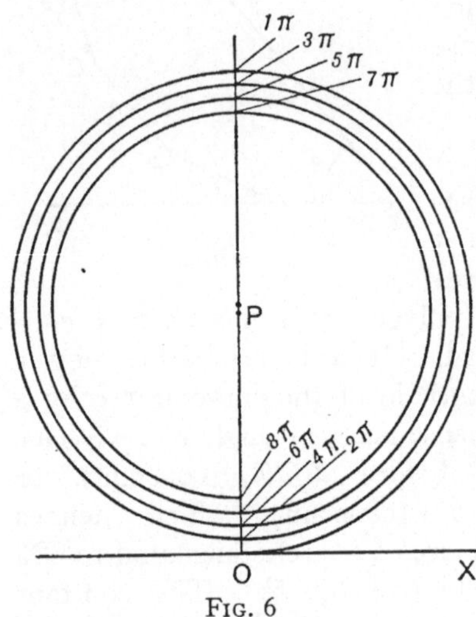

Fig. 6

$$R = \frac{R_1}{2} \pm \frac{R_L}{2},$$

where R_1 and R_L are the resultants corresponding to the first and the last (L^{th}) half-turn of the spiral. As L increases R_L diminishes and ultimately becomes zero when L is infinite.

Case II – (a) *Vibrations of Different Frequency, and in the same Straight Line.* In this case

$$x = a_1 \cos(n_1 t + \epsilon_1) + a_2 \cos(n_2 t + \epsilon_2) \quad . \quad . \quad (11)$$

If n_1 and n_2 are incommensurable the value of x never recurs, consequently the resulting motion is non-periodic. If, however, n_1 and n_2 be in the ratio of two whole numbers, $\dot{x}$ will repeat after a time equal to the L.C.M. of n_1 and n_2, but the vibration will not be *simple harmonic* although it is periodic. The case when n_1 and n_2 are commensurate will be considered under Fourier's theorem (see p. 25).

(b) *Slightly different Frequencies. Beats* – A special case arises when the frequencies of the component vibrations are almost equal. This case is of frequent occurrence and of considerable importance. If the two vibrations represented on the right-hand side of equation (11) have the same amplitude, frequency, and initial phase it is evident that the resultant amplitude will be double that due to either of the forces individually. If now one of the vibrations has a slightly higher frequency than the other the combined vibration will at first be nearly the same as if the frequencies were equal—*i.e.* the resultant amplitude will be double the 'single' amplitude. As the higher frequency vibration gains on the lower, however, thereby changing the relative phase, a point will be reached when they are in opposition and will neutralise each other (amplitude zero). After a further interval they will again be in phase and the amplitude doubled and so on.

Let the two frequencies be $n/2\pi$ and $(n+m)/2\pi$, m being small compared with n. The corresponding vibrations are

$$\left.\begin{aligned} x_1 &= a_1 \cos (nt+\epsilon_1) \\ \text{and} \quad x_2 &= a_2 \cos (nt+mt+\epsilon_2) \end{aligned}\right\} \quad . \quad . \quad . \quad (12)$$

Since m is small compared with n these equations are similar to (1) above, if we regard $(mt+\epsilon_2)$ as the phase term in the second equation. Consequently as in equations (6) and (7) the resultant amplitude and phase are given respectively by

$$r = [a_1^2 + a_2^2 + 2a_1a_2 \cos \{\epsilon_1 - (mt+\epsilon_2)\}]^{\frac{1}{2}} \quad . \quad . \quad (13)$$

and

$$\tan \theta = \frac{a_1 \sin \epsilon_1 + a_2 \sin (mt+\epsilon_2)}{a_1 \cos \epsilon_1 + a_2 \cos (mt+\epsilon_2)} \quad . \quad . \quad (14)$$

the resultant displacement being

$$x = r \cos (nt+\theta) \quad . \quad . \quad . \quad (15)$$

The resultant vibration of frequency $n/2\pi$ may thus be regarded as harmonic with amplitude r and phase θ, which vary slowly with a frequency $m/2\pi$. The amplitude is a maximum when

$$\cos (\epsilon_1 - \epsilon_2 - mt) = +1, \qquad r = a_1 + a_2 \,;$$

and a minimum when

$$\cos(\epsilon_1 - \epsilon_2 - mt) = -1, \qquad r = a_1 - a_2.$$

If we take for simplicity the case when

$$a_1 = a_2 \quad \text{and} \quad \epsilon_1 = \epsilon_2 = 0,$$

$$x = r \cos nt \qquad . \qquad . \qquad . \qquad . \qquad (16)$$

and

$$r = \{2a^2(1 + \cos mt)\}^{\frac{1}{2}} \qquad . \qquad . \qquad . \qquad . \qquad (17)$$

i.e. *the resultant amplitude varies between* $2a$ *and* 0, *the time interval between two successive maxima being* $2\pi/m$.

The phenomenon of '*beats*,' illustrated in fig. 7, is observed

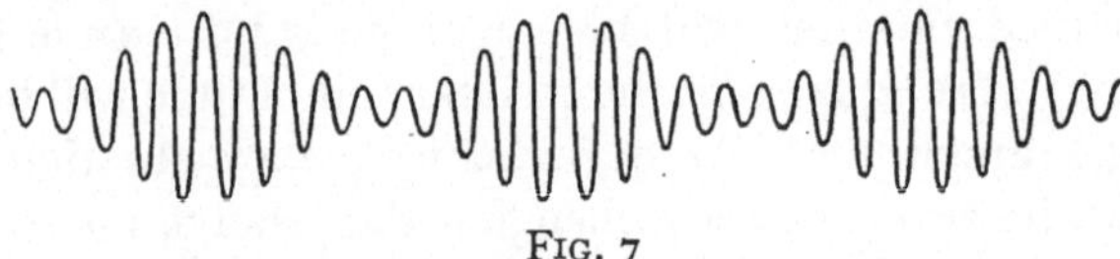

FIG. 7

when two notes nearly, but not quite, in tune are heard together. In such a case a periodic rise and fall of intensity is noticed. The explanation is afforded by the above analytical treatment, which indicates also the importance of arranging approximately equal component amplitudes to secure the greatest contrast between maxima and minima. When two notes of nearly equal pitch are sounded together the frequency of the beats is $m/2\pi$ if $n/2\pi$ and $(n+m)/2\pi$ are the frequencies of the two notes. This principle may be employed to tune two notes to unison with considerable accuracy. Under favourable conditions beats as slow as 1 in 30 seconds may be recognised.

The phenomenon of 'beats' has also a very important significance in the case of electrical oscillations of radio-frequency (of the order 10^6 cycles per second) and in the corresponding case of high-frequency sounds. The beat effect in these examples is generally known as '*heterodyne*.' Briefly, two very high and quite inaudible frequencies n and $(n+m)$ are rendered audible, after rectification, by slight mistuning—the number of beats per second m falling within the range of frequencies perceptible by the ear.

Case III - *Vibrations of the same frequency in directions at right angles*. In this case we are dealing with the motion of a particle in a plane. The projections of this motion on rectangular axes x and y each represent a simple harmonic motion. Let these motions be indicated by the equations

$$x = a\cos(nt+\epsilon_1) \quad \text{and} \quad y = b\cos(nt+\epsilon_2) \quad . \quad (18)$$

Since the axes are rectangular and the amplitudes of the motion a and b, the path of the particle must lie within a rectangle of sides $2a$ and $2b$. Eliminating t from these equations, we obtain the equation to the curve of resultant motion of the particles, viz.

$$\frac{x^2}{a^2}+\frac{y^2}{b^2}-\frac{2xy}{ab}\cdot\cos(\epsilon_1-\epsilon_2)-\sin^2(\epsilon_1-\epsilon_2)=0 \quad . \quad (19)$$

This is the equation of an ellipse whose position and dimensions depend upon the amplitudes a and b and the initial phases ϵ_1 and ϵ_2. *If* ϵ_1 *is equal to* ϵ_2 *or they differ by* π (half a period) the ellipse takes the form of two coincident straight lines

$$\left(\frac{x}{a}-\frac{y}{b}\right)^2=0.$$

If the phases differ by $\pi/2$ (quarter period), $\cos(\epsilon_1-\epsilon_2)=0$ and the equation becomes

$$\frac{x^2}{a^2}+\frac{y^2}{b^2}=1,$$

an ellipse whose axes coincide with the axes of co-ordinates x and y. If, in addition to this condition the amplitudes a and b are equal, the orbit of the particle becomes a circle with centre at the origin, viz. $x^2+y^2=a^2$.

Frequencies nearly Equal – When the two S.H. vibrations are exactly alike in frequency the elliptic paths referred to above remain perfectly steady. If, however, there is a slight difference between the frequencies, the form of the ellipse will slowly change. This effect is due to the gradual change in the relative phase $(\epsilon_1-\epsilon_2)$ of the two components. The boundary of all the possible elliptic orbits is a rectangle of sides $2a$ and $2b$, since the extreme values of x and y are $\pm a$ and $\pm b$ respectively. We shall now consider the effect of maintaining a and b constant whilst slowly varying $(\epsilon_1-\epsilon_2)$. Starting with the phases in agreement, $\epsilon_1=\epsilon_2=0$, the ellipse coincides with the diagonal $\frac{x}{a}-\frac{y}{b}=0$ of the rectangle. As $(\epsilon_1-\epsilon_2)$ increases from 0 to $\pi/2$ the ellipse opens out to the form $\frac{x^2}{a^2}+\frac{y^2}{b^2}=1$, afterwards closing up again and ultimately coinciding with the other diagonal $\frac{x}{a}+\frac{y}{b}=0$, as $(\epsilon_1-\epsilon_2)$ increases from $\pi/2$ to π. As the phase difference varies from π to 2π the reverse process takes place until the ellipse again coincides with the first diagonal. The cycle of changes is illustrated in fig. 8 (*a*). The frequency of this cycle is clearly equal to the difference of frequency of the component simple harmonic motions.

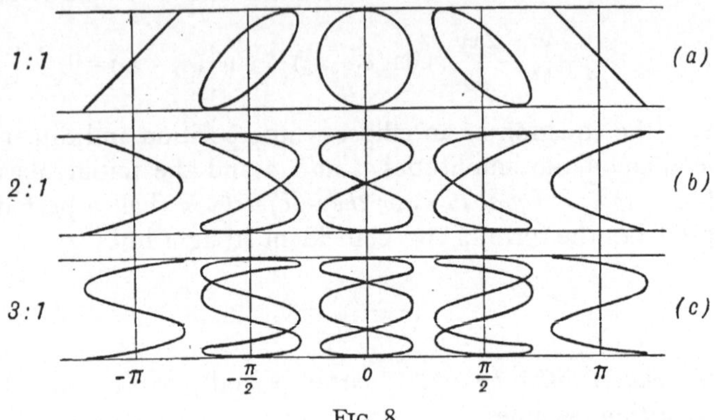

FIG. 8

Vibrations of this character may be demonstrated in a very convincing manner by means of the spherical pendulum. A heavy bob suspended on a thread or a wire describes an elliptic path with period $T=2\pi\sqrt{l/g}$ as it swings under the action of gravity. By varying the amplitudes of displacement along the x and y axes the bob can be made to describe the series of ellipses, circles, and straight lines shown in fig. 8 (*a*).

Case IV – *Vibrations, having frequencies commensurate (or nearly so), in directions at right angles.* When *the ratio of frequencies is* 2 : 1 the component vibrations are represented by

$$x=a\cos(2nt+\epsilon_1) \quad \text{and} \quad y=b\cos(nt+\epsilon_2) \qquad . \quad (20)$$

which, on elimination of t, give the resultant orbit of the particle, viz.

$$\left\{\frac{x}{a}-\sin(\epsilon_1-\epsilon_2)\right\}^2+\frac{4y^2}{b^2}\left\{\frac{y^2}{b^2}+\frac{x}{a}\sin(\epsilon_1-\epsilon_2)-1\right\}=0 \ . \quad (21)$$

This is the general equation for a curve having two loops which (for $(\epsilon_1-\epsilon_2)=m\pi/2$) degenerates into two coincident parabolas, represented by the equation

$$\left(\frac{2y^2}{b^2}+\frac{x}{a}-1\right)^2=0.$$

If the frequencies depart slightly from the ratio 2 : 1 the form of the curve slowly varies in a manner indicated by variations of $(\epsilon_1-\epsilon_2)$ in equation (21). Typical curves of this character are shown in fig. 8 (*b*).

When the ratio of frequencies is 3 : 1 an equation of the sixth degree represents the resultant orbit. Taking the case where

$$\epsilon_1=\epsilon_2=0, \quad x=a \cos 3nt \quad \text{and} \quad y=b \cos nt,$$

the equation obtained by eliminating t is

$$\left(\frac{4y^3}{b^3}-\frac{3y}{b}+\frac{x}{a}\right)^2=0,$$

representing two coincident cubic curves. When $(\epsilon_1-\epsilon_2)=\pi/2$ the equation is one of the sixth degree,

$$\left(1-\frac{y^2}{b^2}\right)\left(1-\frac{4y^2}{b^2}\right)^2-\frac{x^2}{a^2}=0 \quad . \quad . \quad . \quad (22)$$

which represents a curve having three loops. As in the previous cases, when the ratio of frequencies is not exactly 3 : 1, the form of the curve changes slowly in accordance with the change of phase relationship $(\epsilon_1-\epsilon_2)$. Typical phases of the curve are shown in fig. 8 (*c*). If the ratio of frequencies is a whole number 'N' the number of loops in the figure will also be N. This analytical process of determining the resultant motion of a particle acted on by simple harmonic forces at right angles becomes rather cumbrous when the ratio of frequencies is greater than 3, and graphical methods become more suitable.

Blackburn's Pendulum – A very simple experimental method of investigating the combination of rectangular vibrations is that known as Blackburn's pendulum. Taking the general case of two motions at right angles,

$$x=a \cos (knt+\epsilon_1) \quad \text{and} \quad y=b \cos (nt+\epsilon_2) \quad . \quad (23)$$

where k is a constant representing the ratio of the frequencies of the component vibrations, it is possible by means of this pendulum to determine the resultant motion. Blackburn's pendulum is of simple construction. It consists of a weight P hanging on a string OP attached at O to another string AB fixed at the ends (see fig. 9). Neglecting the inertia of the strings, the point P will always be in the same plane as AOB. It will be seen that the motion of P may be either that of a simple pendulum of length OP (vibrating *in* the plane of the diagram), or of a simple pendulum of length NP (vibrating *at right angles* to the plane of the diagram), or of a combination of these two vibrations. Thus if OP$=l_1$ and NP$=l_2$, the two component frequencies will be

$$f_1=\frac{1}{2\pi}\sqrt{\frac{g}{l_1}} \quad \text{and} \quad f_2=\frac{1}{2\pi}\sqrt{\frac{g}{l_2}},$$

respectively. The resultant motion of the bob will in the general case be represented by equation (23) above.

Various modifications of this pendulum, to give simple controls of the relative frequencies, amplitudes, and initial phases, have been described* and the reader is referred to these descriptions for further details of this fascinating and instructive apparatus.

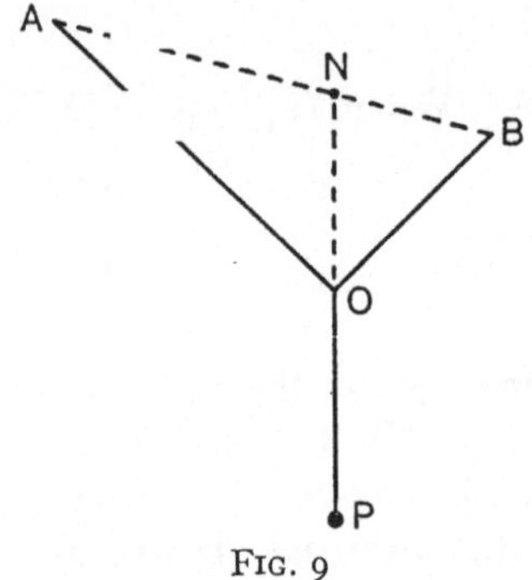

FIG. 9

The numerous figures obtainable by means of Blackburn's pendulum are generally known as *Lissajous' figures*. The figures were first produced by Lissajous using an optical method as an accurate means of comparison of vibration frequencies of two systems in motion at right angles. The experiment is most directly carried out by means of two tuning-forks vibrating in planes at right angles. One prong of each fork carries a small mirror, and a spot of light is reflected successively from the two mirrors on to a screen where the combined motion of the two forks is clearly shown. Many other optical and mechanical devices have been described for producing Lissajous' figures. Wheatstone's "kaleidophone" (1827) gives some of these figures in a very simple way. It consists of a straight rod clamped in a vice and carrying an illuminated bead at its upper end. If the rod is circular in section the end will describe a circle, ellipse, or a straight line—but on account of slight differences in elasticity in different directions the figure usually changes and revolves. By bowing the rod at different points, vibrations of overtones are superposed on the fundamental. In this way the frequencies of the first and second overtones were found by Tyndall to be 6·25 and 17·36 times the frequency of the fundamental (see p. 113).

* (1) *Étude Optique des mouvements Vibratoires.* J. A. Lissajous. Published 1873. (2) *Harmonographs and Vibration Figures.* Gould, Benham, Kerr, and Wilberforce (Newton & Co., London). See also W. Pupp (*Phys. Zeits.*, **30**, p. 427, July 1929).

By means of the cathode-ray oscillograph * the formation of Lissajous' figures at any frequency (*e.g.* of the order 10^6) is very simply demonstrated. In this case the 'vibrations' at right angles are electrical and produce proportionate displacements of a beam of cathode rays which subsequently fall on a photographic plate or a phosphorescent screen, where the resultant motion, *i.e.* the Lissajous figure, is indicated.

As we have already seen, the frequency of repetition of the cycle of Lissajous' figures is equal to the difference of frequency of the component vibrations when these are nearly equal—corresponding to the effect of beats in the case where the component vibrations are co-linear.

The production of stationary, or slowly-changing, Lissajous' figures has an important application in the comparison of frequencies of vibration. Thus König studied the effect of temperature on the frequency of a tuning-fork by means of the Lissajous' figures formed by this fork and another 'standard' fork. In a similar manner he investigated the effect on frequency of a fork, produced by proximity of a resonator having varying degrees of slight mistuning. Recently D. W. Dye has employed the cathode ray oscillograph for the harmonic comparison of very high frequencies (10^5 cycles/sec.)—the high frequency being adjusted to an exact multiple of a 1000 p.p.s., the frequency of a standard fork. The figures reproduced in Dye's paper † at these high frequencies bear a striking resemblance to those in Tyndall's R.I. lectures on Sound, ‡ at low frequencies of the order of 10 to 1000 p.p.s.

Fourier's Theorem

Summation of any number of simple harmonic vibrations of commensurate frequencies—the vibrations being in the same straight line Synthesis and analysis of complex forms of vibration. In Case II (a) above, a brief reference was made to the combination of S.H. vibrations of commensurate frequencies. A very important theorem which deals with this case is due to J. B. J. Fourier. § The theorem is stated in various forms by different writers, and is generally quoted in relation to the study of the transverse vibration of strings. Its application is, however, of much wider scope than this. Briefly, the theorem states that *any single-valued periodic function whatever can be expressed as a summation of simple*

* A. B. Wood, *Proc. Phys. Soc.*, **35**, p. 109, 1923.

† *Proc. Phys. Soc.*, **37**, part 3, April 1925, pp. 158–168.

‡ Pp. 133–135.

§ *Theorie de la Chaleur* (Paris, 1822).

harmonic terms having frequencies which are multiples of that of the given function.

The theorem not only deals with the synthesis of a complex periodic vibration from simple harmonic terms, but also indicates a method by which a complex vibration can be analysed into its component simple vibrations.

It should be noted that the theorem has two provisos regarding the form of the complex vibration, viz. (1) that the displacement must be single-valued and continuous—it is obvious that this condition is fulfilled in all cases of mechanical vibrations, *e.g.* a particle cannot actually have two different displacements at the same instant of time; (2) that the displacement must always have a finite value—this condition also is clearly fulfilled in the case of sound. Fourier's theorem may be expressed by the series *

$$\left.\begin{aligned} x=f(nt)&=a_0+a_1\cos(nt+\epsilon_1)+a_2\cos(2nt+\epsilon_2)+\ldots \\ &\qquad +a_r\cos(rnt+\epsilon_r) \\ \text{that is,}\quad & \\ x=f(nt)&=a_0+\sum_{r=1}^{r=\infty} a_r\cos(rnt+\epsilon_r) \end{aligned}\right\} \quad . \quad (1)$$

In this equation x is the displacement of the complex periodic vibration of frequency $n/2\pi$; a_1, a_2, a_r are the amplitudes of the component simple harmonic vibrations, ϵ_1, ϵ_2, ϵ_r being the respective initial phases. It is sometimes convenient in analysis to express x as the sum of a sine and a cosine series. This can easily be done by writing $A=a\cos\epsilon$ and $B=-a\sin\epsilon$, when we obtain

$$\left.\begin{aligned} x=f(nt)=A_0+A_1\cos nt+A_2\cos 2nt+\ldots+A_r\cos(r\,.\,nt) \\ +B_1\sin nt+B_2\sin 2nt+\ldots+B_r\sin(rnt) \end{aligned}\right\} \quad (2)$$

Integrating this equation, with respect to t, for a complete vibration of period $T=2\pi/n$, we find

$$A_0=\frac{1}{T}\int_0^T x\,dt \quad . \quad . \quad . \quad . \quad (3)$$

To obtain the amplitude-values A_r of the cosine series we multiply all terms of equation (2) by $\cos(r\,.\,nt)$ and integrate for a complete cycle (0 to T), whence

$$A_r=\frac{2}{T}\int_0^T x\cos(r\,.\,nt)\,dt \quad . \quad . \quad . \quad (4)$$

---

* The analytical proof of this theorem is beyond the scope of this book. Reference should be made to Fourier's original work or to various mathematical treatises on the subject. H. S. Carslaw, *Fourier's Series and Integrals*, London, 1906. Lamb (*Dynamical Theory of Sound*) deals with the theorem in a very helpful manner, but does not give a rigid proof.

Similarly the amplitude-values B_r of the sine series are obtained by multiplying all terms of equation (2) by $\sin(r\,.\,nt)$ and integrating as before

$$B_r = \frac{2}{T}\int_0^T x \sin(r\,.\,nt)dt \quad . \quad . \quad . \quad (5)$$

By means of equations (2) to (5) it is possible to analyse or synthesise any complex form of vibration. The constant amplitude term A_0 or a_0 is a measure of the displacement of the axis of the vibration curve from the axis of co-ordinates. If the abscissa of the curve coincides with the axis of the curve the constant vanishes. Equation (3) shows that A_0 represents the mean displacement. Measuring from the lowest point of the displacement curve to the mean value of x we obtain the *axis of the curve.*

The terms required to construct a given curve may be limited or may be infinite in number. Certain harmonics of the series may be missing entirely. As an example consider the case of a combination of two S.H.M.'s of nearly equal frequency resulting in 'beats' (see fig. 7). If the two frequencies are, say, 300 and 302, there will be two beats per second, *i.e.* the fundamental frequency of the resultant is 2 p.p.s. Analysis of the 'beat' vibration will therefore indicate that only the 300th and 302nd terms of Fourier's series are present—the fundamental of the series will have a frequency of 2 p.p.s. but zero amplitude.

Examples illustrating Analysis and Synthesis by Fourier's Theorem

EXAMPLE (1) – *Displacement x constant (equal to c) from $t=0$ to* T/2, *and zero from* T/2 *to* T (a 'square' wave), see fig. 10.

$$x=f(nt)=c \quad \text{from} \quad t=0 \text{ to } T/2,$$
$$x=f(nt)=0 \quad \text{from} \quad t=T/2 \text{ to } T,$$

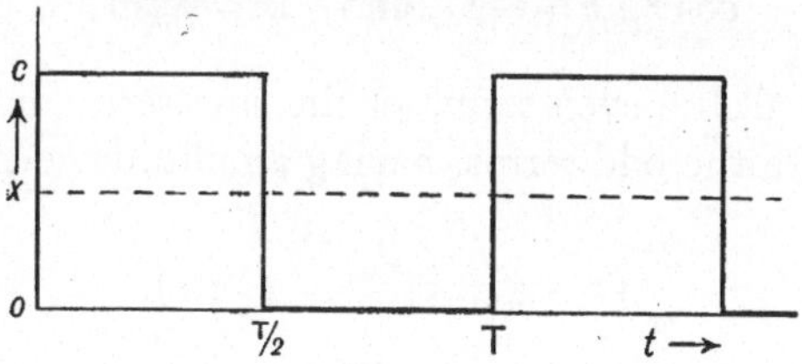

c
x
0
T/2
T
$t \rightarrow$

FIG. 10

when the time axis of co-ordinates is taken through the lowest point of the displacement curve. From equation (3) we have for A_0,

$$A_0=\frac{1}{T}\int_0^T xdt,$$

which may be written

$$A_0=\frac{1}{T}\int_0^{T/2} cdt=\frac{c}{2},$$

i.e. the axis of the displacement curve is the line $x=c/2$.

From equation (4)

$$A_r=\frac{2}{T}\int_0^T x\cos(rnt)dt,$$

which reduces to

$$A_r=\frac{c}{\pi r}\left[\sin\frac{r\,.\,2\pi t}{T}\right]_0^{T/2}=0,$$

when we remember that $T=2\pi/n$, *i.e.* all the cosine terms of the Fourier series have zero amplitude, and consequently disappear altogether.

Similarly, from equation (5),

$$B_r=\frac{2}{T}\int_0^T x\sin(rnt)dt,$$

which becomes

$$=\frac{c}{\pi r}\left[-\cos r\,.\,\frac{2\pi t}{T}\right]_0^{T/2}=\frac{c}{\pi r}[-\cos r\,.\,\pi+1].$$

When r is even

$$\cos r\,.\,\pi=+1 \quad\text{and}\quad B_r=0;$$

when r is odd

$$\cos r\,.\,\pi=-1 \quad\text{and}\quad B_r=2c/\pi r.$$

Consequently all the even terms of the sine series disappear, and we are left with the odd terms, having amplitude coefficients

$$\frac{2c}{\pi}(1,\ \tfrac{1}{3},\ \tfrac{1}{5},\ \tfrac{1}{7},\ \tfrac{1}{9},\text{ etc}).$$

The complete series therefore becomes

$$x=f(nt)=\frac{c}{2}+\frac{2c}{\pi}\left[\sin nt+\frac{1}{3}\sin 3nt+\frac{1}{5}\sin 5nt+\ \ldots\ +\frac{1}{r}\sin r\,.\,nt\right].$$

In fig. 11 the addition of these terms is indicated graphically. In (*a*) the first three terms of the series are shown independently, and as a resultant curve. It will be seen that even in this case

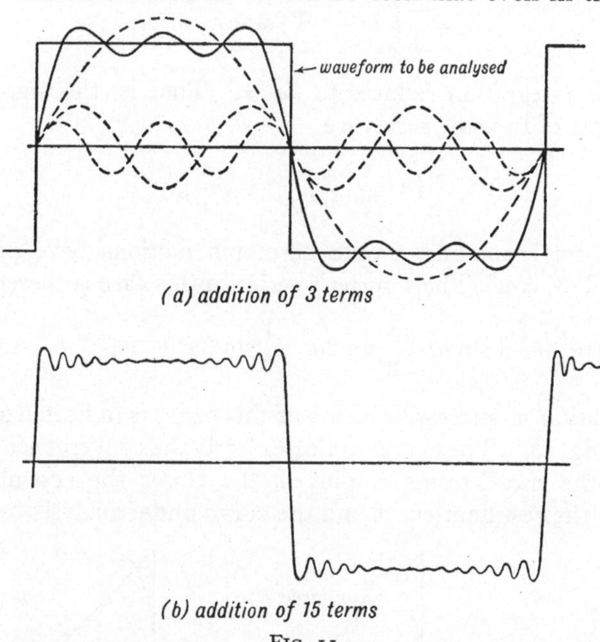

FIG. 11

the resultant bears a rough resemblance to the curve to be analysed. By the addition of 15 terms of the series shown in (*b*) the curve is almost perfectly reproduced.

EXAMPLE (2) – *Displacement linear with time from $x=c$ when $t=0$ to $x=0$ when $t=\mathrm{T}$* (see fig. 12). A 'saw-tooth' wave.

$$x=f(nt)=c(1-t/\mathrm{T})$$

when the time axis passes through the lowest points of the periodic curve.

FIG. 12

$$\mathrm{A}_0=\frac{1}{\mathrm{T}}\int_0^{\mathrm{T}} x\,dt=\frac{c}{\mathrm{T}}\int_0^{\mathrm{T}}(1-t/\mathrm{T})\,dt=\frac{c}{2},$$

which gives the ordinate of the axis of the curve.

Again

$$\mathrm{A}_r=\frac{2c}{\mathrm{T}}\int_0^{\mathrm{T}}(1-t/\mathrm{T})\cos(rnt)\,dt,$$

which, on integration between the required limits, reduces to zero for all values of r, *i.e.* the cosine series becomes zero.

$$B_r = \frac{2c}{T}\int_0^T (1 - t/T)\sin(r\,.\,nt)dt,$$

which on integration reduces to $2c/\pi r$. That is, the amplitude coefficients of the sine series are

$$\frac{2c}{\pi}\left(1\,.\,\frac{1}{2}\,.\,\frac{1}{3}\,.\,\frac{1}{4}\,.\,.\,.\,\frac{1}{r}\right),$$

and the corresponding simple harmonic motions have periods T, T/2, T/3, etc. The Fourier's series in this case is therefore

$$x = f(nt) = \frac{c}{2} + \frac{2c}{\pi}\left(\sin nt + \frac{1}{2}\sin 2nt + \frac{1}{3}\sin 3nt + \;.\;.\;.\; + \frac{1}{r}\sin rnt\right).$$

The addition of successive terms of this series is indicated graphically in fig. 13. The curves indicate clearly the convergence of the series; the more terms employed, the closer the resemblance between the resultant curve and the curve under analysis.

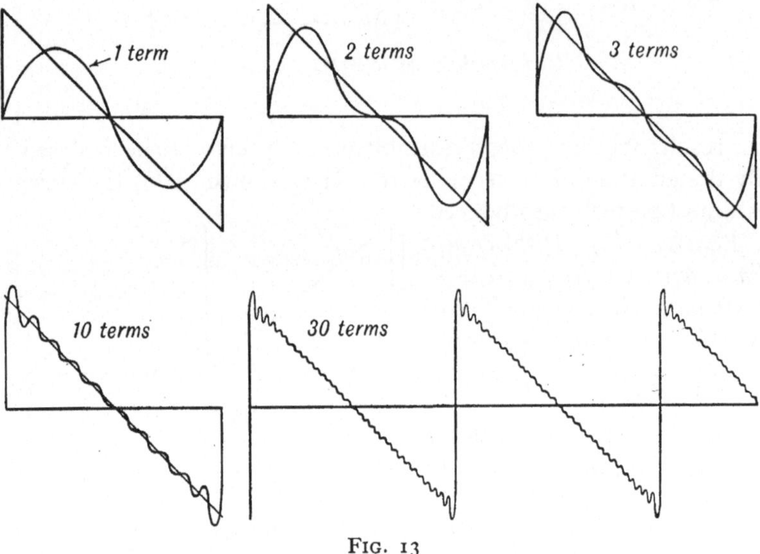

FIG. 13

EXAMPLE (3) – *Two straight lines* meeting at $x = c$, $t = T/2$ and passing through $x = 0$ at $t = 0$ and $t = T$ respectively (see fig. 14 (a)). In this case we have

$$x = 2ct/T \quad \text{from} \quad t = 0 \text{ to } t = T/2$$

and

$$x=2c(1-t/\mathrm{T}) \quad \text{from} \quad t=\mathrm{T}/2 \text{ to } t=\mathrm{T}.$$

$$A_0=\frac{1}{\mathrm{T}}\int_0^{\mathrm{T}/2}\frac{2c}{\mathrm{T}}\,.\,t\,.\,dt+\frac{1}{\mathrm{T}}\int_{\mathrm{T}/2}^{\mathrm{T}}2c\left(1-\frac{t}{\mathrm{T}}\right)dt=\frac{c}{2}$$

$$A_r=\frac{2}{\mathrm{T}}\left[\int_0^{\mathrm{T}/2}\frac{2ct}{\mathrm{T}}\cos(rnt)dt+\int_{\mathrm{T}/2}^{\mathrm{T}}2c\left(1-\frac{t}{\mathrm{T}}\right)\cos(r\,.\,nt)dt\right]$$

$$=\frac{2c}{\pi^2r^2}[(-1)^r-1].$$

For even values of r the values of A_r become zero. When r is odd $A_r=-\dfrac{4c}{\pi^2r^2}$.

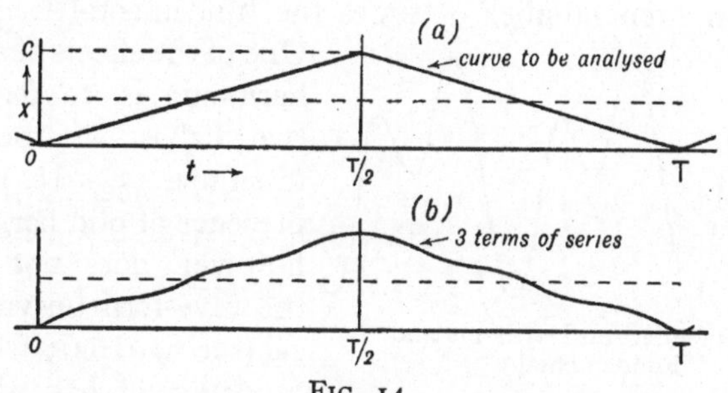

FIG. 14

$$B_r=\frac{2}{\mathrm{T}}\left[\int_0^{\mathrm{T}/2}\frac{2ct}{\mathrm{T}}\,.\sin(r\,.\,nt)dt+\int_{\mathrm{T}/2}^{\mathrm{T}}2c\left(1-\frac{2t}{\mathrm{T}}\right)\sin(r\,.\,nt)dt\right]=0,$$

i.e. all the sine terms disappear. The Fourier series thus reduces to a series of cosine terms of amplitude coefficients,

$$-\frac{4c}{\pi^2}\left(1,\ \frac{1}{3^2},\ \frac{1}{5^2},\ \frac{1}{7^2},\ \text{etc.}\right)$$

and periodic times T, T/3, T/5, T/7, etc. Thus

$$x=f(nt)=\frac{c}{2}-\frac{4c}{\pi^2}\left(\cos nt+\frac{1}{3^2}\cos 3nt+\frac{1}{5^2}\cos 5nt+\ldots+\frac{1}{r^2}.\cos r\,.\,nt\right).$$

The resultant obtained by adding three terms of this series is shown in fig. 14 (*b*). It will be observed that the series is rapidly convergent, and therefore the curve under analysis is closely represented by the sum of comparatively few terms of the series. In this respect, compare with fig. 13, where the series is not so rapidly convergent. Writing $n=2\pi/\mathrm{T}$ in the above equation for

x, we may obtain a useful check on the accuracy of the result. Thus when t is an odd multiple of T/4, the displacement x is always $c/2$, *i.e.* the particle is then passing through the mid-point of its vibration—this is evident from a consideration of the displacement curve (fig. 14). Again, when t is an odd multiple of T/2, the displacement x has always reached its maximum value c, and we have

$$x=\frac{c}{2}-\frac{4c}{\pi^2}\left(1+\frac{1}{3^2}+\frac{1}{5^2}+\ \ldots\ \text{to infinity}\right),$$

writing $x=c$ this becomes $\pi^2/8=1+\frac{1}{3^2}+\frac{1}{5^2}+\ \ldots$ to infinity, which is known to be correct.

Even and Odd Harmonics – An even harmonic has a frequency which is an even number of times the fundamental frequency, One of the effects of an even harmonic is to make the two halves of the wave dissimilar (fig. 15A). The presence of odd harmonics, however, does not render the wave-form unsymmetrical (see fig. 15B); when the fundamental has advanced through half a period the odd harmonics have advanced through a number of complete periods plus half a period, hence their instantaneous values are in the same direction relative to the instantaneous value of the fundamental.

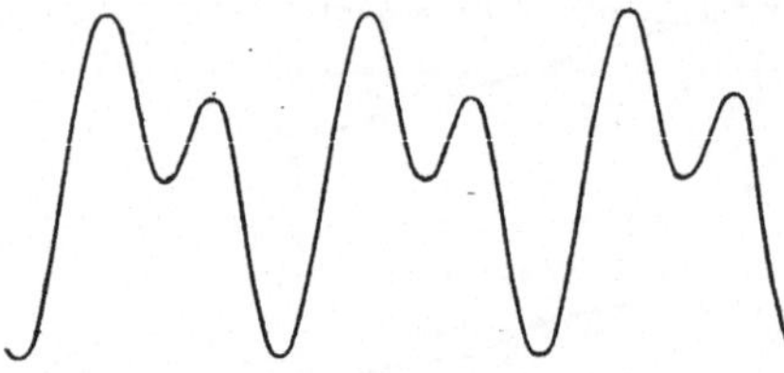

FIG. 15A—Second Harmonic and Fundamental

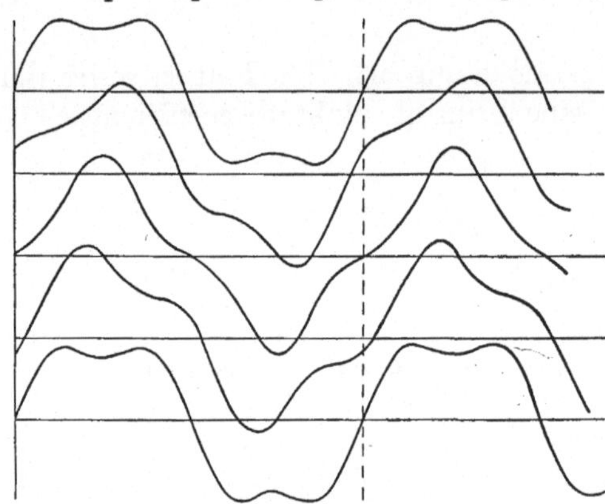

FIG. 15B—Third Harmonic and Fundamental (illustrating effect of varying relative phase)

The relative phase of the harmonics and the fundamental affects the resulting character of the wave to a marked extent. This is illustrated in the series of curves shown in fig. 15 (*b*), which result from the substitution of different values for the angle ϵ in the expression $x=B_1 \sin \theta + B_3 \sin (3\theta+\epsilon)$, where the amplitude B_3 of the third harmonic is one quarter B_1, the amplitude of the fundamental.

The above examples of the application of Fourier's theorem to the analysis and synthesis of comparatively simple types of periodic vibration, also serve to indicate the complexity of the calculation when the curve between 0 and T is very irregular. In such cases the analysis by calculation is a long and tedious process requiring perhaps several days' work for a single curve. In order to simplify the process when many complex periodic curves have to be examined, various types of mechanical devices have been constructed. These devices, known as '*harmonic analysers*,' are designed to perform mechanically the integrations indicated in equations (3), (4), and (5) above. Instruments of this nature have been devised by Kelvin (*Proc. Roy. Soc.*, **27**, 371, 1878), Henrici (*Phil. Mag.*, **38**, 110, 1894), Michelson and Stratton (*Amer. Journ. Science*, **5**, 1–13, 1898), and many others.* Fourier's theorem and 'harmonic analysis' has a wide field of application, not only in the study of sound, but in astronomy, meteorology, tide prediction, mechanical and electrical engineering.

Damped Vibrations

Frictional Forces – It has been assumed up to this point that all vibrations have a constant amplitude for an indefinite time. Dissipative, or frictional forces, involving losses of energy in heat and other forms of mechanical motion, have been disregarded. There is no simple law which indicates the variation of frictional forces when a body moves through a medium offering resistance to its motion. Experiment has shown that the law of resistance changes with the velocity of the body. For small velocities, however, it is generally assumed that the retarding force, due to the medium, is proportional to velocity. The problem of the vibration of a particle subject to a restoring force proportional to its displacement and a resistance varying as the velocity is very important. All vibrating bodies are subject to such forces,

* An interesting description of harmonic analysis and synthesis, with a valuable list of references, is given by D. C. Millar in *The Science of Musical Sounds* (Macmillan : New York, 1916).

otherwise there would be no loss of energy by the body and consequently no radiation of sound energy.

The equation of motion of a vibrating particle of mass m subject to a damping force proportional to the velocity is

$$m\frac{d^2x}{dt^2}+r\frac{dx}{dt}+sx=0 \quad . \quad . \quad . \quad (1)$$

where s is the 'spring factor' (stiffness factor) or the restoring force per unit displacement, and r is the 'resistance constant' or retarding force per unit velocity. Equation (1) may conveniently be solved as follows. Writing

$$x=ye^{-kt} \quad . \quad . \quad . \quad . \quad (2)$$

we have

$$\frac{dx}{dt}=\left(\frac{dy}{dt}-ky\right)e^{-kt} \quad . \quad . \quad . \quad (3)$$

and

$$\frac{d^2x}{dt^2}=\left(\frac{d^2y}{dt^2}-2k\frac{dy}{dt}+k^2y\right)e^{-kt} \quad . \quad . \quad (4)$$

Substituting in equation (1) we have

$$\frac{d^2y}{dt^2}+\left(\frac{r}{m}-2k\right)\frac{dy}{dt}+\left(k^2-\frac{r}{m}k+\frac{s}{m}\right)y=0 \quad . \quad . \quad (5)$$

Writing $s/m=n^2$ (where $2\pi/n$ is the period of the free oscillations of the unrestrained particle), and $r/2m=k$ (since we are at this stage free to choose any convenient value for the constant k), equation (5) becomes

$$\frac{d^2y}{dt^2}+(n^2-k^2)y=0 \quad . \quad . \quad . \quad (6)$$

Equation (1) is therefore satisfied by

$$x=e^{-kt}y \quad . \quad . \quad . \quad . \quad (7)$$

provided the value of y is defined by equation (6). Three important cases arise:

Case I – *Frictional Forces Small.* $n>k$. This stipulation makes (n^2-k^2), in equation (6), a positive quantity. Consequently if we write $\sqrt{(n^2-k^2)}=n'$ the solution of (6) is $y=a\cos(n't+\epsilon)$. Substituting this value of y in equation (7) we obtain

$$x=ae^{-kt}\cos(n't+\epsilon) \quad . \quad . \quad . \quad (8)$$

This relation for x represents geometrically a simple harmonic curve of which the amplitude ae^{-kt} diminishes exponentially to zero with increasing time. Since the periodic term $\cos(n't+\epsilon)$ alternates between the limits ± 1, the space-time curve of x lies between the curves $x=\pm ae^{-kt}$ (see fig. 16). The measure of the decay of amplitude is the constant k, called the '*damping coefficient.*' This constant depends on the ratio of resistance to mass $(r/2m)$—it should be noted, however, that the rate of dissipation of energy $(r\dot{x}^2)$ in the system due to friction depends only on the resistance factor r.

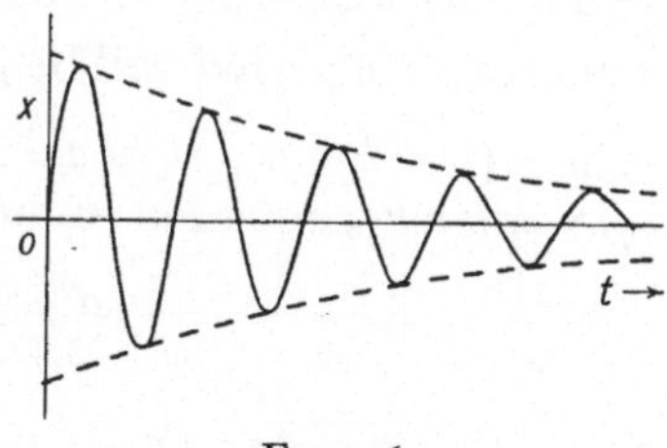

FIG. 16

The ratio of successive maximum amplitudes (on opposite sides of zero displacement) is $e^{k\pi/n}$, and the logarithm of this ratio,

$$\frac{k\pi}{n}\log_{10} e = 0{\cdot}217\frac{k\pi}{n},$$

is known as the '*logarithmic decrement*' of the oscillations. Reference is sometimes made to the '*modulus of decay*' of damped oscillations; this is defined by the time taken for the amplitude to fall in the ratio $1/e$.

A simple harmonic vibration, undamped, can be represented by the orthogonal projection of a uniform circular motion. Similarly the damped vibration which we have just been considering is obtained from the orthogonal projection of an equiangular spiral, described at constant angular velocity with the radius vector diminishing at a uniform rate.

Effect of Damping on Frequency – In equation (8) above, the period of the damped oscillation is $2\pi/n'$, where $n'=\sqrt{n^2-k^2}$ and $2\pi/n$ is the undamped period. The frequency is therefore diminished by the damping. In most cases occurring in practice, however, the difference between n and n' is a very small quantity of the second order, and may usually be neglected, *i.e.* the effect of damping on the period is negligible, except in a few extreme cases where the vibrations are strongly damped. As an example of the effect of damping on frequency take the case of a tuning-fork of frequency 25 p.p.s. ($n=157$), and damping coefficient $k=0{\cdot}1$ (found experimentally from observation of the time required for the amplitude to fall to half its initial value).*

$$n' = \sqrt{n^2 - k^2} = \sqrt{24{,}649 - 0{\cdot}01} = 157 - 3{\cdot}2 \times 10^{-5}.$$

That is, the damping has lowered the frequency by one part in five millions, a quantity which in practice is usually quite negligible. In general, the period will be given by $2\pi\sqrt{m/s}$ $(=2\pi/n)$.

Case II - *Large Frictional Forces.* $k>n$. In this case of *very heavy damping*, equation (6) becomes

$$\frac{d^2y}{dt^2} - (n')^2 y = 0,$$

where $n' = \sqrt{k^2 - n^2}$. The solution is $y = Ae^{n't} + Be^{-n't}$, whence

$$x = Ae^{-k_1 t} + Be^{-k_2 t} \text{ (by equation (7))} \quad . \quad . \quad (9)$$

if

$$k_1 = k - \sqrt{k^2 - n^2} \quad \text{and} \quad k_2 = k + \sqrt{k^2 - n^2}.$$

The particle does not 'vibrate,' but the displacement after passing its first maximum decays asymptotically to zero. This is the case of an '*overdamped*,' '*deadbeat*,' or *aperiodic* displacement, illustrated in a moving coil galvanometer shunted by a very low resistance. It has, in common with the following Case III, important applications in the electrical recording of sound by means of an oscillograph, and in the design of certain forms of sound receivers.

Case III - $k=n$. Equation (6) now becomes $\frac{d^2y}{dt^2} = 0$, whence

$$y = At + B \quad \text{and} \quad x = e^{-kt}(At + B) \quad . \quad . \quad (10)$$

This is the transition case of *critical damping*, the motion being just aperiodic or non-oscillatory. The displacement at first increases by virtue of the factor $(At+B)$, but the exponential term becomes relatively important as t increases. The displacement can only become zero for one finite value of t.

The case of critical damping has an important bearing on the problem of recording sound vibrations either mechanically or electrically.

Forced Vibrations. Resonance

So far we have dealt only with 'free' vibrations where no external forces, apart from frictional forces, are applied to the

* See A. B. Wood, "Electrically Maintained Tuning-Forks. Some Factors affecting Frequency," *Journ. Sci. Instrs.*, **1**, p. 330, 1924.

vibrating system. Such vibrations, as we have seen, diminish in amplitude in accordance with certain simple laws, and ultimately die away.

We now come to consider the very important case where a mass is maintained in a state of vibration by a periodic force which may or may not have the same period as the free vibration. As we shall see, the mass in such a case will ultimately vibrate with the same period as that of the force, its vibrations being designated *forced vibrations*. In the special case where the period of the free vibrations coincides with the period of the force we have the phenomenon of *resonance*.

Let the external periodic force be represented by $\mathrm{F}\cos pt$, where F is the maximum value of the force of frequency $p/2\pi$, then the equation of motion of a particle of mass m is given by

$$m\frac{d^2x}{dt}+r\frac{dx}{dt}+sx=\mathrm{F}\cos pt \quad . \quad . \quad . \quad (1)$$

If we write $s/m=n^2$ and $r/2m=k$ as before, and $\mathrm{F}/m=f$, equation (1) becomes

$$\frac{d^2x}{dt^2}+2k\frac{dx}{dt}+n^2x=f\cos pt \quad . \quad . \quad . \quad (2)$$

Assume $x=a\cos(nt-\epsilon)$ and writing $\cos pt=\cos\{\epsilon+(pt-\epsilon)\}$, we obtain on substitution in (2)

$$a(n^2-p^2)\cos(pt-\epsilon)-2kpa\sin(pt-\epsilon)=f\cos\epsilon\,.\,\cos(pt-\epsilon)$$
$$-f\sin\epsilon\sin(pt-\epsilon) \quad . \quad (3)$$

Now equation (3) must hold for all values of t, consequently we may equate coefficients of $\sin(pt-\epsilon)$ and of $\cos(pt-\epsilon)$ and obtain

$$\left.\begin{aligned} a(n^2-p^2)&=f\cos\epsilon \\ \text{and}\qquad 2kpa&=f\sin\epsilon \\ \text{whence}\qquad a&=\frac{f\sin\epsilon}{2kp} \\ \text{and}\qquad \tan\epsilon&=2kp/(n^2-p^2) \end{aligned}\right\} \quad . \quad . \quad . \quad (4)$$

The solution is therefore

$$x=\frac{f\sin\epsilon}{2kp}\cos(pt-\epsilon) \quad . \quad . \quad . \quad (5)$$

where

$$\tan\epsilon=2kp/(n^2-p^2)\ (\text{from (4)}) \quad . \quad . \quad (6)$$

Equation (5) represents the *forced vibration, i.e.* the vibration maintained by the external force. It will be seen that the amplitude $a=f\sin\epsilon\,.\,/2kp$ is proportional to f the magnitude of the force, and the period is the same as that of the force. In the case where there is *no friction*

$$k=0,\quad \sin\epsilon=0,\quad \epsilon=0 \text{ or } \pi,\quad \text{and}\quad a=f/(n^2-p^2).$$

Thus

$$x=\frac{f}{n^2-p^2}\cos pt \qquad . \quad . \quad . \quad (5a)$$

Equation (5) does not represent the *complete* solution of the fundamental equation (2). We may add any value of x which makes the expression on the left of equation (2) vanish. Such values of x have been obtained (*e.g.* see equation (8), p. 34) in the case of 'free' oscillations subject to a damping force. Hence, in the case of $n>k$, we have the complete solution

$$x=ae^{-kt}\cos(n't+\epsilon')+\frac{f\sin\epsilon}{2kp}\cos(pt-\epsilon) \qquad . \quad (7)$$

The first term on the right of this equation represents a 'free' damped oscillation which is superposed on the forced oscillation represented by the second term. On account of the exponential factor, the free oscillations gradually die out and, after a time, only the effects of the forced vibration are appreciable.* The same conclusion holds also in the cases when $n<k$ and $n=k$.

Amplitude of Forced Vibrations – In the steady state of maintained oscillations, we may regard equations (5) and (6) as representing the motion produced by the external force. There

* An interesting theoretical and experimental study of this *initial* stage of forced vibrations has recently been made by L. W. Blau (*Journ. Frank. Inst.*, **206**, p. 355, Sept. 1928). He writes equation (2) in a different form which exhibits a number of important characteristics of the initial vibrations. Writing $s=(n-p)$ and $\Delta n=n-\sqrt{n^2-k^2}$ it is found (*a*) when $s \neq \Delta n$, there are 'beats,' and the first beat *maximum* is greater than any later maximum, whilst the first beat *minimum* is less than any later beat minimum; (*b*) when $(n^2-p^2)\doteq k^2$ there are no beats and the resultant amplitude grows monotonically from zero to the steady amplitude of forced vibrations; (*c*) at resonance when $n=p$, there are still maxima which occur with a frequency $\Delta n/2\pi$ in a damped system; (*d*) the absence of beats is neither a sufficient nor a necessary condition of resonance in a damped system. Blau obtained photographic records of the forced oscillations of a pendulum with varying degrees of damping, these records show the initial 'beats' and confirm the theoretical deductions.

are several important points to be noticed in regard to the forced vibration. The *amplitude A of the forced vibration* is $f \sin \epsilon . /2kp$. Now

$$\sin \epsilon = 2kp/\{(n^2-p^2)^2+4k^2p^2\}^{\frac{1}{2}} \text{ from (6).}$$

Therefore

$$A = f/\{(n^2-p^2)^2+4k^2p^2\}^{\frac{1}{2}} \quad . \quad . \quad . \quad (8)$$

giving the amplitude of the forced vibration in terms of (1) the impressed force f ($=F/m$), (2) the free frequency $n/2\pi$ ($n^2=s/m$), (3) the forced frequency $p/2\pi$, and (4) the damping k ($=r/2m$). This expression is of fundamental importance.

Now the amplitude A is a maximum when the expression in brackets in equation (8) is a minimum, as regards variation of the forced frequency. Differentiating this expression with respect to p and equating to zero, we find for the minimum condition $p^2=n^2-2k^2$ *provided* $n^2>2k^2$. Substituting this value of p^2 in equation (8) we obtain for the maximum value of amplitude, when k is small compared with n, that is, $n=p$ nearly,

$$A_{max} = f/2kn \quad . \quad . \quad . \quad . \quad (9)$$

In the case where there is no damping ($k=0$) and the free frequency is the same as the forced frequency ($n=p$), it will be seen that the amplitude A becomes infinite. Such a case, however, never occurs in practice, for the damping k is never zero. For small values of damping ($n^2>2k^2$), the forced amplitude is greatest when the forced frequency coincides with that of the free vibration in the absence of damping. This condition is known as *resonance*. Frequency amplitude curves for varying degrees of damping are shown in fig. 17.

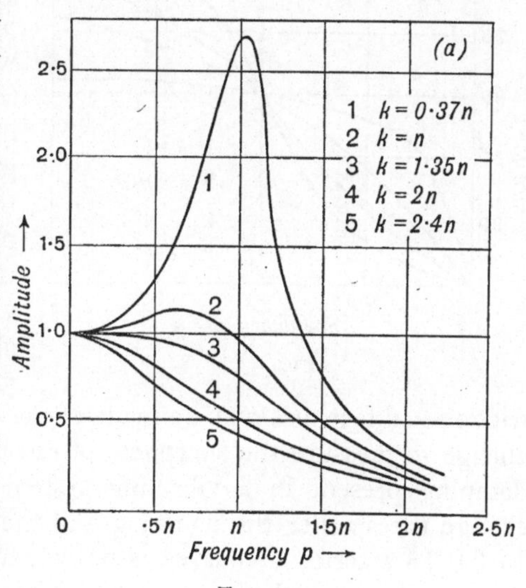

FIG. 17

Phase of Forced Vibrations – In general, the force and the resultant forced vibration will not be in phase. Equation (5)

indicates a lag ϵ between the displacement and the force, the magnitude of this lag being determined by equations (4) and (6). From (4) we see that $\sin \epsilon$ is always positive, whilst $\cos \epsilon$ may be positive or negative according as n is greater or less than p. The angle ϵ therefore lies between 0 and $\pi/2$ when the frequency (p) of the force is less than the frequency (n) of the 'free' vibration, and between $\pi/2$ and π when the forced frequency is greater than the free frequency. The lag ϵ (given by equation (6)) will, in general, *if the damping be small*, be nearly equal to 0 or π in these cases, i.e. *the displacement will be in phase or out of phase with the force according as the frequency of the force is less or greater than the frequency of the free vibration. As resonance is approached,* however, the effect of even slight damping becomes more noticeable, and the phase lag ϵ approaches $\pi/2$ at resonance in a manner determined by equation (6). At exact resonance when $\tan \epsilon$ becomes infinite, $\epsilon=\pi/2$, *i.e.* the force and displacement are in quadrature (that is, differing 90° in phase), whilst the force and *velocity* are in phase.

When there is no damping, $k=0$, $\sin \epsilon=0$, $\epsilon=0$ or π, and the displacement x is expressed by equation (5*a*). Thus when p is less than n the displacement x is always in phase with the force, when p is greater than n the displacement reverses in sign and is in opposite phase to the force. As the frequency of the force passes through resonance, the phase *suddenly* reverses from 0 to π. This change of phase is more gradual when there is friction—in fact, use may be made of the abruptness of this phase change to measure the sharpness of resonance and the degree of damping present in a vibrating system—the more abrupt the change the smaller the damping and the sharper the resonance. In fig. 18 a series of curves is drawn, to an arbitrary scale, indicating the general nature of the variation of phase lag ϵ as the frequency of the forced vibration (p) passes through the resonance frequency (n), each curve of the series corresponding to a particular value of k the damping factor. All the phase curves pass through $\epsilon=\pi/2$ when $p=n$.

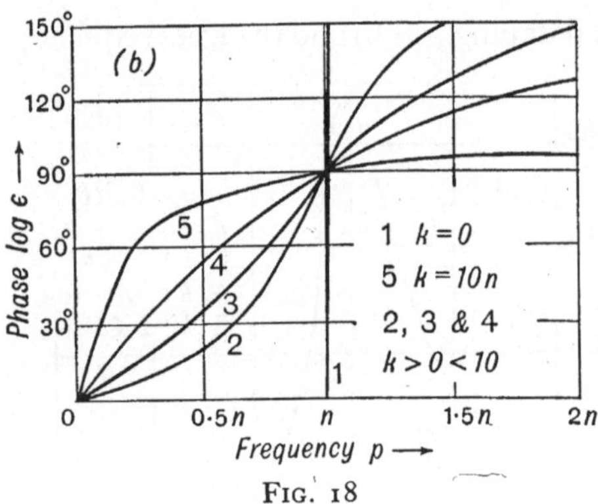

FIG. 18

The relationships of phase between the applied periodic force and the motion can be illustrated by means of simple pendulums under the action of a periodic driving force. This driving force may, for example, be supplied by a massive 'compound' pendulum C which carries at convenient positions the supports of two 'simple' pendulums A and B whose frequencies are respectively higher and lower than the frequency of C. When the large pendulum C is set swinging it will be found that the motions of A and B are at first irregular (see equation (7) and subsequent remarks, p. 38), giving the appearance of 'beats,' but at a later stage the simple pendulums will swing regularly, the short one 'A' being *in phase* with C, and the long one 'B' being in opposition, or *out of phase* with C. The critical case of resonance may also be studied in this way, the simple and compound pendulums being then in quadrature ($\epsilon = \pm 90°$). Resonance phenomena at higher frequencies (20 to 1000) may be very strikingly demonstrated by means of a frequency indicator of the reed type (see fig. 19). If such an indicator is supplied with alternating, or

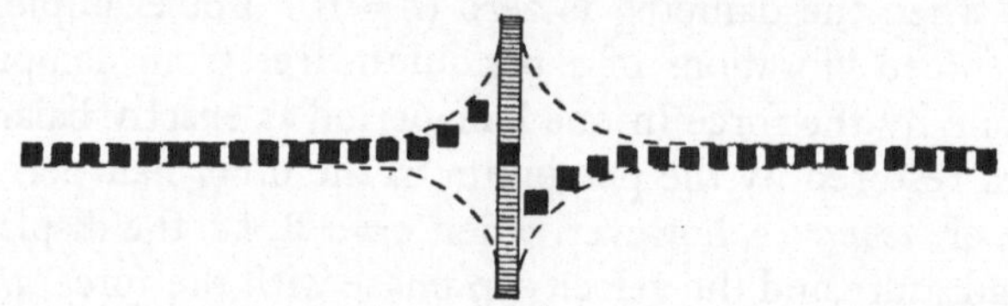

FIG. 19—Instantaneous appearance of Reeds of Frequency Meter (viewed stroboscopically)

interrupted, current of suitable frequency one of the graduated series of reeds will usually be found to vibrate vigorously, whilst those on either side of it are only moving slightly. If the reeds are viewed stroboscopically either by intermittent illumination or by intermittent viewing (by means of a moving-slit stroboscope or a glimpse-apparatus such as the Ashdown 'Rotoscope'), the vibrations may be seen in 'slow-motion.' It will then be noticed that the reeds on opposite sides of the resonant reed are in exactly opposite phase, whilst the phase of the resonant reed is intermediate —as indicated in fig. 19.

Energy Considerations. Power Dissipation – An important question relating to the maintenance of forced vibrations is that of power or rate of supply of energy. This question is closely connected with that of the phase relationships which we have just considered. Now mechanical power, or the rate of doing work, is the product of force and velocity. The force is $f \cos pt$, and the velocity dx/dt is obtained by differentiating equation (5). Thus the 'instantaneous' power supplied to maintain the oscillations is $f \cos pt \,.\, dx/dt$, that is,

$$=-\frac{f^2}{2k} \,.\, \sin \epsilon \,.\, \sin (pt-\epsilon) \cos pt$$

$$=-\frac{f^2 \sin \epsilon}{4k}\{\sin \epsilon - \sin (2pt-\epsilon)\} \quad . \quad . \quad (10)$$

The average value of the second term is zero, consequently the mean power supplied is

$$W=\frac{f^2}{4k} \,.\, \sin^2 \epsilon \quad . \quad . \quad . \quad . \quad (11)$$

the value of $\sin \epsilon$ being obtained from equation (6).

The mean power supplied by the impressed force to maintain the oscillations is therefore zero when $\epsilon=0$ or π. This case is realised when the damping is zero ($k=0$). For example, in the case of forced vibrations of a pendulum free from damping, the work done by the force in one half-period is exactly balanced by the work restored by the pendulum in the other half-period. In the case of *resonance*, however, when $\epsilon=\pi/2$, *i.e.* the displacement is in quadrature and the velocity in phase with the force, *the mean power supplied by the force is a maximum* and equal to

$$W_{max}=f^2/4k \quad . \quad . \quad . \quad . \quad (12)$$

a quantity increasing as the damping diminishes. This result may be obtained more directly from the equation of motion (equations (1) and (2), p. 37). Regarding this as analogous to the electrical case, at resonance the effects of inductance and capacity neutralise, and the force is expended in overcoming friction (corresponding to resistance) only, *i.e.*

$$2k\frac{dx}{dt}=f \cos pt, \quad \text{whence} \quad \frac{dx}{dt}=\frac{f}{2k} \cos pt,$$

and

$$\text{Power}=f \cos pt\frac{dx}{dt}=\frac{f^2}{2k} \cos^2 pt,$$

the mean value of $\cos^2 pt$ is $\frac{1}{2}$ over a cycle. Therefore

$$\text{Mean Power}=f^2/4k \qquad \text{as before.}$$

Sharpness of Resonance – As we have already seen, when the 'forced' and 'free' frequencies are equal, it is only the presence of damping which prevents the amplitude of the vibration from becoming infinite. The damping becomes relatively more and more powerful in its influence as resonance is approached. At resonance the amplitude and power dissipation reach their maximum values. On account of the very frequent occurrence of resonance phenomena in sound it is important to consider how these quantities vary with frequency in the neighbourhood of resonance.

The amplitude A under any specified conditions of k, p, and n may be obtained from equation (8), or combining this equation with (9) A may be expressed as a ratio of A_{max}, the maximum amplitude (at resonance $n=p$). Thus

$$\frac{\mathrm{A}}{\mathrm{A}_{\mathrm{max}}}=\frac{2kn}{\{(n^2-p^2)^2+4k^2p^2\}^{\frac{1}{2}}}\quad\left(\text{or }=\frac{n}{p}\sin\epsilon\right)\quad . \quad (13)$$

Similarly, the power W dissipated under the specified conditions is given by (11); combining this with (12), giving maximum power dissipation, we obtain

$$\frac{\mathrm{W}}{\mathrm{W}_{\mathrm{max}}}=\sin^2\epsilon\quad . \quad . \quad . \quad (14)$$

Substituting the value of sin ϵ obtained earlier (p. 39), we find

$$\frac{\mathrm{W}}{\mathrm{W}_{\mathrm{max}}}=\frac{4k^2p^2}{(n^2-p^2)^2+4k^2p^2}\quad . \quad . \quad . \quad (15)$$

a relation indicating how the power required to maintain vibrations against frictional and radiation losses varies in the neighbourhood of resonance. It will be seen from this relation that the energy dissipated at a certain frequency $(p/2\pi)$ near resonance is half that dissipated at resonance (*i.e.* $\mathrm{W}/\mathrm{W}_{\mathrm{max}}=\frac{1}{2}$) when $(n^2-p^2)^2=4k^2p^2$. That is when

$$\frac{p}{n}=1\pm\frac{k}{n}\text{ (approximately)}\quad . \quad . \quad . \quad (16)$$

Thus *the energy-dissipation at resonance falls to half value when the forced frequency departs from the natural or free frequency by the fraction k/n.* This indicates that the smaller the damping and the greater the natural frequency, the more rapidly will the dissipation fall off on either side of resonance—that is, the sharper will be the resonance. The ratio k/n *is therefore a measure of the sharpness of resonance.*

Typical 'resonance curves' are shown in fig. 17. These curves indicate a diminishing 'sharpness' as k increases. This important principle has many striking and important applications in sound. When a vibrating system is only lightly damped, very careful tuning is required to attain the resonance condition of maximum amplitude and power dissipation. In frequency standardisation very sharp tuning, and consequently minimum damping, is desirable in order to obtain an accurate determination of the peak of the resonance-curve. In other circumstances resonance is undesirable, *e.g.* in the faithful recording or reproduction of sounds extending over a range of frequencies, it is desirable to select 'n' far removed from any possible values of 'p,' or alternatively to employ a system which is very heavily damped (k very large)—in both cases the system is relatively insensitive.

A well-balanced tuning-fork is a good example of sharp tuning when it may be excited through the medium of the surrounding air by the vibrations of a second fork adjusted to the same frequency. When the two forks are closely in tune there will be a very marked response, otherwise there is relatively little effect. Even sharper tuning is required in the case of certain quartz resonators and oscillators (see p. 142) which have extremely little damping. The damping constant k of a vibrating system may be conveniently determined by measuring the rate of decay of its free oscillations (see p. 35) or from a resonance curve.

Perhaps one of the simplest and most instructive experiments to illustrate the effect of damping on resonance is that in which a massive compound pendulum drives two simple pendulums of the same period as the large pendulum. One of the simple pendulums has a bob of pith, the other of lead. The former is heavily damped by air friction, whereas the latter is not seriously affected (for the damping constant $k=r/2m$, see p. 34). Consequently the amplitude of the pith pendulum at resonance is small compared with that of the lead pendulum. On account of the large damping the light pith pendulum loses during each swing nearly all the energy given to it by the large pendulum, very little energy is stored and the amplitude remains small.

Coupled Systems. Reaction – As in the case of coupled electrical circuits, two mechanical vibrating systems react on one another when connected by some form of mechanical 'link.' When mechanical power is supplied to a vibrating system B from a system 'A' a condition may arise, and this is especially so at resonance, when the drainage of energy is so great that the system A can no longer maintain its maximum amplitude. The interchange of energy between the two systems also involves a change of natural frequency, for the primary system A is now loaded by the system B. The general features of mechanical coupling between two such systems may be simply demonstrated by means of apparatus described by T. R. Lyle * in a paper "On the Mechanical Analogy to the Coupled Circuits used in Wireless Telegraphy." A horizontal beam of mass M, free to move on the direction of its length, carries two simple pendulums, of lengths $l_1 l_2$ and masses $m_1 m_2$ respectively, vibrating in a vertical plane through the axis of the beam. The surging of energy backwards and forwards is very well illustrated by the model. If one pendulum is started swinging it will be seen that the second pendulum begins to swing with increasing amplitude, while the amplitude of the first at the same time diminishes. This goes on to a certain point when the reverse takes place, and the transfer of energy forwards and backwards many times may be observed. The effect is very striking when the two pendulums have equal or nearly equal periods. The 'coefficient of coupling' is shown to be $\sqrt{\frac{m_1 m_2}{(M+m_1)(M+m_2)}}$, analogous to $\sqrt{\frac{M_e^2}{L_1 L_2}}$ in the electrical case where L_1 and L_2 are the self-inductances and M_e the mutual inductance of the coupled circuits. The mechanical coupling of the pendulums obviously diminishes as M the mass of the connecting beam increases. As in the electrical case, the coupled system has two natural frequencies.

The vibrations of any mechanical system are communicated to a greater or less degree to other systems in contact with it. In certain cases such an effect may have important consequences, *e.g.* the frequency of an accurate electrically maintained tuning-fork may be seriously affected by the manner in which the fork is supported.†

Relaxation Oscillations ‡

The type of vibration represented by equation (1), p. 34, or by equation (2), p. 37, is the most common form of periodic

* T. R. Lyle, *Phil. Mag.*, **25**, pp. 567–592, April 1913. E. H. Barton and H. M. Browning, *Phil. Mag.*, **44**, pp. 573–576, Sept. 1922. See also C. H. Lees, *Phil. Mag.*, **48**, p. 129, 1924.

† A. B. Wood, *Journ. Sci. Instrs.*, **1**, p. 330, Aug. 1924.

‡ See B. van der Pol, *Phil. Mag.*, **2**, p. 978, Nov. 1926, and p. 763, Nov. 1928.

oscillation. The periodic time of free vibration is dependent solely on the ratio of the inertia and elastic factors, whilst the resistance factor controls only the resultant amplitude. Another type of vibration which is of fairly common occurrence has, however, very different characteristics. Thus, in equation (1), p. 34, if the sign of the resistance coefficient r be *negative*, we have

$$\ddot{x}-\frac{r}{m}\dot{x}+\frac{s}{m}x=0 \quad . \quad . \quad . \quad (1)$$

the solution of which, writing

$$\frac{r}{m}=2k \quad \text{and} \quad \frac{s}{m}=n^2,$$

is

$$x=ae^{+kt}\cos(n't+\epsilon) \quad . \quad . \quad . \quad (2)$$

cf. with equation (8), p. 34. In this case the amplitude, instead of dying away, *increases* to infinity. Such a condition is physically impossible, of course, but equation (1) may be applicable for values of x up to a certain limit. Thus we may rewrite equation (1) as follows:

$$\ddot{x}-2k(1-x^2)\dot{x}+n^2x=0 \quad . \quad . \quad . \quad (3)$$

representing an oscillatory system in which the resistance is a function of the displacement. When k is a positive quantity the system has a resistance which, for a small amplitude, is negative. Therefore the position $x=0$ is unstable. When, further,

$$k^2>>n^2 \quad . \quad . \quad . \quad . \quad (4)$$

it is obvious that as long as $x^2<<1$ the variable x will initially leave the value $x=0$ aperiodically, but when later $x^2>1$ the resistance has changed sign and has become positive, x will tend to return towards the zero value. It may be shown * that the periodic time T of a vibration typified by equation (3) is

$$T=1{\cdot}61\frac{2k}{n^2} \quad \text{or} \quad 1{\cdot}61\frac{r}{s} \quad . \quad . \quad . \quad (5)$$

The fundamental period of the vibration is, apart from the numerical constant, defined by a quantity involving the resistance and elastic factors only. B. van der Pol describes such oscillations as 'relaxation oscillations.' They differ very markedly from the sinusoidal oscillations of a mass controlled by a spring, for the resistance may now be negative (as in analogous electrical cases),

* *Loc. cit.*

increasing to a positive value as the displacement increases. The displacements take place in sudden jumps which recur periodically. As examples of relaxation oscillations, van der Pol mentions: the æolian harp, the pneumatic hammer, the scratching noise on a plate, the waving of a flag in the wind, the squeaking of a door, the Wehnelt interrupter, the intermittent discharge of a condenser through a neon tube, etc., etc. In all these examples the frequency is not dependent on the customary ratio of elasticity and mass, but is controlled by some form of resistance which, on reaching a certain critical value, suddenly relaxes, builds up again, relaxes, and so on. The properties of relaxation oscillations are:

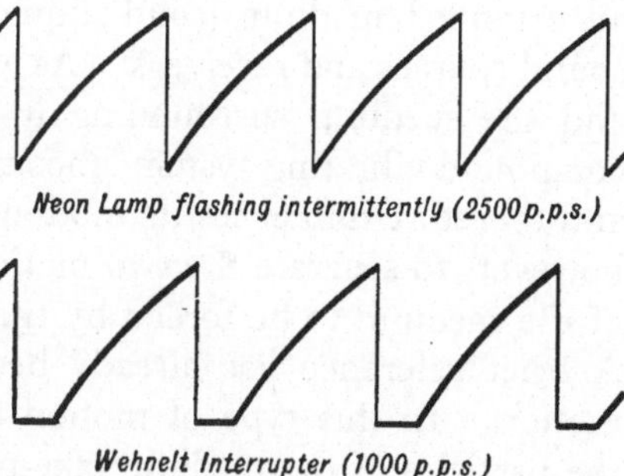

FIG. 20

(1) Their *time period* is determined by a time constant, or relaxation time, dependent on resistance and elastic forces only.

(2) Their *wave-form* is far from simple harmonic. Very steep parts occur on 'relaxation' of the resistance, and as a consequence many harmonics of large amplitude are present.

(3) A small impressed periodic force can easily bring the relaxation system into step with it (automatic synchronisation), whilst under these circumstances the amplitude is hardly influenced at all.

If equation (3) is regarded as an electrical oscillation, $2k = R/L$, $n^2 = 1/LC$, and consequently $T = 1{\cdot}61RC$. The fundamental period of the relaxation oscillation is therefore dependent on the product of resistance R and capacity C. An example of this may be found in the Anson-Pearson * neon lamp intermittent 'flashing' circuit, in which a neon tube and a condenser in parallel are connected in series with a high resistance and source of current. Copies of cathode ray oscillograph records of a Wehnelt interrupter and of a flashing neon lamp are shown in fig. 20.

VIBRATIONS IN AN EXTENDED MEDIUM WAVE-MOTION

In the foregoing treatment consideration has been given mainly to the vibrations of a particle either 'free' or acted upon by

* *Phys. Soc. Proc.*, **34**, p. 204, 1922.

external periodic forces and friction. This treatment finds particular application in the study of sound production and reception. It is important now to consider what takes place in an extended medium (solid, liquid, or gas), which may contain sound sources and receivers. As we have seen, the vibrating body and the medium surrounding it may be regarded as a single compound vibrating system since the medium reacts on the body and therefore influences its motion. It is often more convenient, however, to separate the two motions and to regard the vibration of the medium to be forced by the *actual* vibrations of the body. A brief reference has already been made to longitudinal wave motion. In this type of motion the vibratory displacements of the particles of the medium take place along the same line as the direction of propagation of the wave. It should be noted at the outset that the 'particles' to which we refer are not necessarily individual molecules, but may be small volumes or thin layers of the medium. The vibratory motions transmitted by the medium are therefore superposed on the irregular motion of the molecules. Sound waves are the inevitable result when a vibrating body is immersed in an elastic medium. We are therefore concerned with the physical properties (density, elasticity, viscosity, etc.) of the medium when considering the mechanism of wave transmission.

The problem in its simplest form, viz. that of transmission of 'plane' (or non-spreading) waves, is closely analogous to that of transmission of longitudinal waves in a long helical spring. If one end of such a spring, suitably suspended, is subjected to a single to-and-fro displacement alternate waves of compression and tension will be propagated along it, the velocity of transmission depending on the relative stiffness and inertia of the spring. If any particular point of the spring is observed during the passage of the wave it will be seen to perform a vibration similar to that imposed at the free end. In the case of a very long spring we can visualise the effect of a simple harmonic displacement of the end. As before, the velocity of transmission of the wave will depend on the inertia and stiffness of the spring, and the motion of any point will be simple harmonic like the forced motion of the free end. Taking an 'instantaneous view' of the spring as a whole when a number of such waves have been started along it, we should see a series of alternate compressions and extensions of the spring, the maxima of compression, or extension, being separated by equal intervals. It requires no proof here to show that the length of these intervals is equal to the velocity c of wave propagation divided by the number of vibrations of the end of the spring per second (the frequency N). We shall refer to these equal intervals between successive maxima, and successive minima, as the *wave-length* λ. Consequently we may write $N\lambda=c$.

Similarly, in the case of transmission of simple harmonic waves through an extended elastic medium the linear density of the particles, or their state of condensation or rarefaction (compression or expansion), is at any instant represented by a simple harmonic curve, which repeats itself at regular intervals of a wave-length, this wave-length being the ratio of the velocity of wave transmission to the frequency of the impression vibration.

Plane Waves

Suppose a progressive wave of simple harmonic type to proceed in a positive direction along the axis of x.* Then the displacement ξ at any time t of a point whose mean position is x will be given by

$$\xi=a \sin 2\pi\left(\frac{t}{T}-\frac{x}{\lambda}\right) \quad . \quad . \quad . \quad (1)$$

or

$$\xi=a \sin 2\pi(Nt-x/\lambda) \quad . \quad . \quad . \quad (2)$$

where a is the amplitude, T and N the period and frequency, and λ the wave-length of the vibration. Writing $c=N\lambda=\lambda/T$ we obtain other forms of equation (1), viz.

$$\xi=a \sin \frac{2\pi}{\lambda}(ct-x) \quad . \quad . \quad . \quad (3)$$

and

$$\xi=a \sin n(t-x/c) \quad . \quad . \quad . \quad (4)$$

where n has the usual significance and is equal to $2\pi N$. These forms of expression for the particle displacement ξ in a progressive plane wave will be found convenient for most purposes. Such a system of waves could, of course, be represented by a displacement curve similar to that of fig. 2, where the ordinates represent 'ξ' and the abscissa 't'. Consideration of the above expressions shows that, in the case of progressive waves, phases may be expressed in terms of fractions of a wave-length—one wave-length λ corresponding to a phase angle of 2π. Thus the difference of

* We assume in what follows, that the effects of viscosity and other attenuation factors in the medium are negligible.

phase between the vibration of two particles at distances x_1 and x_2 from the origin is $2\pi(x_2-x_1)/\lambda$. It will be evident that a system of waves travelling in the negative direction of x will be represented by the introduction of a positive instead of a negative sign inside the brackets in equations (1) to (4).

Particle-Velocity - Differentiating equation (4), which gives the displacement of a particle in the path of the wave, we obtain the particle-velocity $\dot{\xi}$,

$$d\xi/dt = na \cos n(t-x/c) \quad . \quad . \quad . \quad (5)$$

which may be written

$$\frac{d\xi}{dt} = -c\frac{d\xi}{dx} \quad . \quad . \quad . \quad . \quad (6)$$

giving the particle-velocity in terms of the wave-velocity and the slope of the displacement curve.

Acceleration of Particles in progressive waves. Differentiating (5) we obtain

$$d^2\xi/dt^2 = -n^2 a \sin n(t-x/c) \quad . \quad . \quad (7)$$

or

$$\frac{d^2\xi}{dt^2} = c^2\frac{d^2\xi}{dx^2} \quad . \quad . \quad . \quad . \quad (8)$$

i.e. the acceleration is approximately equal to the product of the square of the wave-velocity and the curvature of the displacement curve. Equation (8) is the differential equation which characterises a wave-motion. Its complete solution is

$$\xi = f(ct-x) + \mathrm{F}(ct+x) \quad . \quad . \quad . \quad (9)$$

which represents two independent systems of waves travelling in opposite directions with the same velocity c. This velocity is, within the limits of the approximate equation (8), entirely independent of the form of the wave. For example, in the case of the simple form of wave represented by equation (3), viz.

$$\xi = a \sin \frac{2\pi}{\lambda}(ct-x),$$

the velocity c is the same whatever the wave-length λ may be. Within certain limits, set by equation (8), the velocity c is also independent of the amplitude a.

In other words, the velocity of wave transmission is independent of the frequency or amplitude (provided this is small), and is determined solely by the physical properties of the medium.

Velocity of Wave Transmission – The physical properties of the medium which influence the velocity are density and elasticity, corresponding to the mass and stiffness factors in the case of the vibrations of a particle previously considered. As the wave of compression and rarefaction passes through the medium the volume and density fluctuate locally about the normal values. The magnitude of these fluctuations is controlled by the properties of the medium and the applied forces. The following definitions * are important in this connection:—

(1) *Dilatation* (Δ) is the ratio of the increment of volume δv to the original volume v_0, thus

$$\Delta=\delta v/v_0 \quad \text{and} \quad v=v_0(1+\Delta) \qquad (10)$$

(2) *Condensation* (s) is the ratio of the increment of *density* $\delta\rho$ to the original density ρ_0, thus

$$s=\delta\rho/\rho_0 \quad \text{and} \quad \rho=\rho_0(1+s) \qquad (11)$$

Since

$$\rho v=\rho_0 v_0, \qquad (1+s)(1+\Delta)=1 \qquad (12)$$

and $s=-\Delta$ if we neglect $s\Delta$ as a small second order quantity.

(3) *Volume or 'cubic' elasticity* (κ)—sometimes known as the bulk modulus of elasticity,

$$\kappa=\frac{\delta p}{-\delta v/v_0}=-v_0\frac{\delta p}{\delta v}=\frac{\delta p}{s} \qquad (13)$$

where $\delta\rho$ is the stress and δv the corresponding strain. Whence we have, for small variations of s,

$$p=p_0+\kappa s \qquad (14)$$

(4) *Compressibility* (C) is the reciprocal of the bulk modulus κ. Thus

$$\mathrm{C}=-\frac{1}{v_0}\frac{\delta v}{\delta p} \quad \text{and} \quad \mathrm{C}=\frac{1}{\kappa} \qquad (15)$$

Consider the case of plane waves travelling along the axis of x, the displacement of a point whose mean position is x being ξ at any time t. The displacements of planes normally at x and $(x+\delta x)$ will be ξ and $\left(\xi+\frac{d\xi}{dx}\cdot\delta x\right)$. The thickness of the layer bounded by the planes x and $(x+\delta x)$ is therefore changed from

* Lamb's notation is adopted as being most suitable.

δx to $(1+d\xi/dx)\delta x$, consequently the dilatation $\Delta=d\xi/dx$, and the condensation

$$s=\frac{\rho-\rho_0}{\rho_0}=-\frac{d\xi}{dx}\bigg/\left(1+\frac{d\xi}{dx}\right)$$

from (11) and (12).

[For *small* displacements we may write $s=-d\xi/dx=-\Delta$.] The mass of unit area of the layer is $\rho_0\delta x$. If the excess pressure on the $(x+\delta x)$ face is $\delta\rho$, we may equate this force to the product of the acceleration $d^2\xi/dt^2$ and the mass $\rho_0\delta x$ of the layer. That is,

$$\rho_0\delta x d^2\xi/dt^2=-\delta\rho,$$

or

$$d^2\xi/dt^2=-\frac{1}{\rho_0}\cdot\frac{dp}{dx}.$$

Now we have $\delta p=\kappa s$ * from (13) and $s=-d\xi/dx$, consequently

$$\frac{d^2\xi}{dt^2}=\frac{\kappa}{\rho}\cdot\frac{d^2\xi}{dx^2} \quad . \quad . \quad . \quad . \quad (16)$$

or writing

$$c=\sqrt{\kappa/\rho} \quad . \quad . \quad . \quad . \quad (17)$$

equation (16) becomes

$$\frac{d^2\xi}{dt^2}=c^2\frac{d^2\xi}{dx^2} \quad . \quad . \quad . \quad . \quad (18)$$

which is identical with equation (8) obtained above. The complete solution is given in equation (9), which represents two independent wave-systems travelling in opposite directions with velocity c. As equations (16) and (17) indicate, *the velocity of the wave is determined solely by the elasticity and density of the medium through which it is passing.*

The treatment just given applies to all cases of transmission of small amplitude plane waves, in solid, liquid, or gaseous media, *provided the appropriate modulus of elasticity is applied.* The same method of treatment also applies to the case of longitudinal waves in solid rods—this will be dealt with in a later section. In the case of a rod, the elasticity factor is Young's modulus, whereas in the case of plane waves in an extended solid, when lateral displacements are impossible, it is necessary to use the longitudinal elasticity $(\kappa+\frac{4}{3}\mu)$ where κ is the bulk modulus and μ the rigidity. In gases, and fluids in general, the elasticity is the

* This applies to an unlimited fluid medium. For a solid in bulk κ must be replaced by $(\kappa+\frac{4}{3}\mu)$ (where μ is the rigidity of the solid) because a solid changes in shape as well as in volume.

ordinary bulk modulus, except in certain cases where the volume of the medium is limited.

Energy of Plane Waves of Sound – The rate of transfer of energy per unit area of cross-section of the wave may be regarded as a physical measure of the intensity * of the sound transmitted. In the first place, we shall determine the kinetic energy of the wave-system per unit area of the plane parallel to the wave-front. Considering the simple harmonic wave expressed by equations (1) and (4), viz.

$$\xi = a \sin n(t - x/c),$$

we obtain the particle-velocity by differentiation with respect to time,

$$d\xi/dt = an \cos n(t - x/c).$$

Now the mass of unit area of a layer of thickness δx is $\rho_0 \delta x$. Consequently, the kinetic energy $\delta \mathrm{E}$ of the layer is given by

$$\delta \mathrm{E} = \tfrac{1}{2}\rho_0 \delta x \left(\frac{d\xi}{dt}\right)^2,$$

and the kinetic energy of the whole wave-system is

$$\left.\begin{aligned} \mathrm{E} &= \tfrac{1}{2}\rho_0 a^2 n^2 \int \cos^2 n(t - x/c)dx \\ &= \tfrac{1}{4}\rho_0 a^2 n^2 \int \{1 + \cos 2n(t - x/c)\}dx \end{aligned}\right\} \quad . \quad . \quad (19)$$

The mean value of the 'cos' term under the integral is zero. Hence the *average kinetic energy per unit area and unit length* (i.e. *per unit volume*) *of the wave* is equal to

$$\tfrac{1}{4}\rho_0 a^2 n^2 \quad . \quad . \quad . \quad . \quad (20)$$

The *maximum* kinetic energy per unit volume of the wave is from (19)

$$\tfrac{1}{2}\rho_0 a^2 n^2.$$

But the sum of the kinetic and potential energies is a constant, and when one form of energy is a maximum the other form is zero, therefore

Total energy of the wave-motion per unit volume $= \tfrac{1}{2}\rho_0 a^2 n^2 \quad . \quad (21)$

This quantity is the total energy contained in unit length of the wave per unit area of wave-front, and may therefore be

* This must not be confused with 'loudness.'

conveniently called *energy density*. We require to know also the rate of flow of energy through the medium. This is at once obtained when we notice that the wave travels a distance c per second. Consequently the flow of energy per unit time per unit area of wave-front, or

Transmission of energy per second per unit area of wave-front $= \frac{1}{2}\rho c a^2 n^2$. (22)

This may be regarded as the *intensity* of the sound wave. It is directly proportional to the density of the medium and the wave-velocity, and to the square of the amplitude and frequency of the vibration. The relation thus obtained for the intensity (or rate of energy flow) is of fundamental importance. As we have just shown, *the intensity is equal to the product of energy density* (*i.e.* energy per unit volume of wave) *and wave-velocity*.

By differentiation of the equation of wave-motion we obtain the maximum particle-velocity $(d\xi/dt)_{max}$ equal to an, and the maximum condensation s_{max} equal to $(-d\xi/dt)_{max}$, or

$$s_{max} = \frac{an}{c} = \frac{\dot{\xi}_{max}}{c},$$

indicating that *the maximum condensation is equal to the maximum particle-velocity divided by the wave-velocity*. Substituting this relation in (22) we obtain an expression for the energy flow, or intensity, in terms of particle-velocity and wave-velocity, viz.

$$\textit{Energy flow} = \tfrac{1}{2}\rho c^3 s^2_{max} \quad . \quad . \quad . \quad . \quad (23)$$

Now

$$c^2 = \kappa/\rho \quad \text{(see (17))}$$

and

$$\delta p_{max} = \kappa s_{max} \quad \text{(see (13))}.$$

Therefore the energy flow or *intensity* is, from (23),

$$= \frac{1}{2} \frac{(\delta p_{max})^2}{\rho c} \quad . \quad . \quad . \quad . \quad (24)$$

an expression giving the intensity of the sound wave in terms of the maximum pressure-variation in the path of the wave.

Power of a Source emitting Plane Waves. Radiation Resistance - The energy thus present in the sound wave must be derived primarily from the vibrating source. The rate at which the source does work, *i.e.* the power of the source in producing sound waves, is equal to the product of pressure-variation and particle-velocity, *i.e.* to $\delta p \times d\xi/dt$ per unit area of wave-front. Now

$$\delta p = \kappa s = \frac{\kappa}{c} \cdot \frac{d\xi}{dt},$$

since $c = \sqrt{\kappa/\rho}$ we have

$$\delta p = \sqrt{\kappa\rho} \,.\, (d\xi/dt) \quad \text{or} \quad \rho c\dot{\xi} \qquad . \qquad . \qquad (25)$$

Hence the instantaneous power expended by the source per unit area of wave-front

$$= \sqrt{\kappa\rho} \,.\, (d\xi/dt)^2 \quad \text{or} \quad \rho c\dot{\xi}^2 \qquad . \qquad . \qquad (26)$$

The *mean* power radiated per unit area of wave-front is therefore given by

$$\tfrac{1}{2}\sqrt{\kappa\rho} \,.\, (d\xi/dt)^2_{\max} \quad \text{or} \quad \tfrac{1}{2}\rho c\dot{\xi}^2_{\max} = \tfrac{1}{2}\rho c n^2 a^2 \qquad . \qquad (27)$$

It will be seen that the relations (25), (26), and (27) are closely analogous to the electrical expressions connecting voltage, resistance, and current, if we regard δp, $\sqrt{\kappa\rho}$ (or ρc), and $(d\xi/dt)$ as corresponding to these electrical quantities respectively. Thus equation (25) is the mechanical analogue of *Ohm's law*, and equations (26) and (27) represent the law of power dissipation in a circuit of resistance $\sqrt{\kappa\rho}$ (or ρc). The quantity $\sqrt{\kappa\rho}$, or ρc, is therefore generally described as the *radiation resistance* (or impedance) of the medium which transmits the sound waves. It is a fundamental characteristic of the medium, to which frequent reference is made. Equation (27) indicates that the mean power radiated by a source of plane waves of area S (large compared with λ^2) and uniform amplitude

$$= \tfrac{1}{2}\rho c n^2 a^2 \mathrm{S} \qquad . \qquad . \qquad . \qquad . \qquad (28)$$

Taking the case of a circular disc or *piston source* (see p. 147) of radius r, frequency N, and uniform amplitude a,

$$(\text{Mean Power radiated}) = 2\pi^3 r^2 \rho c \mathrm{N}^2 a^2 \text{ ergs/sec.}$$

EXAMPLE – A disc 20 cms. radius r, vibrates in water ($\rho = 1$, $c = 1{\cdot}5 \times 10^5$ cms./sec.) with a frequency $\mathrm{N} = 10^5$ p.p.s. The power radiated $= 37 \times 10^{13} a^2$ ergs/sec. or $37 \times 10^6 a^2$ watts. If the amplitude of vibration is 0·001 cm., the power radiated is 37 watts. It is important to notice that the wave-length of the sound in this case is small ($\lambda = 1{\cdot}5 \times 10^5/10^5 = 1{\cdot}5$ cms.) compared with the radius of the disc. Otherwise the waves would cease to be plane and the calculation would have no meaning.

Spherical Waves

The results derived above for the energy in plane waves hold equally well for spherical waves at a sufficient distance from the source. The general case of spherical waves at *any* distance from the source requires further consideration. We have now to deal with a sound wave which is progressively increasing (or decreasing) in diameter. The amplitude of the particles of the medium through which the wave passes is therefore dependent on the distance from the source of the waves. Elementary considerations at once indicate that the energy density in the wave will vary inversely as the square of the distance from a point source of spherical waves, whence it may be inferred that the amplitude (of displacement, pressure, or condensation) will vary inversely as the distance. As in the case of plane waves we shall assume that the displacement is always small at all parts of the wave, and that there is no loss of energy, due to viscous, thermal, or other causes. In order to deduce the fundamental equation for spherical wave-motion it is necessary first of all to refer to the equation of continuity and the equations of motion.

The Equation of Continuity – In effect, this equation is merely a mathematical statement of an otherwise obvious fact that matter is neither created nor destroyed in the interior of the medium. Consider an elementary volume $\delta x \delta y \delta z$ of the medium. Then, in any small time interval δt, the difference in the quantity of matter entering and leaving through its faces must be equal to the change in the quantity of matter inside the small volume during this time. Let u, v, and w denote the velocity components of the particle whose co-ordinates are x, y, and z at the instant t. Then the quantity of matter entering the face in the yz plane is $u\delta y \,.\, \delta z \,.\, \rho\delta t$. Again, the quantity passing out at the opposite face of the elementary volume will be

$$\left(u \,.\, \rho + \frac{\partial \,.\, u\rho}{\partial x} \,.\, \delta x\right)\delta y \delta z \,.\, \delta t.$$

Thus the excess passing out of the small volume by the two faces is

$$\frac{\partial \,.\, u\rho}{\partial x} \,.\, \delta x \,.\, \delta y \,.\, \delta z \,.\, \delta t.$$

Similar expressions are obtained for the other faces. Consequently the total mass lost in the time δt is

$$\left\{\frac{\partial(u\rho)}{\partial x} + \frac{\partial(v\rho)}{\partial y} + \frac{\partial(w\rho)}{\partial z}\right\}\delta x \,.\, \delta y \,.\, \delta z \,.\, \delta t.$$

Now the mass lost is also equal to

$$-\frac{\partial\rho}{\partial t}\delta x\,.\,\delta y\,.\,\delta z.$$

Equating these two expressions we have

$$\frac{\partial\rho}{\partial t}+\frac{\partial(u\rho)}{\partial x}+\frac{\partial(v\rho)}{\partial y}+\frac{\partial(w\rho)}{\partial z}=0 \quad . \quad . \quad (1)$$

which is the general form * of the equation of continuity. Since $\rho=\rho_0(1+s)$, $\partial\rho/\partial t=\rho_0\partial s/\partial t$; and for small-amplitude vibrations $\partial(\rho u)/\partial x=\rho_0\partial u/\partial x$ approximately. Equation (1) therefore becomes

$$\frac{\partial s}{\partial t}+\frac{\partial u}{\partial x}+\frac{\partial v}{\partial y}+\frac{\partial w}{\partial z}=0 \quad . \quad . \quad . \quad (2)$$

a form of the equation of continuity more convenient for application to the case of small-amplitude vibrations.

Equations of Motion – Referring again to the elementary volume $\delta x\delta y\delta z$, the total pressures on the two yz faces (δx apart) are $(p\pm\frac{1}{2}\delta x\,.\,\partial p/\partial x)\delta y\,.\,\delta z$, giving a resultant force along the x-axis of $-(\partial p/\partial x)\delta x\,.\,\delta y\,.\,\delta z$. Now the mass in question is $\rho\delta x\delta y\delta z$ and its acceleration $\partial u/\partial t$. Consequently we have

$$\rho\frac{\partial u}{\partial t}=-\frac{\partial p}{\partial x},$$

and similarly

$$\left.\rho\frac{\partial v}{\partial t}=-\frac{\partial p}{\partial y} \text{ and } \rho\frac{\partial w}{\partial t}=-\frac{\partial p}{\partial z}\right\} \quad . \quad . \quad (3)$$

But $\delta p=\kappa s$ (see equation (13), p. 51), and $c^2=\kappa/\rho_0$, therefore

$$\frac{\partial u}{\partial t}=-c^2\frac{\partial s}{\partial x},\qquad \frac{\partial v}{\partial t}=-c^2\frac{\partial s}{\partial y} \text{ and } \frac{\partial w}{\partial t}=-c^2\frac{\partial s}{\partial z} \quad . \quad (4)$$

neglecting variations of $\rho\partial u/\partial t$, etc., and regarding $\rho=\rho_0$ to a sufficient degree of accuracy for small oscillations. Equations (4) are fundamental when dealing with the propagation of sound waves through an elastic medium.

General Equation – We have now to combine the equation of continuity (2) with the equations of motion (4). Differentiating (2) with respect to t we obtain

$$\frac{\partial^2 s}{\partial t^2}=-\left\{\frac{\partial^2 u}{\partial x dt}+\frac{\partial^2 v}{\partial y\partial t}+\frac{\partial^2 w}{\partial z\partial t}\right\} \quad . \quad . \quad (5)$$

* See Rayleigh, *Theory of Sound*, 2, p. 3.

Differentiating the equations in (4) with respect to x, y, and z respectively, and adding, we have

$$-c^2\left\{\frac{\partial^2 s}{\partial x^2}+\frac{\partial^2 s}{\partial y^2}+\frac{\partial^2 s}{\partial z^2}\right\}=\frac{\partial^2 u}{\partial x\partial t}+\frac{\partial^2 v}{\partial y\partial t}+\frac{\partial^2 z}{\partial z\partial t} \quad . \quad (6)$$

From (5) and (6) we obtain

$$\left.\begin{aligned}\frac{\partial^2 s}{\partial t^2}&=c^2\left\{\frac{\partial^2 s}{\partial x^2}+\frac{\partial^2 s}{\partial y^2}+\frac{\partial^2 s}{\partial z^2}\right\}\\ \text{or}\quad \\ \frac{\partial^2 s}{\partial t^2}&=c^2\nabla^2 s\end{aligned}\right\} \quad . \quad . \quad . \quad (7)$$

which is the general equation of sound waves in a three-dimensional homogeneous medium. It will readily be seen in the case of plane waves travelling in the direction of the axis of x that equation (7) reduces to

$$\frac{\partial^2 s}{\partial t^2}=c^2\frac{\partial^2 s}{\partial x^2} \quad . \quad . \quad . \quad . \quad (8)$$

which is similar to that previously obtained (equation (18), p. 52).

Spherical Sound Waves – In order to apply the general equation (7) to the case of spherical waves, we must transform to polar co-ordinates, writing

$$x=r\sin\theta\cos\phi, \qquad y=r\sin\theta\sin\phi, \qquad z=r\cos\theta.$$

We now find

$$\nabla^2 s=\frac{\partial^2 s}{\partial r^2}+\frac{2}{r}\cdot\frac{\partial s}{\partial r}+\frac{1}{r^2\sin\theta}\cdot\frac{\partial}{\partial\theta}\left(\sin\theta\frac{\partial s}{\partial\theta}\right)+\frac{1}{r^2\sin^2\theta}\cdot\frac{\partial^2 s}{\partial\phi^2} \quad . \quad (9)$$

For a symmetrical spherical wave, s will be a function of r and t only, and independent of θ and ϕ. Equation (9) may then be simplified

$$\nabla^2 s=\frac{\partial^2 s}{\partial r^2}+\frac{2}{r}\cdot\frac{\partial s}{\partial r}=\frac{1}{r}\frac{\partial^2(rs)}{\partial r^2} \quad . \quad . \quad . \quad (10)$$

Consequently by substitution in (7) the equation of motion becomes

$$\frac{\partial^2(rs)}{\partial t^2}=c^2\frac{\partial^2(rs)}{\partial r^2} \quad . \quad . \quad . \quad (11)$$

The general solution of this equation is

$$rs=f(ct-r)+F(ct+r) \quad . \quad . \quad . \quad (12)$$

representing independent diverging and converging waves of any type, having a common velocity c. This velocity $c=\sqrt{\kappa/\rho}$ is, of course, identical with the velocity in the case of plane waves.

Velocity Potential* – It is convenient at this point to introduce a new function into our equations. This function, the velocity-potential ϕ, was first employed by Lagrange in dealing with hydrodynamical problems. The name 'velocity-potential' was given on account of its analogy with electrostatic potential. The relation it bears to the particle-velocity at any point in the medium is given by

$$u=-\frac{\partial\phi}{\partial x},\qquad v=-\frac{\partial\phi}{\partial y},\qquad w=-\frac{\partial\phi}{\partial z}\quad . \quad . \quad (13)$$

provided there is no rotational motion. This latter condition is of course fulfilled when the particles of the medium are in vibration when transmitting sound waves. (The expressions for u, v, w in (13) represent components of current if ϕ denote electric potential, provided the resistivity of the substance be unity.) If equipotential surfaces, ϕ=constant, are drawn throughout the sound-field, then *the velocity* is at any point in the direction in which ϕ *decreases* most rapidly and, as in (13), *is equal to the gradient of* ϕ.

We require first of all to determine ϕ in terms of the condensation s. This may be done by integrating equation (4), thus

$$u=-c^2\frac{\partial}{\partial x}\int_0^t s\,dt+u_0 \quad \text{and two similar,}$$

or, if the medium is initially at rest, $u_0=0$, and

$$u=-c^2\frac{\partial}{\partial x}\int_0^t s\,dt \quad \text{and two similar} \quad . \quad . \quad (14)$$

Comparing equations (13) and (14) we obtain

$$\phi=c^2\int_0^t s\,dt \quad \text{or} \quad \frac{\partial\phi}{\partial t}=c^2 s \quad . \quad . \quad . \quad (15)$$

which is the required expression for the velocity-potential ϕ in terms of the condensation s. The general equation of sound waves (7) now becomes

$$\frac{\partial^2\phi}{\partial t^2}=c^2\nabla^2\phi \quad . \quad . \quad . \quad . \quad (16)$$

* See pp. 201–205, Lamb's *Dynamical Theory of Sound.*

and *the equation of motion* (11) *of the medium for spherical waves is now*

$$\frac{\partial^2(r\phi)}{\partial t^2}=c^2\frac{\partial^2(r\phi)}{\partial r^2} \quad . \quad . \quad . \quad (17)$$

the solution being

$$\phi=\frac{A}{r}f(ct-r)+\frac{B}{r}\,.\,F(ct+r) \quad . \quad . \quad (18)$$

Considering only the divergent wave

$$\phi=\frac{A}{r}f(ct-r) \quad . \quad . \quad . \quad . \quad (19)$$

and regarding ξ and $\dot{\xi}$ as the displacement and velocity of a particle along the radius r, we have from (15) and (19)

$$s=\frac{A}{cr}f'(ct-r) \quad . \quad . \quad . \quad . \quad (20)$$

where $f'(ct-r)$ is the derivative of f with respect to $(ct-r)$. This equation indicates that the condensation in a diverging spherical wave is inversely proportional to the distance r as it travels outwards. The particle-velocity along the radius r is

$$\dot{\xi}=-\frac{\partial\phi}{\partial r}=\frac{A}{r}f'(ct-r)+\frac{A}{r^2}f(ct-r) \quad . \quad . \quad (21)$$

As the distance r increases with the spreading of the wave the first term on the right of this equation becomes large compared with the second term. In such a case therefore the velocity tends to become proportional to $1/r$, and the energy density to $1/r^2$.

The pressure-variation p at any point distant r from the source may be obtained from equation (15), viz. $\partial\phi/\partial t=c^2s$. Substituting $c^2=\kappa/\rho$ and $s=p/\kappa$ in this equation, we find

$$\frac{\partial\phi}{\partial t}=\frac{p}{\rho} \quad \text{or} \quad p=\rho\frac{\partial\phi}{\partial t} \quad . \quad . \quad . \quad (22)$$

Consequently from (19) and (22)

$$p=\rho\frac{A\,.\,c}{r}\,.\,f'(ct-r) \quad . \quad . \quad . \quad (23)$$

When r is sufficiently great to render the second term of equation (21) negligible, then

$$p=\rho c\dot{\xi}=\sqrt{\kappa\rho}\,.\,\dot{\xi} \quad . \quad . \quad . \quad (24)$$

That is, *the pressure is equal to the product of the radiation resistance*

(ρc) *and the particle-velocity*, when the distance from the source is large. This result, which we obtained for plane waves (see equation (25), p. 55), is analogous to the electrical relation between voltage, resistance, and current.

The only distinction between plane and spherical waves, when r may be regarded as large, is that due to the spreading of the spherical wave. Thus we must replace $\dot{\xi}$ and p for the plane waves by $(r\dot{\xi})$ and (rp) for diverging spherical waves.

Simple Harmonic Spherical Waves – The equations of motion obtained in the foregoing treatment apply to all forms of small-amplitude vibrations. The form of particular interest is, of course, the simple harmonic vibration. In this case we may write equation (19) in the form

$$\phi=\frac{A}{r}\cos n\left(t-\frac{r}{c}\right) \quad . \quad . \quad . \quad (25)$$

Then the particle-velocity $\dot{\xi}$ along the radius is

$$\dot{\xi}=\frac{\partial\phi}{\partial r}=-\frac{An}{cr}\sin n\left(t-\frac{r}{c}\right)+\frac{A}{r^2}\cos n\left(t-\frac{r}{c}\right) \quad . \quad (26)$$

The rate of flow of medium, i.e. *the flux*, through a sphere of radius r, will therefore be

$$4\pi r^2\dot{\xi}=-\frac{4\pi An}{c}\,.\,r\sin n\left(t-\frac{r}{c}\right)+4\pi A\cos n\left(t-\frac{r}{c}\right) \quad . \quad (27)$$

Regarding the source as a *very small sphere* of radius r_0 (a *point source*) and having a maximum surface velocity of $\dot{\xi}_0$ we may imagine that fluid (the medium) is introduced or abstracted at a certain rate, the '*strength*' *of the source being defined as the maximum rate of volume variation at the source.* This fictitious source provides the necessary train of sound waves. Then at the surface of the source we have

$$\dot{\xi}=\dot{\xi}_0\cos nt \quad . \quad . \quad . \quad . \quad (28)$$

Also from equation (27) we have, when $r=r_0=0$ nearly,

$$(4\pi r_0^2)\dot{\xi}_0\cos nt=4\pi A\cos nt \quad . \quad . \quad (29)$$

The quantity $(4\pi r_0^2\,.\,\dot{\xi}_0)$ on the left of this equation represents the maximum rate of introduction (or abstraction) of fluid at the source, i.e. *the strength of the source.** Denoting this quantity by A′ we have from (29)

$$A'\cos nt=4\pi A\cos nt \quad . \quad . \quad . \quad (30)$$

* Rayleigh, 2, p. 109 *et seq.*; Lamb, Arts. 69–76.

The velocity potential of a small source of strength A′ is therefore from (30) and (25)

$$\phi=\frac{A'}{4\pi r}\cos n\left(t-\frac{r}{c}\right) \quad . \quad . \quad . \quad (31)$$

This equation is generally expressed in the form

$$\phi=\frac{A'}{4\pi r}\cos(nt-kr) \quad . \quad . \quad . \quad (32)$$

where

$$k=\frac{n}{c}=\frac{2\pi}{\lambda} \quad . \quad . \quad . \quad . \quad (32a)$$

The particle-velocity being

$$\dot{\xi}=\frac{\partial\phi}{\partial r}=-\frac{kA'}{4\pi r}\sin(nt-kr)+\text{(a negligible term involving) }1/r^2) \quad . \quad (33)$$

Equation (32) is of fundamental importance in dealing with sound radiation from any type of source. Although it applies essentially to a 'point' source (*i.e.* of dimensions small compared with λ), it is quite evident that it must also be applicable to sources of any shape or size, regarding each elementary area of the source as a 'point source' and integrating over the whole of the vibrating surface. It is therefore possible to deduce the fundamental quantities, particle-velocity $\dot{\xi}$, condensation s, and pressure p, at any point in the medium provided we know the motion at all points of the source itself. Numerous applications of the velocity-potential function are to be found in the treatises on *Sound* by Lamb and Rayleigh. A number of examples of practical importance are dealt with in a book by I. B. Crandall.*

Power of a Source radiating Spherical Waves.– The rate of doing work, or power, of the sound waves from a 'point' source must be the same at any spherical surface surrounding the source, and is therefore independent of distance r from the source. Neglecting steady pressures which contribute nothing to the average effect, the rate of doing work at the surface of a sphere of radius r is the product of alternating pressure p, area of surface and radial velocity, *i.e.* the power W of the source

$$W=\frac{dE}{dt}=p\,.\,4\pi r^2\frac{d\phi}{dr}=4\pi r^2\rho\frac{d\phi}{dt}\cdot\frac{d\phi}{dr} \quad . \quad . \quad (34)$$

since $p=\rho d\phi/dt$ (see equation (22)).

* *Theory of Vibrating Systems and Sound.* (Macmillan.)

Obtaining $d\phi/dt$ and $d\phi/dr$ from (32) we find

$$W=4\pi r^2\rho\frac{An}{4\pi r}\sin(nt-kr)\,.\,\frac{Ak}{4\pi r}\sin(nt-kr).$$

Since the mean value of the $\sin^2$ term is $\frac{1}{2}$, the average rate at which the source does work, or

$$\textit{Mean Power}=\frac{\rho A^2nk}{8\pi}\quad\text{or}\quad\frac{\rho n^2A^2}{8\pi c}\quad\text{or}\quad\frac{\rho cA^2k^2}{8\pi}\quad . \qquad (35)$$

where *A is the strength of the source defined as the maximum rate of volume variation at the source, i.e.* $A=S\dot{\xi}_0$ where S is the area of the vibrating surface and $\dot{\xi}_0$ $(=n\xi_0)$ is the maximum velocity (assumed uniform) over the surface. Writing $n=2\pi N$,

$$\textit{Mean Power}=2\pi^3\rho N^4S^2\xi_0{}^2/c \text{ ergs/sec.} \quad . \quad (36)$$

If the source is a circular piston of radius R (small compared with λ) this becomes

$$=2\pi^5\rho N^4R^4\xi_0{}^2/c \text{ ergs/sec.} \quad . \quad . \quad . \quad (37)$$

In the *case of a small circular diaphragm* radiating sound waves, the amplitude ξ and velocity $\dot{\xi}$ are *not* uniform over the whole surface. We must therefore substitute for ξ_0 in (37) the *average* amplitude $\bar{\xi}$ over the surface which, in the fundamental mode, is approximately $\xi_0/3$, where ξ_0 is the maximum amplitude at the centre of the diaphragm.* We have in this case

$$W=2\pi^5\rho R^4N^4\xi_0/9c \text{ ergs/sec. (approximately)} \quad . \quad (38)$$

The relations (35) to (38) for the mean power radiated in spherical waves are of course only applicable to a small isolated source. The emission of energy would be modified by discontinuities in the medium near the source. Thus *a rigid plane near the source* would double the amplitude by reflection, but would limit the radiation to one hemisphere only. Therefore the net power radiated, as given in the above expressions, would have to be doubled on one side of the sound source and reduced to zero on the other. Rayleigh deals with this case of sound radiation from a small piston source forming part of an infinite rigid wall.† The expression given in (38) can be checked experimentally by simultaneous observations of the frequency, amplitude, and motional impedance (see p. 73) of a diaphragm vibrating in water. The amplitude may be measured optically by means of

* See also equation (2), p. 158.

† *Sound*, **2**, p. 162.

a small mirror attached at a point half-way along a radius of the diaphragm. The power radiated is most conveniently indicated by a wattmeter in circuit with the alternating electrical supply which excites the diaphragm. The change of 'watts' on passing through resonance is a measure of the sound-power radiated (10^7 ergs/sec. $=1$ watt). The sound *intensity* at any point in the medium will of course be the power divided by $4\pi r^2$, where r is the distance of the point from the source (assumed small).

Double Sources - The simple 'point source' which we have been considering is a theoretical abstraction. All vibrating bodies have a finite size which may be very large (the case of plane waves) or very small (in case of spherical waves) compared with a wave-length of the sound radiated. Further, many sources of sound, unless special precautions are taken, will act as *double sources*. Thus a vibrating diaphragm exposed to the free atmosphere on both sides is, at any instant, sending out a compression pulse on one side and a rarefaction pulse on the other, behaving as *two sources near together and in opposite phase*. A vibrating wire is another example of this. Lamb * has developed the theory of the double source, regarding it as two simple sources near together but in opposite phase. It requires little proof to show that such a source will radiate less energy than the corresponding simple source, for one component will almost neutralise the effect produced by the other. The velocity potential ϕ at a point P distant r from a double source of strength C cos nt is, for large values of r,

$$\phi=-\frac{kC}{4\pi r}\sin n(t-r/c)\cos\theta \qquad (39)$$

where θ is the angle made between a line joining the two simple sources and a line from the midpoint to the point P. Also the rate of emission of energy by the double source is

$$W=\frac{dE}{dt}=\frac{\rho ck^4C^2}{24\pi} \text{ or } \frac{\rho n^4C^2}{24\pi c^3} \qquad (40)$$

the intensity at any point in the medium depending on the orientation θ as in (39). It is instructive to compare equations (35) for a simple source, and (40) for a double source. Regarding these as having the same strength, we observe that

(*a*) In both cases the power radiated is proportional to ρc, the radiation resistance of the surrounding medium, and

* *Sound*, p. 226.

(*b*) the emission from a simple source varies as n^2 (*i.e.* as $1/\lambda^2$), whereas it varies as n^4 (*i.e.* as $1/\lambda^4$) from a double source. The double source, therefore, rapidly becomes more efficient as a radiator, relative to the simple source, with increasing frequency and diminishing wave-length of the sound.

In the case of the double source (see equation (40)), if the frequency N is fixed, the energy emitted in different media will vary directly as the density ρ and inversely as the cube of the wave-velocity c. The energy emission *for sources of the same strength* in gaseous media, where this velocity varies inversely as $\sqrt{\rho}$, will, at constant frequency, vary inversely as the *fifth* power of the wave-velocity. This accounts for the apparent feebleness of a bell or a tuning-fork (both double sources) when vibrating in hydrogen as compared with air. The wave-velocity in hydrogen is 3·9 times that in air. Consequently the energy emission is $(3 \cdot 9)^5$ or 900 times as great in air as in hydrogen. A comparison might also be made between air and water as the transmitting media, the relative densities and wave-velocities being 1/770 and 1/4·4 respectively. Apart from other considerations, the energy emission from sources of equal strength would, for the reasons given above, be about 3400 times as great in water as in air. The powerful damping effect of water, as compared with air, on the vibrations of a solid body are thus explained.

In both cases of a simple and a double source (see equations (35) and (40)) the energy emission is proportional to the square of the strength of the source, *i.e.* to $S^2\xi_0{}^2$, where S is the area of the vibrating surface. It will be clear, therefore, why a vibrating tuning-fork or wire emits its energy at a much greater rate when it is brought into contact with a solid body of large area such as a table top. The sounding-board of a piano or of a violin radiates practically all the vibrational energy of its strings. As we have seen, the increase in energy-emission with frequency is most rapid ($\propto n^4$) in the case of a double source. When the vibrations are very slow (n small) the fluid surrounding the source behaves as if it were incompressible and simply *flows locally* from one face of the double source to the other which is in opposite phase. In such a case no sensible vibration is *transmitted* by the medium. As the frequency increases, and λ becomes comparable with, or greater than, the distance apart of the two simple sources which form the 'double-source,' this 'local flow' is reduced and true waves of alternating pressure, *i.e.* sound waves, are set up in the medium. The local flow increases with increase in the velocity of sound in the medium, when the latter more nearly approaches incompressibility, and less energy is radiated as sound (as in the case of hydrogen and air cited above).

This reduction of 'local flow' has important practical application in the design of sound generators to work in media such as air or water. It is obviously better, for example, to employ a vibrating diaphragm with one side enclosed, when it behaves as a simple source and radiates much more energy than if its efforts were partially neutralised by the simultaneous out-of-phase vibrations from the opposite side. For a similar reason a diaphragm enclosed on one side is much more efficient as a radiator when sounding its fundamental tone, than vibrating with one nodal circle when there is local flow between the two areas vibrating in opposite phase. In the case of a vibrating string the local flow in the surrounding air accounts for the feebleness of the sound emitted.

In certain circumstances, however, more particularly in the directional transmission of sound, the double (or possibly a multiple) source (see p. 217) is very important. When the distance apart of the two equivalent simple sources is comparable with or greater than a wave-length, the directional term $\cos\theta$ in equation (39) becomes important.

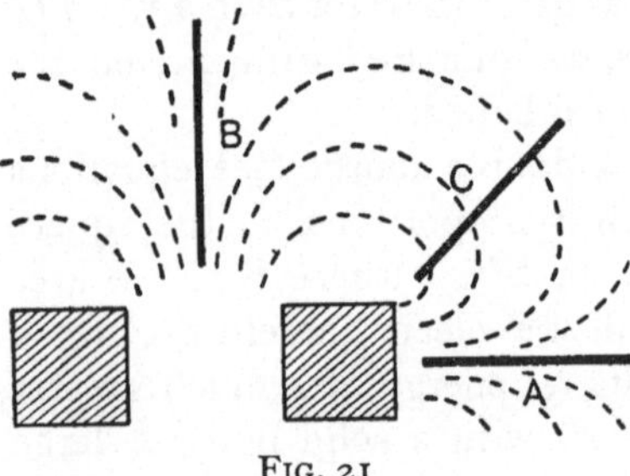

FIG. 21

Rayleigh refers to a simple experiment performed by Stokes* to illustrate the increase of the sound emitted from a tuning-fork when local flow around the prongs is prevented. A piece of cardboard is held with one edge parallel and close to one prong of the vibrating fork. When the card is placed as at A or B in fig. 21 relative to the prong there is little or no effect, but when placed at C, making an angle with the plane of the prongs, the sound is considerably intensified.

The Principle of Superposition – Huyghens first drew attention to the important fact that the passage of one beam of light through an aperture is in no way affected by the passage of another beam through the same aperture. The light waves cross each

* See Rayleigh, *Sound*, **2**, p. 306; and Stokes, *Phil. Trans.*, **158**, p. 447, 1868.

other at the aperture without interfering at all with each other's course. The same is true also of sound waves. This independence of the separate waves is explained on the *principle of superposition.* This principle implies that the resultant displacement, velocity, or condensation at any point in the medium is equal to the *vector* sum of the corresponding quantities due to each wave-train independently. The principle is valid only for small amplitudes, when it can be assumed that the stresses are linear functions of the strains, a condition which is fulfilled in the case of ordinary sound waves of small amplitude. It is important to notice that the principle is only applicable to quantities which may reverse in sign (viz. displacements, velocities, and condensations); it would lead to erroneous results if applied to the addition of energies or intensities, involving the squares of these quantities. From a purely physical standpoint there cannot, of course, be two wave-systems at once in the same part of the medium, since an element of the medium cannot have two different densities or velocities at the same instant. The *actual* motion of the medium is different from any of the individual motions which would be produced by the wave-trains independently; when we speak of two or more wave-systems passing at the same instant through the same medium, the actual or resultant effect is a single wave-system different from any of the imaginary ones. It is important to remember when applying the principle that there is no restriction as to the relation between the component frequencies.

A special case of superposition, which may appear to be a contradiction, is known as interference (see p. 293). Two independent wave-trains of the *same* frequency mutually cross each other. At certain places there is reinforcement (*i.e.* increased amplitude), whilst at others there is reduced, or possibly zero, amplitude. Such 'interference,' as it is called, is a direct outcome of the principle of superposition.

Huyghens' Secondary Waves – Another important principle due to Huyghens is the following: The wave-front of a disturbance may at any instant be obtained as the envelope of the secondary waves proceeding from all points of the wave-front at some preceding instant.

Thus a disturbance diverging with velocity v from a point source may at any time t_1 be represented by a thin spherical shell. According to Huyghens' principle, we may regard this shell as the disturbed region, and determine the disturbance at a subsequent time t_2 by drawing spheres of radii $v(t_2-t_1)$ round each point of the shell. The outer spherical envelope of these spheres will be the new wave-front at the instant t_2. By this construction it will be found that the direction of advance of the wave is normal to the wave-front. As the distance from the source becomes large, a small area of the wave-front approximates to a plane surface at right angles to the direction of advance of the wave. It is often convenient, and sufficiently accurate, to regard a spherical wave at a large distance from the source as a plane wave. A wave-front may be defined as a surface such that the disturbance over it commenced at the same source at the same instant. Such a wave-front may of course be entirely in one medium or in two or more media. The wave-front due to a point source in a homogeneous medium will always be spherical. The definition of course involves the assumption that waves of all frequencies are propagated with equal velocities.

Making use of the theory of the simple point source of sound, and combining it with Huyghens' principles of superposition and secondary waves, it is often possible to deduce the sound distribution in a medium containing a source of sound of appreciable size. The reflection and refraction of sound waves at the boundary of two different media follow the ordinary laws applicable to light waves, *provided* the dimensions of the boundary are always large compared with the wave-length of the sound incident upon it. In the majority of sound problems with which we shall have to deal, however, this condition is not fulfilled, the dimensions of obstacles and apertures being frequently of the same order as the wave-length of the sound. Diffraction problems in sound are consequently of much more frequent occurrence, and are of greater relative importance, than they are in the case of light. Only when sounds of very high frequency (and therefore of short wave-length) are employed is it possible to demonstrate the ordinary optical laws of reflection, etc., with apparatus of moderate dimensions. In order to obtain effective focussing of sound waves by curved reflectors or lenses it is important that the dimensions of the reflector should be many times the wave-length of the incident sound. The same statement is applicable to sound screens if 'optical' shadows are required. In the majority of cases, therefore, where diffraction effects predominate or cannot be neglected, the principle of secondary waves as proposed by Huyghens may be of great assistance. The method is often applied experimentally by means of spark-pulse and ripple photography (see p. 340) where the progress of a wave-front can be followed from point to point in a medium containing obstacles of any required form and distribution. In Section III such questions will be examined in greater detail.

ELECTRICAL VIBRATIONS

Attention has already been drawn to the close analogy which exists between mechanical and electrical vibrations. On this account alone it would seem justifiable to introduce the subject of alternating electrical phenomena at this point, but there is also another important reason why further reference should be made to this subject. Alternating electrical and magnetic effects provide a very convenient means of exciting the corresponding mechanical vibrations. An alternating current supply of variable frequency and power may be employed in the production of sound vibrations of corresponding frequency and intensity.

The earlier forms of alternating current generators (of sonic and supersonic frequency), such as the Duddell singing arc, were very effective but rather inconvenient in use. The advent of the thermionic three-electrode-valve during recent years has, however, supplied an extremely convenient and inexpensive source of alternating electrical power. By means of the valve it is a comparatively simple matter to produce alternating current of almost any conceivable frequency, from zero to many millions of cycles per second. In addition to this the power output at any particular frequency may be easily controlled and measured. It is not surprising, therefore, that a rapid development has taken place in the design of electrical sound sources (or generators as they are sometimes called) and in electrical methods of receiving and recording sound waves. The transfer of electrical to mechanical energy and *vice versa* is of frequent occurrence in many problems concerned with the production, transmission, and reception of a sound wave.

We shall therefore consider briefly the more salient facts in the theory and methods of production of electrical oscillations.

Free Oscillations in a Circuit containing Inductance, Capacity, and Resistance – When a condenser of capacity C is discharged through a coil of inductance L and resistance R, the electrical charge Q in the condenser at any time t after closing the circuit may be deduced from the equation of electromotive forces operating in the circuit, viz.

$$L\frac{d^2Q}{dt^2}+R\frac{dQ}{dt}+\frac{Q}{c}=0 \quad . \quad . \quad . \quad (1)$$

A comparison of this equation with equation (1) on p. 34 is instructive. It will be seen at once that they are of the same form, consequently the solutions will be similar. It is important to note the corresponding mechanical and electrical quantities involved, viz.

Inductance L≡Inertia (mass) m.

Resistance (electrical) R≡Resistance (mechanical) r.

Capacity C≡Compliance or $\frac{1}{\text{stiffness}} \cdot \frac{1}{s}$.

Quantity (electrical displacement) Q≡Displacement x.

Current dQ/dt≡Velocity dx/dt.

Rate of current change d^2Q/dt^2≡Acceleration d^2x/dt^2.

The method of solution of the above equation, follows exactly the same lines as that on p. 34, etc., in the mechanical case. When R is less than a certain critical value the solution therefore takes the same form as equation (8), p. 34.

$$Q = Q_0 e^{-kt} \cos(n't + \epsilon)$$

where Q_0 is the initial charge on the condenser,

$k = R/2L$ the electrical damping coefficient

and

$$n' = \sqrt{\frac{1}{LC} - \frac{R^2}{4L^2}} \qquad . \quad (3)$$

This equation represents a vibratory S.H. motion with amplitude dying away exponentially. The condition for oscillations is that n' should be real, *i.e.*

$$\frac{1}{LC} > \frac{R^2}{4L^2} \quad \text{or} \quad R < 2\sqrt{L/C} \qquad . \quad . \quad (4)$$

Provided R is considerably smaller than the critical value in (4), the frequency of oscillation is

$$N = \frac{n}{2\pi} = \frac{1}{2\pi\sqrt{LC}}$$

or the periodic time

$$T = 2\pi\sqrt{LC} \qquad . \quad . \quad . \quad (5)$$

substituting m for L and $1/s$ for C, we obtain the corresponding relation $T = 2\pi\sqrt{m/s}$ for the mechanical system.

Forced Oscillations – In the previous case we considered the effect of discharging a condenser through a circuit having inductance and resistance. Now take the case of an alternating e.m.f. E cos pt applied to such a circuit. The equation of e.m.f.'s will be

$$L\frac{d^2Q}{dt^2}+R\frac{dQ}{dt}+\frac{Q}{c}=E\cos pt \quad . \quad . \quad (6)$$

Comparing with the analogous equation for the forced vibrations of a mechanical system (see equation (1), p. 37), we obtain the solution equivalent to equation (5), p. 37, viz.

$$\left.\begin{aligned} &Q=\frac{f\sin\epsilon}{2kp}\cos(pt-\epsilon) \\ &\tan\epsilon=\frac{2kp}{(n^2-p^2)} \quad \text{and} \quad \sin\epsilon=\frac{2kp}{\{(n^2-p^2)^2+4k^2p^2\}^{\frac{1}{2}}} \\ &\text{where} \\ &f=E/L, \quad k=R/2L, \quad \text{and} \quad n^2=1/LC \end{aligned}\right\} \quad . \quad (7)$$

The current i in the circuit at any instant t is obtained by differentiating Q thus,

$$i=\frac{dQ}{dt}=-\frac{f\sin\epsilon}{2k}\sin(pt-\epsilon) \quad . \quad . \quad (8)$$

On substitution of the corresponding electrical quantities for f, k, and n^2, this becomes

$$i=\frac{E\sin(pt-\epsilon)}{\left\{R^2+\left(Lp-\frac{1}{Cp}\right)^2\right\}^{\frac{1}{2}}}=\frac{E}{Z}\sin(pt-\epsilon) \quad . \quad (9)$$

The denominator of this expression for the current is known as the *electrical impedance* Z of the circuit, the quantity $\left(Lp-\frac{1}{Cp}\right)$ being designated the *reactance* X. The phase lag ϵ is given by

$$\tan\epsilon=\frac{R}{(Lp-1/Cp)}=\frac{R}{X} \quad . \quad . \quad . \quad (10)$$

Resonance – When the frequency of the force coincides with that of the free oscillation (resistance damping being small), we have from (9) and (7),

$$i=\frac{E}{R}\sin(pt-\pi/2)=\frac{E}{R}\cos pt \text{ [since } n=p, \tan\epsilon=\infty \text{, and } \epsilon=\pi/2],$$

the maximum current being therefore E/R, in phase with the applied voltage. When R is zero, which of course never occurs in practice, the current becomes infinite. In ordinary cases, the current at resonance is limited solely by the resistance in the circuit, the effect of inductance being neutralised by the negative effect of the capacity. This is obviously analogous to the case of resonance in a mechanical circuit where the amplitude and particle velocity are limited only by the frictional damping.

The question of electrical power dissipation is also similar to the mechanical case dealt with on p. 42. It is unnecessary to repeat the discussion already given.

The Circle Diagram – The phase relations between e.m.f. E, impedance Z, and current i (or the corresponding mechanical quantities force, mechanical impedance, and particle-velocity) may be indicated simply by a vector diagram.* If an alternating voltage be applied to a circuit consisting of a resistance R and a reactance X (inductance and capacity), the voltages across the two component parts of the circuit will always be 90° out of phase with each other, this being independent of the magnitudes of the resistance and the reactance. Fig. 22 (a) shows a vector diagram

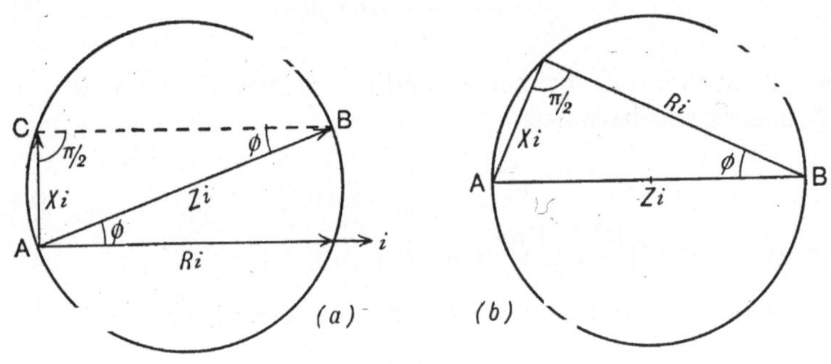

FIG. 22

of the circuit where AB $=Zi=$voltage applied to the circuit. A circle described on AB as diameter will therefore pass through the point C since the angle ACB is always a right angle. If the relative values of the resistance and reactance be altered, the angle ϕ, the phase lag between current and applied e.m.f., is altered also, but the angle ACB is still a right angle. The diagram generally takes the form shown in fig. 22 (b), in which AB indicates the phase of the applied e.m.f. and BC the phase of the

* *E.g.* see Kemp, *Alternating Current Electrical Engineering* (Macmillan).

current. A similar diagram may be constructed for a vibrating mechanical system, substituting f for E/L, k for R/2L, and n^2 for 1/LC.

Motional Impedance. Efficiency of Electrically-driven Sound Sources – On account of the extensive use of electrodynamic or electromagnetic forms of sound generators and receivers, it is a matter of some importance to know their characteristics and efficiency. In any form of machine which converts electrical energy into motion, the moving mechanism reacts on the electrical circuit. In the case of an ordinary electric motor, for example, a back e.m.f. is developed in the armature when the latter rotates. The electrical efficiency of the motor is equal to the ratio of this back e.m.f. E to the voltage V of the supply, since Vi is the gross power supplied and Ei is the net power utilised. The current i through the armature is equal to $(V-E)/r$ which consequently becomes smaller, the greater the efficiency of the motor, *i.e.* the nearer E approaches V. The back e.m.f. therefore behaves like an added resistance in the circuit. Denoting this apparent additional resistance due to the rotation of the armature by R, we may write the current i equal to $V/(r+R)$. Now, the power supplied is $i^2(r+R)$ and the power utilised is i^2R, whence the efficiency is $R/(r+R)$. The *efficiency* is therefore the *ratio of the 'motional' resistance* R *to the total effective resistance* $(r+R)$. The net power utilised by the motor is a maximum when $d(i^2R)/dR$ is zero, *i.e.* when $R=r$, in which case the efficiency is 1/2.

In a similar manner, the mechanical vibrating element of an electrical sound generator or receiver reacts on the electrical circuit. The back e.m.f. produced by the vibration manifests itself as a change of impedance in the circuit. It is customary, therefore, to determine the characteristics of such a vibrator from the analysis of its impedance at different frequencies of excitation. This method of analysis was first introduced by Kennelly and Pierce * who applied it to study the vibrations of a telephone diaphragm. The principle involved with an alternating current motor, *e.g.* a telephone receiver, is analogous to that of the direct current motor to which we have just referred. The only difference is that we have now to deal with impedances and phase angles instead of the ordinary D.C. resistance. Thus, in such an

* See Kennelly and Pierce (*Amer. Acad. Science Proc.*, **48,** p. 113, 1912); *Electrical Vibration Instruments*, A. E. Kennelly (Macmillan); and *Dictionary of Applied Physics*, **2,** 'Electricity,' pp. 844–846, 971, and 977.

electrically maintained vibrating system, if R′L′ are the 'vibrating' and RL the 'stationary' values of resistance and inductance, then (R′−R) is the *motional resistance*, (L′*p*−L*p*) or (X′−X) the *motional reactance*, whilst the *motional impedance* is the vector sum of (R′−R) and (X′−X).

Motional Impedance Circle – When the impressed frequency differs widely from the resonant frequency of the vibrating system, the amplitude of vibration will be small for a given input of electrical energy. Under such circumstances, therefore, the motional impedance will be small and the efficiency very low. Near resonance, however, in a well-designed 'sound transmitter,' the motional impedance may be large and the efficiency correspondingly high. To determine the characteristics of a transmitter, therefore, a series of impedance measurements (*i.e.* of R and X) is made at constant current over a range of frequencies in the neighbourhood of resonance. Strictly speaking, two series of measurements should be made with the vibrator, (1) 'free' to move, and (2) clamped, or heavily damped, in the normal position. In practice, however, it is generally unnecessary to make the measurements (2) as they can be inferred from the results of series (1). In fig. 23 (*a*) and (*b*) the resistance and reactance of a

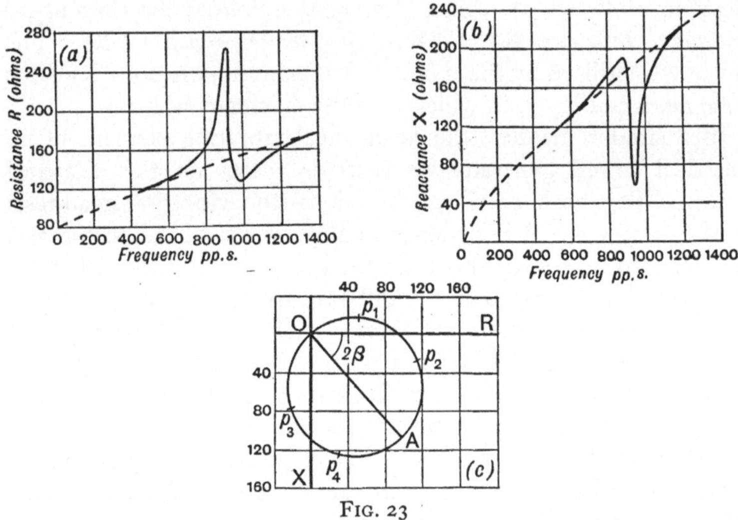

FIG. 23

vibrating telephone (diaphragm type) are plotted in terms of frequency near resonance.

If now the corresponding values of motional resistance (R′−R) and motional reactance (X′−X) obtained from these curves are plotted as in fig. 23 (*c*), each point being marked with its corresponding frequency, it will be found that the locus of the points is a circle passing through the origin. The motional impedance at any given frequency will then be the chord Op_n of the circle. At *resonance*, the motional impedance is represented by the chord of maximum length, *i.e.* the diameter OA of the circle. In this case the motional impedance is a *pure resistance* in phase with the current. Consequently the diameter of the circle, expressed in ohms, multiplied by the square of the current (amps.) gives a measure of the mechanical power developed (watts) by the motion of the vibrator. The angle 2β between the diameter OA and the R axis is a measure of the 'iron loss' in a moving iron transmitter, such as a telephone earpiece. As in the case of the electric motor, *the ratio of the maximum motional resistance to the total effective 'free' resistance is a measure of the efficiency at resonance.* If the resonance is very sharp, the damped (or clamped) impedance may be regarded as constant over the range of the circle. In such a case it is often sufficiently accurate to obtain the motional resistance at resonance by simply subtracting the minimum from the maximum observed impedances near resonance.

Wattmeter Method – If the power of a sound transmitter is sufficiently large, the output and efficiency may be determined with sufficient accuracy by a more direct method. Measuring the electrical power consumed in the circuit by a wattmeter at a series of frequencies near resonance, it will be found that a marked increase occurs at resonance. Such a curve for a small diaphragm sound transmitter under water is shown in fig. 24. It will be seen that the total power consumed at resonance is 80 watts, whereas the power absorbed on account of the motion of the diaphragm is 35 watts. Assuming this power represents energy radiated per second as sound, the efficiency is 44 per cent. A similar result is obtained if an ammeter in series and a voltmeter in parallel with the transmitter are used instead of a wattmeter. In this case the impedance of the circuit at any frequency is given by the ratio volts/amps.

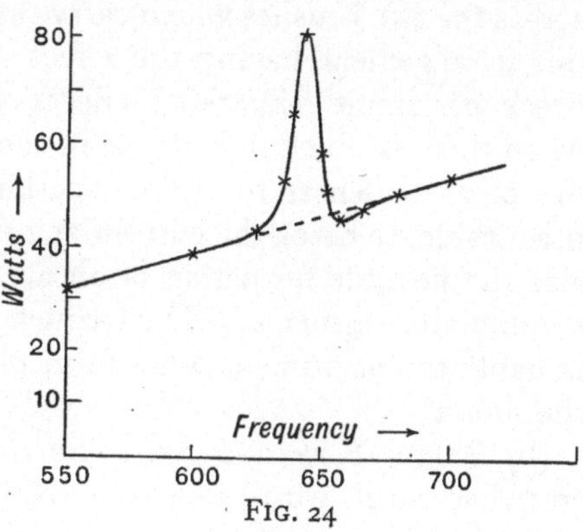

FIG. 24

Sources of Alternating Current

Any attempt to give details of the numerous ways in which alternating currents for acoustic purposes may be produced is manifestly outside the scope of this book. It is important, nevertheless, to refer briefly to a few which are in general use at the present time. For further information the reader is referred to standard treatises in A.C. electrical engineering. A useful descriptive list of suitable generators is given in the *Dictionary of Applied Physics*.* We shall refer briefly to a few of the more important.

(1) **Interrupters** – Perhaps the simplest type of 'alternator,' if it may be so described, is an interrupter of direct current. 'Interrupted' current may be used either directly or through the secondary of a transformer. In the latter case the current may be more properly described as alternating. Direct current may be rendered intermittent at any desired frequency within certain limits in numerous ways.

(a) *Motor-driven Commutator* – This is a very convenient form of interrupter in which the frequency is controlled by the speed of the motor. The number of segments in the commutator will of course be chosen to suit the frequency-range required, having regard to the speed-range of the motor. If much power is required from such a device it is usually the practice to connect a suitable condenser and series resistance across the two brushes which carry the current to the commutator, this arrangement having the effect of reducing sparking at the break of circuit. Sparking effects are particularly troublesome when there is much inductance in the circuit. In addition to the use of a condenser to reduce sparking it is advisable, as far as practicable, to break the current at a high speed in order to reduce also the possible formation of an arc between the brush and the commutator segments. The frequency of this form of interrupter is liable to vary unless some form of speed governor is fitted to the motor.

(b) *Tuning-Fork* and *Wire Interrupters* – Tuning-forks and stretched steel wires may be maintained in vibration on the electric-bell principle, by means of suitable contact-breakers —wires dipping into mercury cups or solid spring-mounted metal contacts—combined with a maintaining electromagnet. Auxiliary contacts may of course be provided to interrupt the

* Vol. 2, 'Electricity,' pp. 393–401 and pp. 897–903.

current in the main circuit, without interfering with the maintaining circuit of the fork or the wire. Vibrating wires and forks provide a reliable source of interrupted current of constant frequency. By altering the tension of the wire the frequency may be varied continuously from, say, 30 to 300 cycles/sec. The tuning-fork interrupter, on the other hand, is essentially fixed in frequency. If different frequencies are required it is necessary to have a suitable range of forks. Interrupted current of frequency as high as 200 per second may be obtained by means of tuning-forks in this way, but the difficulties of making and breaking mechanical contacts carrying current become very great at higher frequencies than this. Reference will be made later to valve-maintained tuning-forks of frequency of the order 1000 p.p.sec.

(c) *Buzzers* – Various forms of buzzers have been devised, the principle of operation being in most cases essentially that of the ordinary electric bell. The vibrating diaphragm or reed makes and breaks a contact in the circuit of the maintaining magnet. An improved form, due to Dr. Dye of the N.P.L., has an additional secondary winding in which alternating current is induced by the intermittent primary current winding which maintains the buzzer. The diaphragm buzzers made by the General Electric Company are made with frequencies lying within the range 300 to 1000 cycles per second.

(d) *Electrolytic Interrupters* (Wehnelt Type) – An interrupter of an entirely different type is one due to Wehnelt, which depends on the interrupted flow of direct current through an electrolyte. If a small platinum point is used as the anode and a lead or aluminium plate as the cathode in an electrolytic cell (containing a concentrated solution of almost any salt, *e.g.* ammonium phosphate, or even common salt), then an interrupted flow of current will take place at a regular frequency provided the circuit (at, say, 100 volts) contains sufficient inductance. The wave-form is rich in harmonics. The purity of the wave-form may be improved by using a tuned circuit (inductance and capacity) across the electrodes. It is of interest to note in passing that a Wehnelt interrupter not only forms a convenient means of obtaining interrupted current of frequency-range 100 to 5000 p.p.s. or so, but also may serve as a fairly powerful source of sound. The sudden production of gas bubbles at the same frequency as the interruptions results in corresponding pressure vibrations being set up in the surrounding liquid.

(2) **Microphone Hummers** – The wave-form of interrupted

current is obviously far from sinusoidal, since the current is varied suddenly between zero and maximum. In the microphone hummer a vibrator (tuning-fork, bar, telephone diaphragm, etc.) is maintained in vibration by the fluctuating current through a microphone attached to the vibrator. This current, acting through an electromagnet, reacts on the vibrator which in turn shakes the microphone—the cycle being continuous. A moderately pure sinusoidal alternating current may then be obtained from a secondary winding on the maintaining electromagnet. The same principle has been applied by A. Campbell * at moderately high frequencies, 1000 to 5000 p.p.s., by means of a steel bar with a microphone and exciting magnet suitably mounted. As before, a transformer is used to couple the vibrator to the external circuit. The humming telephone † is a microphone hummer in which a vibrating telephone diaphragm reacts through the intervening air on a microphone coupled electrically to the telephone.

(3) **Singing Arcs** – Duddell's singing arc ‡ provides a source of alternating power of a fairly wide range of frequency, but in its original form is not very convenient. A modification, due to C. L. Fortesque, employs a 'pointolite' (tungsten arc) instead of the carbon arc. This results in much steadier working conditions. As in the Duddell arc the pointolite is tuned by means of inductance L and capacity C across the arc electrodes, a suitable coupling coil being employed to connect to the external circuit. The frequency is simply determined by $1/2\pi\sqrt{LC}$ (or $159{\cdot}2/\sqrt{LC}$ if C is measured in microfarads and L in henries).

(4) **Alternating Current Machines** – Low-frequency alternators working up to 200 p.p.s. are well known and are described in electrical engineering textbooks. For audiofrequencies, other designs have been developed.

(a) *Dolezalek Toothed Wheel* – In its simplest form this consists of a toothed iron wheel rotated, by means of a motor of controlled speed, close to the poles of an electromagnet carrying secondary coils near its poles. The pole pitch is equal to an exact whole multiple of the tooth pitch of the iron wheel. As the teeth pass the poles the flux through the magnet, and consequently the secondary coils, is varied, the frequency being determined simply by the product of the number of teeth and revolutions per second of the wheel. To reduce losses due to eddy currents, it

* *Roy. Soc. Proc.*, **78**, p. 209, 1906.
† F. Gill, *J. Inst. Elect. Eng.*, **31**, p. 388, 1901.
‡ *J. Inst. Elect. Eng.*, **30**, p. 232, 1900.

is an advantage to laminate the magnet and the teeth of the wheel. Frequencies up to 4000 p.p.s. may easily be obtained in this way. A very simple form of Dolezalek alternator may be constructed by using a two-pole telephone receiver with its bobbins, and mounting it near the teeth of a suitable motor-driven wheel. The wave-form is rich in harmonics.

(b) *Duddell Alternator* – This machine is well designed and provides a reasonably pure sine-wave current over a frequency range up to about 2000 p.p.s. The rotor of the machine carries the field winding on 30 poles. The stator consists of a laminated iron ring wound like a gramme ring—the air gaps between stator and rotor being fairly long to reduce armature reaction. The output of the machine varies from about 10 watts at low speeds to about 500 watts at high speeds.

(c) *Alexanderson Alternator*. – In very high-speed alternators it is necessary to employ a specially strong mechanical construction for the rotor. This has been achieved in the Alexanderson machine, which runs at 30,000 r.p.m., *i.e.* ten times the speed of an average small motor. A rotating disc has 300 slots set round its edge to leave 300 steel teeth. Each of these teeth acts in turn to close the magnetic circuit between two of the poles of the stator, which holds the magnetising field coils and a peculiar form of slot winding in which the alternating e.m.f. is generated. These alternators were generally used for purposes of wireless telegraphy. A good description may be found in the *Dictionary of Applied Physics*, **2**, 'Electricity,' pp. 1051–1053.

(d) *Valve Oscillation Generators* – The introduction of the thermionic valve has undoubtedly provided the best source of alternating current for sound production. Simplicity combined with wide ranges of power and frequency make the valve, with its associated circuits, an ideal means of electrically exciting mechanical vibrations of any conceivable frequency. It has the advantage over nearly every other type of alternator, of silence and steadiness of operation combined with great flexibility of frequency control. The three-electrode valve with its grid, plate, and hot metal filament is now so well known in its associations with radiotelephony that a further description here is superfluous. Many different methods are in regular use for maintaining electrical oscillations, and the reader is referred to treatises on radiotelephony for details of these. Perhaps one of the simplest and most convenient forms of oscillation generator circuit is that shown in fig. 25. When due allowance is made for the difference in frequencies of oscillation the arrangement of an audiofrequency valve generator is identical with that employed at radiofrequencies. In the arrangement shown an inductance coil L_g connecting the filament to the grid is coupled inductively to another coil L_p connecting the plate to the filament through a high-tension battery B. A variable condenser C is connected in parallel with the plate coil L_p. In alternative arrangements this condenser may be connected across the grid coil, or condensers may be used with both grid and plate coils. A third coil L_e, coupled to the coils L_p and L_g, provides the alternating current for the external circuit. Briefly, the system functions as follows. A small oscillation set up in the tuned circuit L_pC, *e.g.* by switching on the filament current or the high-tension battery, induces a corresponding change in the grid coil which consequently causes the potential of the grid relative to the filament to fluctuate in the same manner. In turn, this varies the plate current, giving an amplified voltage fluctuation in the plate L_pC circuit, thereby increasing its oscillations. The grid is again affected, and the cycle continues to repeat itself until a steady state of oscillation is maintained. Regarded in this very elementary manner, the frequency of oscillation is dependent solely on the values of L_p and C in the plate circuit, *i.e.* the frequency is equal to $1/2\pi\sqrt{L_pC}$, provided the resistance R of the plate coil is not too great. A more complete view of the matter includes the effect of mutual inductance M between the plate and grid coils, as well as the amplification factor σ and 'plate' resistance ρ of the valve itself. The valve adds to the resistance R in the oscillatory circuit a term of the form $(L_p+\sigma M)/\rho C$, the equation of oscillation * being

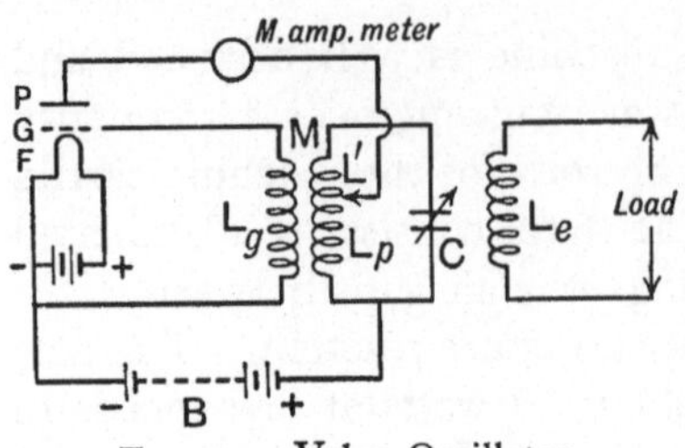

FIG. 25—Valve Oscillator

$$L_p\frac{d^2i}{dt^2}+\left(R+\frac{L+\sigma M}{\rho C}\right)\frac{di}{dt}+\frac{i}{C}\left(1+\frac{R}{\rho}\right)=0,$$

where i is the current flowing in the plate inductance coil L_p. The damping constant

$$k=\frac{1}{2L}\left(R+\frac{L+\sigma M}{\rho C}\right)$$

* See 'La Lampe à trois Electrodes,' C. Gutton, **5**, p. 59, edited by *Le Société 'Journal de Physique.'*

and the frequency of oscillation, when damping is small,

$$N=\frac{n}{2\pi}=\frac{1}{2\pi}\sqrt{\frac{(1+R/\rho)}{LC}}$$

$$N=\frac{1}{2\pi\sqrt{LC}},$$

when R is small compared with ρ, as is usually the case in practice. The condition for oscillation requires that n^2 should be always greater than k^2.

If it is required that an oscillator should give a maximum power-output *at a particular frequency* it is an advantage to divide the inductance L_p in the tuned circuit into two parts, L_p and L', one of which, L', is introduced in series with the capacity as shown in fig. 25. This amounts to connecting the plate to a suitable tapping point on the inductance L_p in the tuned circuit.

The frequency of an oscillator is conveniently adjusted by means of a set of condensers, from a variable air condenser 0·001 microfarad to large capacities of the order of 10 microfarads in mica or paraffin paper condensers. It is convenient also to have a series of pairs of inductances suitable for different ranges of frequency, *e.g.* 1 henry for frequencies from 200 to 1000 or 2000, 10 millihenries for frequencies of the order 10,000, and so on. A set of coils made for radiotelephony purposes will cover the higher frequency ranges if required. There are numerous commercial forms of valve which are suitable for oscillators having power outputs from a few watts to many kilowatts. 'Loud speaker' valves are very convenient where a moderate output, 10 or 20 watts, is required for general laboratory purposes. If larger powers are necessary it is advisable to refer for special information to treatises on radiotelephony.

Valve Amplifiers – The use of amplifying valve circuits to increase the intensity of relatively feeble alternating currents and voltages is of considerable importance in connection with electrical methods of sound reception. Feeble sound waves, of whatever frequency, may be picked up by means of a receiver which converts the mechanical into electrical vibrations and the latter multiplied many hundreds of times, if required, by means of a valve amplifier. As in the case of valve oscillators, many forms of amplifier exist. The design of the inter-connections of a multi-valve amplifier depends essentially on the frequency range in which it is to be used, and on the amount of permissible distortion of wave-form.

Resistance-capacity coupled amplifiers are very useful and, with suitable valves and grid bias, give a faithful reproduction of wave-form. At audiofrequencies it is possible to use commercial forms of transformer without serious distortion. In figs. 26A and

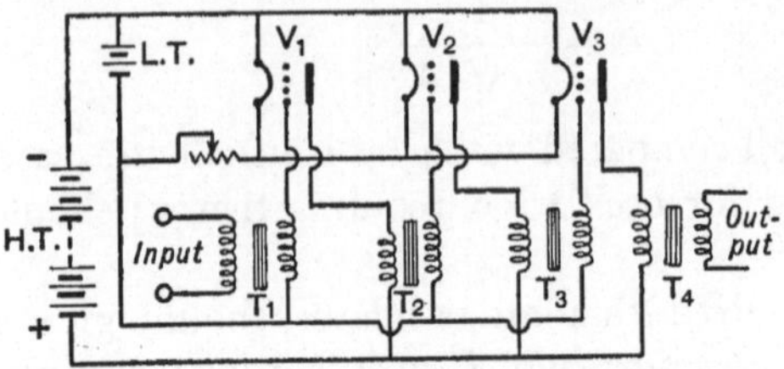

FIG. 26A—Amplifier Transformer Coupling

26B typical valve amplifier circuits of this nature are indicated. The design of the input and output ends of a multi-valve amplifier must, of course, depend on the particular circumstances under which it is to be used. Commercial types of amplifying valves give voltage amplifications varying from 6 to 36 or more, according

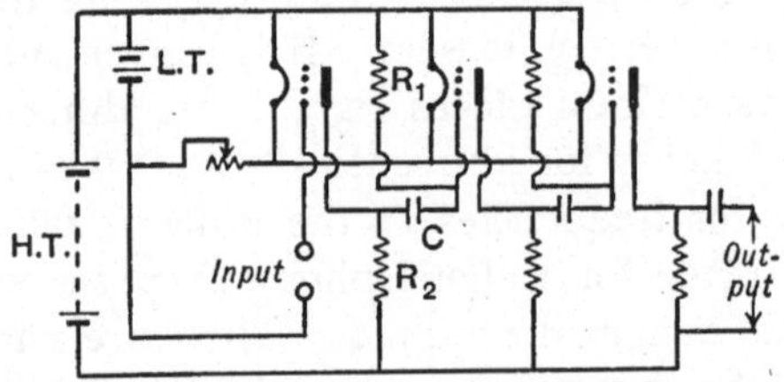

FIG. 26B—Amplifier Resistance-Capacity Coupling

to the design. Usually a high voltage amplification implies a high plate impedance. Descriptions of valves and valve circuits with their amplifying, rectifying, and oscillatory properties are to be found in numerous publications dealing with radiotelephony. These should be consulted if further information is desired.

SECTION II

VIBRATING SYSTEMS AND SOURCES OF SOUND

HITHERTO we have dealt with simple 'ideal' systems, point sources, homogeneous media, and simple harmonic motion. It is necessary now to consider how the results thus obtained are applied to the more complex vibrations of actual sources of sound. Solid bodies are particularly liable to vibrate in more than one mode, and in each of these possible modes there may be overtones present in addition to the fundamental. By adopting suitable precautions, it is possible in certain cases to confine the vibration mainly to one particular type. The vibrations of a stretched string, for example, are predominantly transverse, but it is not difficult to excite longitudinal or torsional vibrations also. As the diameter of a wire increases relatively to the length the importance of the longitudinal mode of vibration increases, and in a rod the transverse and longitudinal modes are both easily excited. As the diameter of the rod is still further increased relatively to the length, the rod may ultimately be regarded as a plate which vibrates in the transverse mode most easily. In all these cases, however, if the method of excitation is suitably chosen, it is possible to encourage one particular mode, whilst discouraging the others. The vibrations of the body are also affected by the nature of the medium surrounding it, and by the presence of other bodies in the neighbourhood, particularly if these bodies have natural periods of vibration equal to, or near to, that of the vibrating source.

In considering actual cases it is generally necessary to make simplifying assumptions and to deal with vibrators conforming approximately to certain simple types. Vibrations of the intermediate types which do not conform with the mathematical assumptions and approximations must be considered mainly from the purely 'physical' aspect, the mathematical treatment, when possible, being very complex. For example, the string vibrator is regarded as a perfectly uniform and flexible filament of solid matter stretched between two fixed supports. In practice, allowance has to be made for lack of uniformity and flexibility and for movement of the points of support. The method of excitation of vibration in a body has a considerable influence on the wave-form or quality of the sound it emits. A string, for example, vibrates transversely in different ways according as it is bowed, plucked, or struck. A steel rod struck at one end will vibrate both transversely and longitudinally, but by means of an electro-magnet it is possible to make either mode predominate.

In regard to the nature of the motion of solid vibrating systems, the more important factors with which we shall have to deal are, broadly speaking, (1) the resonant or 'natural' frequencies of the vibrating system, (2) the segmental arrangement of the various forms of vibration, and (3) the extent to which these vibrations may be communicated to the surrounding medium. The applications of vibrating systems as sources of sound will be discussed.

TRANSVERSE VIBRATIONS OF STRINGS

The type of vibration of stretched strings or wires with which we are most familiar is that in which the particles of the string vibrate in a plane perpendicular to the line of the string. This form of vibration is described as the *transverse* or lateral type. Another mode of vibration which is of comparatively little interest is the longitudinal vibration in which the string remains straight but the tension varies.

In the first instance we shall assume the 'theoretical' string which (*a*) is perfectly uniform, having a constant mass per unit length, (*b*) is perfectly flexible, its stiffness being supposed negligible, and (*c*) is not subjected to appreciable changes of length whilst in vibration. A long, thin, tightly stretched wire held between massive and well-clamped supports fulfils these conditions fairly well. After dealing with the 'ideal string' we shall consider what modifications are necessary to conform with *actual* conditions, in which the wire has stiffness and the supports yield as the wire vibrates. When any point of a wire stretched between two fixed points is displaced and released the wire commences to 'vibrate.' This vibration in effect results from transverse motions travelling in opposite directions along the string, and successive reflections of the 'wave' from the opposite fixed ends. In order to visualise such reflection of a transverse wave the student is recommended to make a few simple experiments on waves travelling along a stretched rope, one end of which is held in the hand whilst the other end is fixed. It will be observed that the movement of each particle of the rope whilst it forms part of the reflected wave is in the opposite direction to its motion whilst it formed part of the original wave. If such a transverse wave arrives at the '*free*' end of a rope it is also reflected, but the direction of motion of the particles is not reversed. The rope assumes the form of a sine wave if the end is moved up and down harmonically.

Velocity of a Transverse Wave along a String – A long string of mass m per unit length is stretched by a force of T dynes. Assume, in the first place, that the gradient of the curve formed by the string in its displaced position is always so small that the tension T may be regarded as constant. Taking the x axis of co-ordinates in the direction of the undisplaced string, and the y axis in the direction of displacement at right angles to the string, we require to determine the motion of an element δx whose displacement is y at the point x at the instant t. If $(\phi+\delta\phi)$ and ϕ are the angles of inclination of the tangents to the curve at the ends of the element δx, the difference in the tensions acting on the ends of the element in the direction of the y axis will be

$$\mathrm{T}\,\{\sin(\phi+\delta\phi)-\sin\phi\}=\mathrm{T}\cos\phi\,.\,\delta\phi=\mathrm{T}\delta(\sin\phi).$$

Also the mass of the element δx is $m\delta x$, and its acceleration in the direction of y is d^2y/dt^2, whence

$$m\delta x\,.\,d^2y/dt^2=\mathrm{T}\delta(\sin\phi) \quad . \quad . \quad . \quad (1)$$

Writing $\sin\phi=\tan\phi=dy/dx$ when the gradient of the curve is small, we have

$$\frac{d^2y}{dt^2}=\frac{\mathrm{T}}{m}\cdot\frac{d^2y}{dx^2} \quad . \quad . \quad . \quad . \quad (2)$$

which characterises a wave-motion travelling with a velocity

$$c=\sqrt{\mathrm{T}/m} \quad . \quad . \quad . \quad . \quad (3)$$

The solution of equation (2) is

$$y=f(ct-x)+\mathrm{F}(ct+x) \quad . \quad . \quad . \quad (4)$$

representing transverse waves travelling in opposite directions with the same velocity c.

Cf. with equations (8) and (9), p. 50.

A novel method of proof of the wave-velocity formula (3) has been given by Tait * in which the string is drawn through an

* *Encyclopædia Britannica*, 9th ed., Art. 'Mechanics.'

imaginary smooth tube with velocity c. The tube is straight except for the isolated curved portion which represents the wave on the string. If R is the radius of curvature at any point of the tube, the force acting in the direction of the normal to an element δs is $T\delta s/R$, where T is the tension in the string. Now the centrifugal force of the element δs of mass m per unit length and velocity c will be $m\delta s \,.\, c^2/R$, and this must balance the force $T\delta s/R$ if there is to be no reaction on the tube. Consequently, if $T\delta s/R = m \,.\, \delta s \,.\, c^2/R$, $c=\sqrt{T/m}$, and there is no reaction on the tube, *i.e.* the tube may be regarded as absent, and the wave travelling along the string with the velocity $c=\sqrt{T/m}$. In this method of proof, no assumptions need be made regarding the smallness of the displacement.

Reflection. Formation of Stationary Waves – If both ends of the string of length l are fixed, the wave is reflected successively from each end, and the resultant motion is determined by the superposition of the direct or *incident* wave and the reflected waves. The equation of motion expressed in (4) above must now be modified to suit the end-conditions of the string where the displacement y must always be zero. Thus we have, when $x=0$, $y=0$, *i.e.* $f(ct)=-F(ct)$; we may then rewrite (4),

$$y=f(ct-x)-f(ct+x) \quad . \quad . \quad . \quad (5)$$

Since $y=0$ when $x=l$ we have also

$$f(ct-l)=f(ct+l) \quad . \quad . \quad . \quad (6)$$

or writing

$$z=(ct-l)$$
$$f(z)=f(z+2l),$$

indicating that z is a periodic function repeating at intervals of $2l$. Consequently the motion of the string is periodic, the period $T=2l/c$ being the time taken for a wave to travel along the full length of the string and back again. The frequency of this form of vibration is consequently

$$N=\frac{c}{2l} \quad \text{where} \quad c=\sqrt{\frac{T}{m}},$$

i.e.

$$N=\frac{1}{2l}\sqrt{\frac{T}{m}} \quad . \quad . \quad . \quad . \quad (7)$$

If the displacement y varies sinusoidally with the frequency we may obtain *stationary waves* on the string. Thus equation (5) may be written

$$y=a\cos n(t-x/c)-a\cos n(t+x/c),$$

which reduces to

$$\left.\begin{aligned} y&=2a\sin nt\sin nx/c \\ &=2a\sin(2\pi \mathrm{N}t)\,.\,\sin(2\pi \mathrm{N}x/c)\end{aligned}\right\} \quad . \quad . \quad (8)$$

At any point x along the string the amplitude is therefore varying sinusoidally *with time* between *zero* and $2a$. The amplitude of successive particles is also varying *with distance* x according to a sine law. The result is a series of loops formed on the string, the amplitude varying between 0 and $2a$, *i.e.* the maximum amplitude is double that in the initial wave. The condition for the formation of loops on a string of finite length l is clearly that in which l is a whole multiple of the length of a loop. At $x=0$ and $x=l$ we must have $y=0$, therefore, from (8),

$$\sin(2\pi \mathrm{N}l/c)=0 \quad \text{or} \quad 2\pi \mathrm{N}l/c=\pi,\ 2\pi,\ 3\pi,\ \text{etc.}$$

Consequently the frequencies of the various partials are given by

$$\mathrm{N}_s=\frac{sc}{2l} \quad \text{or} \quad \frac{s}{2l}\sqrt{\frac{\mathrm{T}}{m}}, \quad \text{where } s=1, 2, 3, \text{etc.} \quad . \quad (7a)$$

representing *a complete harmonic series o tones.* The number of loops, *i.e.* half wave-lengths, into which the string divides is equal to s. A few of the modes of vibration are indicated in fig. 27. There are a number $(s-1)$ of points which are at rest between the fixed ends of the string. These points marked N are called *nodes*, whilst intermediate points marked A, at which the amplitude of vibration is a maximum, are known as *antinodes* or *loops*. Any node might be clamped without interfering with the motion of the remainder of the string.

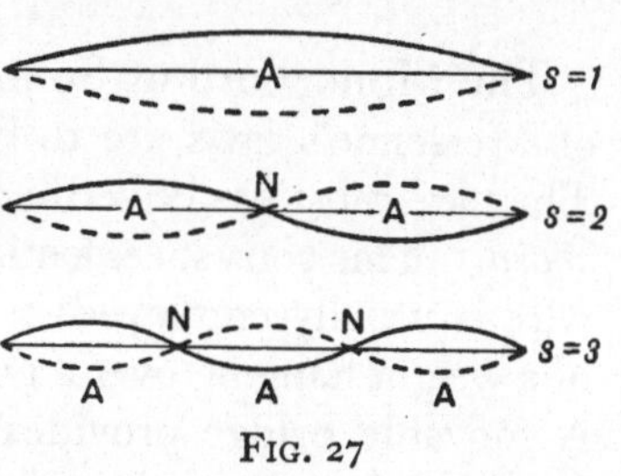

FIG. 27

The string may, of course, vibrate with any number of partials at the same time; the resultant sound being characterised by the overtones present in addition to the fundamental. The harmonic relation of the overtones is regarded as essential to the good musical 'quality' of a vibrating string.

The laws of vibration of strings, embodied in equation (7*a*) were discovered experimentally in 1638 by Mersenne. They may be stated as follows:—

1. For a given string and a given tension the frequency varies inversely as the length. (*Note.*—This principle was known to Aristotle.)

2. For a string of given length and material the frequency varies as the square root of the tension applied.

3. The frequency of vibration of strings of the same length and subjected to the same tension, varies inversely as the square root of the mass 'm' per unit length of the string, *i.e.* varies inversely as the square root of the area of cross-section A and the density ρ of the string (since $m=\rho A$).

These results may be simply deduced by the method of dimensions. Assuming that the frequency of vibration N is dependent only on the length l, mass per unit length m, and the tension T_1 of the string,

$$l\equiv(L) \quad \rho\equiv(ML^{-1}) \quad T_1\equiv(MLT^{-2}) \quad N\equiv(T^{-1}).$$

Thus

$$\text{Frequency } N\equiv T^{-1}\equiv l^a\rho^b T_1{}^c,$$

and we have

$$\left.\begin{array}{ll}\text{(length)} & a-b+c=0\\ \text{(mass)} & b+c=0\\ \text{Time} & -2c=-1\end{array}\right\}\text{giving } c=\tfrac{1}{2},\ b=-\tfrac{1}{2},\ a=-1.$$

Consequently, the frequency $N\propto\frac{1}{T}\propto\frac{1}{l}\sqrt{\frac{T_1}{\rho}}$ as before.

The Monochord or Sonometer - Examples of the application of Mersenne's laws are to be found in all stringed instruments. They are most easily verified by means of the *sonometer* or *monochord*. This consists essentially of a thin metallic wire (steel piano wire is usually employed) stretched over two 'bridges' by means of a weight hanging over a pulley, or by a spring tensioning device. A movable bridge provides a convenient means of varying the vibrating length of the wire. It is important that the base on which the string and bridges are mounted should be strong and massive, yielding as little as possible to the forces of tension applied to the string. The vibrations may be excited in any convenient manner, *e.g.* by plucking, striking, bowing, or by electromagnetic means. The monochord is very useful as a means of comparing frequencies (see also p. 103). Fixing the tension of the string, the length which tunes to a fork of standard frequency is determined. The corresponding length for the 'unknown' note is also found, and the ratio of frequencies is the inverse ratio of the lengths. Sonometers are often provided with a second wire of fixed length and tension to provide a standard frequency as a basis for comparison. The use of the frequency formula, equation (7), for the absolute determination of pitch does not permit of great accuracy on account of the difficulty of measuring the tension, which is liable to variation on opposite sides of a bridge. The 'effective' length of a string is rendered somewhat uncertain by the stiffness of the wire, and the frequency is further affected by yielding of the bridges.

The various overtones, in this case harmonics, of a string may be very simply demonstrated by the monochord. If the string is lightly damped at the midpoint, by touching it gently with the finger, and then plucked, it will be found to vibrate in two loops. The antinodes, if of small amplitude, may be easily located by means of small paper riders arranged at short intervals along the wire. If the string is touched at a point one-third of its length from the end, it will vibrate in three loops. Similarly it can be made to vibrate in four, five, or more sections by touching it at one-fourth, one-fifth, etc., of its length from the end.

Small Motion at the Nodes of a String – It is generally agreed that the nodes of a string which is maintained permanently in vibration in two or more loops cannot be points of *absolute* rest, as the energy requisite for the maintenance of the vibrations is transmitted through these points. The problem has been investigated by Raman,* who has shown that these small motions at the node differ in phase by a quarter period from the vibrations of the rest of the string. By attaching a thread to an obliquely-set electrically-maintained tuning-fork, and to various points of the vibrating string, it was found that points of the string in the loops describe 'figure of eight (8)' curves, whilst the so-called nodes describe very flat parabolas, the chief motion being longitudinal. S. Ray † has shown that the velocity of a transverse wave on a string is a function of the ratio a/λ, where a is the amplitude of vibration and λ the wave-length. (The ordinary theory assumes the wave-form to be very flat, and consequently the velocity is independent of a and λ.) When a/λ is constant, or the waves are geometrically similar, the velocity of propagation is constant, but not otherwise. The velocity of a wave therefore cannot be independent of the amplitude of vibration. On this basis ‡ S. Ray explains the motion of the nodes of strings as investigated by Raman.

* *Nature*, **82**, p. 9, Nov. 4, 1909. † *Phys. Rev.*, **28**, p. 229, July 1926. ‡ *Ibid.*, p. 1321, Dec. 1926.

Stiffness of Strings – The definition of a string given above requires that the string shall have no 'stiffness,' *i.e.* shall require no force to bend it. A stretched steel wire, at the point where it passes over the bridge of a monochord, reveals the weakness of this theoretical assumption; especially if the wire is moderately thick. At such a point the 'wire' is tending towards a 'bar' or rod, in which elastic forces are called into play when bending takes place. A short length of steel wire clamped at one end has a definite period of transverse vibration of its own due to this inherent stiffness which we have hitherto ignored. In the case of a wire having finite stiffness, the restoring force acting on an element displaced from its normal position consists of the component of the applied tension acting 'inwards,' *plus* an elastic force called into play by the bending of the wire. Thus the total restoring force is increased a with corresponding increase in frequency. Savart, who first investigated this effect, obtained the result $N_1^2 = N^2 + N_0^2$, where N_1 is the observed frequency, N the calculated frequency for an 'ideal' string (see equation (7)), and N_0 is the frequency with which the string would vibrate at zero tension (*i.e.* by the virtue of its own stiffness). Rayleigh * has determined the effect of stiffness and finds that the frequency of a wire is increased in the ratio $1:(1+Bs^2\pi^2/2Tl^2)$, where B is a quantity depending on the material and form of section of the wire. Donkin † has calculated the frequencies of the harmonic series of a 'stiff' wire of circular section, viz.

$$N=\frac{s}{2l}\sqrt{\frac{T}{m}}\cdot\left\{1+\frac{\pi^3 s^2 r^4 E}{8l^2 T}\right\}. \qquad (9)$$

where r is the radius of section and E is Young's Modulus of the material of the wire. The second term in brackets represents the correction factor for the 'stiffness' of the wire. It will be seen that the correction is small for the fundamental ($s=1$) of a wire of small radius of section, but becomes more and more important as the number of loops s increases. An increase in the number of loops has, from this point of view, a similar effect to a decrease in length l of the wire.

Taking the case of a steel wire of length 50 cms., 0·5 mm. diameter, subjected to a tension of 10^7 dynes (about 25 lbs. weight), the fundamental frequency is found to be approximately 1·00012

* *Sound*, **1**, p. 207; and E. Allan (*Phil. Mag.*, p. 1324, Dec. 1927); see also Lamb, *Sound*, p. 133.

† Donkin, *Accoustics*.

times that of the ideal wire. The same wire vibrating in 5 loops ($s=5$) would have a frequency 1·003 times that of the 'ideal' string—the correction is here more important. When the stiffness of the string is taken into account, not only is the fundamental raised in pitch but the overtones are raised in pitch in increasing proportion. Consequently the fundamental and overtones of a 'stiff' string no longer form an *exact* harmonic series. With the strings usually employed in music these slight divergences from the ideal string are not of serious importance.

Yielding of the Supports – Rayleigh * has calculated the effect on frequency produced by yielding of the supports, or bridges, carrying the tensioned wire. The principal cases are examined in which the supports have (*a*) negligible mass, but a very large spring factor μ, and (*b*) negligible spring factor but very large mass. In the former case the effect of the yielding is the same as that due to an increase of length of the string in the ratio $1:(1+2T/\mu l)$. In the latter case, where mass predominates, the effect is equivalent to shortening the string in the ratio

$$1:(1-2Tl/Ma^2s^2\pi^2).$$

Whether the frequency of the string will be raised or lowered by the yielding of the supports depends therefore on the sign of the correction factor, *i.e.* whether the natural frequency of the yielding support (of mass M and stiffness μ) is lower or higher than that of the string. In case (*a*) there is a proportionate lowering of frequency for all overtones, which therefore still form a harmonic series. In case (*b*), however, the proportionate rise in frequency is greater the lower the harmonic component (*i.e.* the smaller s).

Loaded Strings – It will be clear that any normal mode of vibration of a string having a node at the point to which a load is attached will be unaffected by the presence of the load. The fundamental will, of course, be reduced in frequency. Thus if a load is attached to the midpoint of a string the even harmonics, having a node at the centre, will be unaffected, whereas the odd harmonics will be lowered in frequency. The harmonic relationship is therefore disturbed by the addition of a load at any point of a uniform string. Rayleigh (**1,** p. 205) calculates the period of the fundamental vibration of a string of mass m per unit length

* *Sound*, **1,** p. 203.

carrying a load M (large compared with mass of string) at its mid-point, viz.

$$\text{Period}=\pi\sqrt{\frac{\text{M}l}{\text{T}}}\cdot\left(1+\frac{ml}{6\text{M}}\right) \quad . \quad . \quad (10)$$

If the mass of the string ml is small compared with the added load M this becomes

$$\text{Period}=\pi\sqrt{\text{M}l/\text{T}},$$

a result which may be deduced directly from simple dynamical considerations. For example, if the mass M is displaced a distance y from its normal position and released we have

$$\text{M}d^2y/dt^2=-2\text{T}\sin\theta,$$

where θ is the inclination of the displaced string relative to its zero position. Now $\sin\theta$ is equal to $2y/l$ (approx.), consequently

$$\text{M}\frac{d^2y}{dt^2}+\frac{4\text{T}y}{l}=0,$$

whence $n^2=4\text{T}/\text{M}l$. The frequency $\text{N}=\frac{n}{2\pi}=\frac{1}{\pi}\sqrt{\frac{\text{T}}{\text{M}l}}$, and the period$=\pi\sqrt{\frac{\text{M}l}{\text{T}}}$. A load distributed uniformly along the whole length of the string merely adds to the effective mass per unit length m of the string without adding to its stiffness (if this can be assumed), the result being a lowering of frequency of the fundamental and all the overtones in the same ratio. Use is made of this type of loading in the lower strings of a pianoforte. In order to obtain a sufficiently low frequency within the limits of length and tension possible in the instrument, the lower strings are loaded with a closely wound layer of wire. This increases the mass without appreciably increasing the stiffness.

Methods of Producing Vibration in Strings. Quality – A stretched string may be set in transverse vibration in a variety of ways. Plucking, bowing, and striking are the more familiar methods, exemplified in the harp, the violin, and the piano respectively. A string may also be set in vibration by forced oscillation of a point of support or of some point intermediate between the fixed ends. *Melde's* classical *experiment*, in which one end of a string is attached to a vibrating tuning-fork, is an example of the former, whilst the electromagnetically maintained wire described on pp. 76 and 102 exemplifies the case of forced excitation at an intermediate point. It will be appreciated that the method of excitation has a very important influence on the form of the wave which travels along the string. A Fourier analysis of this wave reveals the relative amplitudes of the various harmonics present in the disturbance. Now the quality of the note emitted by the wire is dependent essentially on the relative amplitudes of the various overtones which are added to the fundamental tone. It is just this addition of overtones which makes it possible for the ear to distinguish between the sounds of a piano, a violin, and a tuning-fork emitting the same note. If the pure fundamental tones could be separated in each of these cases they would be indistinguishable from one another. The overtones introduce the necessary *quality* which is so important to the ear. It is therefore necessary to consider the influence of the mode of excitation on the production of overtones.

The general form of the displacement of a stretched string is given by equations (2), (3), and (4) on p. 85. Equation (4) represents two waves travelling with equal velocity $\sqrt{T/m}$ in opposite directions. When these waves are reflected at the fixed ends of the string and are superposed on the incident waves, stationary waves are formed and the wire is divided into nodes and loops in a manner dependent on the form of the incident waves, *i.e.* dependent on the method of displacing the string. In the case of the fundamental type of vibration in which the string forms one loop only ($l=\lambda/2$), the displacement y_1 in the stationary waves is given by equation (8), p. 87, where $N=c/2l$, that is,

$$y_1=A\sin\frac{\pi x}{l}\,.\,\sin\frac{\pi ct}{l} \quad . \quad . \quad . \quad (11)$$

In the case of the sth harmonic ($l=s\lambda/2$) and $N=sc/2l$,

$$y_s=A_s\sin\left(s\,.\,\frac{\pi x}{l}\right)\sin\left(s\,.\,\frac{\pi ct}{l}\right) \quad . \quad . \quad (12)$$

The resultant vibration of the string will generally be complex and dependent on the mode of excitation, consequently the resultant displacement will be represented by the summation from $s=1$ to $s=\infty$ of a complete Fourier series involving sine and cosine terms (see p. 26), thus

$$y=\sum_{s=1}^{s=\infty}\sin\left(s\frac{\pi x}{l}\right)\left\{a_s\cos\left(s\frac{\pi ct}{l}\right)+b_s\sin\left(s\frac{\pi ct}{l}\right)\right\} \quad . \quad (13)$$

This is the most general solution of the motion of a string. The equation was first obtained by Bernouilli in 1755. The constants a_s and b_s will of course depend on the special circumstances of the vibration, and may be expressed in terms of initial values of y and $\dot{y}$. In equation (13), when $t=0$, representing the string in its initial or 'starting' position, the coefficients of $b_s=0$, and

$$\left.\begin{aligned} y &= \sum_{s=1}^{s=\infty} a_s \sin\frac{s\pi x}{l}. \\ &\text{Similarly} \\ \dot{y} &= \frac{\pi c}{l}\sum_{s=1}^{s=\infty} s \,.\, b_s \sin\frac{s\pi x}{l} \quad \text{(the coefficients of } a_s \text{ being zero)} \end{aligned}\right\} \quad . \quad (14)$$

The values of the coefficients a_s and b_s in (14) may be determined as in the case of Fourier's Theorem (p. 26 *et seq.*). Multiplying by $\sin(s\pi x/l)$ and integrating from 0 to l, we find

$$\left.\begin{aligned} a_s &= \frac{2}{l}\int_0^l y \sin\frac{s\pi x}{l}\,dx \\ \text{and} \qquad b_s &= \frac{2}{\pi c s}\int_0^l \dot{y} \sin\frac{s\pi x}{l}\,.\,dx \end{aligned}\right\} \quad . \quad . \quad . \quad (15)$$

Equations (13) to (15) may now be used to calculate the motion of the string in particular cases, assuming the condition of the string at one particular instant is known.

Strings Excited by Plucking or Striking - If the initial displacement of a string be produced by 'plucking,' *i.e.* by pulling aside at one point and suddenly releasing, the wave-form which travels along the string will depend on the nature of the bend in the string at the point of plucking. We shall assume that the displacing force acts at a particular *point* C of the string AB (fig. 28). Actually, of course, the force must act on an element of appreciable length. Let the coordinates of the 'plucked' point P of the string at the instant $t=0$ be $x=d$ and $y=\beta$. Between A and P, $y=\beta x/d$; between B and P, $y=\beta(l-x)/(l-d)$, consequently from equation (14) we obtain

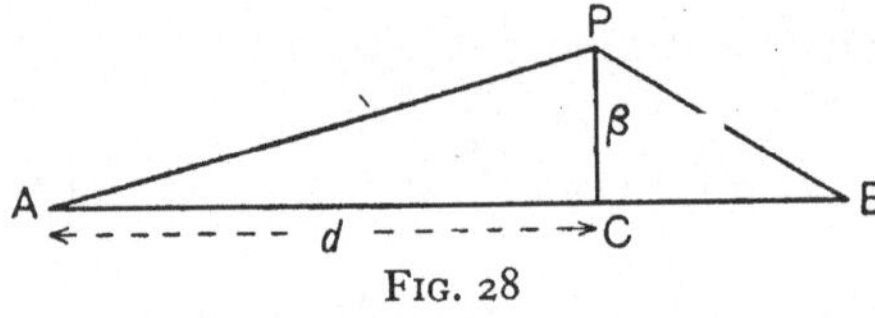

FIG. 28

$$a_s = \frac{2\beta}{l}\left[\int_0^d \frac{x}{d}\sin\frac{s\pi x}{l}\,dx + \int_d^l \frac{l-x}{l-d}\sin\frac{s\pi x}{l}\,dx\right] = \frac{2\beta l^2}{\pi^2 s^2 d(l-d)}\sin\frac{s\pi d}{l} \quad (16)$$

This relation indicates that *the sth harmonic will disappear* when $\sin(s\pi d/l) = 0$, that is, when sd/l is any integer, or *whenever there is a node of the sth harmonic situated at* P. If the string is divided into s equal parts and is plucked at any dividing point, the sth harmonic will be absent from the resultant vibration. This principle also shows, in the case of a string at rest in its equilibrium position, that it is impossible for *any* force applied at the midpoint of the string to produce any *even* harmonics. If, after the application of a force at the midpoint, this point be damped by touching it lightly, the string will be brought to rest; for the odd harmonics cannot persist with a node at the midpoint, and the even harmonics were absent initially for the reason given above. Thomas Young * (1841) proved experimentally when any point of a string is plucked, struck, or bowed, all the overtones (or partials) which require that point for a node will be absent from the resultant vibration.

The relation for a_s given in (15) indicates that a string plucked in the above manner has the full harmonic series of overtones, the amplitude varying inversely as s^2, *i.e.* the amplitude of the partials falls off rapidly towards the higher frequencies. The quality of the resultant tone depends on the ratio d/l, and consequently varies with the point of plucking the string.

It must be remembered that the foregoing treatment refers to an 'ideal' string. As we have seen, the 'stiffness' of an actual string causes a slight departure from the harmonic series. Damping due to internal friction is another factor which has a similar tendency—its effect becoming more important at higher frequencies. Since the higher partials are damped most rapidly, the tone of a plucked string improves in purity (*i.e.* tends towards the fundamental S.H. type) as it diminishes in loudness. Well-known examples of plucked string instruments are the harp, guitar, mandolin, banjo, and various 'jazz' instruments.

When a *string is excited by striking*, complications arise due to the finite time of contact of the striking hammer with the string. The general features of the vibration produced by striking are much the same as those in which the string is plucked, but the motion is complicated by the fact that the point of the string which receives the blow of the hammer is displaced before the remainder of the string. The resultant wave-form is therefore dependent on a number of factors, including the duration of the blow, the velocity of the hammer, and the relative masses of the hammer

* See Tyndall's *Sound*, pp. 118–124.

and string. During recent years a considerable amount of research has been carried out in connection with this complex question.

Various theories have been developed, notably those of Helmholtz and of Kaufmann.* In both theories the assumptions made relative to the mode of contact between hammer and string are of a somewhat artificial character when compared with the actual conditions experimentally determined.† As in the case of the plucked string, the sth harmonic is absent if the string be plucked at one of the nodes corresponding to this harmonic. The more abrupt and localised the blow, the greater the relative intensity of the higher harmonics. Harmonics above the sixth are considered undesirable in a pianoforte string, consequently the hammers are covered with felt to reduce the suddenness of the impact, and the blow is struck at a point from one-seventh to one-ninth the length of the string from the end.‡ As Helmholtz has pointed out, all overtones above the sixth are dissonant tones which must be suppressed by making the striking-point coincide with a node for these overtones.

Recent work on the elastic impact of a pianoforte hammer has thrown some light on this difficult problem. Bhargava and Ghosh § have shown that the elasticity of the hammer felt must be taken into account in calculating the duration of impact. With this addition, Kaufmann's theory, which refers to the unyielding hammer, agrees well with experimental results. Ghosh and Dey || have subsequently verified the theoretical deductions by a photographic method. W. H. George ¶ has investigated by an oscillographic method the duration and mode of contact between metallic hammers and strings, and has found in some cases momentary separation between hammer and string. The existence of important pressure variations and the instants of their occurrence during the impact have been established. The results obtained cannot be explained on the theories of Helmholtz, Lamb, or Bhargava-Ghosh, but may be reconciled with those of Kaufmann, Das,** or Raman-Banerji. Using a suitably designed

* Helmholtz, *Sensations of Tone*; and Kaufmann, *Wied. Ann.*, **54**, (1895).

† See papers by G. H. Berry (*Phil. Mag.*, **19**, p. 648, April 1910); G. H. Bryan (*Phys. Soc. Proc.*, **25**, 1912–13); C. V. Raman and B. Banerji (*Proc. Roy. Soc.*, **97**, p. 108, 1920).

‡ Tyndall, *Sound*, p. 122.

§ *Phil. Mag.*, **49**, p. 121, Jan. 1925; and *Phys. Rev.*, **24**, p. 456, Oct. 1924.

|| *Ind. Assoc. Cult. Proc.*, **9**, 1925.

¶ *Proc. Roy. Soc.*, **108**, p. 284, 1925; and *Phil. Mag.*, **49**, p. 97; and **50**, p. 491, 1925.

** P. Das, various papers in *Ind. Assoc. Cult. Proc.*, **7**, **9**, and **10**.

metal hammer, equivalent to a pendulum, George and Beckett * have shown that the loss of energy on rebound from a string struck transversely is a measure of the energy communicated to the string. As the position of the impact moves away from a bridge the energy lost by the hammer increases exponentially until a point of maximum energy-loss is reached. This point is nearer the bridge the heavier the hammer. Measurement of the duration of the impact shows that for maximum energy-loss by the hammer the conditions are such that the hammer rebounds just before the wave reflected from the farther bridge reaches the point of impact.

In a recent paper P. Das † gives a theory involving an elastic pianoforte hammer, explaining why a hard hammer imparts a piercing quality to the general tone of the vibrating string. Again, if the elastic hammer has a large initial velocity, and hence undergoes a large compression, its effective hardness increases. Consequently, if the string be struck very hard, the tone loses its softness and acquires a metallic ring.

Strings Excited by Bowing – Many attempts have been made to develop a theory for the vibrations of a bowed string, but up to the present none of these can be regarded as completely satisfactory. Helmholtz in 1861 gave an explanation of the mode of vibration of a violin string which is in tolerably good agreement with observation. He observed the motion of a bowed string by means of the vibration microscope. The latter was used to view an illuminated particle attached to the string, the 'microscope' (objective) being mounted on a prong of a tuning-fork (electrically maintained) vibrating in a direction at right angles to the vibrations of the string. The Lissajous figures thus observed served to indicate the nature of the vibration of the string. Helmholtz ‡ concludes: "We learn therefore by these experiments:

"(1) The strings of a violin when struck by the bow, vibrate in one plane.

"(2) That every point of a string moves to and fro with two constant velocities.

"These two data are sufficient for finding the complete equation of the motion of the whole string. It is the following:—

$$y=A\sum\left\{\frac{1}{s^2}\sin\left(\frac{s\pi x}{l}\right)\sin\left(\frac{2\pi st}{T}\right)\right\},$$

* *Proc. Roy. Soc.*, **114**, Feb. 1927.
† *Proc. Phys. Soc.*, **40**, p. 29, 1927; and *Phil. Mag.*, **6**, p. 479, 1928.
‡ *Phil. Mag.*, **21**, 393–396, 1861.

T is the duration of the vibration and A an arbitrary constant. A comprehensive idea of the motion represented by this equation may be given in the following way: Let *ab*, fig. 29 (*a*), be the equilibrium position of the string. During the vibration its forms will be similar to *acb*, compounded of two straight lines *ac* and *cb*, intersecting at *c*. Let this point of intersection move with a constant velocity along two flat circular arcs, lying symmetrically on the two sides of the string, and passing through its ends, as represented in fig. 29 (*a*). A motion the same as the actual motion of the whole string is thus given. As for the motion of every single point it may be deduced from the above equation that the two parts *ab* and *bc* (see fig. 29 (*b*)) of the time of every vibration are proportional to the two parts of the string which are separated by the observed point. The two velocities, of course, are inversely proportional to the times *ab* and *bc*. In that half of the string which is touched by the bow, the smaller velocity has the same direction as the bow; in the other half of the string it has the contrary direction. By comparing the velocity of the bow with the velocity of the point touched by it, I found that this point of the string adheres fast to the bow and partakes of its motion during the time *ab*, then is torn off and jumps back to its first position during the time *bc*, till the bow again gets hold of it."

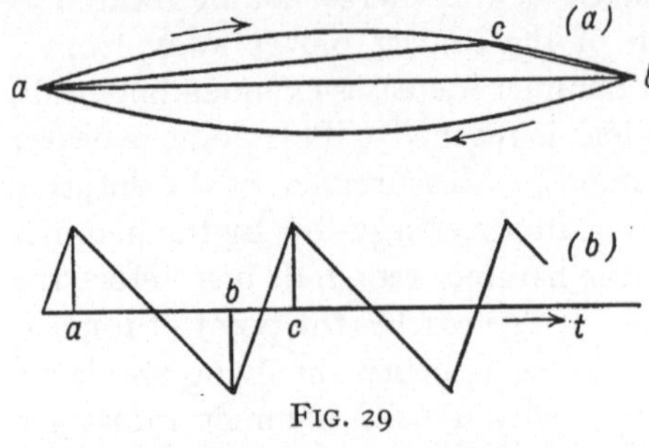

Fig. 29

After remarking on Young's law that all component vibrations are absent which have a node at the point of excitation, Helmholtz refers to the suppression of high overtones which are dissonant. "Near the end of the string, where the bow is commonly applied by players, the nodes of different harmonics are very near to each other, so that the bow is nearly always at, or at least very near to, the plane of a node."

Helmholtz's observations indicate that the note produced by bowing has the same pitch as the natural note of the string. The vibrations, apparently forced by the bow, may therefore still be regarded as free. *Although the energy for maintenance comes from the bow, the note is determined by the string.* The intermittent dragging action of the bow serves to maintain the natural vibration of the string in the short intervening periods. Rayleigh suggests that *the sustaining power of the bow* depends on the fact that solid friction is less at moderate than at small velocities. Consequently when the part of the string acted on is moving with the bow (not improbably with the same velocity) the mutual action is greater than when the string is moving in the opposite direction with a greater relative velocity. "The accelerating effect in the first part of the motion is not entirely neutralised by the subsequent retardation, and an outstanding acceleration remains capable of maintaining the vibration in spite of other losses of energy."

The foundations of a mechanical theory of bowed strings have been laid during recent years principally by the researches of C. V. Raman.* One of the outstanding questions which is not yet fully cleared up is the manner in which the overtones of the bowed string are influenced by the position of the bowing point. The problem is complicated by other variable factors which influence the character of the overtones, viz. the bowing pressure and speed, and the width of the region of contact between the bow and the string. Raman has found in these investigations that the vibration-curve of a point on the string departs slightly from the Helmholtz two-step zigzag; one of the motions being uniform and the other fluttering. Young's theorem is verified, the amplitude of a partial being zero for a bowing at its node, and the phase of that partial passes discontinuously from $+\pi/2$ to $-\pi/2$ as the bowing point passes that node.

Wolf-Note – On all stringed instruments of the violin type a certain note can be found which it is difficult, and often impossible, to produce by bowing. When this "wolf-note," as it is called, is sounded, the whole body of the instrument vibrates in an unusual degree. At this pitch the bow refuses to 'bite' and a soft pure tone cannot be obtained. If the pressure of the bow on the string is increased, the tone resulting is often of an unsteady nature with variable intensity. G. W. White,† examining the point experimentally by a photographic method, came to the conclusion that the pitch of the wolf-note was about that of best resonance of the belly of the cello on which the experiments were made. Further, the fluctuations of intensity were shown to correspond to waxing and waning of the vibrations of the belly. C. V. Raman ‡

* (1) *Ind. Assoc. for Cult. of Science*, **15**, pp. 1–158, 1918; (2) *Phil. Mag.*, **38**, pp. 573–581, Nov. 1919.

† *Camb. Phil. Soc. Proc.*, **18**, pp. 85–88, June 1915.

‡ *Nature*, **97**, pp. 362-363, June 29, 1916.

gives a somewhat different explanation. He suggests the effect depends on the fact that, when the pressure of the bow is less than a certain critical value proportionate to the rate of dissipation of energy from the string, the principal mode of the string's vibration, in which the fundamental predominates, is incapable of being maintained and passes over into one in which the octave is prominent. When the bow sets the string in vibration, the belly of the instrument is strongly excited by sympathetic resonance, and the rate of dissipation of energy increases beyond the limit to which the bow can maintain the fundamental pre-eminent; the vibration of the string, therefore, changes to that in which the fundamental is feeble. These views are confirmed by a photographic record of simultaneous vibrations of string and belly.

Strings as a Source of Sound. The Sounding-Board – In the foregoing treatment we have, in general, regarded the string as *isolated* by its supports or bridges, whereas in reality it is *connected* by these supports to some form of base. The string, in performing its vibrations, reacts on its end supports, which in turn react on the base. We have seen (p. 65) that on account of the local reciprocating flow in the air surrounding a vibrating wire, the latter acts as a double source of very feeble strength. Consequently a vibrating wire *rigidly* supported would radiate extremely little sound energy into the surrounding medium. When it is desired, therefore, to employ a vibrating string as a source of sound, *e.g.* in all stringed musical instruments, it is essential that its vibrations should be 'transformed'—*i.e.* transferred to another body which is more suitable to transmit vibrations into the surrounding medium. This office is performed by the sounding-board, which must be connected to the string by suitable flexible supports. In this manner a much larger vibrating area is brought into contact with the medium (air usually) and the rate at which energy is radiated is greatly increased. The sounding-board of a piano and the belly of a violin radiate practically all the energy of strings. It is important in a good stringed instrument that the sounding-board should possess no predominant resonance frequencies of its own, otherwise these will reinforce disproportionately corresponding frequencies of the strings. In certain cases of such resonance, *e.g.* the 'wolf-note' quoted above, the energy absorbed from the string and reradiated by the sounding-board may become so great as to be comparable with the total energy supplied, in which case the reaction on the string becomes excessive.

The manner in which a string communicates its vibrations to the sounding-board is illustrated in its simplest form in fig. 30, where AB represents the vibrating string, AC and BD the flexible supports or bridges, and CD the sounding - board. The effect of the motion of the string is to bend the upper ends of the supports inwards when it is at the upper *and* at the lower limits of its vibration. The supports therefore vibrate at *twice* the frequency of the string. The sounding-board, however, vibrates with the *same* frequency as the string, the corresponding positions 1 and 2 being indicated in fig. 30. The process will be evident from consideration of the diagram. The amplitude of vibration of the base will of course depend to a large extent on its mass and stiffness relative to the corresponding quantities for the wire. The nature of the supports AC and BD will also influence the result (see also p. 91). If the base-board and supports of the wire did not yield there would be no appreciable emission of sound from the wire. The exact tuning of a stretched string and a tuning-fork is sometimes tested by pressing the stem of the fork on one of the end supports of the wire—in this process it is tacitly assumed that the end support of the string is in motion when the string vibrates transversely. W. H. George * has determined the amplitude of this end-motion of the string and the variations of tension which the string undergoes during vibration. The length l of the displaced string is given by

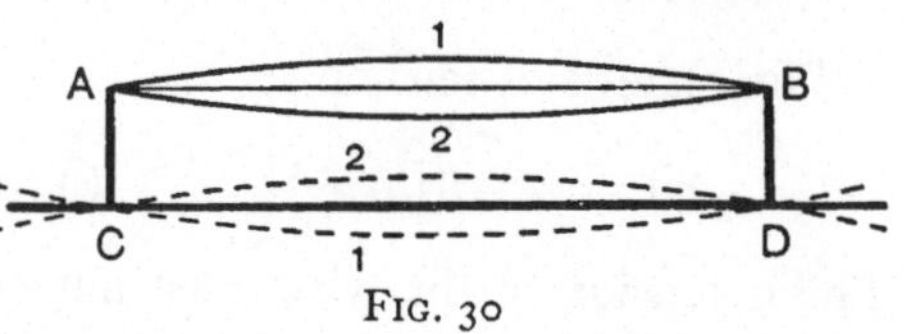

FIG. 30

$$l=\int_0^{l_0}\sqrt{1+(dy/dx)^2}\,.\,dx,$$

in which y is given by equation (13) on p. 93.

In the case of the fundamental form of vibration

$$y=\mathrm{A}\sin(\pi x/l_0)$$

and

$$l=\int_0^{l_0}\sqrt{1+\frac{\mathrm{A}^2\pi^2}{l_0^2}\,.\,\cos^2\frac{\pi x}{l_0}}\,.\,dx$$

$$=\frac{2l_0}{\pi}\sqrt{1+\kappa^2}\int_0^{\pi/2}\sqrt{(1-\frac{\kappa^2}{1+\kappa^2}\,.\,\sin^2\phi)}\,.\,d\phi,$$

* W. H. George, *Phil. Mag.*, **49**, p. 97, Jan. 1925.

where

$$\kappa = A\pi/l_0 \quad \text{and} \quad \phi = \pi x/l_0.$$

The change of tension

$$\delta T = \pi r^2 Y/(l/l_0 - 1) \qquad (Y = \text{Young's modulus}).$$

The extension of the wire must not exceed the elastic limit, consequently there is a maximum permissible amplitude with a given tension in the wire. For maximum emission of sound energy the tension and amplitude must be as large as possible, *i.e.* a wire with a large breaking stress, or tenacity, is required. Wolfenden * states in this connection: "It is not too much to say that the improvements made during the last century in the tenacity and elasticity of steel wire have rendered the modern piano possible." Greater tenacity in the wire ultimately results therefore in increased sound output with increase of intensity and duration of tones. The design of sound-boards for various types of stringed instruments is a very complex problem which is largely dealt with on an empirical basis by those experienced in the work. Experimental studies have been made by various investigators, notably E. H. Barton, † of the nature of the vibrations of the sounding-boards of the violin, pianoforte, and other stringed instruments. In these investigations simultaneous photographic records are made of the vibrations of the string and that part of the 'body' of the instrument under consideration.

Strings Excited by Electromagnetic Means – We have so far considered only impulsive or discontinuous forms of excitation. The most direct and simple means of obtaining continuous or maintained vibrations in a stretched string is undoubtedly the electromagnetic method. We have already referred to one electromagnetic method of maintaining the vibrations of a steel wire (see p. 76). In this method the string carries a contact breaker which interrupts the current in an electromagnet as the wire vibrates. The form of maintenance of the vibrations is analogous to that of the electric bell. A second form of maintenance of vibrations also employs a contact breaker, or interrupter, but in this case the interrupted current passes through the wire itself. The latter may be of any non-magnetic material having suitable elastic properties. Near its midpoint

* *Treatise on the Art of Pianoforte Construction*, p. 16, 1916.

† E. H. Barton, *Sound*.

the wire passes through the transverse field of a magnet (permanent or electromagnet). The current through the wire reacting on the steady magnetic field causes corresponding motion of the wire at right angles to the field. The force acting on the wire is equal to the product of current, strength of field, and length of wire in the field. The harmonics in the wave-form of the current passing through the wire may, if desired, be suppressed by electrical tuning. An interrupter of this kind was described by Arons.* It gives a very steady and easily adjustable frequency. A third method of maintaining a wire in continuous vibration by electromagnetic means is analogous to the second method, but in this case the wire is not self-maintained. Alternating current of controllable frequency is supplied (*e.g.* from a valve oscillator) to the wire, part of which lies in a strong magnetic field. When the frequency of the current coincides with one of the possible harmonic frequencies of the wire a large amplitude vibration is set up. Alternatively the frequency of the wire may be tuned to that of the current by varying the tension by means of any convenient 'continuous' form of tensioning device (*e.g.* a strong helical spring extended by means of a screw capable of coarse and fine adjustment). This method is extremely convenient and, by suitably disposing the permanent magnet, *i.e.* by controlling the point of mechanical excitation of the wire, it is possible to obtain the higher harmonics. The principle involved is of course that of the Einthoven string galvanometer (or oscillograph), where a very fine conducting fibre is caused to vibrate in a magnetic field by virtue of the small alternating currents passing through it. In the oscillograph, however, resonance frequencies are avoided as far as possible, whereas in the present application, as a source of continuous sound, resonance is encouraged. The principle has been applied by D. W. Dye † at the National Physical Laboratory, Teddington, in a *standard sonometer* or monochord—the instrument being used mainly to determine audiofrequencies without the necessity of employing a large range of tuning-forks. This instrument makes it possible to measure or adjust an electrical frequency with an accuracy of 1 in 1000 over a considerable range. The sonometer is shown in fig. 31. The design is very simple. A phosphor-bronze wire 0·3 mm. diameter has a weight hung on its lower end and is thereby tightly strained

* L. Arons, *Wied. Ann.*, **66**, p. 1177, 1898; and E. Orlick, *Elekt.-Zeits.*, **24**, p. 502, 1903.

† See *N.P.L. Annual Report*, p. 83, 1924.

between a fixed upper nodal point in the form of a V groove (formed by two very small steel balls in contact) and a lower groove formed by a little wheel about 4 mm. diameter on a light steel spindle. This lower nodal point is on a carriage which can be racked up and down on slides covering nearly the whole length of the wire. The carriage has two pointers, which indicate the frequency on direct reading scales. A small alternating current passes through the wire, a portion of which lies between the poles of a permanent magnet mounted so that it can slide along the length of the wire. When the free length of the wire is such that its frequency of vibration in a single loop or in a number of loops is equal to that of the source, a large resonant vibration occurs. The wire can be made to vibrate in any number of loops (up to 10) provided the magnet does not coincide with a node (Young's law). It is thus possible to cover a wide range of frequencies by means of a number of scales. When vibrating in a single loop the range is 200 to 400; for two loops 400 to 800; for three loops 600 to 1200; for five loops 1000 to 3000; for ten loops 2000 to 10,000 cycles per second. The range for one loop also serves for two loops (using a multiplying factor 2), but the other ranges require separate scales because the stiffness of the wire causes these scales to depart

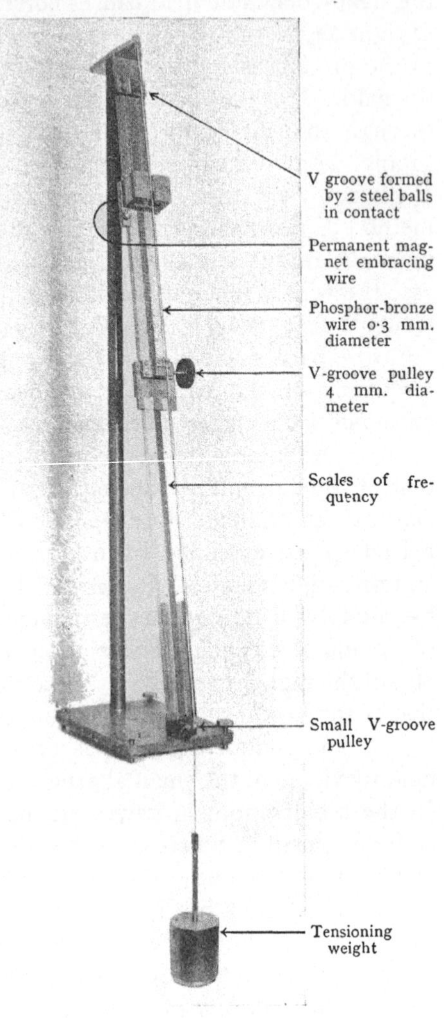

FIG. 31—Dye's 'Sonometer'

from exact multiples of the unit scale. By this system it is never necessary, except on the highest range, to use a length of wire shorter than half the maximum, and so the accuracy of reading can be kept high. Over the whole range a change of frequency of 1 in 1000 corresponds to a length of wire not less than 0·4 mm. The error of a setting is less than half this length under any conditions. The power taken by the apparatus is only about 1 milliwatt to give a good sound above 300 cycles per second or to give visible loops at lower frequencies. Frequencies between 100 and 200 are obtained by using a weight of one quarter that used for the other ranges.

When such a vibrating wire is used as a frequency standard, it is important of course that the supporting base and 'bridges' should be as rigid as possible. When the wire is to be used as a source of sound, however, a certain amount of yielding of the bridges and flexibility of the base—the sounding-board—is essential. The difference in application is exemplified in the monochord, as a frequency-measuring device, and the violin, which is essentially a sound-producing device. The radiation of sound energy from the sounding-board results in increased damping of the vibrations of the wire. Thus the amplitude of the vibrations in a well-constructed monochord will be large and persistent, compared with the corresponding vibrations in an instrument with flexible supports and base.

Reaction of Surrounding Medium on a Vibrating String – Assuming a vibrating wire is rigidly supported, its vibrations are influenced only by the medium and internal friction. Neglecting the latter quantity we have to consider the added mass (due to a layer of the medium moving bodily with the wire), the radiation of sound energy and loss of energy due to viscosity of the medium. The problem was investigated mathematically by Stokes,* who showed that the damping coefficient k is given by

$$k=\sqrt{2n\rho\mu}/\rho_0 a,$$

where $n/2\pi$ is the frequency, μ the viscosity, and ρ the density of the medium, whilst ρ_0 is the volume density of the material of the string of radius of section a. H. Martin † has attacked the problem experimentally. He observed, by means of a microscope, the vibrations of an electrically-maintained wire in various liquids, obtaining resonance curves which indicated both the resonant

* *Camb. Phil. Soc. Trans.*, **9**, p. 8, 1856.
† *Ann. d. Physik*, **77**, p. 627, 1925.

frequency and the damping coefficient. His experimental results lead to the empirical formula

$$k=1{\cdot}45\sqrt{n\rho\mu/\rho_0}a,$$

which is in good agreement with Stokes's calculation.

When massive wires vibrate in air at audible frequencies the damping due to the above causes is relatively small, but when the wire is *very fine* and the frequencies relatively high the value of k may reach serious proportions. This is particularly the case in the Einthoven string galvanometer, where a string of silvered quartz 1μ or 2μ thick ($1\mu=10^{-3}$ mm.) is sometimes required to vibrate at a frequency of several thousand cycles per second. Recently W. F. Einthoven * has tuned such strings to radio-frequencies, of the order 3×10^5 cycles per second, when the viscous damping in air at ordinary pressures would be so great as to render the string extremely insensitive. By enclosing the string (1 mm. long) in a high vacuum ($<10^{-3}$ mm. Hg), however, the sensitivity obtained was quite satisfactory when the frequency of the string was tuned to the incoming 'wireless' signals. When such fine strings are used the damping is of course almost entirely air-damping. The problem of the vibration of Einthoven 'fibres' has been thoroughly investigated by H. B. Williams,† who gives the theory of the vibrations of a string under the action of a uniformly distributed alternating force (the *whole* string, carrying alternating current, is influenced by a strong magnetic field). The loading effect of the air and viscous damping at different frequencies is calculated and experimentally determined. The agreement between these observations with very fine wires (10^{-4} cm. diameter) vibrating in air and the theoretical deductions from Stokes's theory is remarkably good.

A fine electrically conducting fibre, like the Einthoven string, when vibrating in a magnetic field is electrically ‡ as well as mechanically damped. Its vibrations set up a back e.m.f., which reacts on the input alternating e.m.f. and affects the amplitude of vibration. This phase of the problem is, however, outside our present consideration (but see p. 446). Einthoven, when using the 'high-frequency string,' removed almost all mechanical damping and controlled the string by electromagnetic means.

* *Proc. Köningl. Akad. von Wetenschaften te Amsterdam.*

† *Opt. Soc. Amer. Proc.*, **9,** No. 2, Aug. 1924; and **13,** No. 3, Sept. 1926.

‡ S. Butterworth, A. B. Wood, and E. H. Lakey, *Journ. Sci. Instrs.*, **4,** p. 8, 1926.

TRANSVERSE VIBRATIONS OF ELASTIC BARS

If the ratio of diameter to length of a wire be increased, the stiffness becomes increasingly important and the tension relatively unimportant. The transition stage between a wire, affected mainly by tension, and a bar, affected mainly by stiffness, is somewhat complicated,* and we shall deal only with the ideal bar, which is unaffected by tension.

The theory of transverse vibration of bars, even when simplified as far as possible by the omission of quantities of secondary importance, is still very complex in comparison with the theory of perfectly flexible strings. In the case of strings, harmonic waves travel with a velocity independent of the wave-length, but in the case of bars this is not so.* A non-harmonic wave in a bar has no definite velocity, but the motion is the sum of the separate motions which would be due to the harmonic components each travelling with the velocity appropriate to its own wave-length. A thin bar clamped at both ends may, like a string, form stationary waves vibrating in one or more loops, but the laws which regulate the frequencies of the successive modes of vibration are entirely different in the two cases.

In what follows it will be assumed that the vibrating bar is straight and uniform in cross-section and density, and is not subjected to tension or compression. It will also be assumed that the amplitude of vibration is so small that rotary effects can be neglected. Further, we shall suppose that in its bent condition, when vibrating, curvatures are so small that they may be represented by d^2y/dx^2, and the length of the bent bar practically the same as in the unbent condition. In the first place we require to know the 'bending moment' M, or couple required to bend a bar uniformly, so that its axis becomes an arc of a circle of radius R, *i.e.* the curvature of the bar is $1/\mathrm{R}=d^2y/dx^2$, where the axis of x is in the direction of the length of the unbent bar and y is in the direction of transverse displacement. The strain in any part of the bent bar will obviously be proportional to the curvature. Hence by Hooke's law the bending moment M is proportional to curvature, *i.e.*

$$\mathrm{M}=\mathrm{B}/\mathrm{R},$$

where B is a constant depending on the size and shape of the cross-section of the bar and on the elastic qualities of the material

* See Rayleigh, *Sound*, **1**, pp. 256–301.

of which it is made. When the bar is bent there is a *neutral surface* AB at right angles to the plane of bending, which is neither extended nor contracted (see fig. 32). The material of the bar will be stretched or compressed according as it lies on the convex or concave side of the neutral surface in the bent position. Consider a filament ab distant r from the neutral surface. Let the length of the filament be δx and area of cross - section w. After bending, its length becomes $ab' = \delta x + \Delta$. If E is Young's elastic modulus * of the material of the bar, we have

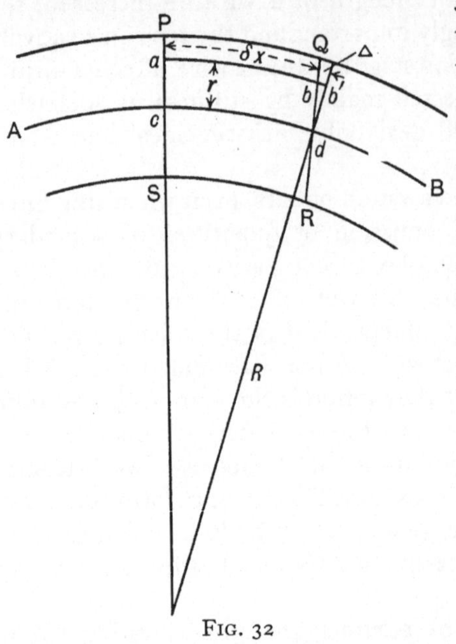

FIG. 32

$$\text{Force stretching the filament} = \text{E}w\Delta/\delta x.$$

Now $$\frac{\delta x + \Delta}{\text{R}+r} = \frac{\delta x}{\text{R}}, \quad \text{whence} \quad \frac{\Delta}{\delta x} = \frac{r}{\text{R}}.$$

Consequently the force stretching the filament ab is

$$= \text{E}wr/\text{R},$$

and the moment of this force about the neutral axis

$$= \text{E}wr^2/\text{R}.$$

The total moment M of all horizontal tensile and compressive stresses acting perpendicular to the section PS about a line cd

$$= \frac{\text{E}}{\text{R}} \sum wr^2 = \frac{\text{EI}}{\text{R}},$$

where I is the moment of inertia of the cross-section about a line through c at right angles to the plane PQRS. We have therefore, since the curvature $1/\text{R} = d^2y/dx^2$,

$$\text{M} = \text{EI} \,.\, d^2y/dx^2 \qquad . \quad . \quad . \qquad (1)$$

* Defined as the ratio of stress/strain, *i.e.* the force per unit area of section divided by the extension per unit length.

We now require to find the relation between this bending moment M, the transverse shearing force F, and the forces applied to the bar. If ρ is the density and S the area of cross-section of the bar, we have, since d^2y/dt^2 is the acceleration of the element δx, $\rho S\delta x \,.\, d^2y/dt^2 = \delta F$, where δF represents the resultant transverse shearing force acting on the element δx, or

$$\rho S d^2y/dx^2 = dF/dx \qquad . \quad . \quad . \quad (2)$$

Considering the angular motion, the transverse slice of thickness δx and area S is rotated, in the bending, through a small angle $\theta = dy/dx$ from its equilibrium position. Consequently we have

$$I d^2\theta/dt^2 = \delta M + F\delta x,$$

writing $I = \rho\delta x \,.\, S \,.\, \kappa^2$ where $\rho\delta x S$ is the mass of the slice and κ its radius of gyration, we have

$$\rho S \frac{d^2y}{dxdt} = \frac{dM}{dx} + F,$$

and combining this with equation (2) we obtain

$$\rho S\left(\frac{d^2y}{dt^2} - \kappa^2 \frac{d^4y}{dx^2dt^2}\right) = -\frac{d^2M}{dx^2}.$$

Substituting the value of M given in equation (1) and writing $I = \rho\delta x S\kappa^2$ we find

$$\frac{d^2y}{dt^2} - \kappa^2 \frac{d^4y}{dx^2dt^2} + \frac{E\kappa^2}{\rho}\frac{d^4y}{dx^4} = 0 \qquad . \quad . \quad (3)$$

the second term, involving the rotary motion of the bar, may generally be neglected and the equation reduces to

$$\frac{d^2y}{dt^2} + \kappa^2c^2\frac{d^4y}{dx^4} = 0 \qquad . \quad . \quad . \quad (4)$$

where $c = \sqrt{E/\rho}$ is the velocity of a *longitudinal* elastic wave in the bar (see p. 136). (*Note.*—$\kappa^2 = a^2/4$ for a bar of circular section and radius a, and $\kappa^2 = t^2/12$ if the bar is rectangular and of thickness t in the plane of vibration.) The constant κc in equation (4) cannot be regarded as a velocity unless we divide it by a quantity of the dimensions of a length, viz. λ. As Rayleigh points out, the mere admission that harmonic waves can be transmitted at all is sufficient to prove that the velocity must vary inversely as the wave-length. Consequently we must have

$$c_{\text{trans}} = \frac{\kappa}{\lambda} c_{\text{long}} = \frac{\kappa}{\lambda}\sqrt{\frac{E}{\rho}}.$$

In elastic bars stationary transverse vibration is produced, as in strings, by the interference of wave-systems which have been reflected from the ends of the bar. The modes of such vibrations depend on the method of clamping or supporting the bar. It should be noted that an end fixed in direction is not a *true* node since it is not a point of maximum change of slope; similarly a free end is not a true antinode, since it is not a place where the bar moves parallel to itself. The point of zero transverse displacement nearest to a free end is also not a *true* node, since the change of slope is not so great there as at the free end. The various 'end conditions' for a bar of finite length are:

(1) *Free End.* The displacement y and the slope dy/dx may have any small values. The curvature d^2y/dx^2 and the third differential d^3y/dx^3 must always be zero.

(2) *Clamped End.* If the clamp is rigid, the displacement y and the slope dy/dx must always be zero.

(3) *Supported End.* A knife-edge support (assumed rigid) prevents displacement, therefore $y=0$, but the slope dy/dx may have any small value. The curvature d^2y/dx^2 is zero.

Assuming $y=a\cos nt$, equation (4) becomes

$$\frac{d^4y}{dx^4}=\frac{\rho n^2}{E\kappa^2}\cdot y=m^4y \qquad . \quad . \quad . \qquad (5)$$

where $m^4=\rho n^2/E\kappa^2$.

If $y=ue^{\alpha x}$ is a solution,* α is a root of m^4, that is, $\alpha=\pm m$ or $\pm\sqrt{-1}\,.\,m$. The complete solution is therefore

$$y=(A\cosh mx+B\sinh mx+C\cos mx+D\sin mx)\cos nt \quad . \qquad (6)$$

the expression involving the constants A, B, C, D giving the amplitude at the point x. The ratios of the constants A : B : C : D and the possible values of m are determined by the end-conditions. When m is determined, we obtain the frequency $n/2\pi$ from equation (5).

Bar Free at both Ends – Consider a perfectly free bar of length l. Taking the origin of x at the midpoint * the end-conditions which must be satisfied are $\dfrac{d^2y}{dx^2}=\dfrac{d^3y}{dx^3}=0$ at $x=\pm\frac{1}{2}l$. The four equations thus obtained give a value for m which satisfies

$$\tan\tfrac{1}{2}ml=-\tanh\tfrac{1}{2}ml \qquad . \quad . \quad . \qquad (7)$$

---

* See Lamb's *Sound*, p. 125.

in the case of the symmetrical vibrations when

$$y=(\mathrm{A}\cosh mx+\mathrm{C}\cos mx)\cos nt.$$

The roots of equation (7) are obtained most simply by graphical construction (see fig. 33). Plotting the curves $y=\tan\theta$ and

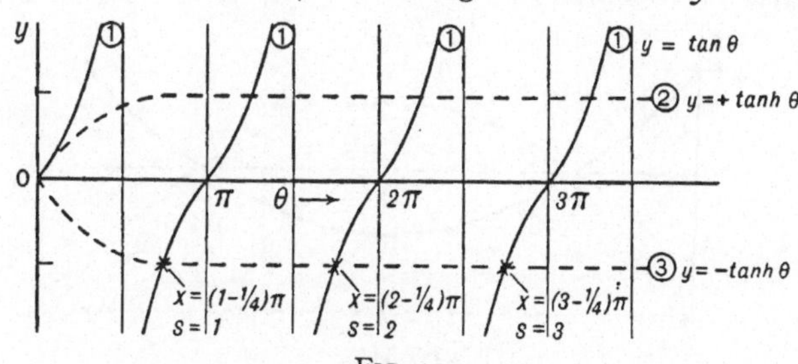

FIG. 33

$y=-\tanh\theta$, the values of $\theta(=ml/2)$ at the points of intersection are found to be closely represented by

$$x=ml/2=(s-\tfrac{1}{4})\pi+\beta \quad . \quad . \quad . \quad (8)$$

where $s=1, 2, 3 \ldots$ and β is a small quantity only appreciable in the case of the fundamental tone when $s=1$ (see fig. 33). Thus

$$m=2\pi(s-\tfrac{1}{4})/l \quad . \quad . \quad . \quad . \quad . \quad (9)$$
$$=\sqrt[4]{\rho n^2/\mathrm{E}\kappa^2} \text{ from equation (5).}$$

Consequently *the frequency of vibration* of the partials is given closely by

$$\mathrm{N}=\frac{n}{2\pi}=\frac{\pi(4s-1)^2}{8}\cdot\frac{\kappa}{l^2}\sqrt{\frac{\mathrm{E}}{\rho}}. \quad . \quad . \quad (10)$$

In the case of the fundamental vibration, the factor $(4s-1)$ in the above relation should be replaced by 3·0112. *It is important to note*:

(*a*) The frequency N is approximately proportional to $(4s-1)^2$ where $s=1, 2, 3 \ldots$ etc., *i.e. the frequencies of the successive symmetrical modes of vibration of the bar are proportional to* 3^2, 7^2, 11^2, 15^2, *etc., approximately*. The overtones are therefore not harmonics as in the case of the vibrations of a string.

(*b*) The frequency N varies inversely as the square of the length of the bar, indicating that $\mathrm{N}\lambda$ is not constant, *i.e. the velocity of transverse vibrations in a bar is dependent on the frequency.*

(*c*) The frequency is proportional to $\sqrt{\mathrm{E}/\rho}$, *i.e.* to the velocity of longitudinal elastic vibrations in the bar.

The form of curve assumed by the bar * vibrating in its fundamental mode can be obtained from equation (6) subject to the necessary end-conditions (see p. 110). The curve is shown in fig. 34. The two 'nodes' occur at distances $0{\cdot}224l$ from the ends of the bar.

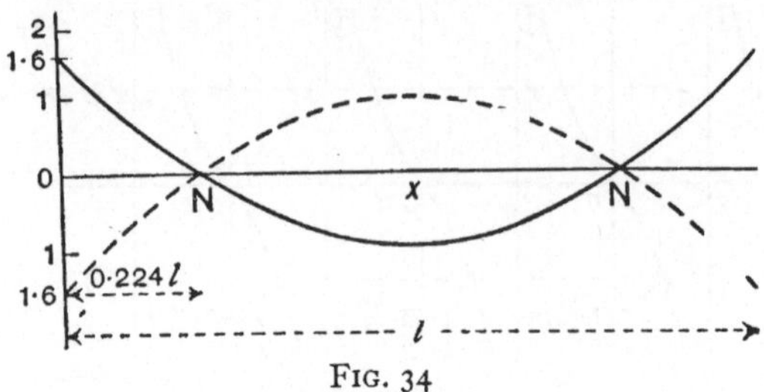

FIG. 34

The next symmetrical mode, frequency $(7/3)^2$ times that of the fundamental, has four nodes, and so on.

It may be shown in a similar manner that the asymmetric modes of vibration of the bar, in which an odd number of nodes are present, have frequencies approximately proportional to $(4s+1)^2$, *i.e.* to 5^2, 9^2, 13^2, etc. Consequently the possible frequencies of the whole series of partials are proportional to the squares of the successive odd numbers commencing at three. The result was discovered experimentally by Chladni,† and has been more accurately verified subsequently by E. H. Barton.‡ The transverse vibrations of a free-free bar in water have been investigated by Moullin and Browne.§ For an iron bar they find $N_{air}/N_{water}=\sqrt{1+0{\cdot}1d/t}$ (d=width and t=thickness of the bar). The following table, which relates to 'free-free' bars in air, gives the accurate values of relative frequency and position of nodes (true or false) for the first few tones (Seebeck and Donkin).

No. of Tone.	No. of Nodes.	Distance of Nodes from one end (in terms of length of bar).	Relative Frequencies.	Frequencies as a ratio of fundamental Frequency.
1	2	0·2242, 0·7758	$3{\cdot}011^2$	1
2	3	0·1321, 0·5, 0·8679	5^2	2·756
3	4	0·0944, 0·3558, 0·6442, 0·9056	7^2	5·404
4	5	0·0734, 0·277, 0·5, 0·723, 0·9266	9^2	8·933

* See Rayleigh, **1**, pp. 282–284.
† Chladni, *Die Akustik*. Leipzig, 1802.
‡ *Phil. Mag.*, p. 578, Nov. 1907.
§ *Proc. Camb. Phil. Soc.*, **24**, p. 400, July 1928.

Bar Clamped at both Ends – Applying the appropriate end conditions to equation (6), it is found that the same series of tones is obtained as for a bar free at both ends.

Bar Clamped at One End only – At the clamped end ($x=-l/2$) we must have $y=0$ and $dy/dx=0$, and at the free end ($x=+l/2$) we have $d^2y/dx^2=0$ and $d^3y/dx^3=0$. Applying these end conditions to equation (6), we find ultimately

$$\tan \tfrac{1}{2}ml = \pm \coth \tfrac{1}{2}ml \quad . \quad . \quad . \quad (11)$$

which is most easily solved graphically. The intersections of the curves $y=\tan\theta$ and $y=\coth\theta$ give $\theta=\frac{1}{2}ml=(s\pm\frac{1}{4})\pi+\beta$, where $s=0, 1, 2, 3$, etc., and β is a small quantity which affects only the fundamental frequency seriously. For the overtones we may write with sufficient accuracy

$$m=2\pi(s\pm\tfrac{1}{4})/l=4\sqrt{\rho n^2/E\kappa^2} \quad . \quad . \quad (12)$$

and

$$N=\frac{n}{2\pi}=\frac{\pi(4s\pm1)^2}{8}\cdot\frac{\kappa}{l^2}\sqrt{\frac{E}{\rho}}\ \text{(approx.)} \quad . \quad . \quad (13)$$

In the case of the fundamental ($s=0$) and the first overtone ($s=1$) a more accurate value of N is obtained by substituting for $(4s-1)$ the values 1·19372 and 2·98836 respectively. The possible frequencies of the overtones are therefore approximately proportional to 3^2, 5^2, 7^2, . . . etc., having respectively 1, 2, 3, etc., nodes in addition to the clamped end. The table gives the accurate theoretical frequencies and position of nodes (true or false, but excluding the clamped end) for the first few tones (Seebeck and Donkin).

No. of Tone.	No. of Nodes.	Distances of Nodes from free end (in terms of length of bar).	Relative Frequencies.	Frequencies as a ratio of fundamental Frequency.
1	0	..	$1{\cdot}194^2$	1
2	1	0·2261	$2{\cdot}988^2$	6·267
3	2	0·1321, 0·4999	5^2	17·55
4	3	0·0944, 0·3558, 0·6439	7^2	34·39

The frequencies are again inversely proportional to the square of the length 'l' of the bar, and the partials do not form a harmonic series.

Bar 'Supported' at both Ends – We now have the end conditions (at $x=\pm l/2$), $y=0$, $d^2y/dx^2=0$, and ultimately from equation (6),

$$\tfrac{1}{2}ml=\pi/2\,.\,(1, 2, 3, 4, \ldots \text{etc.})$$

or

$$m=\pi/l\,.\,(1, 2, 3, 4, \ldots \text{etc.}) \qquad . \quad . \quad (14)$$

and

$$\mathrm{N}=\frac{n}{2\pi}=\frac{\pi s^2}{2}\cdot\frac{\kappa}{l^2}\sqrt{\frac{\mathrm{E}}{\rho}} \qquad . \quad . \quad . \quad (15)$$

where $s=1, 2, 3, 4$, etc. The frequencies of the partials are consequently proportional to the squares of the natural numbers, and the nodes are equidistant as in the case of a vibrating string.

In all cases of bars vibrating transversely it is found that the frequency of vibration is proportional to $\frac{\kappa}{l^2}\sqrt{\frac{\mathrm{E}}{\rho}}$. This relation has been verified experimentally by numerous physicists. Using a 'free-free' bar of rectangular section ($\kappa^2=t^2/12$), Barton * has obtained all the partials up to the seventh, agreeing remarkably well with theory as regards the nodal positions. The theoretical conclusions relative to the vibrations of a clamped-free bar have been similarly verified.

Methods of Excitation – Transverse vibrations in a bar may be excited by similar methods to those used for strings, *e.g.* striking, plucking, bowing, or electromagnetic methods are all applicable. As in the case of the string also, the partials present in the resultant vibration depend both on the method of excitation and on the point of application. Young's law, involving the absence of all partials requiring a node at the point of excitation, is applicable also to bars. The various partials of a bar may therefore be encouraged or otherwise by supporting and exciting it in the required manner. Bars of circular section are liable to vibrate simultaneously in two directions at right angles, *i.e.* the end of the bar vibrates in a closed curve resulting from two rectangular vibrations. *Wheatstone's "Kaleidophone"* † is a good example of the complex motion of a bar vibrating in this way. A steel bar of circular section has a small silvered bead attached to its free end, the bead being suitably illuminated. When the bar vibrates the bright spot reflected from the bead describes a brilliant line which indicates the nature of the vibration. Extremely beautiful

* Barton, *loc. cit.* † See Tyndall's *Sound*, pp. 132–134.

figures are obtained with this simple contrivance. The motion of the bar may also be regarded as a circular combined with a radial vibration. Tyndall shows how, by suitable bowing, such a clamped bar may vibrate with its fundamental mode with superposed partials "$6\frac{1}{4}$ and $17\frac{13}{36}$ times the rate of vibration as a whole." For most purposes, however, it is desirable that the vibrations of the bar should take place in a particular plane. Bars of rectangular cross-section, with the thickness in the direction of bending, are therefore most suitable.

As we have seen above, the only method of supporting a bar which results in 'harmonic' partials is the 'free-free' method. This method has been applied to a musical instrument known as the 'harmonicon' or 'dulcimer,' in which freely supported bars of graded lengths are struck by a hammer, thus producing an agreeable succession of musical tones.

Thin rods, or reeds, clamped at one end and excited by plucking, have been employed in the instrument commonly known as the 'musical box.' In this device a graded series of reeds, like teeth in a comb, are plucked by small pins fixed in the required positions on the surface of a rotating cylinder.

A more important application of the clamped-free reed is in combination with a suitable resonator (*e.g.* an air cavity) which reinforces the fundamental mode and consequently gives a much purer note.* Thin reeds clamped at one end are also employed in wind instruments—the vibrations of the reed serving to open and close the opening which admits air to the instrument (*e.g.* in the oboe, bassoon, and the clarinet group of instruments). The combined action of an air blast and a vibrating reed is illustrated in the harmonium and the concertina—the frequency of the reed here plays a predominant part in determining the pitch of the note. When a metal reed is used in combination with a resonant air column, as in the 'reed pipes' of an organ, the quality of the tone is modified by the presence of the pipe. By varying the construction of the reeds and the shape of the pipes various qualities can be obtained to imitate different orchestral instruments.

Electromagnetic methods of maintenance of the vibrations are, whenever applicable, probably the most suitable for experimental study of the vibrations of a bar. A steel bar is readily set into transverse resonant vibration by means of a suitably disposed electromagnet carrying alternating current—the latter being conveniently supplied by an oscillating valve circuit of controllable

* See Rayleigh, p. 275.

frequency (see p. 79). The principle of excitation is merely that applied in the ordinary telephone earpiece, more particularly in the reed form designed by S. G. Brown (see p. 386). By varying continuously the frequency of the A.C. supply the various resonance frequencies of the fundamental and overtones can very simply be determined. The positions of the nodes can also be obtained by the simple expedient, due to Chladni, of sprinkling fine, dry sand on the bar. The sand will collect in lines across the bar at the nodal positions. The laws of vibration of bars are conveniently verified in this way.

The determination of the frequencies of a steel bar of known dimensions and density, clamped or supported in a definite manner, provides a means of obtaining Young's modulus of elasticity E for the material of the bar. Equations (10), (13), and (15) all show

$$E=\frac{N^2 l^4 \rho}{\kappa^2 C},$$

where C is a constant depending on the method of holding the bar, and on the particular partial excited.

A clamped steel bar electrically maintained is sometimes employed as a rough standard of frequency. An electromagnet, actuated by a battery in series with a contact fairly near the clamped end of the bar, is mounted at a point about $\frac{1}{5}$ to $\frac{1}{4} \cdot l$ from the free end. In this manner the fundamental tone is maintained and the first overtone discouraged. It should be pointed out, however, that the base of such a vibrator should be massive and stiff, otherwise it is liable to vibrate with a comparatively large amplitude, and react on the bar. The frequency and damping are thereby rendered uncertain. These quantities will vary also with the manner in which the base itself is supported. Such a vibrator is consequently not to be recommended in accurate work.

Frequency Meters – Hartmann and Kempf * have employed a graded series of clamped steel reeds (of calibrated frequencies), actuated by a common electromagnet as a means of indicating directly the frequency of alternating currents. They have constructed frequency meters on this principle, covering various ranges from 1 or 2 cycles to 1500 cycles/sec. The tip of the vibrating reed is painted white and the magnet pole is hollowed out to permit of large amplitude resonant vibrations without fear of the reed striking the pole. A photograph of a low-range

* *Ann. d. Phys.*, **13,** 1904; and *Phys. Zeits.*, **11,** Dec. 1910.

frequency meter in operation is shown in fig. 35. The sharpness of tuning of the reeds is roughly indicated in the photograph. A similar instrument may also be used to determine resonant vibrations in machinery or other mechanical structures subject to

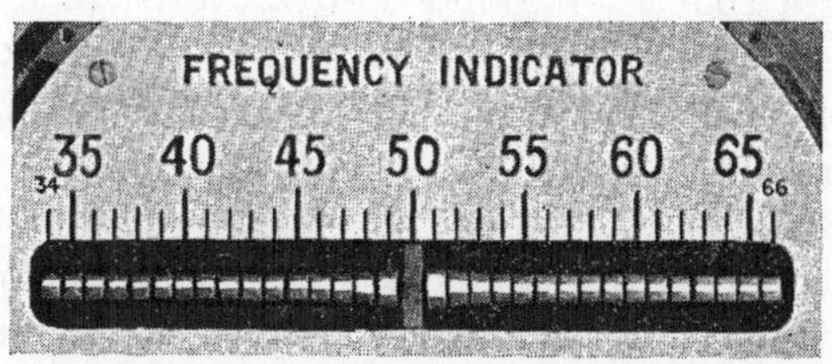

FIG. 35

shocks or unbalanced reciprocating forces. By means of such a frequency determination it is sometimes possible to trace the origin of dangerous or otherwise objectionable vibrations, and take steps to remove it. 'Critical' or 'resonant' speeds in mechanical or electrical machinery often induce large unbalanced forces which may result in serious damage.

Method of Tuning a Bar – A bar clamped at one end may be tuned to a given frequency in a number of ways. The pitch may be lowered either by loading the free end or by reducing the cross-section of the bar near the clamped end. Similarly the pitch may be raised either by shortening the bar or by reducing its cross-section near the free end. These methods of tuning are employed in practice in the construction of reeds of musical instruments (harmoniums, for example) and in frequency meters. A bar may be conveniently tuned for experimental purposes by means of a collar, a sliding fit on the bar, clamped rigidly in the position corresponding to the required frequency. The sliding weight may be fitted with an extra 'rider' with fine adjustment if accurate frequency setting is required. The nearer the sliding weight approaches the free end of the bar the lower will be the frequency.

TUNING-FORKS

The early development of the tuning-fork is mainly due to König, who excelled in making forks of great purity of tone covering a wide range of frequency. Some of König's forks had a frequency as high as 90,000 cycles/sec.* During recent years

* D. C. Millar, *Science of Musical Sounds*; and Winkelmann, *Akustik*, pp. 345–367.

the tuning-fork has increased in importance as a standard of frequency. On account of its great purity of tone and constancy of frequency the tuning-fork has proved of great value as a means of indicating and preserving standard pitches. Not only does it function as a source of sound of definite frequency, but, in its electrically maintained forms, it serves to control electrical circuits in such a manner as to form a *standard of electrical frequency* of great accuracy and of extensive range.

The tuning-fork is regarded by different writers in different ways. The earliest investigators, Chladni for example, regarded it as developed from a 'free-free' bar by bending in the form of an elongated U. Other writers, Rayleigh * for example, have regarded the fork as consisting of two 'clamped free' bars mounted on a heavy, stiff block of metal. Various forms of tuning-fork are made, approximating in appearance to one or other of these forms.†

Consider first of all the bent-bar type. Theory and experiment alike show that the effect of bending a bar is to cause the two nodes of the fundamental tone to approach each other more and more closely and to lower the frequency of vibration. This is illustrated in fig. 36 (a), the various positions of the two nodes n, n being indicated in the progressive stages of bending of the bar. Fig. 36 (b) indicates the form of vibration of such a U-shaped bar. The amplitude of vibration at the 'antinode' at the midpoint of the bend is small compared with that at the ends of the prongs. The addition of a stem at the bend between the two nodes has the double effect of adding mass at an antinode and of increasing the stiffness of that portion of the fork between the two nodes. This results in a further approach of the nodes towards the stem and a reduction in the amplitude of vibration at the midpoint, *i.e.* of the stem. It is this small vibration of the stem of a fork, however, which serves to actuate a sounding-board or a resonance box when pressed into contact with the stem. If the tip of one prong of a vibrating tuning-fork be lightly touched, the vibration is immedi-

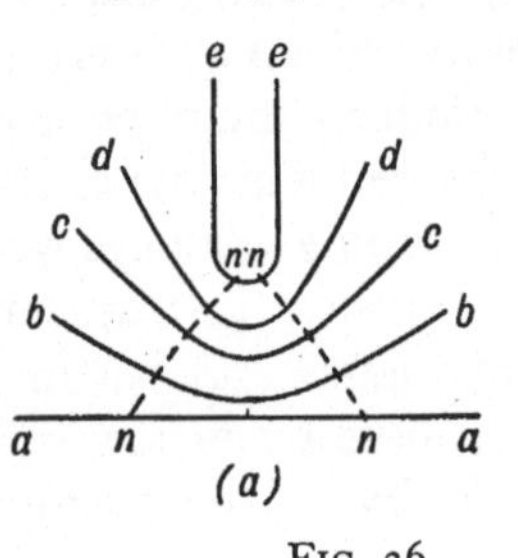

FIG. 36

* Rayleigh, *Sound*, 1, p. 58.

† See (1) D. C. Millar, *Science of Musical Sounds*; (2) Winkelmann, *Akustik*, pp. 345–367.

ately stopped in *both* prongs. The withdrawal of the energy of vibration is in this case extremely rapid. If, however, the stem of the fork be pressed on a sounding-board the vibrations may persist a considerable time, for the amplitude of vibration of the stem is very small and the energy is transferred relatively slowly from the fork to the sounding-board.

Regarding the fork as consisting of a pair of symmetrical clamped-free bars, somewhat similar conclusions may be reached in a different manner. First of all consider a single bar (or prong) clamped to a block of metal. In consequence of the oscillation of its centre of gravity there will be communication of vibration to the block. To reduce the amplitude of such vibration the block must be firm and massive. If now a second exactly similar bar be attached symmetrically to the block and arranged to vibrate with exactly the same amplitude as the first bar, but in opposite direction, the two reactions on the block will exactly compensate each other and the centre of gravity will be unmoved *in the direction in which the prongs vibrate*. On account of the circular path of the centre of gravity of the prongs, however, there must still remain a small vibration of the centre of gravity of the block in the direction of the length of the prongs, the extent of which will depend on the amplitude of the fork and on the relative mass and stiffness of the prongs and of the block. The frequency of a fork of this construction will approximate to that of a single clamped-free rectangular bar of thickness t, *i.e.* the fundamental frequency will be

$$N=(1{\cdot}1937)^2\frac{\pi}{8}\cdot\frac{t}{\sqrt{12}\,.\,l^2}\sqrt{\frac{E}{\rho}} \qquad (1)$$

$$=8{\cdot}24\times10^4 t/l^2$$

for a steel bar in which $\sqrt{E/\rho}=51\times10^4$ cm./sec.

The frequency of a fork will vary approximately as the inverse square of the length of the prongs, directly as the thickness of the prongs and directly as the velocity of sound in the material of which the fork is made. The frequency is independent of the width of the prongs.

The overtones of a fork may be excited by bowing at a suitable point—those overtones requiring nodes at the bowing-point being discouraged. The traces of fork vibrations shown in fig. 37 were obtained by the dropping-plate method. A massive low-frequency fork with a fine bristle attached to the tip of the prong was bowed vigorously, and the bristle allowed to touch gently the smoked surface of a glass plate suspended vertically. Whilst the fork was vibrating the thread supporting the glass plate was burnt through and a wavy line traced on the smoked surface as the plate fell past the fork. This principle is sometimes applied as a rough laboratory method of determining 'g' when the frequency of the fork is known; or conversely. The three records shown in the figure, obtained in this way, show clearly the overtones $6\frac{1}{4}$ and $17\frac{1}{2}$ times the fundamental frequency.* By varying the point of bowing the various partials may be obtained separately or occurring two or three at the same time.

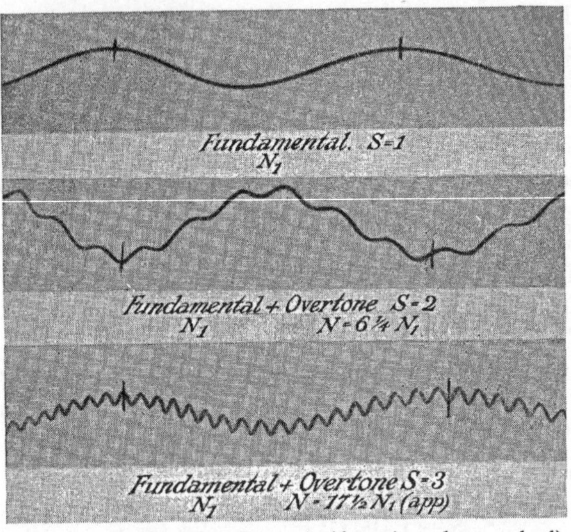

FIG. 37—Tuning-Fork Records (dropping-plate method)

The fundamental tone of a fork may be reinforced by attaching the stem to a resonance box of the same frequency—the overtones of a fork and an air cavity of the same fundamental frequency are widely different, so the overtones of the fork are not reinforced.

Small tuning-forks, of the types devised by König, have long been used as standards of pitch for musical and scientific purposes where a moderate accuracy was required. The method of beats gives results of considerable accuracy in comparing the frequencies of two forks or in comparing a fork with any other vibrating body of approximately the same frequency. The number of beats in a second is equal to the difference of frequencies of the two sounds. Under favourable circumstances beats as slow as one in thirty

* See also F. H. Parker, *Nature*, **90**, p. 361, Nov. 28, 1912; and E. H. Barton, *Nature*, **90**, p. 436, Dec. 19, 1912.

seconds may be recognised. This represents very exact tuning between two forks of say 200 p.p.s. The method of beats provides a very convenient means of investigating the factors which produce small changes of frequency in a tuning-fork. One fork is kept under standard conditions, whilst the other is subjected to the influence, *e.g.* temperature change, which affects the frequency. In this manner it is not difficult to show experimentally that the temperature coefficient of frequency for an ordinary steel fork is approximately -10^{-4} per degree Centigrade rise of temperature. The method of beats is employed in making copies of standard tuning-forks. The substandard is modified until it gives very slow beats when sounded simultaneously with the standard. The addition of a small load to the prongs of the substandard is sufficient to indicate, by the change in the number of beats per second, which of the two forks is gaining on the other.

On account of its behaviour as a double source, a tuning-fork is a relatively poor radiator of sound energy (see p. 66). E. A. Harrington * has found experimentally, for example, that a certain electrically-maintained tuning-fork expended only 3·5 per cent. of its energy in producing sound, the remaining 96·5 per cent. being employed in overcoming internal friction and in vibrating the support of the fork.

Forks are now commonly used as *substandards of time.* The period of vibration of a fork is a very constant quantity and serves as a convenient subdivision of a second when time intervals have to be measured with accuracy. As a consequence of the increased application of the tuning-fork for this purpose, methods have been devised for increasing the accuracy and perfection of electrically-maintained forks. Low-frequency forks of this character, giving an accuracy greater than 1 in 10,000, are in common use. If particular care is taken with the choice of material, design, and control of the fork, an accuracy of 1 in 10^5 or even of one in a million is possible.

Electrically-maintained Forks – Reference has already been made (p. 76) to the method of maintaining the vibrations of a tuning-fork by an electromagnet. The principle involved is that of an electric bell, the vibrations of the prongs of the fork 'making' and 'breaking' a current through the exciting electromagnet. The self-maintaining action of a tuning-fork is not quite as obvious as it may appear.† If it were not for the presence of inductance

* *Journ. Opt. Soc. Amer.*, **17**, p. 224, Sept. 1928.
† See Rayleigh, **1**, pp. 68–69.

in the circuit, the fork would not continue to vibrate under the action of the electromagnet. The action is as follows (see fig. 38 (*a*), (*b*), and (*c*)). The fork is maintained by a battery of e.m.f. E, in series with a pair of contacts (on the *outside* of one prong), and an electromagnet M. The inductance and resistance in the circuit are L and R respectively. As the prong moves from *a* to *b* and back again, the circuit is closed (approximately for half a period T/2), whilst from *a* to *c* and back again the circuit is open. When the circuit closes $t=0$ (fig. 38 (*b*)) the circuit i begins to rise, the rate depending on the values of L, R, and E. At break, $t=T/2$, the current falls suddenly to zero (assuming there is no arcing). From *a* to *b* on the outward movement of the prong, a quantity of electricity Q_1 flows (time 0 to T/4) through the magnet which is then *opposing* the motion. From *b* to *a* on the return movement of the prong, a quantity Q_2 flows (time T/4 to T/2) through the magnet which is *assisting* the motion. The difference (Q_2-Q_1) represents the quantity per cycle effectively used in maintaining the vibration of the fork. The sum (Q_2+Q_1) represents the total quantity per cycle taken from the battery. The current i at any instant 't' after closing the circuit is given by

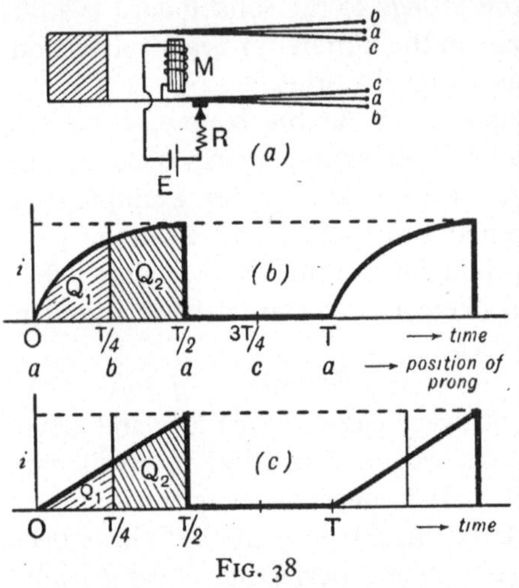

FIG. 38

$$i=\frac{E}{R}\left(1-e^{-\frac{R}{L}t}\right)$$

and

$$Q_1=\int_0^{T/4} idt, \quad Q_2=\int_{T/4}^{T/2} idt,$$

whence

$$\text{Efficiency}=\frac{Q_2-Q_1}{Q_2+Q_1}=\frac{\frac{L}{R}\left\{e^{-\frac{R}{L}\frac{T}{2}}-2e^{-\frac{R}{L}\frac{T}{4}}+1\right\}}{\frac{T}{2}-\frac{L}{R}\left(1-e^{-\frac{R}{L}\frac{T}{2}}\right)} \quad . \quad (2)$$

When the *inductance L is zero* this ratio becomes zero also. No energy is supplied to the fork to compensate for frictional damping and the vibrations are consequently not maintained. When the *inductance* L *is very large* and R is negligible, the expression approximates to $\frac{1}{2}$. This is the maximum value of the efficiency (see fig. 38 (*c*)). When L/R is sufficiently small to allow the current to attain its maximum value in a time less than T/4, the expression for the efficiency reduces to

$$\frac{Q_2-Q_1}{Q_2+Q_1}=\frac{1}{\frac{TR}{2L}-1}.$$

For intermediate values of L and R the efficiency is given by equation (2).

The above considerations illustrate the necessity for inductance in the circuit and indicate that this should be as large as possible

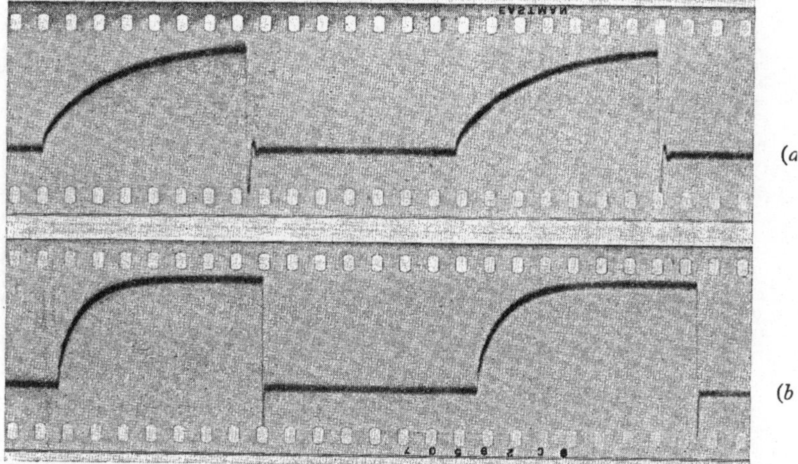

FIG. 39—Oscillograph Records of Current exciting a Tuning-Fork

to secure efficient running. A large inductance implies a large back e.m.f. at 'break,' with consequent arcing at the contacts. It is advisable, therefore, to include a certain amount of resistance in the circuit, and to use a suitable condenser in parallel with the fork contacts. This reduces the risk of arcing whilst making it possible to obtain a good running efficiency. Oscillograph records of the current maintaining a 25 cycles/sec. fork are shown in fig. 39. The effect of added resistance in record (*b*) is clearly shown in the more rapid rise of current at 'make.' The two cases represent respectively 50 per cent. and 10 per cent. of maximum possible efficiency.

When an electrically-maintained fork is to be used as a precision time-standard, it is essential that attention be paid to certain features of the design. In papers by Dadourian * and by the writer † it is shown that the base of the fork must be massive and stiff, in order that mechanical coupling between the fork and its supports (the top of a bench, for example) should be reduced to a minimum. For this reason also it is desirable that the plane of the prongs should be parallel to the base, and that the block supporting the prongs should be very short and rigidly attached to the base (preferably forming part of it). The best forks are milled out of a solid steel block, but very satisfactory forks can be made by silver-soldering two steel bars to the supporting block. The frequency of the fork is also affected slightly by the change in the electrical constants of the electromagnet circuit and on the setting and number of contact pairs operated by the prongs. The frequency depends slightly on the amplitude of vibration, this being partly due to the presence of the contact springs.

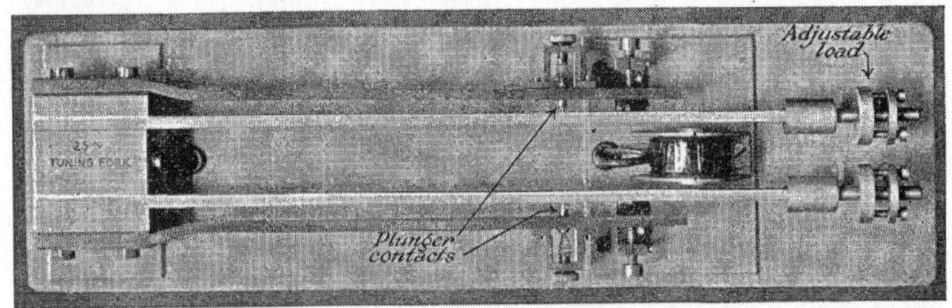

FIG. 40—Electrically-maintained Tuning-Fork (low frequency)

In fig. 40 is shown a low-frequency fork (25~ or 50~) by J. M. Ford and the writer,‡ in which the spring contacts were designed to give very exact timing of the moments of 'make' and 'break' relative to the position of the prongs. The frequency of the fork and balance of the prongs may be regulated by means of adjustable loads near the free ends. The positions of the electromagnet and contacts were carefully selected to give the best running conditions. The magnetic circuit is also more efficient than the more customary form, the flux being confined to

* H. M. Dadourian, *Phys. Rev.*, **13**, No. 5, pp. 337–359, May 1919.
† A. B. Wood, *Journ. Sci. Instrs.*, **1**, No. 11, Aug. 1924.
‡ A. B. Wood and J. M. Ford, *Journ. Sci. Instrs.*, **1**, p. 161, 1924.

the shortest possible path *across* the prongs. See fig. 41. This results in a considerable reduction of the current necessary to maintain the fork in vibration, with consequent increase of 'life' of the platinum contacts. Forks similar to that shown in fig. 40 are often required to run continuously for long periods whilst driving or controlling electric motors for time-marking or recording purposes. (See section on Sound-Ranging, p. 478.)

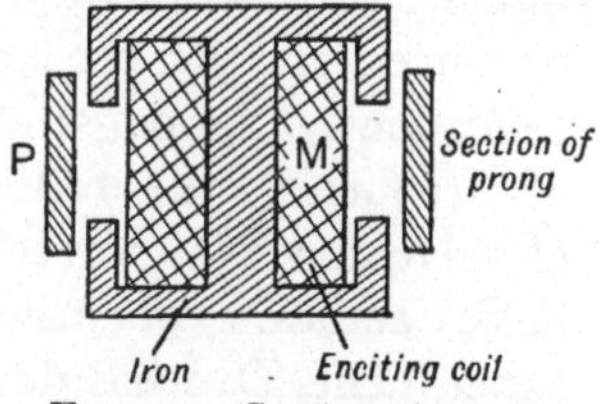

FIG. 41—Section through Magnet and Prongs of Fork

Valve-maintained Forks – The method of maintaining the vibrations of a fork just described ceases to be really efficient when the frequency exceeds about 100 cycles/sec. W. H. Eccles* has described a method which is particularly applicable to forks of higher frequency (*e.g.* 1000 p.p.s.) and which has the additional advantage that no mechanical contact is made with the vibrating prongs. In the form described, two electromagnets act on the prongs, one of these being in the grid circuit and the other in the plate circuit of a valve (see fig. 42). These coils must be wound in opposite directions.* The vibration of the fork induces alternating voltage in the grid circuit, which controls the current in the plate circuit and its magnet, the vibration of the fork being thus maintained. To ensure that the fork responds readily to excitation, it is advisable to introduce a condenser across each of the magnet coils as shown in fig. 42, so as to get approximate tuning in the electrical circuits. S. Butterworth † has given a mathematical analysis of the mode of operation of a valve-maintained fork, and has determined the conditions essential to its success. He deals with the problem by substituting for the mechanical system the fork, its electrical equivalent. In this manner the variation of frequency in the neighbourhood of the natural frequencies of both the dynamical and electrical systems is calculated. Using a pair of valve-maintained forks, one of

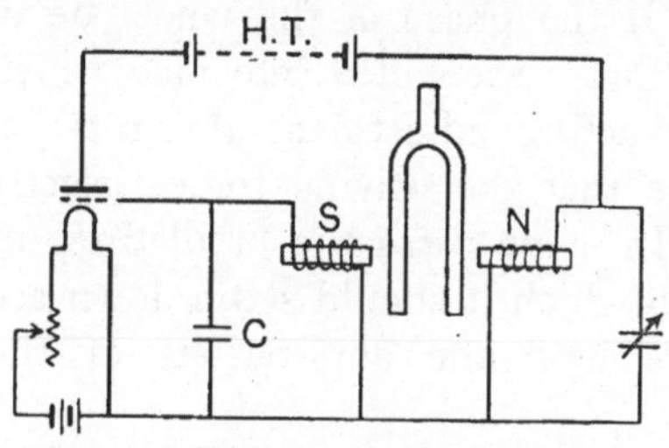

FIG. 42—Valve-maintained Tuning-Fork (Eccles)

* W. H. Eccles, *Proc. Phys. Soc.*, **31**, p. 269, 1919.

† S. Butterworth, *Proc. Phys. Soc.*, **32**, p. 345, Aug. 1920.

which was kept as invariable as possible, D. W. Dye * has determined, by the method of beats, the variations of frequency produced in the second fork by changes in the experimental conditions. Amongst other factors likely to produce changes of frequency the following were studied: temperature, effect of valve controls (current, H.T. voltage, etc.), polarising magnetic field, addition of mass to base, etc. As a result of these experiments Dye has developed the valve-maintained tuning-fork as a precision time-standard which controls a multi-vibrator (see p. 132) and so forms the foundation of the radio-frequency standards of the National Physical Laboratory.

Phonic Motors – The phonic wheel, invented independently by Rayleigh † and La Cour, is of great importance when used in conjunction with an electrically-maintained tuning-fork as a means of converting a vibratory into a rotary motion of constant speed. In describing a phonic motor Rayleigh says: "... the essential feature is the approximate closing of the magnetic circuit of an electromagnet, fed with intermittent current, by one or more soft iron armatures carried by the wheel and disposed symmetrically round the circumference. If in the revolution of the wheel the closest passage of the armature synchronises with the middle of the time of excitation, the electromagnetic forces operating on the armature during its advance and its retreat balance one another. If, however, the wheel be a little in arrear, the forces promoting advance gain an advantage over those hindering the retreat of the armature, and thus upon the whole the magnetic forces encourage the rotation. In like manner, if the phase of the wheel be in advance of that first specified, forces are called into play which retard the motion. By a self-acting adjustment the rotation settles down into such a phase that the driving forces exactly balance the resistances. . . . In some cases the oscillations of the motion about the phase into which it should settle down are very persistent ‡ and interfere with the applications of the instrument."

Early forms of phonic wheel consisted essentially of a toothed iron wheel with a small electromagnet embracing an integral number of teeth of the wheel. These simple phonic motors served fairly well as 'time-markers' in optically recording systems,

* *Proc. Roy. Soc.*, A, **103**, pp. 240–260, 1923.

† Rayleigh, *Nature*, **18**, p. 111, 1878 ; see *Sound*, **1**.

‡ A phenomenon known as 'hunting.' Rayleigh recommends a mercury-filled fly-wheel to damp out such oscillations.

e.g. in the Einthoven string galvanometer recorder, but were useless, however, when called upon to supply a small amount of power, *e.g.* to drive a drum with a pen recorder. In 1919 J. M. Ford

FIG. 43—Phonic Motor (assembled)

and the writer designed an improved type of phonic motor (see fig. 43) which was capable of driving an electrically-operated chronometer * or a moving-tape pen-recorder. The magnetic

FIG. 43—Phonic Motor (parts)

circuit of this motor is so designed that even moderate exciting currents produce a large driving torque. In one pattern the rotor, mounted on ball-bearings, is made from a bar of soft iron with ten slots machined longitudinally, leaving ten radial bars

* A. B. Wood and J. M. Ford, *Journ. Sci. Instrs.*, **1**, pp. 160–173, 1924.

outstanding. The stator consists of a soft-iron cylinder and two soft-iron discs each with ten teeth corresponding to the ten bars of the rotor. The exciting coil, connected through the fork contacts, is a single former-wound coil which lies within the iron shell of the stator and completely surrounds the rotor. The troublesome transverse vibration of the shaft experienced with earlier forms of phonic motor is now entirely eliminated, for the radial forces on the rotor are balanced and the tangential pull is applied uniformly on all the bars of the rotor. The magnetic flux has a short path and the torque is large, consequently the motor is easy to start and relatively small driving currents through the fork contacts are sufficient. The motors are silent when running and are free from troublesome mechanical vibrations. In addition to the more obvious applications to phonic chronometers, chronographs, and stroboscopic devices, these motors have been used to control the speed of much more powerful motors. Thus a phonic motor has been used in this way to drive a Radio-Beacon at constant speed.* Controlled by low-frequency forks (up to 100 p.p.s.) they have a relatively large driving torque. At higher frequencies (100 to 1000 p.p.s.), however, the difficulties increase, and it is necessary to use laminated iron in the rotor and the stator. The motor can quite easily be made self-starting, if required, by the addition of a small commutator on the shaft and a stroboscopic synchronising device, or by a method described by D. C. Gall,† in which a special form of winding is employed. The simple form of motor is, however, easily spun by hand into its synchronous speed.‡

Frequency Determination of Forks

There are many simple laboratory methods of measuring approximately the frequency of a vibrating tuning-fork. The dropping-plate method, illustrated in fig. 37, p. 120 (see also p. 221), is direct but is somewhat inaccurate on account of the fact that the tip of the prong touches the falling plate. In another method, the note of the fork is compared with that of a siren, or of a toothed wheel, of which the speed is accurately known, the frequency of the fork being thus obtained in absolute measure. The final tuning is done

* R. L. Smith-Rose, *Nature*, **121**, p. 745, May 12, 1928; and *I.E.E. Journ.*, **66**, p. 256, 1928.

† D. C. Gall, *Journ. Sci. Instrs.*, **3**, pp. 351–352, July 1926.

‡ The tuning-forks and phonic motors shown in figs. 40 and 43 are manufactured by Messrs. Tinsley & Co., South Norwood, S.E.

by the method of beats. A photographic record, with superposed time marks from a standard clock, is also another possible method. Reference has already been made to the use of the monochord in *comparing* the frequencies of two forks.

Phonic-Motor Method – The phonic motor provides a very convenient and accurate method of determining the frequencies of electrically-maintained forks, a reliable pendulum clock or Greenwich radio time-signals being the time-standard to which reference is made. The following method was employed by the writer in 1919.* A tuning-fork of nominal frequency N controls a phonic motor having n teeth, or bars. The speed of the motor is consequently N/n revolutions per second. If the motor makes m revolutions in t seconds, the frequency N of the fork must be equal to nm/t. Attaching a revolution counter to the shaft of the motor it is a simple matter therefore to count the revolutions m in any time interval t. The greater the time interval the greater the accuracy of the method. The value of N obtained in this way is, of course, the *mean* value during the period of test. As an example of the method, take the case where $N=50$ approximately, and $n=10$. The motor-speed is consequently 5 r.p.s. approximately. In 5 minutes $m=1500$, in 1 hour $m=18{,}000$, and in 24 hours $m=432{,}000$ revolutions. Now the value of m can be determined with certainty to ± 1 by direct observation of the revolution counter and a standard clock. Therefore the accuracy of the method is approximately $\pm 1/m$ for short intervals, and ultimately becomes equal to that of the standard clock for long intervals. An accuracy of 1 in 10,000 is easily attained. Dye † has successfully applied the phonic motor method to valve-maintained forks of higher frequency. The motor-revolutions and seconds-contacts of the standard clock are recorded side by side on a chronograph tape, whence the fractional part of a second lost or gained by the fork during a given period of time can readily be determined. Three runs for a particular fork, at intervals of a month, gave the values $1000{\cdot}21_8$, $999{\cdot}99_7$, $999{\cdot}87_6$ vibrations per second, the corresponding temperatures being 13·36° C., 15·23° C., and 16·27° C. respectively.

The phonic-motor method of *counting* the vibrations of a fork in a given time lends itself readily to the study of the various factors likely to influence the frequency of a fork. Temperature,

* See also A. B. Wood, *Journ. Sci. Instrs.*, March and Aug. 1924.

† D. W. Dye, *loc. cit.*; see also T. G. Hodgkinson, *Proc. Phys. Soc.*, **39**, p. 203, 1927.

amplitude of vibration, spring contact pressure, magnet current, loading of base, etc., have all an important influence on frequency.

Temperature-Coefficient of Frequency of Forks – Change of temperature is one of the most serious causes of change of frequency of a tuning-fork. This is due mainly to the variation of the elastic properties of the material with temperature, the corresponding density changes being relatively negligible. Thus König found that the temperature-coefficient of frequency for a steel fork was $-1{\cdot}12 \times 10^{-4}$ per °C. rise of temperature. The temperature-coefficient of Young's modulus E for ordinary steels is given by

$$E_t = E_{15°}\{1-\alpha(t-15°)\},$$

where

$$\alpha = 2{\cdot}4 \times 10^{-4} \quad \text{and} \quad t = \text{temp. °C.}$$

For a 25-cycle fork with mild steel prongs the writer has obtained a value of $-1{\cdot}16 \times 10^{-4}$ per °C. for the temperature-coefficient of frequency at ordinary room temperatures. Dye gives a value $-1{\cdot}15 \times 10^{-4}$ per °C. between 15° and 21° C., for a fork of 1000 frequency, and Dadourian * observed a change in the coefficient from $-1{\cdot}04 \times 10^{-4}$ at $-25°$ C. to $-1{\cdot}43 \times 10^{-4}$ at $+56°$ C. "Elinvar" (*el*asticity *invar*iable), one of the new nickel steels due to Guillaume (who discovered "invar," which has a very small expansion coefficient), has a very small temperature-coefficient of elasticity, and is used in the best watches and chronometers. Forks made of this material have a very small temperature-coefficient of frequency. A good elinvar fork, valve-maintained and kept at a temperature constant within $\pm 0{\cdot}1°$ C., remains constant to within one or two parts in a million.

Variation of Frequency with Amplitude – All observations with 'free' forks agree as regards a lowering of pitch with increasing amplitude. In large low-frequency forks vibrating with amplitudes of the order of 1 mm. at the tip of the prongs, this effect may become of considerable importance. The effects are more complicated in the case of electrically-maintained forks, the increase in amplitude involving questions such as stiffness and length of travel of contact springs, pull of electromagnet, and so on. These additional complications modify the result.† Thus Dadourian observed with a particular type of fork with spring contacts, that increase of amplitude first lowered and then raised

* Dadourian, *loc. cit.*

† See papers by Dadourian, Wood, and others, *loc. cit.*

the frequency, indicating a best working amplitude (where dN/da is a minimum) for the fork. E. Mallett * has shown that this change of frequency with amplitude results in a distortion of the resonance curve for a tuning-fork vibrating at large amplitudes—similar effects being noted for vibrating reeds, vibration galvanometers and telephone receiver diaphragms.

When a tuning-fork is to be used as a time-standard therefore, it is important that its amplitude (and conditions of maintenance) should be defined. A simple 'V' diagram attached to the tip of the prong is a very convenient indication of amplitude—when in vibration the 'V' assumes a double form W, the point of overlap being well defined. By means of a series of equidistant horizontal lines crossing the 'V' symmetrically, it is a simple matter to determine the amplitude in terms of the angle of the 'V' and the distance apart of the horizontal lines.

Clock-controlled Forks - König † devised a clock in which the pendulum was replaced by a tuning-fork, a small escapement mechanism controlling a train of wheels and imparting a small impulse to the fork at each vibration. Thus the vibrations of the fork were maintained mechanically and were counted by means of dial wheels. D. C. Millar ‡ has employed this 'clock-fork' to determine the frequency of standard forks by the method of coincidences. The variations of frequency of a fork running in this manner are considerable, small errors due to various causes, *e.g.* temperature changes, are cumulative, and the final error may be serious. D. W. Dye § and the writer ‖ independently devised methods of controlling a low-frequency electrically-maintained fork directly from a master pendulum clock. In the writer's method, the fork is arranged to *gain* slightly on the clock. When the error has reached a certain very small fraction of a period of vibration, a relay causes the amplitude of the fork to increase, thereby causing the fork to *lose* slightly on the clock. The process is then repeated, the fork alternately gaining and losing on the clock. The extent of this 'hunting' of the fork-frequency can be adjusted to any desired value, depending on the circumstances. In this manner a fork and phonic motor have been kept in exact synchronism with a pendulum clock for a continuous period of several days.

In Dye's method, the seconds contacts of a pendulum clock

* Mallett, *Proc. Phys. Soc.*, **39**, pt. 4, pp. 334–358, June 1927.
† König, *Quelques Experiences*, p. 173.
‡ Millar, *Science of Musical Sounds*, p. 38, etc.
§ D. W. Dye, Phys. Soc. Exhibition, Jan. 1927.
‖ A. B. Wood, Phys. Soc. Exhibition, Jan. 1927.

give a direct impulse to the fork through the exciting magnet. The fork itself is 'free,' *i.e.* carries no mechanical contacts. Provided the frequency of the fork is sufficiently near to a whole multiple of the frequency of the pendulum a vigorous vibration will be maintained and the fork will be 'corrected' every second by the impulse from the standard pendulum. The accuracy of the fork is therefore equal to that of the master pendulum clock at the National Physical Laboratory.

The Tuning-Fork as a Standard of Frequency (Sonic, Supersonic, and Radio) – *The Multivibrator.* A valve-maintained elinvar fork in a constant temperature enclosure is capable of remaining constant in frequency within one part in a million. This high order of accuracy can be attained for tuning-forks of any frequency up to one or two thousand per second. Making use of a device known as a "Multivibrator," invented by H. Abraham and E. Bloch,* D. W. Dye † has applied the accuracy of the elinvar tuning-fork to standardise supersonic and radio-frequencies.

The multivibrator is an ingenious device which makes use of two thermionic valves to produce a fundamental alternating current *very rich in harmonics.* The circuit is shown in fig. 44.

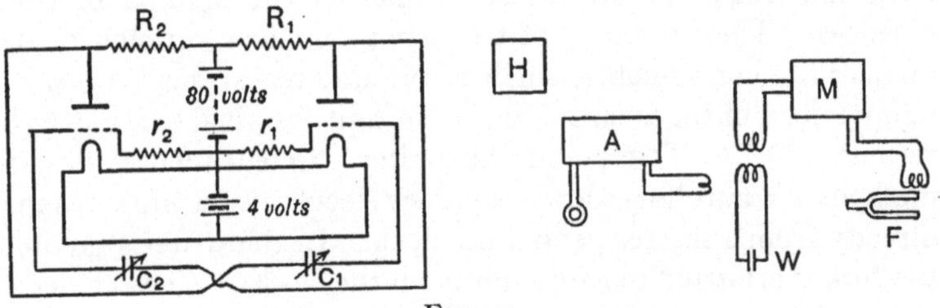

FIG. 44

(*a*) Abraham-Bloch Multivibrator

(*b*) F, valve-maintained fork; M, multivibrator; W, standard wavemeter; A, amplifier; H, heterodyne

In a particular type of apparatus giving a fundamental of 1000 cycles per second, the values of resistances and capacities are as follows: $R_1 = R_2 = 50{,}000$ ohms (inductively wound), $r_1 = r_2 = 75{,}000$ ohms (non-inductive), C_1 and C_2 are subdivided and adjustable condensers giving any value up to 0·008 microfarad.

* Abraham and Bloch, *Comptes rendus*, t. **168**, p. 1105, 1919.

† D. W. Dye, (1) *Trans. Roy. Soc.*, **224**, p. 259, 1924; (2) *Proc. Phys. Soc.*, **38**, pp. 399–458, Aug. 1926; see also *Dictionary of Applied Physics*, **2**, 'Electricity,' pp. 633–634; and W. Hansel, *Phys. Zeits.*, **30**, p. 274, May 1929.

The fundamental frequency of oscillation of the circuit is approximately $N=1/(C_1r_1+C_2r_2)$, all the harmonics up to a very high order being present also. These can be selected up to the 120th by means of a tuned (LC) 'selector' circuit. If, therefore, the multivibrator is controlled by an elinvar fork of frequency 1000·00 (temperature coefficient $+12\times10^{-6}$ per 1° C.) a 'spectrum' of frequencies, all *exact* multiples of 1000, will be obtained. This control is brought about by injecting a small e.m.f. from an electrically-maintained elinvar fork into the multivibrator circuit. Higher frequencies are obtained by the use of a second multivibrator giving impulses at the rate of 20,000 per second. This instrument is kept in harmonic synchronism with the fork-controlled one by the intermediary of a fixed-frequency resonant circuit set to 20,000 cycles/sec. With the second instrument harmonics up to the 75th can be observed, corresponding to a frequency of $1{\cdot}5\times10^6$ cycles/sec. It is therefore possible to obtain standard frequencies as accurate as the frequency of the elinvar fork, at intervals of 1000 up to 1·5 million cycles/sec. Not only can all frequencies which are multiples of 1000 be set exactly, but a very great number of subsidiary frequencies may be obtained within each belt of 1000 cycles. Thus, whenever the source has a frequency very close to $N=1000(f\pm1/r)$, where f and r are integers, a slow-synchronisation beat is heard. With care, further intermediate frequencies can be located. It is never necessary to interpolate by more than about two cycles/sec. from a frequency that is known. Such measurements are made possible by the extreme steadiness of the frequency of the tuning-fork. In this way a wide range of standard frequencies may be covered in very small steps, the degree of accuracy being fixed by the accuracy of the elinvar tuning-fork which stabilises the whole system.

The frequency of the fork is determined by means of the phonic-motor method outlined above. All frequencies are therefore referred ultimately to a standard pendulum clock.

Vibrations of a Ring

The problem of the vibrations of a curved bar may be regarded as an extension of that of a straight bar. Apart from tuning-forks, which we have just considered, the most important case is that of a circular ring.* We shall merely quote the results of theoretical deductions made by Hoppe (1871) and by Rayleigh.

* See Lamb, *Sound*, pp. 133–138.

(a) *Flexural Vibrations in the Plane of the Ring* – The frequency N of any given partial is

$$N=\frac{1}{2\pi}\cdot\frac{s(s^2-1)}{\sqrt{s^2+1}}\cdot\frac{\kappa}{a^2}\sqrt{\frac{E}{\rho}}\quad . \quad . \quad . \quad (1)$$

where s, κ, E, and ρ have the meanings assigned in the case of straight bars. The mean radius of the ring is 'a.' The most important case is that of $s=2$, where the ring oscillates between two slightly elliptical extreme forms. The frequencies are comparable with those of transverse vibration of a bar.

(b) *Flexural Vibrations Perpendicular to the Plane of the Ring* – Mitchell (1889) found, for a ring of circular cross-section,

$$N=\frac{1}{2\pi}\cdot\frac{s(s^2-1)}{\sqrt{s^2+\sigma+1}}\cdot\frac{\kappa}{a^2}\sqrt{\frac{E}{\rho}}\quad . \quad . \quad (2)$$

(σ=Poisson's ratio.)

(c) *Extensional Vibrations* – In this case the ring remains in its own plane, being alternately stretched and contracted,

$$N=\frac{\sqrt{s^2+1}}{2\pi}\cdot\frac{1}{a}\sqrt{\frac{E}{\rho}}\quad . \quad . \quad . \quad (3)$$

When $s=0$ the vibrations are purely radial, the frequency being given by the ratio of the velocity of sound to the circumference of the ring.

The sequences of tones given by equations (1) and (2), for the flexural vibrations of the ring, agree with those found experimentally by Chladni. The flexural vibrations are not purely radial. In case (a) when $s=2$ there are four nodes, the elliptical forms cutting the mean circle in four points. At these four points there is no radial motion, but the tangential motion is a maximum. This is exemplified by the vibrations of a partially filled tumbler when excited by gently drawing a wetted finger along the edge, the nodal points being indicated by the crispations of the surface of the water in the tumbler.

The various modes of vibration of a circular plate and annular rings have been demonstrated recently by W. Hort and M. König.* The annular rings were formed by cutting holes of increasing radius r_1 in a disc of radius r_2, the ratio r_1/r_2 varying from 0 to about 0·9. The numerous radial and circular nodes which are possible were clearly shown by means of sand figures. Even with

* *Zeits. f. Techn. Physik*, No. 10, p. 373, 1928.

the narrowest annular ring ($r_1/r_2=0{\cdot}9$) a circular node was obtained approximately midway across the face of the ring. The results were in general agreement with theory.

The theory of vibration of rings has application in the complicated study of the vibrations of bells (see p. 167).

LONGITUDINAL VIBRATIONS OF RODS

When a rod is set into longitudinal vibration, cross-sections vibrate to and fro in the direction of the axis which remains undisplaced. The quantities involved are the density and elastic properties of the material. When a rod is stretched in the direction of its length the increase of length is accompanied by a contraction in the cross-section. If the rod is very short the motion of a particle near the surface is not only longitudinal but also transverse—the latter motion may be comparable with the former. If, however, the length of the rod is large compared with the diameter, the longitudinal motion accumulates and the transverse motion becomes negligible. We shall, in what follows, assume that this condition holds, viz. that all transverse sections vibrate to and fro along the axis of the rod.

Consider a rod of uniform cross-section A, and an elementary slice bounded by two planes, at x, distant δx apart and at right angles to the axis x of the rod. If a plane at x is displaced at time t to $(x+\xi)$ the plane at $(x+\delta x)$ will be displaced to

$$(x+\delta x+\xi+(d\xi/dx)\delta x),$$

that is, the fractional elongation of the slice δx is $d\xi/dx$. This change of length of the slice implies a difference in the forces acting on its faces. The total force acting on the face at x will be $\mathrm{E}\mathrm{A}d\xi/dx$, where E is Young's modulus * of elasticity. Similarly at the opposite face of the slice the force will be in the opposite sense and equal to

$$\mathrm{EA}\left(\frac{d\xi}{dx}+\frac{d^2\xi}{dx^2}\cdot\delta x\right).$$

Consequently the resultant force acting on the slice will be

$$\mathrm{EA}\cdot\frac{d^2\xi}{dx^2}\cdot\delta x.$$

---

* Young's modulus E is determined by stretching a bar statically, *lateral contraction* being permitted during the extension.

Now the mass of the slice is equal to $\rho A \delta x$, where ρ is the density of the material, and the acceleration is $d^2\xi/dt^2$. Hence we obtain the relation

$$\rho A \delta x \frac{d^2\xi}{dt^2} = EA \frac{d^2\xi}{dx^2} \cdot \delta x,$$

that is,

$$\frac{d^2\xi}{dt^2} = \frac{E}{\rho} \cdot \frac{d^2\xi}{dx^2} \quad . \quad . \quad . \quad . \quad (1)$$

an equation of the same form as that obtained for the transverse motion of a string, indicating waves travelling in opposite directions with equal velocities, $c = \sqrt{E/\rho}$. The general solution is

$$\xi = f(ct - x) + F(ct + x) \quad . \quad . \quad . \quad (2)$$

Assuming ξ varies as $\cos nt$,

$$\left.\begin{aligned} &\frac{d^2\xi}{dt^2} = n^2\xi, \qquad \frac{d^2\xi}{dx^2} = \frac{n^2}{c^2}\xi, \\ \text{and} \\ &\xi = \left(A \cos \frac{nx}{c} + B \sin \frac{nx}{c}\right) \cos nt \end{aligned}\right\} \quad . \quad . \quad (3)$$

which is analogous to equations obtained for stationary waves on a vibrating string. In a rod of finite length, the direct and end-reflected waves interfere to form nodes and loops in accordance with the relation $c = N\lambda \left(\text{that is, } N = \frac{c}{\lambda} = \frac{1}{\lambda}\sqrt{\frac{E}{\rho}}\right)$, as in the case of strings.

Case I – *Rod free at both ends.* Each end must be an antinode, that is, the displacement ξ is a maximum at $x=0$ and $x=l$. This implies in equation (3) that $B=0$ and $\xi=A$, whence $nl/c = s\pi$ (where $s=0, 1, 2, 3$, etc.). The case of $s=0$ is inadmissible, so we may write for the frequencies of the partials

$$N_s = \frac{n_s}{2\pi} = \frac{s}{2l}\sqrt{\frac{E}{\rho}} \text{ (where } s=1, 2, 3, \text{ etc.)} \quad . \quad . \quad (4)$$

which represents *a complete harmonic series.* A similar result is obtained for a rod clamped at both ends ($\xi=0$ when $x=0$ and l).

Case II – *Rod clamped at the midpoint.* All the even harmonics must now be absent, for the midpoint must always be a node and the ends antinodes. The frequencies of the partials are therefore proportional to the odd numbers, *i.e.*

$$N_s = \frac{s}{2l}\sqrt{\frac{E}{\rho}} \text{ (where } s=1, 3, 5, \text{ etc.)} \quad . \quad . \quad (5)$$

The fundamental has the same frequency here as in Case I.

Case III – *Rod clamped at one end only.* Since this end must be a node and the free end an antinode, the length of the rod must be an odd multiple of $\lambda/4$. At the fixed end $x=0$, $\xi=0$, whence A in equation (3) is zero, whilst at the free end $x=l$, ξ is a maximum (equal to B) and $nl/c=s\pi/2$ where s is an odd integer. The partials are therefore represented by

$$N_s = \frac{s}{4l}\sqrt{\frac{E}{\rho}} \text{ (where } s=1, 3, 5, \text{ etc.)} \quad . \quad . \quad (6)$$

The frequencies given by equations (4), (5), (6) are all subject to the proviso that the diameter of the rod is always small compared with $\lambda/2$, and that the lateral bulging and shrinking may be neglected. These relations afford a convenient means of comparing velocities of sound in rods of different materials. Different rods, of the same fundamental frequency, have lengths proportional to the velocity of longitudinal waves in the material. Case II is of special importance in this respect, since it possesses a definite advantage over the other two methods of supporting the rod. When an unbalanced force acts at a 'fixed' point it is essential that this point should be very massive or rigid, otherwise slight motion occurs and the theoretical assumptions for a node are not fulfilled. If a rod is clamped at the midpoint, however, the two symmetrical portions react equally and in opposite directions at the midpoint. Consequently a very slight constraint is sufficient to fulfil the theoretical assumption that it is a point of zero displacement.

The frequency of longitudinal vibrations in bars is usually very high compared with that of the transverse vibrations, the ratio increasing rapidly as the dimensions of the cross-section diminish relative to the length. In the case of a wire (a bar of very small cross-section) under tension T, the ratio of frequencies (longitudinal to transverse) will be equal to the ratio of velocities, *i.e.* $\sqrt{E/\rho}$ to $\sqrt{T/\rho A}$. If the increase of length of the wire under tension T be ϵ, then $T=E\epsilon A$, and the transverse wave-velocity will be $\sqrt{E\epsilon/\rho}$. Hence the ratio of frequencies is $1/\sqrt{\epsilon}$, which is very great.

Excitation and Observation of Longitudinal Vibrations – Longitudinal vibrations may be excited in a bar in a number of different ways, the manner of excitation depending largely on the way the bar is supported and on other requirements. Bars of metal or wood, clamped at one end or at the midpoint, are set into vigorous vibration by the steady frictional drag of a resined cloth drawn with a moderate pressure along the bar towards an antinode. If dry, fine sand be sprinkled on the bar the positions of the nodes may readily be observed. Another method, more suitable for relatively short, stiff bars (as distinct from stretched wires), is to strike the end of the bar a blow with a hammer. This method of excitation usually results in a complex sound due to both transverse and longitudinal vibrations—the one or the other may be rapidly damped out by suitably clamping the bar. Biot, and subsequently Tyndall,* demonstrated to a large audience the stresses at the nodes of a vibrating bar of glass by means of polarised light. The analyser was adjusted to extinction when the bar was stationary, the light (passing through a node) being revived when the bar is set in vibration. Tyndall remarked on the fact that the light transmitted in the latter case is really intermittent. No such effects are observed when the light passes through an antinode where the stresses are negligible. A bar of transparent material vibrating longitudinally might therefore be used as a means of interrupting a beam of polarised light at a definite frequency determined by the length and material of the bar. This principle has recently been revived by Kerr Grant,† who passed a beam of light through a quartz piezo-electric oscillator (see p. 145) placed between crossed Nicol prisms. It was shown by means of a revolving mirror that the frequency of the flashes was double that of the quartz oscillations, which in a particular experiment reached 144×10^3 cycles/sec. Various interesting applications of the method as an optical time-standard were suggested.

Electrical Methods of Excitation – (a) *Electromagnetic*. Powerful longitudinal vibrations can be set up in bars of magnetic material by acting on them with currents of the resonating frequency. A small electromagnet (with a bundle of thin wire or a

* See Tyndall's *Sound*, pp. 168–170.

† Kerr Grant, *Nature*, p. 586, Oct. 22, 1927.

laminated core) mounted close to the end of a steel bar, clamped at the midpoint, serves the purpose very well. The magnet should preferably have a D.C. winding to produce a moderate flux through the free end of the bar, and an A.C. winding (with tuning condenser in parallel) coupled through an auxiliary coil to a valve oscillator. The fundamental and harmonics ($s=1, 3, 5$, etc.) can readily be obtained almost pure, *i.e.* any particular harmonic can be excited free from the others and free from any transverse vibration of the bar. The resonance is extraordinarily sharp (indicating extremely small damping forces in the bar), and careful tuning is required. A thin disc of soft iron firmly cemented to the end of a bar of non-magnetic material (*e.g.* glass) may similarly be employed to excite resonant vibrations.

(b) *Electrostatic* – J. H. Vincent * has succeeded in exciting longitudinal vibrations in bars by an electrostatic method. The plane end of the bar, clamped at the midpoint, is placed close to the plane surface of a massive block of metal, the opposing surfaces forming a condenser which is connected directly to the tuned circuit of an oscillating valve. The alternating attraction of the condenser 'plates' (the bar and the block) sets up oscillations in the bar which responds vigorously at its resonance frequencies. Bars of insulating material, such as fused silica, may be used in this manner if plated with a thin metallic coating.

(c) *Magnetostriction* – Vincent † has also employed a magnetostriction method to excite such vibrations in iron and nickel. In this case a coil carrying alternating current (from the valve oscillator) surrounds the bar. The alternations in magnetic field in the coil induce corresponding alternations in the length of the bar. The best effects are obtained when the bar is initially magnetised by means of a second coil carrying direct current. At resonance the oscillations set up may attain a considerable amplitude. If the exciting coil forms the inductance in the tuned plate-circuit of the valve, and a galvanometer is inserted in the grid circuit containing a coil coupled to the plate coil, the variations in the deflections of the grid galvanometer give definite indications of resonance in the bar. This method is, of course, particularly useful at supersonic frequencies. A comprehensive account and bibliography of magnetostriction phenomena is given by S. R. Williams,‡ in which it is shown that longitudinal, trans-

* J. H. Vincent, *Nature*, p. 952, Dec. 31, 1927.
† *Electrician*, Jan. 1929.
‡ *Journ. Opt. Soc. Amer.*, **14**, p. 383, May 1927.

verse, and torsional effects may be excited. G. W. Pierce * and J. H. Vincent † have employed the longitudinal vibrations of magnetostrictive bars to produce and control electrical oscillations over a frequency range from a few hundreds to more than 300,000 p.p.s. The method involves the interaction of a magnetostrictive bar and the electrical oscillations of a valve-circuit. The constancy of frequency obtained compares favourably with piezo-electric crystal oscillators. Pure nickel and alloys approximating to 36 per cent. Ni and 64 per cent. iron (stoic metal) are stated by Pierce to give very good effects. Nichrome (Ni, Cr, and Fe) and Monel metal (64 per cent. Ni, 28 per cent. Cu+Fe, Si, Mn) also form very powerful oscillators. In all cases it is desirable to use an auxiliary D.C. magnetising coil to polarise the bar magnetically. Pierce shows that the motional impedance circle diagrams (see p. 73) of magnetostriction vibrators indicate a very high efficiency of conversion of electrical energy into mechanical motion. When solid bars are used the diameter of the motion impedance circle dips considerably, indicating hysteresis and eddy-current losses at high frequencies, but these losses almost disappear when the magnetostriction oscillator is a thin tube split longitudinally. Such oscillators are convenient sources of high frequency (super sonic) vibrations. We shall refer later (p. 262) to the application of this method to the measurement of the velocity of high-frequency vibrations in bars of magnetostrictive material.

High-Frequency Vibrations in Piezo-Electric Crystals

The principal phenomena of piezo-electricity are now fairly well known, numerous papers having appeared within recent years dealing with various aspects of the subject. On account of the importance which it has attained, it may not be out of place to refer briefly to the more important features before dealing with the applications in sound.

Piezo Electricity – The discovery of the electrical effects of pressure on asymmetric crystals may be ascribed to J. and P. Curie.‡ They found that certain crystals (particularly the hemihedral variety with inclined faces), when subjected to pressure, exhibited electrical charges on their faces. The converse effect

* *Amer. Acad. Proc.*, **63**, 1, p. 1, April 1928.

† *Electrician*, **101**, p. 729, Dec. 1928 ; and **102**, p. 11, Jan. 1929.

‡ J. and P. Curie, *Comptes rendus*, p. 204, July 1880, and other papers ; see also *Œuvres de Pierre Curie*, Gauthier Villars ; and article on 'Piezo Electricity,' *Dictionary of Applied Physics*, **2**, 'Electricity,' p. 598.

was also shown to be true—that voltages applied to the faces of the crystal produced corresponding changes in dimensions. The effects were best observed in slices cut in certain directions from the natural crystal. The phenomena may be observed in a large number of crystals,* but the best known examples are quartz, tourmaline, and Rochelle salt. Of these, Rochelle salt gives by far the largest piezo-electric effect, but on account of the inferiority of its mechanical properties its practical importance relative to quartz and tourmaline is greatly diminished.

Referring to fig. 45, which represents a transverse section of a quartz crystal cut in a plane perpendicular to the optic axis, the dotted lines E_1E_1, E_2E_2, E_3E_3 are the three electric axes parallel to the bounding faces of the crystal. Plates or bars are cut from the crystal having the length l perpendicular to one of these axes, *e.g.* E_2E_2 in fig. 45, the breadth b parallel to the optic axis, and the thickness t parallel to the electric axis. When such a plate is placed in an electric field, in the direction of the thickness, its length l and thickness t change —the length increasing and the thickness diminishing, or *vice versa*. The two strains are such that there is no change of volume in the crystal.

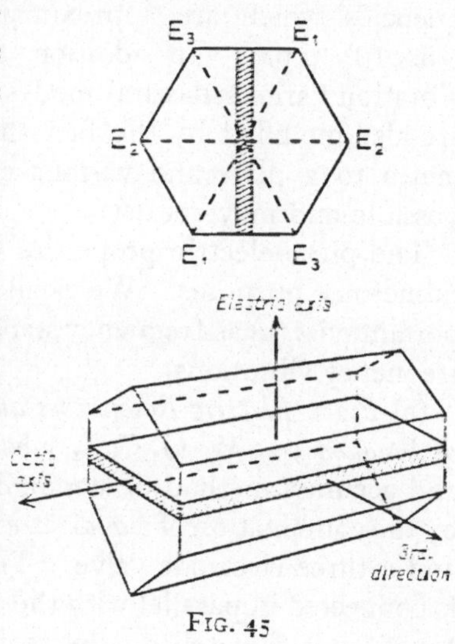

FIG. 45

Resonant Vibrations of Crystal Slices – It will be evident, when alternating voltages are applied to the crystal slice, there will be corresponding alternating stresses and strains in its length and thickness. When the frequency of such alternating voltage coincides with one of the possible modes of vibration of the slice a large resonant vibration will occur. Such vibrations may truly be regarded as longitudinal, whether they are attributable either to the thickness or to the length of the slice.

Even with the largest crystals of quartz obtainable the dimensions of such slices must necessarily be small and the natural

* See article by F. Pockels, *Winkelmann's Handbuch der Physik*, 4, pp. 766–795.

vibrations correspondingly high. Possible frequencies of such a bar will be

$$N=\frac{s}{2l}\sqrt{\frac{E}{\rho}} \quad \text{or} \quad =\frac{s}{2t}\sqrt{\frac{E}{\rho}} \qquad (s=1, 2, 3, \text{etc.}).$$

The value of E/ρ will, of course, depend on the particular axis chosen—for the elasticity varies with direction in a crystal. The difference in the 'l' and 't' axes is, however, very small, and an average value of $c=\sqrt{E/\rho}$ may be taken as $5{\cdot}5\times10^5$ cm./sec. A rod 5 cm. long and 0·5 cm. thick will consequently have frequencies which are approximately multiples of $5{\cdot}5\times10^4$ and $5{\cdot}5\times10^5$ p.p.s. In addition to such longitudinal modes of vibration various flexural modes are possible.* The frequencies are also modified by the fact that the 'bar' of quartz approximates to a plate and various complex modes of vibration are possible and may coexist.

The piezo-electric properties of crystals have been applied to numerous purposes. We shall consider two of the most important, viz. (*a*) as frequency standards, and (*b*) as sources of high-frequency vibrations.

(a) *Piezo-Electric Resonators and Oscillators. Quartz Standards of Frequency* – W. G. Cady † has devised an extremely ingenious and accurate method of standardising high-frequency oscillations by the combination of a *quartz resonator* (of the type just described) and a three-electrode valve. The 'resonator' with its electrodes is connected in parallel with the condenser of an oscillatory circuit tuned approximately to the required resonance frequency of the quartz. A current-measuring device in series with the inductance coil of the oscillatory circuit indicates a sudden change in current at a frequency corresponding to resonance in the quartz. This change is due to a characteristic variation in the effective capacity and resistance of the resonator as the frequency passes through resonance. As a typical example Cady cites a quartz resonator $30\times4\times1{\cdot}4$ m.m.s. vibrating longitudinally in the direction of its length. The fundamental frequency was found to be 89,870 cycles/sec. The effective capacity, normally $4{\cdot}5\times10^{-12}$ farads, becomes $42\ 2\times10^{-12}$ at 89,860, decreasing to the normal value at 89,870 and becoming $-32{\cdot}2\times10^{-12}$ at 89,880 cycles/sec. Further increase of frequency returns the capacity

* See J. R. Harrison, *Proc. Inst. Radio Eng.*, p. 1040, etc., Dec. 1927.

† *Proc. Inst. Rad. Eng.*, **10**, p. 83, 1922; *Phys. Rev.*, **19**, p. 1, 1922; *Journ. Optical Soc. Amer.*, **10**, No. 4, April 1925.

gradually to its normal value again. The changes of oscillating current and effective capacity are shown in fig. 46. It will be seen that the current descends to an extremely sharp minimum at resonance. A variation in frequency of one part in 10^4 produces a marked change of current. The logarithmic decrement of the resonator is only 0·00066 per cycle. The effectiveness of the resonator as a frequency-standard is due entirely to these very

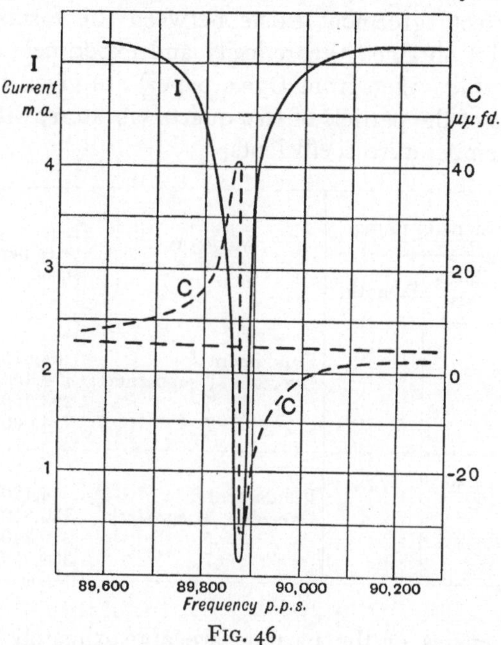

FIG. 46

rapid changes in effective capacity and resistance over a range of frequency comprising only a small number of cycles near resonance. Cady found also that the temperature-coefficient of the longitudinal vibrations (in the direction of the length) was only about 5 in 10^6 per degree C.—in the direction of the thickness (electric axis) this becomes 20 in 10^6. Cady also constructed compound resonators of quartz and steel vibrating at lower frequencies, of the order 10^4 cycles per second. D. W. Dye * has developed the theory of the quartz resonator, using as a basis a theorem due to S. Butterworth,† in which the mechanical quartz

* *Proc. Phys. Soc.*, **38,** p. 399, Aug. 1926.

† *Proc. Phys. Soc.*, **271,** p. 410, 1915. See also Giebe and Schiebe, *Zeits. f. Physik*, **33,** p. 335, 1925; Mallett and Terry, *Wireless World*, **16,** p. 630, 1925; and Schiebe, *Zeits. f. Hochfrequenztechn*, **28,** 1, p. 15, 1926.

system is replaced by an equivalent electrical circuit containing a capacity C, inductance L, and resistance R in series all shunted by a capacity C_1. In a particular case of a quartz bar $7{\cdot}0 \times 0{\cdot}6 \times 0{\cdot}15$ cm., Dye found experimentally that $C = 0{\cdot}08\ \mu.\mu$ fd., $R = 1500$ ohms, $L = 160$ henries, $C_1 = 8\ \mu.\mu$ fd., $2\pi N = 275{,}000$, giving a log. dec. 10^{-4}, and the diameter of the motional impedance circle (see p. 74) no less than 100 megohms. On such a basis almost perfect agreement exists between the forms of current curve (Cady) obtained theoretically and experimentally. In the following table (taken from Dye's paper) are given the observed frequencies of the partials of two quartz vibrators, with the corresponding temperature-coefficients.

	Bar dimensions in m.ms.			Mode of vibration.	Frequency cycles per second.	Temperature-coefficient, parts in a million.
	Length.	Thick-ness.	Breadth.			
A	47·0	3·3	7·50	Fundamental	58,675	$-1{\cdot}0_2$
				Overtone, 3 segments	174,450	$-3{\cdot}0_0$
				,, 5 ,,	293,750	-49
				,, 7 ,,	414,000	-70
B	62·2	1·52	7·54	Fundamental	44,120	$-4{\cdot}0_0$
				Overtone, 3 segments	130,850	$-15{\cdot}0_3$
				5 ,,	219,840	$-5{\cdot}0_2$
				,, 7 ,,	304,650	-66

The frequencies of the partials are approximately in the ratio 1, 3, 5, 7, as would be anticipated from the simple theory. The agreement is not exact, of course, for bars of such cross-sections and lengths are not likely to conform with the simplifying assumptions (see p. 135) that all cross-sections vibrate in the direction of the axis of the bar. Transverse stresses and strains must necessarily exist, these becoming more important the shorter the vibrating segments, *i.e.* the higher the partial. This deduction is confirmed by the values of the temperature-coefficients. The mean temperature-coefficient of frequency for a quartz bar vibrating longitudinally was found to be -5×10^{-6} per 1° C. rise of temperature, whereas for a plate vibrating in the direction of its thickness (the electric axis) the coefficient lies between -30 and -70×10^{-6}. The higher temperature-coefficients for the higher partials therefore suggest motion in the direction of the electric axis as well as longitudinally.

Dye found slight variations of frequency according to the method of mounting and exciting the quartz resonators, but the reader should consult the original paper for further details.

Quartz Oscillators as Frequency Stabilisers – The work of Cady and Dye which we have just considered deals with the quartz *resonator*, which gives a *transient* response as the electrical frequency passes through the resonant frequency of the quartz bar. Cady also devised an oscillator for *maintaining* the quartz at its resonant frequency. A more convenient arrangement has, however, been devised by G. W. Pierce.* Fig. 47 represents a similar circuit in which the quartz 'oscillator' Q reacts in such a manner on the valve circuit that its oscillations are maintained. The writer has found it desirable, for satisfactory working, that the natural frequency of the tuned LC circuit should be slightly *lower* than that crystal mode it is desired to excite. The commencement of oscillations is marked by a sudden drop in the plate current of the valve. The frequency of these oscillations is determined by the frequency of the quartz alone, being practically independent of the variables in the valve circuit. The arrangement may be regarded either as an electrically maintained mechanical vibrator emitting sound waves of a definite frequency, or as a mechanically controlled valve oscillator generating alternating current of this frequency. We shall refer later (see p. 258) to the application of this device in measuring sound-velocity in gases.

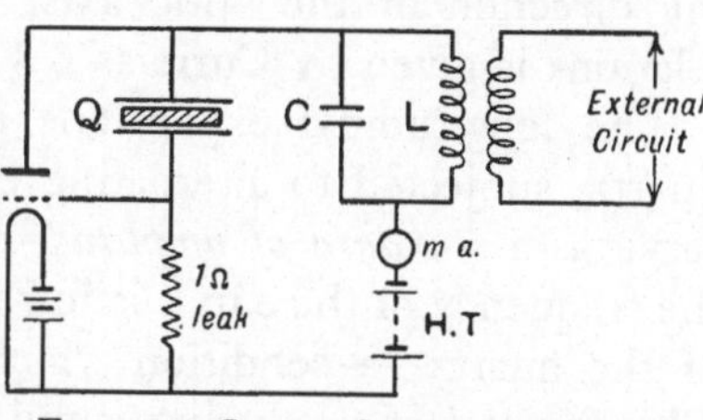

FIG. 47—Quartz Oscillator (Pierce)

(b) *Piezo-Electric Sources of Sound* – The Curies showed that the quantity of electricity q liberated when a total force 'f' is applied to a quartz crystal slice is given by

$$q_1 = \kappa f$$

when the force is applied in the direction of the electric axis,

$$q_2 = 0$$

when the force is applied along the optic axis, and

$$q_3 = -\kappa l f / t$$

* G. W. Pierce, *Proc. Amer. Acad.*, **59**, No. 4, Oct. 1923; and **60**, No. 5, Oct. 1925.

when applied in the third direction (*i.e.* the length, at right angles to both electric and optic axes). Conversely, it was shown that the application of a voltage V to the faces at the ends of the electric axis resulted in changes of length

$$\delta_1 = \kappa V$$

in the direction of the electric axis, and

$$\delta_3 = -\kappa l V/t$$

in the third direction—no change of length being observable in the direction of the optic axis. The constant κ of the above relations is given by Curie as $6 \cdot 3 \times 10^{-8}$ C.G.S. units of charge.

The longitudinal expansions and contractions of a strip of quartz, subjected to alternating e.m.f. of known magnitude may serve as a *standard of amplitude* at audible frequencies provided the frequency of the e.m.f. is low relative to the natural frequency of the quartz—a condition which is easily attained in practice. This method of producing sound vibrations of definite amplitude has been employed in determinations of the minimum amplitude audible at various frequencies. It provides a convenient means also of calibrating sound receivers in terms of known amplitudes of vibration. *Piezo-electric oscillographs* * have been constructed, in which use is made of the alternating expansion and contraction of a strip of quartz or Rochelle salt when subjected to alternating e.m.f.'s. The small displacements of the free end of the strip are mechanically and optically amplified, and photographically recorded.

The piezo-electric properties of quartz and Rochelle salt have been applied to the production of *high-frequency sound transmitters and receivers*. Thus M. Langevin † has employed a slab of quartz, whose faces are perpendicular to the electric axis, as a transmitter and receiver of supersonic vibrations under water—the apparatus being used for depth sounding and the detection of submerged bodies.‡ The arrangement of the Langevin apparatus is illustrated diagrammatically in fig. 48; '*a*' is an assemblage of thin slices of quartz crystal placed between two metal plates '*bb*,' of

* See papers by C. E. Wynn Williams, *Phil. Mag.*, p. 289, Feb. 1925; and A. B. Wood, *Phil. Mag.*, p. 631, Sept. 1925.

† See article on 'Submarine Listening' by A. Troller (*La Nature*, pp. 4 and 12, Jan. 8, 1921); and on 'Echo Depth Sounding' by P. Marti (*La Nature*, Aug. 20, 1921).

‡ See also 'Echo Sounding,' *Nature*, May 9, 1925, pp. 689–690; French Patent No. 505,903 (1918) and British Patent No. 145,691 (1920), Langevin; also *Hydrographic Review* (Monaco), No. 3, 1924; and No. 14, 1926.

equal thickness and calculated according to the frequency of the sound to be emitted. One side of this arrangement is in contact with the sea, the other being shielded by means of the watertight box '*c*.' The quartz-steel 'sandwich' is operated from a source of high-frequency energy—a power-valve oscillator—tuned to

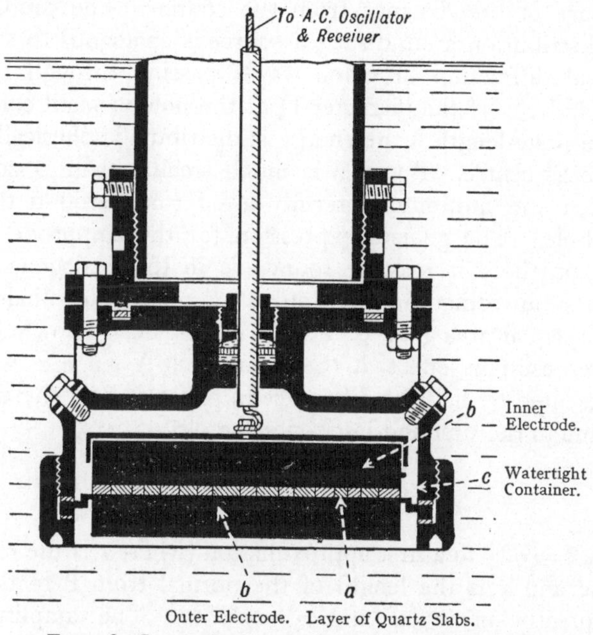

FIG. 48—Langevin Quartz Supersonic Transmitter

the frequency, say 50,000 cycles/sec. of the transmitter. The same device is used to receive the 'echo,' the H.F. alternating pressures falling on the quartz slabs producing corresponding e.m.f.'s. The electrical circuits required for receiving are similar to those used in radiotelephony, the heterodyne (or 'beat') principle being used to render the supersonic echo audible.

Carbon microphones and piezo-electric crystals of Rochelle salt may also be used as detectors of high-frequency sounds if suitably mounted in resonant structures.*

Sound Radiation from a High-Frequency Piston – Langevin's quartz disc transmitter depends for its action on the longitudinal vibrations set up in the direction of the *thickness* of the slab by virtue of its piezo-electric properties. The whole face of the slab vibrates with the same amplitude and phase and behaves in every

* See I. B. Crandall, *Vibrating Systems and Sound*, pp. 138 and 142; Rayleigh, **2**, p. 138 and pp. 162–165.

respect like a *piston* oscillating to and fro at a very high frequency. If the piston has a diameter large compared with a wave-length λ of the sound emitted into the surrounding medium, the waves set up will be approximately plane near the piston, becoming divergent at greater distances, the intensity ultimately varying inversely as the square of the distance from the source. The problem of sound-distribution around such a source is analogous to the case of optical diffraction of plane waves passing through a small circular hole.* If the diameter D of the hole is small compared with the wave-length λ the energy is distributed spherically with the hole as centre. But if λ is much smaller than D zones of maximum and minimum intensity are formed around the axis of the hole. The general expression for the amplitude at any point P on the axis may be found, as in the analogous optical case, by subdividing the disc into a number m of 'half-wave' zones of equal area (see p. 300). It has been shown (p. 17) that the resultant effect at the point P is A sin α/α, where α is represented by half the difference in phase at P of the disturbances due to the first and last zones, *i.e.*

$$\alpha=\frac{1}{2}\cdot\frac{m\lambda}{2}\times\frac{2\pi}{\lambda}=\frac{\pi m}{2}.$$

But $m\lambda/2 \doteq r^2/2x$ to a first approximation (where r is the radius of the disc and x is the length of the normal from P to the disc. Consequently $m=r^2/\lambda x$ and $\alpha=\pi r^2/2\lambda x$. The amplitude A, calculated on the supposition that the disturbances all reach P in phase is $\propto \pi r^2/\lambda x$, *i.e.* the area of the disc divided by λx. The resultant amplitude A_p at the point P on the axis is therefore

$$A_p \propto \frac{A\sin\alpha}{\alpha} \propto \frac{\pi r^2}{\lambda x}\cdot\frac{\sin\left(\frac{\pi}{2}\cdot\frac{r^2}{\lambda x}\right)}{\frac{\pi}{2}\cdot\frac{r^2}{\lambda x}},$$

i.e.

$$A_p \propto 2\sin\left(\frac{\pi}{2}\cdot\frac{r^2}{\lambda x}\right),$$

the intensity being, of course, proportional to the square of this quantity. Up to the point $x=r^2/\lambda$ there will be an approximately parallel beam of radiation, having a succession of maxima and minima of intensity due to interference of rays from the different

* See Schuster's *Optics*, p. 107; also S. Smith, *Trans. Roy. Soc. Canada*, **3**, 1923, p. 173, and 1924, p. 227.

zones into which the disc may be divided. When x is very large (and $\sin \alpha = \alpha$ app.) the intensity, for points along the axis, will be proportional to $\frac{\pi^2 r^4}{\lambda^2 x^2}$—the beam in this case diverging in the form of a cone. As in the optical case, the angle of diffraction θ for the first minimum of intensity is

$$\theta = \sin^{-1} 0{\cdot}61\lambda/r,$$

this representing (for values of x large compared with r) the semi-angle of the beam. The solid angle of the conical beam approximates to

$$\Omega = 0{\cdot}37\pi\lambda^2/r^2,$$

i.e. the divergence of the beam varies directly as the square of the wave-length and inversely as the area of the disc.

When the point P is not on the axis of the vibrating disc, the zones are not concentric with the aperture and the calculation becomes more difficult. The problem may be solved analytically by means of the fundamental equation for the velocity-potential ϕ at any point P (see equation (32), p. 62 *et seq.*), applying it to elementary areas of the disc and integrating over the whole area. The solution has been obtained in this manner by Crandall.* The result may also be derived graphically as in optical treatises on the passage of plane waves through a circular aperture. If the disc comprises only a few zones, the axis is surrounded by diffraction rings of alternate maxima and minima of intensity. Such curves of amplitudes (A $\sin \alpha/\alpha$) and intensities ($A^2 \sin^2 \alpha/\alpha^2$) are shown in fig. 49.

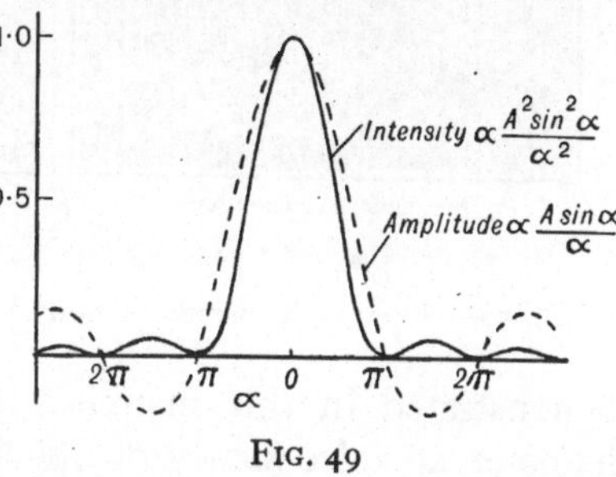

FIG. 49

The sound beam radiated from the disc therefore consists of a primary maximum and a number of subsidiary maxima. If the disc has a large area, most of the energy lies in the primary beam. When the radius r of the disc is small compared with the wave-length λ the primary beam is wide—approximating to the non-directional case.

This directional property of a large vibrating quartz piston is

* Crandall, *loc. cit.*, p. 137 ; Rayleigh, 2, 302 ; see also H. Stenzel, 'Directional Properties of Sound Sources' (*Elecktrische Nachrichten Technik.*, Bd. 4, Heft 6, pp. 239–253, 1927, and p. 165, May 1929).

utilised by Langevin * and R. W. Boyle † in locating marine obstacles, such as ships, icebergs, rocky reefs, etc., and in depth sounding. Boyle, Lehmann, and Reid (*loc. cit.*) have made visible the track of high-frequency beams from such a transmitter by means of powdered coke particles sprinkled in the track of the beam through water (see fig. 50). The primary and subsidiary beams are clearly shown. The reflection of the beam is also

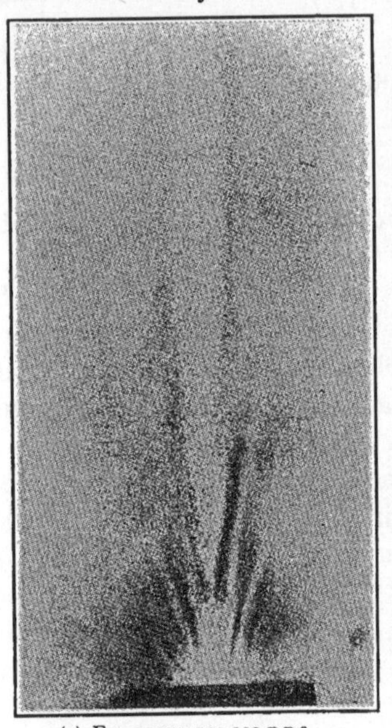

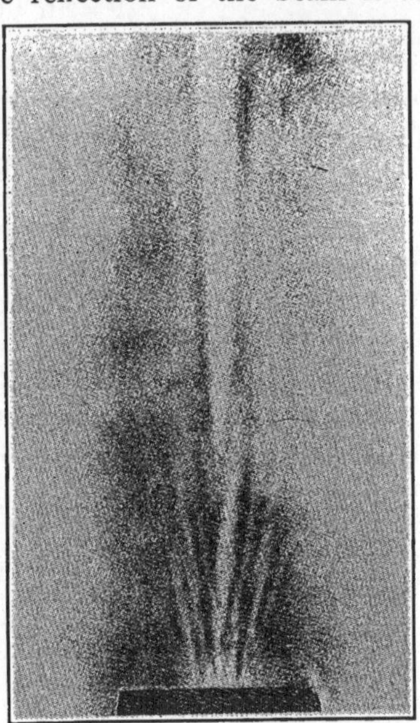

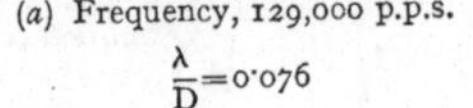

(*a*) Frequency, 129,000 p.p.s. $\frac{\lambda}{D}=0{\cdot}076$

(*b*) Frequency, 178,000 p.p.s. $\frac{\lambda}{D}=0{\cdot}055$

FIG. 50—Supersonic Beams of Sound—showing Primary and Secondaries (Dr. R. W. Boyle)

demonstrated in this manner. Using a transmitter 15·3 cm. diameter at a frequency of 135,000 cycles/sec., the wave-length in water being therefore 1·11 cm. (since $c=1{\cdot}5\times10^5$ cm./sec.), they found that the semi-angle θ of the primary beam was approximately 5° (which agrees with the theoretical relation $\theta=\sin^{-1} 0{\cdot}61\lambda/r$ given above). A high-frequency source of this nature makes it possible to investigate experimentally many problems in 'wave-physics' which cannot be undertaken either with light or with ordinary (audible) sound.

* Langevin, *loc. cit.*

† R. W. Boyle and C. D. Reid, 'Detection of Icebergs by an Ultra-Sonic Beam' (*Trans. Roy. Soc. Canada*, 3rd ser., **20,** sect. iii, p. 233, etc., 1926).

N. W. M'Lachlan* has also dealt with the directional properties of a rigid disc vibrating axially, and has applied his theoretical deductions to the case of a 'loud-speaker' emitting sounds of different frequencies. As the theory indicates, the primary beam becomes more and more sharply directional as the frequency of the sound and the diameter of the disc increase. Thus the 'rigid' cone of a loud-speaker 10 cm. radius may be regarded as non-directional at a frequency of 256 p.p.s., but sharply directional at 4096 p.p.s., frequencies which lie within the range of speech and music.

R. W. Wood and A. L. Loomis,† using a piezo-electric quartz 'oscillator' vibrating with large amplitudes at frequencies of the order 5×10^5 cycles/sec., have obtained some striking results of a physical and biological nature. In certain experiments with a crystal vibrating under oil, they estimated the pressure of the sound-radiation to be equal to 150 grams weight—capable of raising the free surface of the oil into a mound 7 cm. high. Maximum effects were observed when the distance from the quartz to the 'radiometer' disc was a whole number of half wave-lengths. The method of generation of the oscillations was essentially that developed by Langevin.

The uses of Rochelle salt as a sound transmitter and receiver have been demonstrated by Nicholson.‡ In one of the applications, the crystal is attached to a special needle and serves as a gramophone transmitter. The mechanical stresses set up in the crystal as the needle passes over the record produce corresponding electrical effects which may be amplified to operate 'loud-speakers' or telephone receivers. Various commercial applications of the piezo-electric properties of Rochelle salt are dealt with in a 'popular' paper by Russell and Cotton.§ The piezo-electric and mechanical properties of Rochelle salt have been studied by Valasek ‖ and by W. Mandell.¶ The latter shows that the elasticity of Rochelle salt is very small, the crystals consequently having low natural frequencies of vibration. Young's modulus for Rochelle salt is of the order 3×10^8 as compared with 8×10^{11} for quartz and 2×10^{12} for steel.

* *Proc. Roy. Soc.*, A, **122**, p. 604, Feb. 1929; and *Phil Mag.*, **7**, p. 1011, June 1929.

† R. W. Wood and A. L. Loomis, *Phil. Mag.*, **4**, pp. 417–436, Sept. 1927.

‡ Nicholson, *Amer. Inst. Elect. Eng. Proc.*, **38**, p. 1315, Nov. 1919.

§ Russell and Cotton, *Elect. Review*, **92**, No. 2361, Feb. 23, 1923.

‖ Valasek, *Phys. Rev.*, **17**, pp. 475–481, April 1921; *Phys. Rev.*, **20**, pp. 639–664, Dec. 1922.

¶ W. Mandell, *Roy. Soc. Proc.*, A, **116**, p. 623, etc., 1927.

TORSIONAL VIBRATIONS OF BARS

A solid bar or a tube of circular section may be twisted by suitable forces in such a way that each transverse section remains in its own plane. If the section is not circular, however, a warping of the various layers is liable to occur and the analysis becomes very complicated. We shall, in what follows, deal only with the case of bars of circular section, where there is no motion parallel to the axis of the bar. Provided we use the appropriate elasticity coefficient (the rigidity, in this case) the problem is analogous to the case of longitudinal vibrations. The force with which the bar resists a twist depends on 'μ,' the rigidity coefficient of the material (denoting the ratio of stress/strain in a pure shear). The relations between rigidity μ, Young's modulus E, the bulk modulus κ, and Poisson's ratio σ, are

$$\mu=\frac{E}{2(\sigma+1)} \quad \text{and} \quad \kappa=\frac{\mu E}{9\mu-3E} \quad . \quad . \quad (1)$$

The rigidity μ must lie between E/2 and E/3, since σ lies between 0 and $\frac{1}{2}$.

Following Rayleigh's procedure, consider a bar in the form of a thin tube of radius r and thickness δr, and let θ denote the angular displacement of any section distant x from the origin. The 'shear' of the material of the tube is $rd\theta/dx$. The opposing elastic force per unit area is $\mu r \, . \, d\theta/dx$, and since the area of section of the tube is $2\pi r \, . \, \delta r$, the moment round the axis is

$$2\mu\pi r^3 \, \delta r \, . \, d\theta/dx,$$

and the restoring force acting on the slice of thickness δx has the moment

$$2\mu\pi r^3\delta r \, . \, \delta x \, . \, d^2\theta/dx^2.$$

Now the moment of inertia of the slice is $2\pi r\delta r \, . \, \delta x \, . \, \rho r^2$, whence the equation of motion is

$$\rho\frac{d^2\theta}{dt^2}=\mu\frac{d^2\theta}{dx^2} \quad . \quad . \quad . \quad . \quad (2)$$

and this being independent of r, is equally applicable to tubes of all radii or to a solid bar. The velocity of torsional wave transmission is

$$c=\sqrt{\mu/\rho} \quad . \quad . \quad . \quad . \quad (3)$$

i.e. $\sqrt{E} : \sqrt{\mu}$ or $\sqrt{2(\sigma+1)} : 1$ relative to the velocity of longitudinal vibrations. Taking Poisson's ratio σ as 0·25 for a steel bar, the ratio of longitudinal to torsional velocities becomes 1·58. Consequently the ratio of frequencies of longitudinal and torsional vibrations for the same steel bar will have this value also. The frequency of the fundamental, midpoint clamped, will be

$$N=\frac{1}{2l}\sqrt{\frac{\mu}{\rho}} \quad . \quad . \quad . \quad . \quad . \quad (4)$$

Nodes and antinodes for the various partial tones are formed as in the case of longitudinal vibrations.

Torsional vibrations may be set up in solid bars by any convenient means of applying tangential forces to the free end, *e.g.* by 'screwing' a resined leather, pressed tightly round the bar. If the latter is rectangular in section, it is more convenient to 'bow' the opposite sides simultaneously in opposite directions. By comparing the frequencies of the longitudinal and torsional vibrations for the same bar the value of Poisson's ratio σ may be determined. A. O. Rankine and F. B. Young * made interesting observations on the various modes of vibration (torsional, transverse, and longitudinal) of a mild steel bar, observing the alternating magnetic effects produced by the alternating stresses set up in the bar on mechanical excitation. A short brass lever attached rigidly at right angles to one end of the steel bar could be struck by a lead hammer so as to throw the steel bar into torsional, longitudinal, or transverse vibration. Observations with a small search coil, connected to a three-valve amplifier and telephones, indicated frequencies characterising the various possible modes of vibration in the bar.

Piezo-Electric Effects produced by Torsion – Reference has just been made to the use of a Rochelle-salt piezo-electric crystal in a gramophone transmitter. Nicholson † has shown in the case of Rochelle salt crystals of composite structure that a given force produces the greatest piezo-electric effect when it is applied to produce torsion around its principal axis, and advises this method of use for producing electrical or mechanical results. The piezo-electric constant of Rochelle salt subjected to shearing forces in this manner is greater than 100 times the value for quartz subjected to forces of compression.

* Rankine and Young, *Phil. Mag.*, **38,** pp. 528–532, Nov. 1919.
† Nicholson, *loc. cit.*

VIBRATIONS OF MEMBRANES, DIAPHRAGMS, PLATES, AND BELLS

The transverse vibrations of stretched membranes are related to those of diaphragms and plates in a manner analogous to the transverse vibrations of stretched strings and of elastic bars. In the former case the vibrations are conditioned by the applied tension and are independent of elastic forces, whereas in the latter the elastic forces are all-important and tension almost negligible.

Membranes – The theoretical membrane is assumed to be a perfectly flexible, uniform, and infinitesimally thick solid lamina, stretched in all directions by a force which is unaffected by the motion of the membrane. In all cases of practical importance the membrane may be regarded as having a fixed boundary. Denoting the tension by T, and displacements by ξ in a direction at right angles to the plane xy of the membrane, it may be shown by analogy with the case of wave-motion in one dimension (*i.e.* wave-motion on a string) that

$$m\frac{d^2\xi}{dt^2}=\mathrm{T}\left(\frac{d^2\xi}{dx^2}+\frac{d^2\xi}{dy^2}\right) \quad . \quad . \quad . \quad (1)$$

where m is the surface density (mass per unit area $=\rho\times$thickness) of the membrane. The velocity of wave-motion is $c=\sqrt{\mathrm{T}/m}$ as in the case of strings.

To obtain the modes of vibration of a membrane * with a fixed boundary, assume $\xi=a\cos nt$, so that equation (1) becomes

$$\left.\begin{aligned}&\frac{d^2\xi}{dx^2}+\frac{d^2\xi}{dy^2}+k^2\xi=0\\ \text{where}\quad &k^2=n^2m/\mathrm{T}=n^2\rho t/\mathrm{T}\end{aligned}\right\} \quad . \quad . \quad . \quad (2)$$

At a fixed edge $\xi=0$. It is found that the solution of (2), subject to this condition, is possible only for certain values of k, which determine the frequencies of the membrane.

The complete theory of stretched membranes of various forms (rectangular, square, circular, etc.) is mainly of mathematical interest. The reader is referred therefore to the treatises of Rayleigh and Lamb, where the problem is fully discussed. The vibrations of a circular membrane have, however, certain physical points of interest which find practical applications.

* See Lamb, *Sound*, p. 142.

Circular Membranes – From a consideration of forces acting across a circle of radius r, equation (2) for symmetrical modes of vibration * reduces to

$$\frac{d^2\xi}{dr^2}+\frac{1}{r}\frac{d\xi}{dr}+k^2\xi=0 \quad . \quad . \quad . \quad (3)$$

where

$$k^2=n^2\rho t/\mathrm{T}.$$

This leads to a solution of the form

$$\mathrm{J}_0(ka)=0 \quad . \quad . \quad . \quad . \quad (4)$$

[where a=radius of the membrane, and J_0 a Bessel's function of zero order of the first kind, whose roots are

$$ka/\pi=0{\cdot}7655, \quad 1{\cdot}7571, \quad 2{\cdot}7546, \quad 3{\cdot}7534 \ldots . \quad (5)$$

tending towards $(m-\frac{1}{4})$ where m is an integer].

The first of these roots corresponds to the fundamental mode of vibration, of frequency

$$\mathrm{N}=\frac{0{\cdot}765}{2a}\sqrt{\frac{\mathrm{T}}{\rho t}} \quad . \quad . \quad . \quad (6)$$

('a' being the radius of the membrane). In the mth mode there are $(m-1)$ nodal circles in addition to the edge. For the nodal circle $kr/\pi=0{\cdot}7655$, $r/a=0{\cdot}4356$. The partials do not form a harmonic series.

Membranes, approximating in some respects to the ideal, may be made of soap films stretched across a metal ring. The Hilger audiometer employs a stretched circular membrane of extremely thin collodion to record the vibrations of the air in contact with it. Methods of preparing thin tenacious films of collodion suitable for such a purpose have been described by H. Dewhurst.† The vibrations of these membranes, of thickness only a few wave-lengths of light, are examined by optical or photographic methods. In order to examine the effect of tension on a membrane a more robust lamina is usually employed. Thus a sheet of parchment may be stretched until it has a fundamental frequency of several hundreds per second. A thin steel membrane (0·002 inch thick) may be stretched, by applying symmetrical tension *via* a ring, until its natural frequency attains several thousand per second. Wente's ‡ condenser microphone has such a diaphragm, of fre-

* See chap. v, Lamb.

† *Proc. Phys. Soc.*, **39**, p. 39, 1926.

‡ *Phys. Rev.*, **10**, p. 39, 1917; **21**, p. 450, 1918; **19**, p. 498, 1922.

queney about 5000 cycles per second. Attempts to stretch the thin steel membrane still further usually result in disaster, the material passing the elastic limit at greater tensions. 'Loud-speaker' diaphragms have been constructed with highly tensioned aluminium alloy membranes of large area and moderate thickness, having a natural frequency such as to reinforce the lower range of the audible scale of frequencies.*

The various modes of vibration of a stretched steel membrane are very conveniently observed by employing electromagnetic methods of excitation. An electromagnet ('pot type') mounted suitably with respect to the membrane, and supplied with A.C. of variable frequency from a valve oscillator, will excite the fundamental and the various partials as required. The magnet system of an ordinary telephone receiver will serve the purpose. The nodal circles and diameters may readily be observed by sprinkling fine, dry sand (Chladni's method), which collects at the nodes of the vibrating membrane. Millar † has obtained similar results using a glass membrane excited by means of organ-pipes. The electromagnetic method is, however, probably more convenient and instructive on account of the wide range of continuous frequency-variation which is possible. It provides a convenient means also of studying the sharpness of resonance for the various partials. Generally speaking, it will be found that the agreement between theory and observation is only approximate. The ordinary theory takes no account of the large effect due to the damping and loading effect of the surrounding medium, and it is assumed that the stiffness of the membrane is negligible. These assumptions are not realised in practice.

Membranes as Sources of Sound. Directional Properties – The application of stretched membranes to various forms of musical instruments, drums, tambourines, etc., in addition to the uses as a sound receiver already mentioned, brings us to an important point.

In a drum, a membrane is stretched tightly (the tension being adjustable for tuning purposes) over an enclosed cavity, the walls of this cavity being relatively rigid. The air cavity resonates to the fundamental of the stretched membrane and thereby reinforces this note. The wave-length of the note in air will be of the order of 10 feet (at a frequency of 100, the velocity of sound in air being 1100 feet/sec.), whereas the diameter of the drum may be, say,

* Whitaker, *Journ. Sci. Instrs.*, **5**, p. 39, 1928.

† *Science of Musical Sounds*, pp. 151–153.

only 2 feet. Taking this into consideration, with the fact that only one side of the drum radiates sound into the air, the issuing sound will spread spherically. Suppose now the same membrane be stretched on a massive annular ring, permitting *both* sides to radiate sound to the surrounding air. An observer in a position 'edge-on' to the plane of the membrane will receive simultaneously vibrations from opposite sides of the membrane, in opposite phase,* and will consequently hear little or nothing. 'Broadside-on' to the membrane, however, the two sets of vibrations do not neutralise completely owing to a partial screening effect of the remoter vibration by the membrane and ring. As the vibrating membrane is rotated through 360° therefore, the observer will hear two distinct maxima (180° apart) separated by two corresponding minimum (or zero) positions. The maxima will be of smaller intensity than the maximum observed when one side of the membrane is completely screened (*i.e.* in the case of a drum).

This directional property of a membrane stretched on a ring may also be demonstrated in the converse case, when it is used, in conjunction with a small microphone attached to its centre, as a receiver of sound.† The microphone gives zero response when the membrane is 'edge-on' and maximum when 'broadside-on' to the source of sound.

Diaphragms

When the frequency of a membrane can no longer be determined without taking stiffness (or elasticity) into account, we approach the condition of a diaphragm in which stiffness is the primary factor and tension becomes of secondary importance, if not altogether negligible. The complete theory of a vibrating diaphragm is even more complex than that of a membrane. Only approximate solutions have been obtained.

Rayleigh ‡ determined the periods and motions of a thin circular plate (*in vacuo*) rigidly clamped at the edge, regarding the restoring forces involved in its vibration as purely elastic. By means of a series of Bessel functions he found that the fundamental frequency of vibration is given by

$$N=\frac{2{\cdot}96}{2\pi}\cdot\frac{h}{a^2}\sqrt{\frac{E}{\rho(1-\sigma^2)}} \quad . \quad . \quad . \quad (1)$$

* The membrane acting as a 'double source' of very small, or negligible, strength.

† See W. H. Bragg, *World of Sound*, p. 172 (Bell).

‡ Rayleigh, *Sound*, **1**, p. 366, etc.

where 'h' is the thickness, and 'a' the radius of the diaphragm, E Young's modulus, ρ the density, and σ Poisson's ratio for the material of the diaphragm. This value of the frequency is only slightly affected when the diaphragm vibrates in air, since the loading effect due to the added mass of air produces a negligible lowering of frequency. Lamb * has recently dealt mathematically with the problem of frequency and damping of thin circular diaphragms in air and in water, more particularly in relation to the reception and transmission of sound under water. A thin circular plate or diaphragm is regarded as filling an aperture in an infinite plane wall. To simplify the theoretical treatment the wall is supposed to be rigid and the diaphragm to be firmly clamped to it along the circumference, although these conditions are only imperfectly fulfilled in practice. Even with these limitations, a complete solution is not possible, but a sufficient approximation is obtained by Rayleigh's method of an "assumed type" † of displacement. It is known that the frequency thus obtained is slightly too high. The type of displacement assumed

$$\xi = A\left(1-\frac{r^2}{a^2}\right)^2 \quad . \quad . \quad . \quad . \qquad (2)$$

(where ξ is the normal displacement of the diaphragm at a distance r from the centre, and A the displacement at the centre, $r=0$), is justified by the fact that, in the case of the diaphragm vibrating *in vacuo*, the agreement between the calculated and the observed values is very good.

Diaphragm in Air – The frequency of the fundamental '*in vacuo*' is thus found to be almost exactly given by

$$N = 0{\cdot}4745\frac{h}{a^2}\sqrt{\frac{E}{\rho(1-\sigma^2)}} \quad . \quad . \quad . \qquad (3)$$

which agrees closely with Rayleigh's value in equation (1). Lamb also determines the frequencies of the higher partials, the numerical multiplier in (3) being replaced by 1·006 in the case of the vibration with one nodal diameter, and by 1·827 for one nodal circle. The partials are therefore not harmonic. Equation (3) may be written

$$N = 0{\cdot}4745\, hc/a^2,$$

where c is the velocity of elastic waves in an infinite thin plate of the same material and thickness.

* Lamb, *Roy. Soc. Proc.*, A, **98**, p. 205, 1920.

† See Rayleigh, *Sound*, **1**, p. 113, etc.

For an iron diaphragm, assuming

$$E=2\times10^{12},\quad \rho=7{\cdot}8,\quad \sigma=0{\cdot}28\quad \text{C.G.S. units,}$$

we find

$$c=5{\cdot}27\times10^{5}\ \text{cm./sec.,}$$

and equation (3) becomes

$$N=2{\cdot}50\times10^{5}\ h/a^{2} \qquad . \quad . \quad . \qquad (4)$$

Thus if $h=0{\cdot}1$ cm. and $a=5$ cm., the fundamental frequency *in vacuo* or in air will be 1000 cycles per second.

Added Load – Lamb estimates the effective mass of an unloaded diaphragm of total mass M as equivalent to a mass M/5 concentrated at the centre. Consequently a small additional mass m attached to the centre of the diaphragm may be regarded as a uniformly distributed load of mass $5m$. The frequency of a diaphragm (mass M) in air is therefore diminished by a load m (at the centre) in the ratio $(1+5m/M)^{-\frac{1}{2}}$, *i.e.*

$$N_{L}=0{\cdot}4745\frac{hc}{a^{2}}\cdot\frac{1}{\sqrt{1+5m/M}} \qquad . \quad . \qquad (5)$$

Diaphragm vibrating in Contact with Water or Other Medium – The presence of the medium (*e.g.* water) in contact with the diaphragm has two effects. Firstly, the frequency is lowered on account of the virtual increase of inertia of the diaphragm due to the added mass of water; and secondly, the vibrations are damped owing to the energy radiated into the water in the form of sound waves. Lamb has found, in the case of a diaphragm with one side only in contact with water, that the inertia of the diaphragm is increased in the ratio $(1+\beta)$ where

$$\beta=0.6689\frac{\rho_{1}}{\rho}\cdot\frac{a}{h},\text{*}$$

where ρ_1 is the density of the water. The frequency in this case therefore becomes

$$N_{w}=0{\cdot}4745\frac{hc}{a^{2}}\cdot\frac{1}{\sqrt{1+\beta}} \qquad . \quad . \quad . \qquad (6)$$

When *both* sides of the diaphragm are in contact with 'infinite'

* In arriving at this result Lamb assumes that the wave-length in the water is large compared with the diameter of the diaphragm.

water, the value of β in this expression for frequency must be doubled.

The frequency of a loaded diaphragm with water on one side only, and a mass m at the centre, is clearly

$$N_{L.w}=0{\cdot}4745\frac{hc}{a^2}\frac{1}{\sqrt{1+\beta+5m/M}} \quad . \quad . \quad (7)$$

The above relations for the frequency are, of course, equally applicable whatever the medium in contact with the diaphragm, provided the appropriate value of β is introduced. Thus, in the case of a thin diaphragm of large diameter vibrating in air, where the area of the vibrating surface is large and the mass of the diaphragm small, the added air-load may produce an appreciable lowering of frequency.

Damping – Equating the rate of decay of the energy of the vibrating diaphragm (plus the load due to the medium) to the energy emission as sound in the medium, Lamb finds that the decay-constant, in e^{-kt}, is given by

$$k=\frac{5}{36}\cdot\frac{\rho}{\rho_1}\cdot\frac{n^2a^2}{(1+\beta)hc_1} \quad . \quad . \quad . \quad (8)$$

(where $n=2\pi\times$ frequency, and $c_1=$ velocity of sound in the medium).

The *degree of persistence of the vibrations* is indicated by the number of vibrations which elapse whilst the amplitude diminishes in the ratio $1/e$, viz. $n/2\pi k$, whence from equations (6) and (8) we find

$$\text{Persistence}=\frac{n}{2\pi k}=0{\cdot}385(1+\beta)^{3/2}\frac{\rho c_1}{\rho_1 c} \quad . \quad . \quad (9)$$

Thus for a steel diaphragm 7 inches diameter and $\frac{1}{8}$ inch thick (frequency 1013 in air) vibrating in water $n/2\pi k=5{\cdot}16$. In spite of this small persistence, the vibrations are still far from being 'dead-beat.' The damping, though considerable, is not sufficient to affect the frequency appreciably, this influence being of the order k^2/n^2 (see p. 35), which is less than 1 in 1000 in this case. There is also a considerable degree of selective resonance to external sources of sound, when the diaphragm is used as a 'receiver.' The energy of resonance falls to one half the maximum when the frequency of the source differs from the natural frequency of the resonator by the fraction k/n approximately (see

p. 43). Thus in the above example, where the frequency (as modified by the water) is about 550, the range over which the energy will exceed half the maximum is confined between the limits 530 and 570 p.p.s. approximately.

The persistence of vibrations (measured in periods) will be greater, and the resonance will attain a greater maximum (and concentrated over a narrower range) the thinner the diaphragm. The vibrations increase in persistence the greater the load on the diaphragm, i.e. *the sharpness of tuning increases with the load.*

Experimental – Lamb's theoretical deductions relative to the frequency and damping of circular diaphragms have been verified experimentally by Powell and Roberts.* They obtained values in good agreement with theory for the frequencies in air and in water of diaphragms of various dimensions. The theoretical lowering of frequency produced by a water load and by a load concentrated at the centre of the diaphragm were also confirmed experimentally. The resonance frequencies of the various diaphragms were indicated by means of a small 'button' microphone attached to a boss on the centre of the diaphragm. In a further paper, Powell † describes experiments on the damping of such diaphragms in relation to their sensitiveness as receivers of under-water sounds. The values of resonance-amplitude and persistence of vibration were found to be consistent with the mathematical deductions of Lamb quoted above.

The vibrations of diaphragms in air have been studied extensively by A. E. Kenelly,‡ who has also, using the 'motional impedance' principle (see p. 73), developed methods of measuring the efficiency of telephone receivers in converting mechanical into electrical vibratory energy. Kenelly's method of studying the motion of a vibrating diaphragm is outlined on p. 222. The vibrations of a diaphragm may be excited in a number of different ways, *e.g.* by means of sound waves in tune with it, or electromagnetically. A novel method of excitation is described by C. W. Hewlett.§ The diaphragm (of non-magnetic metal) lies between two flat coils through which a constant direct current flows, in such a way as to produce a radial magnetic field in the diaphragm. Alternating current is superposed on the direct current, and circular currents are induced in the diaphragm, which

* J. H. Powell and J. H. T. Roberts, *Proc. Phys. Soc.*, **35**, April 15, 1923.

† J. H. Powell, *Proc. Phys. Soc.*, **37**, Feb. 15, 1925.

‡ See A. E. Kenelly, *Electrical Vibration Instruments*.

§ *Phys. Rev.*, **19**, p. 52, 1922.

behaves as the secondary of a transformer. The field due to these currents reacts on the steady field and the diaphragm is set in vibration with the frequency of the alternating current. Conversely, vibration of the diaphragm, as a receiver of sound waves,

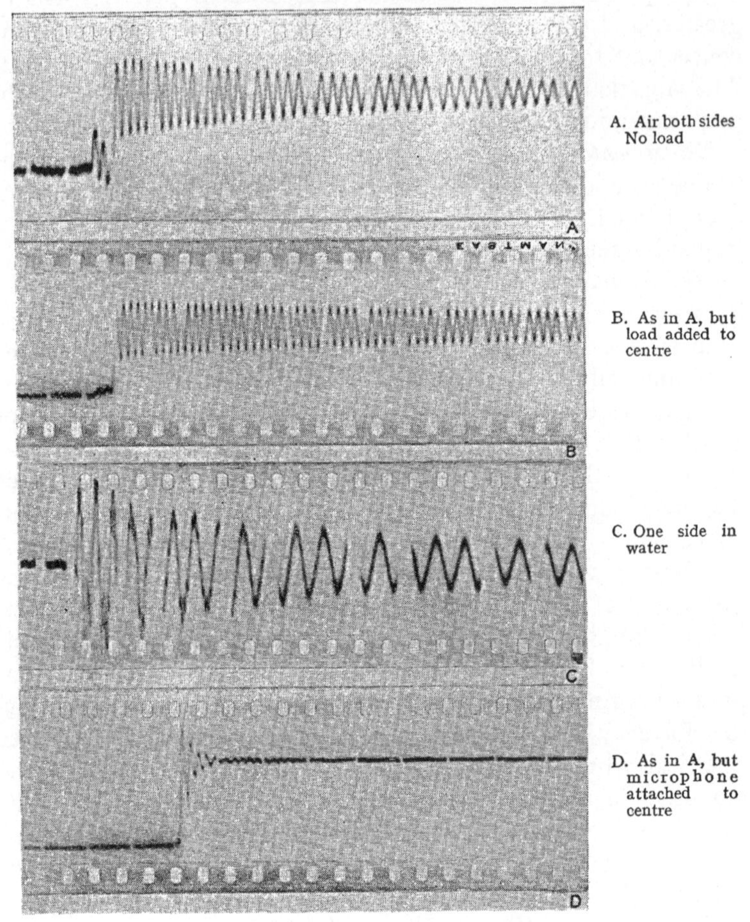

(Time Marks 0·01 second.)

Fig. 51—Damping of a Steel Diaphragm

reacts on the circuit and generates corresponding alternating currents. The arrangement therefore serves as a sound generator and receiver. It gives a pure tone of constant and measurable pitch and intensity over a wide range of frequencies. Kerr Grant * describes a method of exciting a diaphragm by means of

* *Proc. Phys. Soc.*, **34**, p. 104, 1922.

an air blast. The effect is based on the Bernouilli principle by which the diaphragm is attracted towards the jet from which the air issues.

The records shown in fig. 51 were obtained by the writer in an experimental study of the vibration and damping of diaphragms under different conditions. The effects of adding a load in reducing, and a microphone in increasing, the rate of decay of vibrations are illustrated in records B and D. The lowering of frequency produced by immersing one side of the diaphragm in water is shown in C as compared with A when both sides are in air.

L. V. King,* by applying a controlled air pressure to one side of a diaphragm, has succeeded in tuning it continuously over a moderate range of frequency. The method has application in the reception of signals of a definite frequency.

The *mean power radiated from a vibrating diaphragm* has been discussed on p. 63. *The mean amplitude* over the surface of a vibrating diaphragm, in its fundamental mode, is approximately one-third of the amplitude at the centre. A more accurate value may be derived from equation (2), p. 158.

Diaphragms provide one of the most convenient means of producing and receiving sounds. Numerous forms of diaphragm-type telephones and microphones are in daily use in the transmission and reproduction of speech sounds. Large diaphragms, *e.g.* in the Fessenden oscillator (see p. 390), are used as sources of sound-power, in the transmission of sound over large distances under water. A diaphragm operated electromagnetically at its fundamental resonance frequency may become a very efficient sound generator. The choice of diaphragm will depend of course on the purpose in view—the most important controlling factor being the medium through which the sound is to be transmitted. Thus a diaphragm which is suitable for transmitting sound of a particular frequency into air would be quite unsuitable for use under water and *vice versa.* The efficiency of a sound generator, driven electrically, is determined most simply from the analysis of its motional impedance over a range of frequencies including resonance. The method has already been described on pp. 73 to 75. Further reference will be made to the many applications of diaphragms as sound receivers and generators for use in air and under water.

* *Proc. Roy. Soc.*, **99**, p. 163, 1921; and *Journ. Sci. Instrs.*, **3**, p. 241, 1926.

Plates

The vibrations of plates are easily demonstrated experimentally, but the mathematical treatment is very complex. The simplest case of a vibrating plate, viz. a circular disc clamped at the edge (a diaphragm) has already been considered. As we have seen, Rayleigh and Lamb give $0{\cdot}47\ hc/a^2$ as the fundamental frequency of vibration of such a disc of radius a and thickness h. When the plate is free at the edge, however, the mathematical treatment is extremely difficult.* Near the edge of the plate a peculiar state of strain exists which sometimes results in abnormally large shearing forces on sections perpendicular to the edge. A perfectly free rectangular plate cannot vibrate like a bar, with nodal lines parallel to one pair of opposite edges, since couples are required about the other pair of edges to counteract the tendency to anti-clastic curvature. The following general remarks by Rayleigh are important: "If all the dimensions of a plate, including the thickness, be altered in the same proportion, the period is proportional to the linear dimensions, as in every case of a solid body vibrating by virtue of its own elasticity. The period also varies inversely as the square root of Young's modulus, if Poisson's ratio be constant, and directly as the square root of the mass per unit volume of the substance."

The experimental study of the vibrations of plates begins with Chladni (1787). By the simple expedient of sprinkling fine sand on the surface of a plate he succeeded in demonstrating very clearly the positions of the nodal lines when the plate was set in vibration. The figures are best obtained on a plate clamped at the centre of symmetry, by bowing and touching certain points on the edge. The marvellous variety of the patterns obtained in this way is sufficient evidence of the complexity of the problem. An illustration of a large number of possible modes of vibration of a square plate is given in Tyndall's *Sound* (p. 143). Similar results may be obtained with circular plates clamped at the centre. Chladni preferred to use glass plates as they are easily obtained to the required dimensions, and their transparency permits of the fingers being used to damp points underneath which are shown to be nodal by the sand above. It is desirable to hold the plate where two nodal lines intersect, or at the centre of symmetry. Simple figures usually correspond to low-frequency tones, and the more complicated figures to the higher tones. The number

* See Rayleigh, *Sound*, **1**, p. 372; and Lamb, *Sound*, p. 150.

and variety of Chladni's figures are almost unlimited. Chladni gives illustrations of 52 figures with a square plate, 43 circular, 30 hexagonal, 52 rectangular, 26 elliptical, 15 semicircular, and 25 triangular. Doubtless the possibilities are not exhausted by this list.

It was found by Savart that very fine powder, such as lycopodium, behaves differently from sand, collecting at the antinodes instead of at the nodes. This effect was ascribed by Faraday * to the influence of currents in the surrounding air (also in vibration). In a vacuum, all powders move to the nodes.

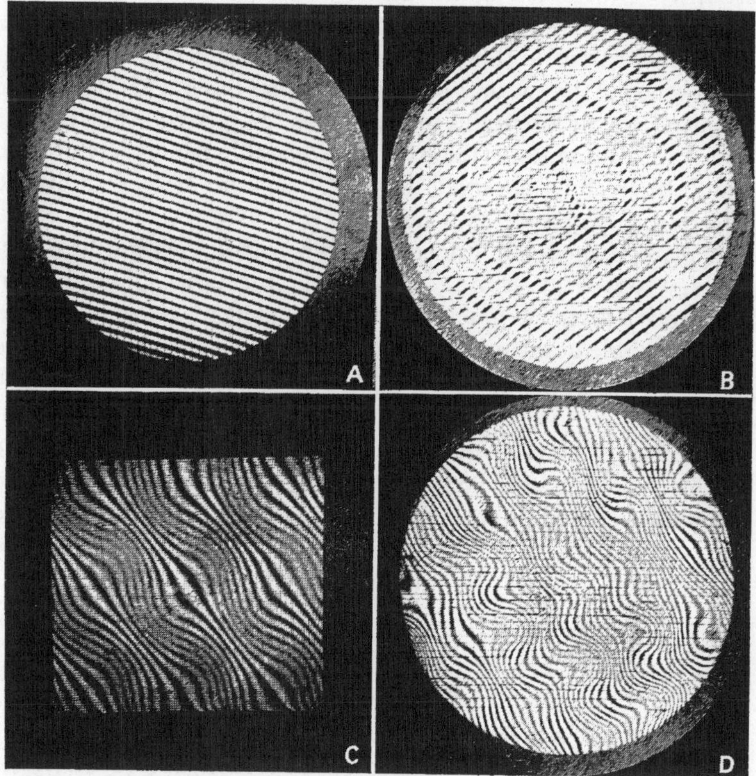

FIG. 52—Vibrations of Quartz Plates (stroboscopic interferometer method)
A. Circular disc at rest. B and D. Same disc as A in vibration. C. Rectangular plate in vibration.
(By courtesy of Dr. D. W. Dye, F.R.S.)

A general idea of the manner of formation of Chladni's figures may be obtained by regarding them as due to the simultaneous

* Faraday, *Phil. Trans.*, p. 299, 1831.

existence of two simple vibrations of the same period.* For example, in the case of a square plate vibrating with three nodal lines parallel to one edge of the plate, the x axis, the same type of vibration is also possible parallel to the other edge, the y axis. The two vibrations usually take place at the same time, but the relative phases may vary and different figures result.

Recently D. W. Dye † has demonstrated the formation of vibration figures at radiofrequencies (of the order 10^5 p.p.s.)—using small pieces of quartz (2 cm. square approximately) excited piezo-electrically in the manner previously indicated (see p. 141 *et seq.*). The motion of the surface of the quartz was observed by means of interference bands formed by reflection between it and a fixed glass surface. The illumination was rendered intermittent (stroboscopic) by means of a neon lamp flashing at the same frequency as the vibrations impressed on the quartz. The various nodes and antinodes of the numerous possible modes of vibration could be readily observed by the resultant distortion of the interference pattern. (See fig. 52.)

A plate of moderate dimensions, when struck with a hammer, emits a powerful sound. This finds practical application in numerous forms of gong. The same method of producing a sound audible at a considerable distance has been applied by the Submarine Signalling Company to a gong or 'disc bell' for use under water.

Vibrations of Curved Plates. Cylinders and Bells – The complex problem of the flat plate is still further complicated when the plate is curved, for it becomes increasingly difficult to separate the various possible modes of vibration. Rayleigh gives for the fundamental frequency of the flexural vibrations of a thin cylindrical shell

$$N=\frac{s(s^2-1)}{\pi\sqrt{s^2+1}}\cdot\frac{h}{a^2}\sqrt{\frac{1}{\rho}\cdot\frac{(k+\mu/3)}{(3k+4\mu)}} \quad . \quad . \quad (1)$$

where s is the number of wave-lengths into which the circumference is divided, $2h$ is the thickness of the shell, a its radius, k the bulk modulus of elasticity, μ the rigidity, and ρ the volume density. As in the case of a diaphragm the frequency N varies as h/a^2 and $\sqrt{1/\rho}$. Fenkner verified this result experimentally. He showed also that the frequency is independent of the length of the cylinder, and the vibrations may therefore be regarded as approximately

* Zeissig, *Ann. d. Physik*, **64**, p. 360, 1898.
† *N.P.L. Reports*, 1928 and 1929.

two-dimensional. Fenkner found that the frequencies of the partials vary as 1, 2·67, 5·00, 8·00, 12·00. The principal mode of vibration corresponds to $s=2$, in which there are four modes $\pi/2$ apart.

Bells – A bell may be regarded as a progressive development of a curved plate or, in certain forms, it may be treated as a cylindrical shell with one end closed. Bells vary in many respects, *e.g.* in regard to relative depth and diameter, shape of cross-section, and thickness-graduations. Various forms are approximately hemispherical, conical, or cylindrical in shape, with cross-sections either circular or rectangular—in fact the variety in form is almost unlimited. One has only to think of such bells as church bells, hand bells, electric bells, submarine bells, table bells, cycle bells, mechanical door bells, cow bells, Chinese bells, etc., etc., in order to realise the extensive use and variety of this means of sound production. In practically all cases the bell is supported at the centre of symmetry and is excited by striking near the free edge. It is unnecessary to refer here to the numerous methods of striking, which must obviously be adapted to the particular type of bell employed. Until recently, the only recorded systematic experimental investigation of the vibrations of bells is that by Rayleigh * who gave particular attention to the characteristics of church bells. Experimenting on two church bells, weighing 6 cwt. and 4 cwt. respectively, he found the various tones produced as follows: *Lowest tone*, 4 nodal 'meridians,' no nodal circle; *second tone*, 4 nodal meridians, 1 nodal circle; *third tone*, 6 nodal meridians (sound best produced when the 'clapper' struck the bell on the lower thick part termed the 'sound bow'); *fourth tone*, 6 nodal meridians (best elicited by striking half-way up); *fifth, and highest tone*, 8 nodal meridians.

English bell-founders (*e.g.* Taylors of Loughborough) recognise five chief tones in a church bell; reckoning from the *highest* they are termed the 'nominal,' 'fifth,' 'tierce,' 'fundamental,' and 'hum-note.' By suitable distribution of the metal in the bell, the bell-founder aims at making the hum-note, fundamental, and nominal successive octaves, but in practice this result is seldom achieved. J. Biehle † and A. T. Jones,‡ in their investigations on the frequencies and modes of vibration of bells, refer to *the*

* Rayleigh, *Sound*, **1**, p. 235.

† J. Biehle, *Phys. Zeits.*, **20**, pp. 429–431, Sept. 15, 1919.

‡ A. T. Jones, *Phys. Rev.*, **16**, p. 247, Oct. 1920; and **31**, p. 1092, June 1928.

striking note which gives its name to the bell. This note cannot be picked up from the bell by a resonator, it cannot be elicited from the bell by resonance, and it does not beat with a tuning-fork of nearly its own pitch. No partial tone of the bell has the same pitch as the striking note, but the fifth partial (the 'nominal') is an octave above it.

It is possible that the 'striking note' is due to a 'chattering' or intermittent contact between the clapper and the bell. A similar phenomenon has been observed by Bond * and others when the stem of a vibrating tuning-fork is held lightly in contact with a fixed surface, *e.g.* a table, viz. the production of so-called 'sub-harmonics' due to intermittent contact between the stem of the fork and the table. Low-frequency, audible chattering may readily be obtained between a piezo-electric crystal, vibrating vigorously at inaudible frequencies, and a metal plate in loose contact with its surface—the frequency of the intermittent contact being perhaps 1000, whilst the frequency of the crystal vibrations may be 100,000. A. T. Jones, however, considers that this 'striking note' is an 'aural illusion,' and such an explanation is not unlikely. Rayleigh's remarks relating to frequency observations on bells are of interest in this connection. "The highly composite, and often discordant, character of the sounds of bells tends to explain the discrepancies sometimes manifested in estimations of pitch . . . when a bell is sounded alone, or with other bells in slow succession, attention is likely to be concentrated on the graver and more persistent elements of the sound rather than upon the acuter and more evanescent elements, while the contrary may be expected to occur when the bells follow one another rapidly in a peal. In any case the *false octaves* with which the table (**1**, p. 393) abounds are simple facts of observation. . . ."

The phenomenon of beats frequently heard in the vibrations of a church bell is explained by Rayleigh in the following manner: "If there be a small load at any point of the circumference, a slight augmentation of period ensues, which is different according as the loaded point coincides with a node of the normal or of the tangential motion, being greater in the latter case than in the former. The sound produced depends therefore on the plane of excitation; in general both tones are heard, and by interference give rise to beats, whose frequency is equal to the difference between the frequencies of the two tones."

* Bond, *Proc. Phys. Soc.*, **36**, p. 340, June 1924.

The Submarine Signalling Company use massive bronze bells * under the sea in charted positions (under lightships and buoys) as a means of signalling to ships fitted with hydrophones to receive the sound. Sound travels much better under water than in air, for the medium is more homogeneous and attenuation losses, due to various causes, are far less serious. The sound of a submarine bell can be heard at ranges of many miles under favourable conditions. The frequency of a bell is, of course, lowered by immersion in water, the loading effect of the water being analogous to that already considered in the case of diaphragms. This loading effect is, however, different for the different partials, as might be expected, and the general character of the 'bell sound' is therefore completely altered.

AIR CAVITIES

Air Columns – The simplest case of a vibrating mass of air in a solid enclosure is that of a parallel cylindrical pipe, the ends of which may be closed or open. The longitudinal vibrations of such a column of air (or other fluid, liquid, or gas) are analogous to the corresponding vibrations in a solid bar (see p. 135 *et seq.*). As in the case of bars, it is necessary to make certain assumptions in order to formulate a simple theory. In the first place, the diameter of the pipe is assumed sufficiently great to justify neglecting viscous effects (see p. 317), and sufficiently small compared with the length of the pipe and the wave-length of the sound. In the simple theory also it is usually assumed that the walls of the pipe are rigid. This is only approximately true, and a considerable correction may be involved when the pipe is liquid-filled. With these assumptions we may regard the motion at any instant to be uniform in any particular cross-section of the pipe. That is, we have to deal with *plane waves* of sound. This plane-wave condition inside the pipe is manifestly incorrect at an open end, where the sound wave diverges spherically into the surrounding medium. Sound energy is radiated into the medium, damping of the pipe oscillations taking place, due to loss of energy to the medium, at every reflection from an open end. Otherwise the vibrations in the pipe would persist indefinitely. The open-end effect results in an increase in the effective length of the pipe, with a consequent lowering of frequency. The simple theory, which

* A description of these bells and the submarine signalling system for safety of ships at sea is described in a booklet issued by the Submarine Signalling Company.

we shall now consider, also assumes small-amplitude adiabatic changes of pressure without rotary or eddy motion. When a cylindrical air column is set into resonant vibration, *e.g.* by blowing or by means of a tuning-fork, stationary waves are set up in it due to the combined effects of the direct and end-reflected waves, just as in the cases of solid bars and strings, already considered. The equation of wave-motion in the pipe is therefore

$$\frac{d^2\xi}{dt^2}=\frac{\mathrm{E}}{\rho}\cdot\frac{d^2\xi}{dx^2} \quad . \quad . \quad . \quad . \quad (1)$$

which characterises a system of waves travelling in opposite directions in the column of fluid with velocity $c=\sqrt{\mathrm{E}/\rho}$, E and ρ being the appropriate elasticity and density of the fluid contained in the pipe. Assuming $\xi=a\cos nt$ this equation becomes

$$\frac{d^2\xi}{dx^2}+\frac{n^2}{c^2}\xi=0 \quad . \quad . \quad . \quad . \quad (2)$$

of which the solution is

$$\xi=\left(\mathrm{A}\cos\frac{nx}{c}+\mathrm{B}\sin\frac{nx}{c}\right)\cos nt \quad . \quad . \quad (3)$$

where A and B are arbitrary constants.

This equation represents a system of stationary waves in the pipe (*cf.* with the case of strings p. 86, and bars p. 136). *At a node* the displacement $\xi=0$, and *at an antinode* the condensation $d\xi/dx=0$.

Pipes Open at both Ends. 'Open' Pipe – A pulse of compression starting from the mouth of the pipe is reflected from the opposite open end *with reversal of phase*, *i.e.* as a wave of rarefaction, and again traverses the pipe to the mouth. Here another reflection takes place with reversal of phase, *i.e.* the reflected pulse is now a compression like the initial pulse. Thus, the initial state of condensation is repeated after *two* complete traverses of the length of the pipe, and the frequency N of the fundamental vibration will be $c/2l$ where l is the length of the pipe. This vibration has therefore a wave-length $\lambda=2l$. Since the condensation $d\xi/dx$ is zero at the open ends (at $x=0$ and $x=l$), we find from (3) that $\mathrm{B}=0$ and $\sin(nl/c)=0$. This gives $nl/c=s\pi$ where $s=1, 2, 3$, etc. ($n=2\pi\mathrm{N}$), representing a *complete harmonic series* of frequencies

$$\mathrm{N}_s=sc/2l \quad . \quad . \quad . \quad . \quad (4)$$

A similar result is obtained in the case of a pipe closed at both ends ($\xi=0$ at $x=0$ and $x=l$).

Pipe Closed at One End and Open at the Other. 'Closed' Pipe – In this case the initial pulse at the open end must traverse the length of the pipe four times before the cycle is repeated; for the reflection at the closed end takes place without change of phase, and two reflections at the open end are necessary before the reflected pulse is restored to the initial phase. The fundamental frequency is therefore given by $c/4l$ and the wave-length $\lambda=4l$. At the closed end ($x=0$) $\xi=0$, and at the open end ($x=l$) $d\xi/dx=0$, whence from equation (3) $A=0$ and $\cos(nl/c)=0$. Thus $nl/c=s\pi/2$ where s is *odd*, 1, 3, 5, etc. ($n=2\pi N$), representing *a series of odd harmonics* of frequencies

$$N_s=sc/4l \text{ (where } s \text{ is an odd integer)} \quad . \quad . \quad (5)$$

Tubes of length l, $3l$, $5l$, etc. (differing by $\lambda/2$), will consequently respond, in different modes, to the same note.

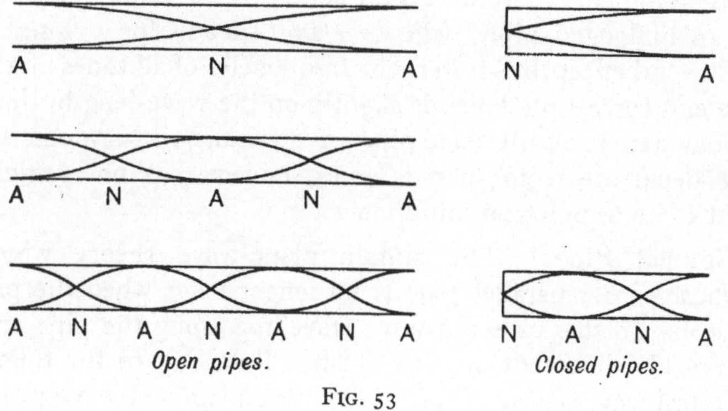

FIG. 53

A few of the possible modes of vibration of open and closed pipes are indicated in fig. 53. It will be seen that the length of the open pipe is twice that of a closed pipe of corresponding frequency. The open pipe gives a complete harmonic series, whereas the closed pipe only gives the odd harmonics of the series, accounting for the difference in characteristic 'quality' of the two types of pipe.

End Correction – In the above theory we have supposed that the open end of a pipe is a true antinode, and that the sound energy in the pipe is perfectly reflected at the open ends. This is not strictly correct, for some sound energy escapes at each reflection and is radiated in the form of spherical waves from the end of the pipe, *e.g.* the sound from an organ pipe rapidly dies away when the air blast which excites it is cut off. The air beyond the open end of the pipe is therefore in vibration and the *effective* length of the pipe is greater than the actual length. The precise calculation of the difference of length is a matter of considerable difficulty. Helmholtz deduced a value $R.\pi/4$ as the open end correction where R is the radius of the pipe, and Rayleigh obtained 0·82R theoretically for a pipe with an infinite flange. Values of about 0·8R have been obtained experimentally for a pipe with a large flange and 0·6R for an unflanged pipe. The end correction is found experimentally by means of a pipe closed at one end by an adjustable piston or water column, the successive lengths l_1, l_2, etc., which resonate to the note of a tuning-fork being determined. If there were no end correction, these lengths would be exactly proportional to the odd numbers, *i.e.* we should have $3l_1=l_2$, $5l_1=l_3$, etc., exactly. Denoting the open end correction by δ, we have approximately $3(l_1+\delta)=(l_2+\delta)$, and so on, whence δ may be determined. Careful experiments give a value about 0·58R for an unflanged pipe; whence $\lambda/4=l+0{\cdot}58R$ for a closed pipe.

The end effect thus lowers the frequencies of all tones of a pipe. The end correction depends slightly on the wave-length, thus the various partials of the same pipe are not *exactly* a harmonic series. The departure from such a series, however, is not sufficiently great to be of practical importance.

Conical Pipes – The simple plane-wave theory which is applicable to a parallel pipe is no longer valid when the pipe is conical. In this case the wave travelling along the pipe will be spherically divergent or convergent. Referring to the theory of spherical waves (see p. 58, eqn. 11), the equation of wave-propagation is

$$\frac{d^2(rs)}{dt^2}=c^2\frac{d^2(rs)}{dr^2} \quad . \quad . \quad . \quad . \quad (6)$$

where r is the radius vector from the centre of disturbance, s the condensation, and c the velocity of the sound wave ($c^2=E/\rho$). Assuming $rs=a\cos nt$, this equation becomes

$$\frac{d^2(rs)}{dr^2}+\frac{n^2}{c^2}rs=0 \quad . \quad . \quad . \quad . \quad (7)$$

the solution being given by

$$rs=\left(A\cos\frac{nr}{c}+B\sin\frac{nr}{c}\right)\cos nt \quad . \quad . \quad (8)$$

This equation represents a system of stationary waves. The constants A and B are determined by the end conditions of the pipe. We shall consider the two cases in which the cone is continued to the vertex, the base being either open or closed.

(a) *Open Cone* – At the open end $r=l$ (the slant length), the condensation $s=0$; at the vertex ($r=0$) the condensation s *must be finite* whether the vertex is closed or open, *therefore* $rs=0$. Hence from equation (8) we have $A=0$ and $B \sin(nl/c)=0$, which gives $nl/c=m\pi$, where m is 1, 2, 3, etc. Since $\lambda=2\pi c/n=2l/m$, the frequency of the mth partial is given by $N_m=n/2\pi=mc/2l$. We see therefore that the harmonic series for a cone closed at the vertex and open at the base is the same as in the case of an *open* parallel pipe of length equal to the slant length of the cone.

(b) *Closed Cone* – At the closed ends where $r=0$ and $r=l$, the condition to be fulfilled is $ds/dr=0$. This leads to the result,

$$\tan(nl/c)=nl/c \qquad . \quad . \quad . \quad (9)$$

the solution of which is obtained graphically by plotting the graphs of $y=x$ and $y=\tan x$, where $x=nl/c$. The intersections of the straight line $y=x$ with the tangent curves give the roots required. In this way it is found that $(nl/\pi c)=0$, 1·43, 2·46, 3·47, 4·47, 5·48, 6·48 . . . etc. The frequency N_m of the mth tone is therefore $N_m=mc/2l$, where m has the corresponding value of the numerical series just given. The overtones are therefore not harmonic. The antinodes, but not the nodes, are equidistant along the pipe.

(c) *Truncated Cone* ($r=l_1$, and $r=l_2$ at the ends) – A conical truncated pipe with open ends has a fundamental frequency and harmonic overtones like those for an open parallel pipe of length equal to the slant length (l_2-l_1) of the conical pipe. The antinodes are equidistant, but the nodes are not. If the cone is closed at both ends, it may be shown that

$$\tan^{-1}\left(\frac{nl_2}{c}\right)-\tan^{-1}\left(\frac{nl_1}{c}\right)=\frac{n}{c}(l_2-l_1) \qquad . \quad . \quad (10)$$

when $l_1=0$, *i.e.* the cone of length l_2 is continued to the vertex, $\tan(nl_2/c)=nl_2/c$ which is the case of the closed cone already mentioned. When l_1 and l_2 are very great (the case of a truncated cone of small angle), $\tan^{-1}(nl_2/c)$ and $\tan^{-1}(nl_1/c)$ are both odd multiples of $\pi/2$, so that (l_2-l_1) is a multiple of $\lambda/2$, in accordance with the theory of parallel pipes.

Experiments on conical pipes have been made by Blaikley *

* *Phil. Mag.*, **6**, p. 119, 1878.

and more recently by Webster.* The latter investigated the pressure distribution in various types of conical horn consisting of a parallel part closed at one end and attached to an open cone at the other end. The experimental results were found to be in approximate agreement with theory. Webster initiated the idea of 'acoustic impedance' analogous to electrical impedance, and developed the theory of pipes and cones from that standpoint.

Liquid-filled Pipes – As in the case of parallel columns of gas, stationary waves may be set up in pipes filled with liquid. The compressibility of a liquid is much less than that of a gas, and is more nearly comparable with that of a solid. Consequently the yielding of the walls of the tube cannot be neglected when considering liquid-filled tubes. At a node in the liquid, where the pressure variation is a maximum, there will be a swelling of the walls of the tube, whilst at the antinode there is no such effect. This swelling at the nodes produces a corresponding drop of the pressure and a lowering of the velocity of wave-propagation in the column of liquid, the lowering being smaller the thicker the tube. Such an effect was first observed by Wertheim in 1847. Lamb † has investigated the question theoretically, assuming that the stresses in the tube adjust themselves so rapidly that the deformation has the statical value corresponding to the instantaneous distribution of pressure in the liquid. Thus if c_0 is the theoretical value of sound velocity in a large bulk of the liquid and c the actual velocity in a tube of small thickness h, then

$$c_0/c=\sqrt{1+2\kappa a/h\mathrm{E}} \quad . \quad . \quad . \quad (11)$$

where a is the internal radius of the tube, κ the bulk modulus of the liquid, and E Young's modulus for the material of the tube. In the case where the walls are very thick $c_0/c=\sqrt{(\kappa+\mu)/\mu}$, where μ is the coefficient of rigidity of the material of the tube. Thus if water ($\kappa=2{\cdot}2\times10^{10}$) is contained in a steel tube ($\mathrm{E}=2\times10^{12}$) 3 cm. radius and 3 mm. thick, we find $c_0=1{\cdot}10c$; if a similar glass tube ($\mathrm{E}=6\times10^{11}$) were used $c_0=1{\cdot}32c$. Kundt and Lehmann succeeded in obtaining 'dust' figures with fine iron filings in resonant liquid columns just as in gases (see pp. 178 and 251). In this way they determined the velocity of sound in liquids in tubes of different diameters and thickness. Application of Lamb's formula to their experimental values gives a mean value of 1436

* *Nat. Acad. Sci. Proc.*, **6**, p. 316, 1920.

† Lamb, *Dynamical Theory of Sound*, p. 173.

metres/sec. at 19° C. for the velocity c_0 of sound in open water, in good agreement with other observations (see p. 246). Green * has recently measured the velocity of sound in resonating liquid columns contained in tubes, and has obtained results in agreement with theory. Resonant vibrations may readily be maintained in a liquid column contained in a tube which is closed at one end by a steel diaphragm. The latter may be excited at any desired frequency by means of a small electromagnet supplied with alternating current from a valve oscillator. The resonance of the liquid column will be observed directly by the ear, and the yielding of the walls of the tube at the nodes in the liquid may be examined by means of a stethoscope tube.

Hubbard and Loomis † have shown that the correction for elasticity of the containing vessel, the tube, can be entirely eliminated by using *high-frequency sounds* so that the wave-length is small compared with the diameter of the pipe and the radiating source. Using a piezo-electric quartz crystal (see p. 140) 10 cm. in diameter and 1·2 cm. thick, sound waves of frequency 200,000 to 400,000 p.p.s. were generated, producing plane waves 3 to 8 mm. in length. The velocities obtained in various liquids (distilled water, solutions of NaCl, KCl, etc.) at different temperatures were in good agreement with velocities determined by other observers for such liquids in bulk. No difference in velocity could be detected when the material and dimensions of the containing vessel were varied (see also p. 260).

Examples of Vibrating Air Columns. Organ Pipes – Among the more familiar examples of resonant air columns is the organ pipe. This usually takes the form of a cylindrical metal tube or a wooden pipe of square section. One end of the tube is specially constructed so that a suitable blast of air will set up resonant vibrations in the air column. In the open '*flue*' *organ pipe* the blast of air impinges on a thin lip which forms the upper end of the slit opening into the tube. When the blast is correctly adjusted, a stable system of alternately spaced vortex filaments is formed, one on each side of the air stream as it reaches the edge of the slit (see also p. 205 on Æolian tones). At a certain minimum blast pressure impulses due to these vortices, constituting the 'edge tone,' are of such a frequency that resonance is set up in the air column and the pipe 'speaks' with its fundamental tone. The higher harmonics of the pipe are excited by increasing the

* Green, *Phil. Mag.*, **45**, May 1923.

† *Nature*, Aug. 6, 1927 ; and *Phil. Mag.*, **5**, p. 1177, 1928.

power of the blast. In another form, known as the '*reed pipe*,' the blast of air impinges on a reed of metal which controls the amount of air entering the pipe. The latter is usually conical, with the reed clamped at the narrow end. The reed and the pipe are tuned to the same fundamental frequency; the inharmonic overtones of the reed consequently do not coincide with the harmonic overtones of the pipe. The air blast sets the reed in vibration and puffs of air are admitted to the pipe which is thereby set into resonance. It should be noted, however, that the reed cannot be regarded as a simple vibrating system apart from the air column, for its motion is influenced considerably by the air blast and by the resonance in the pipe. As we have seen, the length of a vibrating column of gas is subject to a correction at an open end. In the case of actual organ pipes, however, Cavaillé-Coll has shown that the simple formulæ (see equations 4 and 5) for parallel pipes, after application of a correction 0·6R at open ends, do not agree with observation of actual pipes. It was found, for example, that a correction as much as 3·3R was required to the length of a pipe open at both ends. This discrepancy is partly due to the nature of the blast which excites the pipe into resonant vibration, but the principal contribution is due to the fact that the lower or 'blown' end of the pipe is very much less 'open' than the upper end. Rayleigh * remarks that if a sensible correction (0·6R) on account of deficient 'openness' is required at the upper end, a much more important correction is necessary at the lower end, which is partially closed. His experimental observations indicate that at practical pressures the pitch of a pipe as sounded by wind is *higher* than its natural resonant note, and that the large correction to the length found by Cavaillé-Coll is not attributable to the blast, but to the contracted character of the lower end of the pipe, which is regarded as 'open' in the elementary theory. The rise of pitch due to the blast increases with pressure, a certain pipe having its frequency raised by 10 p.p.s. due to a rise of blowing pressure from 1 inch to 4 inches of water. With much larger pressure-increases the pipe was 'over blown' and emitted the octave of its fundamental tone.

Pipes varying in length from 20 metres to 0·5 mm., and in frequency from 8 p.p.s. to 100,000 p.p.s., have been constructed for various purposes. Fifes, flutes, oboes, whistles, etc., are all examples of musical instruments employing resonant air columns.†

* *Phil. Mag.*, **3**, p. 462, 1877, and **13**, p. 340, 1882; and *Sound*, **2**, p. 218.

† See Millar, *Science of Musical Sounds*.

An interesting technical application of resonance in pipes is found in an apparatus for detecting fire-damp in mines. Two similar pipes are blown simultaneously, one with pure air (from a container), the other with the 'doubtful' air of the mine. When the 'mine air' is pure the pipes are in tune, but a slight impurity affects the velocity and therefore the resonant frequency slightly, and beats are observed when the pipes are sounded together. The number of beats per second serves as a measure of the extent of impurity in the air of the mine. We shall have occasion to refer later to the application of the phenomena of resonance in pipes to the measurement of velocity of sound in gases and liquids, and the determination of γ the ratio of specific heats of gases (see Kundt's Tube, p. 240).

High-Frequency Sounds from Pipes. Galton's Whistle. Hartmann and Trolle's Air-Jet Generator – In order to determine the upper limit of audibility in human beings and in animals, Galton devised a miniature organ pipe in the form of a whistle. By means of a screw piston the length of the air column in the pipe, and consequently the pitch of the note, could be varied. In a later form of this whistle designed by Edelman * the pipe is blown from an annular nozzle fitted with a screw to vary its distance from the edge of the pipe. By suitable adjustment of this distance and the pressure of the air blast the little 'pipe' is set into resonant vibration at a frequency corresponding to its length and diameter. Frequencies well above the limit of audibility can be produced by means of this whistle, such sounds being easily detected by a sensitive flame (see p. 371). Hartmann and Trolle † have produced a powerful supersonic source of sound on this principle. The high-pressure air-jet they employ has a velocity greater than that of sound. The air breaks up into a regular series of pulses as it leaves the jet. When these pulses fall on the open mouth of the cylindrical pipe 'oscillator,' resonance is set up at the correct distance (see p. 207 on Jet-tones) and the pipe emits

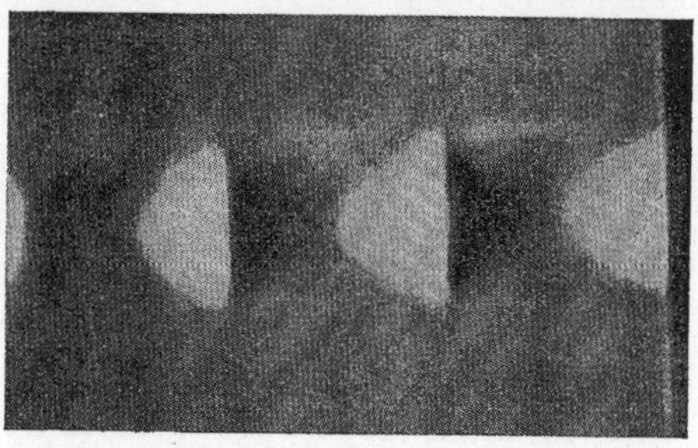

FIG. 54—Schlieren photograph showing structure of jet of Hartmann and Trolle's Air-Jet Generator

* *Ann. der Physik*, p. 2, 1900. † *Journ. Sci. Instrs.*, p. 101, etc., Jan. 1927.

powerful supersonic vibrations as high as 100,000 p.p.s. The wave-length of the vibrations was measured directly by spark photographs revealing the shadows of the high-pressure nodes, and also by means of the dust figures in Kundt's tubes (see p. 239). With jets of hydrogen, the frequencies are found to be about 3·5 times as high as in air, in accordance with theory. The air-jet generator with its resonant air column is a very suitable device for producing supersonic vibrations of high intensity, either for experimental purposes or for signalling. The

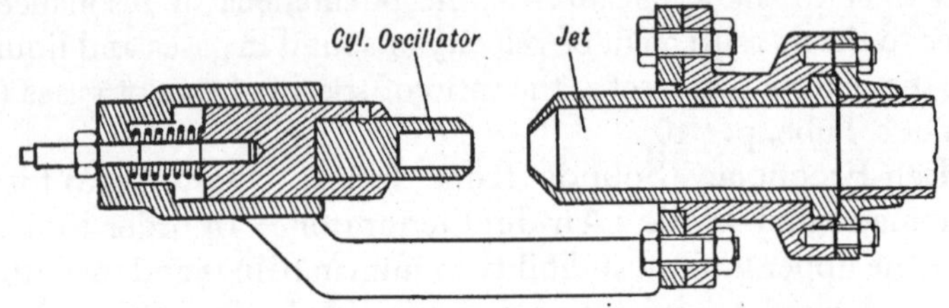

FIG. 55—Hartmann and Trolle's Air-Jet Generator

structure of the jet is shown in fig. 54, obtained by the spark-shadow (Schlieren) method. Fig. 55 indicates the arrangement of jet and pipe. Even at the highest frequencies, greater than 100,000 p.p.s., it was found that the relation $\lambda/4 = l + 0{\cdot}6R$ was valid for a closed pipe (see p. 172).

Experimental Observation of the Vibrations in Pipes – The manner in which air vibrates in a pipe can be examined in a number of different ways, depending on the physical quantity under consideration, *e.g.* displacement, pressure, density, or temperature variation.

(a) *Kundt's dust figures.* One of the earliest methods, and one which has proved most fruitful of results, is due to Kundt (1866). To demonstrate the presence of stationary waves in a horizontal glass tube he devised the method of sprinkling the interior of the tube with fine, dry dust, *e.g.* lycopodium seed, fine sand, or cork dust. This dust is immediately thrown into a recurring pattern when the sound waves are excited in the gas contained in the tube. The vibrations may be produced in a number of ways, *e.g.* by setting up longitudinal vibrations in the glass tube itself by stroking with a wet cloth, or by means of an independent vibrating rod with a free end inserted in one end of the glass tube. A more convenient method employs as a source of vibration a telephone or 'loud speaker' connected to a valve oscillator of controllable frequency (see fig. 56). At resonance, *i.e.* when the length of the gas column corresponds to a whole multiple of half wave-lengths of the sound employed, the dust is thrown into violent vibration and collects into equidistant little heaps at the nodes.

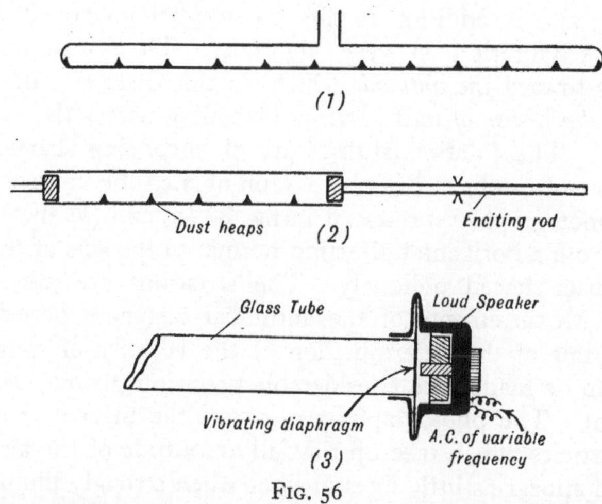

FIG. 56

As we shall see later (p. 240), this simple method not only serves to demonstrate the presence of nodes and antinodes in the gas, but also gives a measure of the velocity of sound in the gas if the frequency of vibration is known. For various reasons, which we shall have to consider when dealing with the measurement of sound velocity by this method, a number of modified forms of Kundt's tube have been developed. The simple apparatus to which we have referred is, however, very effective in demonstrating the state of vibration of the resonating gas column contained in the tube.

In addition to the regular 'half-wave' pattern of dust heaps at the nodal points, the dust tube may exhibit a regular series of fine *striations*, like ribs across the bottom of the tube. This rib-like structure is seen to consist of thin laminæ formed of small dust particles in vigorous vibration. *The effect increases in magnitude towards the antinodes* and disappears entirely at the nodes. König's explanation of the formation of these striations is given on p. 434. More recently E. N. da C. Andrade and S. K. Lewer * have obtained greatly enhanced effects of this nature which are of considerable interest. They employ a telephone diaphragm (a loud-speaker unit) to excite the dust tube into resonance. The loud speaker is supplied with alternating current of fairly pure

* *Nature*, 9th Nov. 1929; see also *Journ. of Sci. Instrs.*, 7, p. 52, Feb. 1930.

wave-form from a valve oscillator, the frequency of which is adjusted to coincide with one of the resonant frequencies of the tube. The striations to which we have just referred are readily obtained, and in addition to this a new phenomenon is noticed when the vibration is very vigorous. The striation effect increases toward *the antinode* which, in this case, *is marked by a clearly defined disc of dust particles* extending across the tube like a barrier. These antinodal discs are of surprising sharpness, as will be seen from fig. 57 (a photograph of the tube in resonance).

The photographs reproduced in fig. 57 (*a*) and (*b*) show a disc as seen from a horizontal direction normal to the axis of the tube, and a disc viewed obliquely. The striations are also clearly shown. Measurements of the antinodal distances between the discs permit of the determination of the velocity of sound (see p. 242) to be made with considerable precision (to one part in a thousand). The photographs also reveal the behaviour of very small particles which take up the full amplitude of the air vibration, and appear as little lines of light when strongly illuminated. These lines, shown in fig. 57 (*c*), have a length of several millimetres when the tube is sounding its fundamental with one end open. The photograph shows the vibration of both large and small particles, the larger ones not attaining the full amplitude of the air vibration. The effect of a drift of some of the heavier particles is clearly visible as a sinusoidal ' oscillograph ' record of the vibration. Measurement of the amplitude of the vibration of the smallest particles may yield important information relative to the motion of the air in the tube.

(b) *König's Manometric Capsule. Pressure amplitude.* The condition of the air in the interior of a vertical organ pipe was investigated experimentally by Savart (1823), who lowered into the tube a small stretched membrane on which a little fine, dry sand was sprinkled. Near a node the sand remained undisturbed, but as a loop was approached it danced with increasing vigour. The use of a thin membrane to detect the presence of vibrations in the air column was developed by König. In this method, the vibration is indicated by a small gas flame fed through a tube in communication with a small cavity called a manometric capsule. One wall of this cavity is formed by a tightly stretched membrane of parchment or rubber on which the sound waves are allowed to fall. The vibrations of the membrane vary the volume of the capsule and set up unsteadiness in the gas supply. The flame, which serves as indicator, therefore rises and falls with the vibrations of

(*a*) Antinodal disc

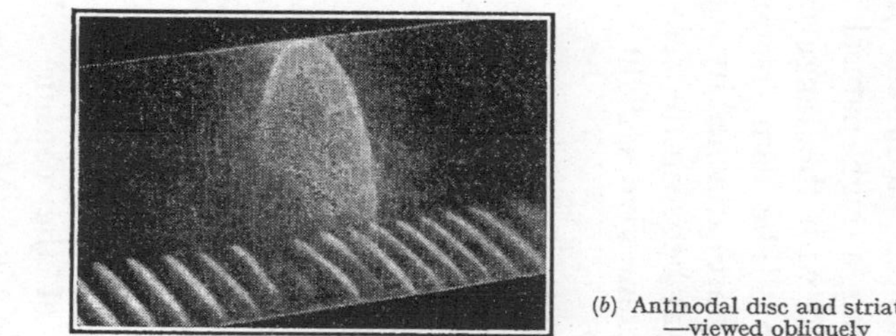

(*b*) Antinodal disc and striations
—viewed obliquely

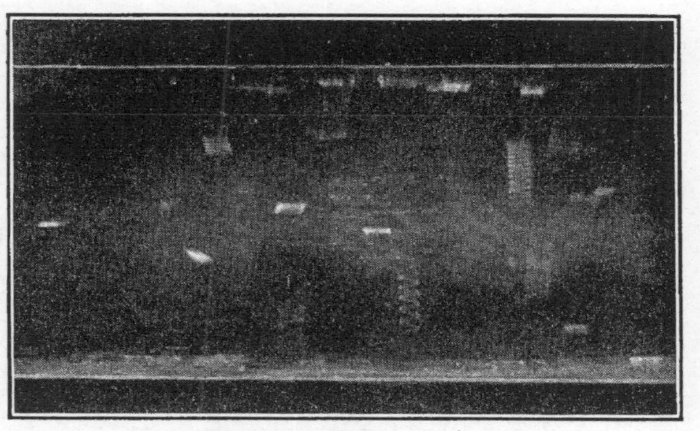

(*c*) Individual dust particles in vibration

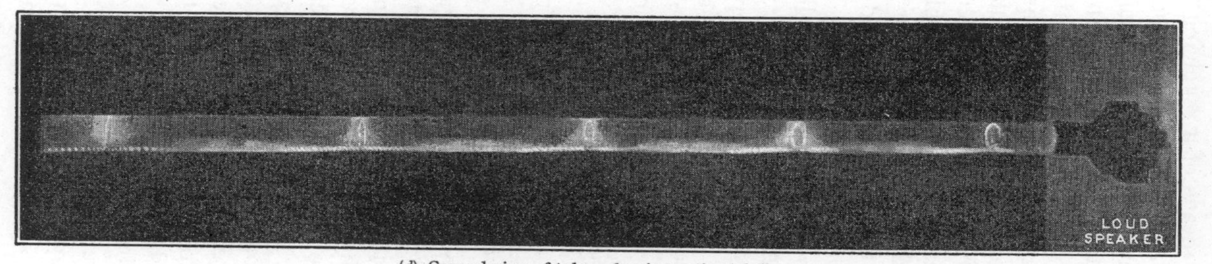

(*d*) General view of tube—showing antinodal discs

FIG. 57—Dust Tube Phenomena

the membrane. The movements of the flame are readily observed by means of a revolving mirror, the appearance being a serrated band of light when in vibration. In the application to organ pipes, a number of small manometric capsules are mounted at intervals in the wall of the pipe in such a manner that the membranes are in contact with the vibrating gas contained in the pipe. The change in the flame, observed in the revolving mirror, is very pronounced as the capsule is moved from a node to an antinode.

König also made direct aural exploration inside the pipe by means of a narrow 'exploring' tube, which could be pushed to and fro along the axis of the pipe. This narrow tube, connected to the ear, was designed so as not to interfere with the normal vibrations of the pipe. The presence of nodes and loops in the pipe were observed aurally without difficulty.

A simplified and more useful form of König's manometric capsule has recently been used by E. G. Richardson * to explore the pressure vibrations in pipes and other forms of resonating air cavities. The capsule shown in fig. 58A widens out conically from a short cylindrical portion to a flange which supports a tightly stretched membrane of sheet rubber having a high natural frequency compared with the range of frequencies of the pipe under investigation. The flame is now replaced by a small mirror (about 1 mm. square) attached by rubber cement to a point of the membrane midway between the centre and the edge. The vibrations of the membrane give a slight angular motion to the mirror and consequently to a beam of light reflected from it to a distant scale. Pressure variations inside the capsule cause the spot of light to be drawn out into a band. The device is calibrated either statically, by applying various small, steady pressures (measured by a water manometer), or dynamically by causing a little piston to oscillate with appropriate frequency and amplitude in the cylindrical portion of the capsule. The manometric

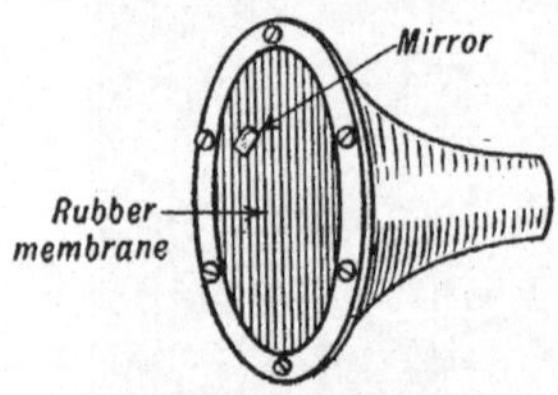

FIG. 58A—Manometric Capsule (Richardson)

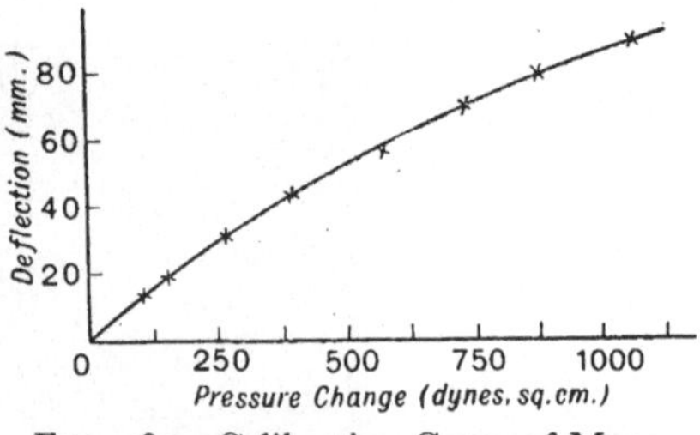

FIG. 58B—Calibration Curve of Manometric Capsule (Richardson)

* *Proc. Phys. Soc.*, May 1928.

capsule thereby becomes a useful metrical device for measuring small alternating pressure amplitudes (see fig. 58B). The narrow cylindrical portion is inserted at the desired point in the pipe or resonator and the pressure amplitude read off as the width of the band on the scale. With this simple apparatus the nodes and antinodes in the pipe may be easily demonstrated.

(c) *Displacement and Velocity Amplitudes.* To obtain complete information relative to the condition of the air in a pipe we must know not only the pressure amplitude but also the displacement or velocity amplitude. Measurements of the latter quantity have

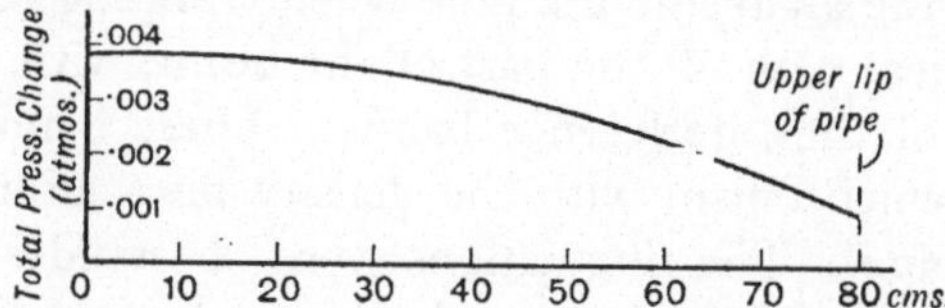

FIG. 59A—Pressure Amplitude along a Stopped Pipe (Richardson)

been made by means of a Tucker hot wire 'microphone' (see p. 402) or similar device. An electrically heated platinum wire (say 0·001 cm. diameter) is cooled by a current of air, whether direct or alternating. The cooling of the wire, as indicated by its change of electrical resistance, is a measure of the velocity of the air current. By measuring the change of resistance of a small heated wire placed at successive points along the axis of the pipe, Richardson * obtained curves (see fig. 59B) showing the velocity and displacement amplitudes in pipes and Helmholtz resonators. Combining these results with measurements of the pressure amplitude (see fig. 59A), using the manometric capsule mentioned above, he was able to determine the acoustic impedance of the mouth of such resonators and compare it with theoretical deductions.

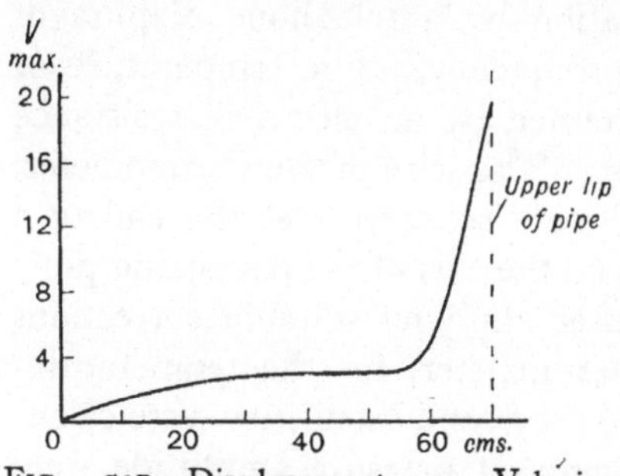

FIG. 59B—Displacement or Velocity Amplitude along a Stopped Pipe (Richardson)

(d) *Density and Temperature changes in a vibrating pipe.* The pressure and displacement fluctuations in a vibrating air column are, of course, attended by corresponding changes of density and temperature. The compressions and rarefactions take place

* *Proc. Roy. Soc.*, **112**, p. 521, 1926.

adiabatically, and the temperature at any point in the pipe must therefore fluctuate in a similar manner. The *changes of density* at the node in a vibrating gas column closed at one end were demonstrated directly by Töpler and Boltzmann * (1870), using the principle of the Jamin interferometer. The upper, closed end of an organ pipe was fitted with sides of optical glass which projected beyond the closed end. A beam of light passed through the glass sides partly inside the pipe (near the closed end) and partly outside. By well-known optical arrangements interference was produced between the two parts of the beam. The changes of density of the air inside the pipe when vibrating resulted in a change of optical path of one part of the beam, with consequent displacement of the interference bands. These bands therefore oscillated in synchronism with the density fluctuations near the node in the pipe. The fluctuations were observed stroboscopically and the amplitude measured, from which the change of density could be inferred. Raps † employed the method later to investigate the partials of an organ pipe, and recorded the displacements of the interference bands photographically. The photographs show clearly how the first overtone, which is initially absent, gradually increases in strength as the wind pressure is increased. Careful *measurements of the temperature fluctuations* in the stationary waves of a resonating pipe have been made recently by Friese and Waetzmann.‡ An air column a metre long was set in resonant vibration by a telephone diaphragm vibrating at the corresponding frequency. The temperature of the nodes and loops was ascertained by an electrical resistance thermometer and valve amplifier. The wire of the thermometer, 0·004 mm. thick and 18 mm. long, was carried at the end of a thin glass tube arranged to slide on the axis of the resonating pipe. Absolute values were found, after applying suitable corrections for the lag in the electrical thermometer, for the temperature-amplitudes at the nodes, whence the sound intensities were calculated. Simultaneous measurements of pressure amplitude were made by a membrane manometer. The temperature amplitudes calculated from $\delta\theta=(\gamma-1)\theta\,.\,\delta p/\gamma p$ agreed well with those observed directly with the resistance thermometer, over a range of frequencies 400 to 1000 p.p.s.

* *Pogg. Ann.*, 141, p. 321, 1870.

† *Ann. d. Physik*, **50**, p. 193, 1891; see also C. Barus, *Interferometry Applied to Acoustics, etc.*, Carnegie Institution, Washington.

‡ *Zeits. f. Physik*, **29**, 2, pp. 110–114, 1924; and **34**, 2–3, pp. 131–141, 1925.

Air Cavity Resonators (Helmholtz Type)

Another type of resonator which may be regarded as distinct from a pipe is that due to Helmholtz and which bears his name. In this type the air cavity is almost completely enclosed, with the result that only a very small proportion of its energy is radiated into the medium. The damping is therefore very small and the tuning is very sharp. For these reasons it is particularly suitable as a detector of sound waves of a definite frequency. It is quite unsuitable as a source of sound, as the mechanical coupling between the air contained in the cavity and the air outside is very weak, whereas in the organ pipe the coupling is much stronger. It must be regarded therefore as particularly suitable as a sharply tuned detector of sound waves of feeble intensity. A small vibration once started in the air cavity persists for a considerable time on account of the small damping, and the amplitude of vibration in the 'neck' of the resonator may be many times greater than that in the surrounding medium. Helmholtz showed that *the vibrations throughout the air cavity were practically uniform and negligible compared with the vibrations in the 'neck'*—*i.e.* at the opening connecting the air cavity to the outer air. Two forms of Helmholtz resonator illustrated in fig. 60 (*a*) and (*b*) are in general use. In one of these (*a*) the length of the neck is negligible compared with its diameter, whereas in the other (*b*) the length is comparable with, or greater than, the diameter. The exact form of the cavity is unimportant, it may be spherical or cylindrical, provided the smallest dimension is considerably greater than the dimensions of the aperture. Rayleigh, in developing the theory of such resonators, regards the air in the aperture as acting like a reciprocating piston compressing and rarefying the air contained in the cavity of the resonator. It is assumed also that the wave-length λ of the vibration in free air is large compared with the dimensions of the cavity. This implies that at any instant the condensation will be uniform throughout the cavity. The system in its simplest form is equivalent to a mass attached to a spring, *the air-piston in the neck being regarded as the mass and the air in the cavity the spring.* If the length and sectional area of the neck are l and S respectively, and ρ the normal density of the gas in

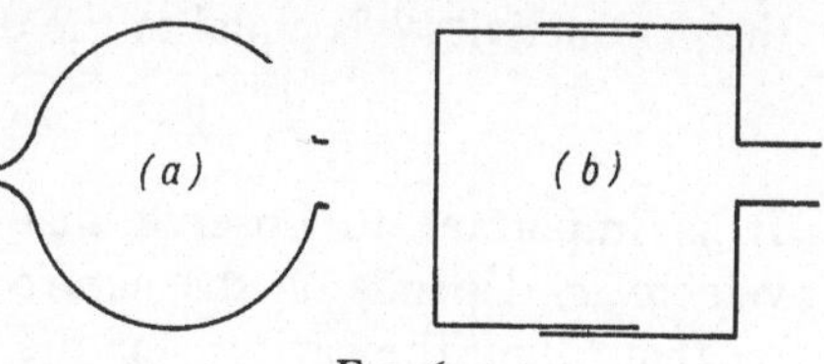

FIG. 60

the neck, then the mass M of the 'piston' will be $\rho l S$. For adiabatic changes of volume in the cavity, we have pv^γ=const., where p and v are the pressure and volume respectively and γ is the ratio of the specific heats. Then

$$dp/dv = -\gamma p/v \quad \text{or} \quad \delta p = -\gamma p \delta v/v.$$

The total force acting on the piston for a small change of volume is δp . S, and the change of volume δv for unit displacement of the piston is S. Thus the restoring force f per unit displacement of the piston is equal to $-\gamma p S^2/v$ or ES^2/v since E, the elasticity of the gas, is equal to γp. The equation of undamped motion of the piston is

$$M d^2\xi/dt^2 + f\xi = 0,$$

which indicates a frequency of oscillation $N = \frac{1}{2\pi}\sqrt{\frac{f}{M}}$,

$$\textit{i.e.} \qquad N = \frac{1}{2\pi}\sqrt{\frac{ES^2}{v \cdot \rho l S}} \quad \text{or} \quad N = \frac{c}{2\pi}\sqrt{\frac{S}{lv}} \qquad (12)$$

since the velocity $c = \sqrt{E/\rho}$. The ratio S/l, of the dimensions of a length, is called the 'conductivity' κ of the neck. For a circular aperture in a thin wall (l=0 or very small) Rayleigh shows that the conductivity κ is equal to the diameter, whence from (12)

$$2\pi N = c\sqrt{2a/v} \qquad (13)$$

It is important to observe that the frequency, involving the velocity c, depends on the nature of the gas *in the neck* and not on that filling the cavity; for the inertia of the gas in the cavity does not come into play and the compressibility of all gases is approximately the same. Usually, of course, the gas is the same throughout, and this distinction does not arise. Sondhauss and Helmholtz obtained experimentally values of frequency which were always slightly lower than the theoretical values given above. The difference may be ascribed to the 'open end' effect to which we have referred in the case of pipes. Thus in the above expression for the frequency, l must be increased to $(l+0{\cdot}6R)$, where R is the radius of the neck.

It is understood, of course, that the frequency N refers only to the fundamental tone of the resonator. The overtones are relatively very high, and the simple theory given above would not apply since it would no longer be possible to neglect the inertia of the gas contained in the cavity.

The modulus of decay (or damping coefficient) of the vibrations in a Helmholtz resonator (due to radiation) is readily shown to be equal to $8\pi v/\kappa^2 c$ where κ is the 'conductivity' of the neck. In the case of a resonator with a circular opening of radius a (when $\kappa=2a$) *the modulus of decay therefore varies directly as the volume of the cavity and inversely as the area of the neck.* This indicates that the vibrations have the greatest persistence, least energy being radiated when the volume of the cavity is large and the area of the neck is small, *i.e.* when the frequency of the resonator is low.

Relation between Pipe and Helmholtz Resonator – Hitherto we have regarded the resonant conditions of the pipe and the resonator as entirely different. In the pipe the length is comparable with λ and the displacement varies sinusoidally along the column of gas due to stationary waves. In the resonator the dimensions of the cavity are regarded as small compared with λ, and the displacement amplitude is everywhere negligible except in the neck, which is small in diameter compared with the dimensions of the cavity. Recently E. G. Richardson * has drawn attention to the fact that these 'classical' theories of the pipe and resonator are mutually exclusive, and has proposed an alternative method of treatment which results in complete reconciliation. He shows that it is possible to obtain a formula which covers all cases of resonant air cavities, by employing a method based on acoustic impedance. The fundamental frequency N of a pipe of length L is given by NL=constant, and for a cylindrical resonator, $v \propto$ L, we have $N^2L=constant$ (see equation (12)). These formulæ fit the extreme conditions, provided a suitable end correction 0·6R is applied to the length of the pipe, but transition cases between a pipe and a 'resonator' are not included. From deductions relating to the 'impedance' of the cylindrical pipe and the orifice, Richardson arrives at the formula

$$\tan k\mathrm{L}=\kappa/k\mathrm{A} \quad . \quad . \quad . \quad . \quad (14)$$

where $k=2\pi\mathrm{N}/c$, κ is the conductivity of the orifice ($=\mathrm{S}/l$) and A is the area of the section of the cylinder. For the ideal pipe the conductivity κ of the orifice is infinite, whence $\tan k\mathrm{L}=\infty$, which gives NL=constant. When κ is small, however, so that $\tan k\mathrm{L}=k\mathrm{L}$ nearly, we obtain the resonator formula $\mathrm{N}^2\mathrm{L}=$ constant. Equation (14) therefore embraces *all cases of resonant cylindrical air cavities*, of whatever length and diameter of opening

* E. G. Richardson, *Proc. Phys. Soc.*, **40**, p. 206, May 1928 ; see also E. T. Hanson, *Proc. Phys. Soc.*, **42**, p. 43, Feb. 1930.

compared with the wave-length λ. When the orifice is equal to the cross-section of the pipe itself, the difference in the value of L from equation (14) (with the appropriate values of κ and k), and the 'ideal' value given by $\tan kL = \infty$ is the *end-correction of the pipe*. The assumption made by Helmholtz that the motion is confined to the neck was verified experimentally by Richardson by means of a 'hot wire' detector (2 cm. of 0·001 cm. platinum electrically heated in a Wheatstone's bridge circuit). Near the neck of the resonator a large cooling effect was observed, indicating a large amplitude of vibration, whereas inside the cavity the cooling effect, and consequently the amplitude, was negligible. Combining the displacement-amplitude measurements, using the hot wire 'anemometer,' with a corresponding set of pressure measurements, using a calibrated manometer capsule, direct values of the acoustic impedance of various orifices were obtained. The maxima and minima of displacement and pressure amplitudes were observed in a similar manner inside a sounding organ pipe.*

Use of Helmholtz Resonators in Sound Analysis - The smallness of the damping and sharpness of tuning in these resonators renders them extremely sensitive detectors of sounds of a particular frequency. Helmholtz used a graduated series of resonators of various volumes and areas of neck in his investigations on the quality of musical notes, in particular to determine which partials were present in a complex musical sound. To cover a moderate range of frequency, however, a large number of resonators is required. It is customary, therefore, to use resonators of continuously variable volume. This may be achieved by means of a sliding piston or a column of water of variable height, forming the closed end of a cylindrical cavity.† Resonance may be detected in a number of ways. In the simplest arrangement the small open 'pip' at the base of the resonator is connected either directly, or by means of a tube, to the ear. The resonator 'speaks' when a sound of relatively feeble intensity falls on the open end. Alternatively the open end of a small manometric capsule, previously described (see p. 182), is inserted into the 'pip' of the resonator. A very sensitive method of detection, giving metrical results, employs the hot-wire microphone of Tucker and Paris. This is described in detail on p. 402. Another method of observing the resonance of an air cavity is that employed by

* E. G. Richardson, *Proc. Roy. Soc.*, **112**, p. 521, 1926 ; and *Proc. Phys. Soc.*, **40**, p. 206, May 1928.

† A. Fage, *Proc. Roy. Soc.*, p. 107, 1925.

Fournier d'Albe. A thin 'reed,' consisting of a 'slip' of mica clamped at one end, is mounted across the mouth; the 'reed' being tuned carefully to the frequency of the resonator. At resonance, the vibrations of the reed become very vigorous and are indicated by a spot of light reflected from a silvered spot on the free end of the mica strip. A series of such resonators of graded frequencies may be used in conjunction with the same source of light and scale. On account of the double-tuning the resonance may be very sharp. The arrangement responds only to the fundamental of the cavity, the overtones of the cavity and the reed being always in disagreement.

Double Resonators – In order to obtain a further increase of sensitiveness Boys * suggested the use of a double resonator consisting of a closed pipe at the end of which was fitted a Helmholtz resonator in tune with it. Rayleigh † worked out the theory of such a double resonator, and Paris ‡ has recently extended the theory to include other forms, *e.g.* (1) a pair of tunable Helmholtz resonators, (2) a conical horn and a Helmholtz resonator, making allowance for damping of the vibrations in the various parts of the system. As generally used, the resonator consists of two tunable cylindrical vessels connected by a short neck. When the separate frequencies are equal, the sensitivity of the double resonator is considerably increased by making the outer resonator of large volume compared with the inner one. Thus Rayleigh § shows that the 'air flow' per unit area in the two necks is in the ratio $(v/v')^{\frac{2}{3}}$, where v and v' are the volumes of the outer and the inner cavities. The double resonator has an important application when used in conjunction with the hot-wire microphone or the Rayleigh disc (see pp. 402 and 426) in the measurement of sound intensities. The device is very sensitive and the tuning very sharp. The intensity of sounds which are hardly perceptible to the ear can be measured, provided they are in tune with the resonator. The selectiveness of the resonator is an important feature when it is desired to measure the intensity of a sound of a particular frequency amidst a medley of sounds of other frequencies. This selectiveness carries with it, however, the corresponding disadvantage that a large number of resonators are required to cover a moderate range of frequency.

* Boys, *Nature*, **42**, p. 604, 1890.

† Rayleigh, *Sound*, **2**, p. 310, and **1**, p. 45; *Phil. Mag.*, **3**, p. 338, 1902.

‡ Paris, *Proc. Roy. Soc.*, A, **101**, 1922; and *Phil. Mag.*, p. 769, Nov. 1924, and p. 751, Oct. 1926.

§ Rayleigh, *Phil. Mag.*, **2**, 751, 1918.

By a suitable choice of frequencies for the two components of the double resonator, however, and by sacrificing a little sensitivity, it is possible to construct resonators which respond strongly to a relatively wide range of frequencies. This is of considerable practical importance in cases when the frequency of the source is liable to small fluctuations (see p. 406).

Absorption of Sound by Resonators. Quincke Filters – The action of a resonator in the neighbourhood of a source of sound in tune with it is very important. A well-tuned resonator may absorb energy from a considerable area of wave-front in the sound field in which it is placed, diverting energy from regions which would otherwise receive the sound waves. This 'absorption' may extend over an area of the order λ^2/π for a resonator of small dimensions compared with a wave-length λ of the incident sound.* Quincke (1866) employed resonators to stop tones of definite pitch from reaching the ear, the arrangement acting as a sound filter. The complex sound to be 'filtered' is led through a short main pipe which carries a side tube fitted with a sliding piston. The piston serves to tune the side tube (length $l=\lambda/4$) to any frequency N it is desired to absorb from the main pipe. Using a succession of such tuned side-tubes a corresponding number of tones may be filtered out of the sound passing through the main pipe. We shall have occasion to refer later (p. 457) to the general question of sound filters, which has recently developed into a subject of considerable practical importance.

Reinforcement of Sound by Resonators – When a vibrating tuning-fork is brought near to the open end of an air column or cavity tuned to the same frequency, the intensity of the sound reaching the ear is greatly increased. It is important to consider why the sound apparently emitted by the resonator should exceed that which would be produced by the tuning-fork alone. Clearly the sound-energy does not originate in the resonator. Any increase of intensity produced by the resonator must be due to some reaction between it and the source of vibrations. If, as in the case of the tuning-fork, the source is not maintained, the presence of the resonator will cause a more rapid drainage of energy from the source, *i.e.* the vibrations of the fork will be more rapidly damped when the resonator is present than otherwise. The increased energy output is therefore the *indirect* consequence of the presence of the resonator, which clearly can supply no energy

* See p. 376; also Lamb, *Sound*, p. 275.

itself. The energy emission W from 'simple' and 'double' sources is given in equations (35) to (40) (pp. 63 to 64). A simple source, for example a vibrating diaphragm, exposed to the medium on one side only, emits sound energy at a greater rate than the corresponding double source, a diaphragm exposed to the medium on *both* sides. Lamb * shows that the presence of the resonator produces an increase of *energy emission* $1/k^2b^2$ *times that of a simple source*, where $k=2\pi/\lambda$ and b is the distance (assumed small compared with $\lambda/2\pi$) of the source from the aperture of the resonator. In the case of a double source the emission induced by the resonator is $1/k^4b^4$ times that due to the source alone. That is, the augmentation of sound emission is proportionately much greater in the case of the double source. This is due to the relatively small efficiency of a double source as compared with a simple source in radiating sound energy (see p. 65). Considering a simple 'point' source near the mouth of a resonator in tune with it, we see that the intensity at a distant point will be increased in the ratio $1/k^2b^2$ (where kb is a small quantity). The introduction of the resonator near the source provides a means whereby the pressure-variation δp can produce a large particle velocity $\dot{\xi}$ *with which it is in phase*, thus increasing the radiating power of the source.† In other words, the tuned resonator decreases the radiation resistance R and neutralises the inertia reactance of the source. The power radiated $(\delta p)^2/2R$ (see equation 24, p. 54) may therefore be greatly increased by diminishing R by means of the resonator. The most simple and effective demonstration of the reinforcement of sound by means of a resonator is that of the tuning-fork and tuned air cavity (a bottle containing the requisite amount of water). In considering the use of resonators in conjunction with *maintained* sources of sound, it is important to bear in mind certain points mentioned above, *e.g.* the action of a resonator is not to magnify the sound *after* it has been produced by the source, but to enable the latter to develop a greater sound output. This increased output must either be supplied directly to the source or must originate in an increased efficiency of radiation. The advantage to be gained by using a resonator may be considerable if the source is initially inefficient (*e.g.* a tuning-fork, or a diaphragm electromagnetically excited in air); with an efficient source, however, there are better methods of increasing radiation efficiency.‡

* Lamb, *Sound*, pp. 273–274 ; see also Rayleigh, **2**, pp. 211–215.
† Rayleigh, *Sound*, **2**, pp. 212–213.
‡ Hart and Smith, *Principles of Sound Signalling*, pp. 30–32 and 91–92.

Trumpets and Horns. (a) *Intensifying Properties* – A little consideration will show that the difference between a resonator and a reflector is not so great as might be supposed. A resonator receiving sound waves in tune with it absorbs energy from the surrounding medium and *re-emits* a part of this energy, thus behaving as a secondary source of sound. In this respect all resonators may be regarded as reflectors, for without such reflection (or re-emission) resonance could not take place. A tuned conical pipe or trumpet behaves, in a similar manner, like a resonator and reflector. Again, a parabolic reflector will resonate to sounds of suitable frequencies.* It will be seen, therefore, that the transition from a resonant air cavity to a reflector is gradual, and that a trumpet or a horn resembles both in that it resonates, reflects, and possesses directional properties when the aperture is sufficiently large.

When a trumpet or horn is used in the transmission of sound from a small source, as in the megaphone, gramophone, or electrical 'loud-speaker,' the source is placed in a small opening near the vertex. In this position it works at maximum efficiency, since the radiation is emitted in the form of *plane waves*, the radiation resistance of a piston source being a maximum under these conditions. The function of the flared trumpet is to replace the small source by a large, nearly flat, source working at the same rate, this large source of nearly plane waves radiating into the surrounding medium with an efficiency approximating to that for plane waves. Thus in a well-designed horn there is reflection from the open end only at low frequencies.† Rayleigh ‡ shows in the case of transmission from a conical trumpet, that the amplitude of the sound received at a distant point varies inversely as the solid angle ω of the cone, and the intensity as ω^2. If the diameter of the open end is small compared with a wave-length a large proportion of the sound energy is reflected (as at the open end of a pipe), with a consequent reduction in the intensity of the sound radiated and the formation of stationary waves in the horn itself. By increasing the length of the cone, and consequently the diameter of the open end, this reflection will be reduced and will become negligible when the diameter of the open end is large compared with a wave-length. Neglecting viscous effects, therefore. it would be possible by diminishing the solid angle ω of the cone to obtain sound of

* Hart and Smith, *Principles of Sound Signalling*, pp. 30–32 and 91–92.

† The theory of conical and exponential horns is developed by Crandall in *Theory of Vibrating Systems*, pp. 152–174.

‡ *Sound*, **2**, p. 113.

any desired intensity and at the same time, by lengthening the cone, to transfer this energy without loss to the surrounding medium. Unfortunately, the practical utility of this form of horn is limited, for a conical horn of finite length and small angle will have little 'flare' and will therefore have serious end reflection, and consequent resonances similar to those of parallel pipes. The difficulty is surmounted in practice by the use of the 'flared' horn of exponentially increasing cross-section. The characteristics of the *exponential horn* are very good even at comparatively low frequencies, giving ample justification for its use in 'loud-speakers,' gramophones,* etc.

(b) *Directive Properties* – It is often supposed that a conical speaking-trumpet prevents the sound which enters at the apex from diverging outside the limits of a cone forming a continuation of the trumpet, and that the whole of the sound energy falling on the open mouth of a conical receiving trumpet is concentrated at the apex. In general, both these suppositions are erroneous. The reflecting properties of a trumpet which, as we have seen, are dependent on the dimensions of the trumpet are sufficient evidence that only a part of the incident energy reaches the apex of a receiving trumpet. The directive properties of a transmitting trumpet are also dependent on its dimensions. As Rayleigh has shown, the problem is essentially one of diffraction, analogous to the propagation of plane waves of light through a circular aperture, regarding the open end of the trumpet as a source of sound in accordance with Huyghens' principle. The problem is almost identical with that of the radiation of sound from a piston which we have already discussed (pp. 147 to 151). On the axis of the trumpet the sound intensity is always a maximum, for all the disturbances from the various elementary areas constituting the plane of the opening arrive in the same phase. In directions inclined to the axis the intensity is less, diminishing steadily to zero when the difference in distance between the nearest and farthest elements of the trumpet opening is rather more than half a wave-length. In a direction still more inclined, the sound increases again to an intensity 0·017 of that on the axis, passing through successive zero and diminishing maxima values as the inclination increases. The primary and secondary maxima, with intermediate zero values of intensity, correspond identically with the various bright and dark diffraction rings in the analogous optical case. The angle θ at which the first silence occurs, *i.e.*

* Crandall, *loc. cit.*

the semi-angle of the cone which delimits the primary beam, is $\sin^{-1}(0{\cdot}61\lambda/R)$ (see p. 149). Thus the primary or central beam will be confined to a cone of small angle when the radius R of the opening of the trumpet is large compared with a wave-length λ of the sound emitted, i.e. *the directionality will be sharp when the frequency of the sound is high.* The polar distribution of amplitude and intensity is shown in fig. 49, p. 149.

When the radius R is less than a quarter wave-length the disturbances from the elementary areas of the opening combine without much opposition in phase, and the intensity is nearly the same in all directions--that is, the source, including the trumpet, approximates to a point source and is practically *non-directional.* The directional condition, that R/λ is large, is not usually achieved in speaking-trumpets, the advantage of which, as we have seen, lies mainly in the increase of the output of sound-energy. Rayleigh obtained general verification of the theory of directional transmission of high-pitched sounds (such as a hiss), observing a considerable concentration of sound on the axis of a trumpet of moderate size.

The Voice – The voice is produced by an expiratory current of air being forced through the narrow slit between the membranous reeds in the larynx, known as the vocal cords. These are set into vibration, and the vibrations are communicated in turn to the resonant air cavities formed by the larynx, the front and back parts of the mouth (separated by the tongue), the nose and associated cavities. During ordinary respiration the glottis, or opening between the vocal cords, remains about half open, being rhythmically widened at each inspiration. For the production of voice, the free edges of the vocal cords are brought close together so as to form a narrow slit with parallel sides. The width of the slit and the tension of the 'cords' are controlled by muscles, which thereby vary the frequency of vibration. The intensity or loudness of the voice depends on the strength of the current of air through the glottis; the more powerful the blast of air, the greater the amplitude of vibration of the vocal cords. The precise action of the vocal cords in setting up resonant vibrations in the upper resonant air cavities is not definitely known. Many writers, following Helmholtz, have regarded the vocal cords as reeds or strings stretched across the larynx and vibrating transversely with a frequency depending on the tension; this vibration being communicated directly to the resonant air cavities above. More recently the view has been put forward that the function of the vocal cords is to induce vortex formation in the stream of air as it passes through the slit. The phenomenon is similar in this respect to the production of Æolian tones (see p. 205), but in the voice the vibration of the tensioned cords also plays an important part. It is significant that the vocal cords of men are about 1·5 times the length of those of women; whilst among men, those with a tenor voice have shorter vocal cords than those with a bass voice. The human voice has a range of about three and a half octaves, although it is rare to find any individual with a range extending over two octaves. In men, two kinds of voice can be distinguished, the 'chest register' and the 'head register' or falsetto. Between these two registers there is a break in the voice. At the lower frequencies the chest voice is employed, the slit between the cords being very narrow and long. For the higher notes, the head voice, the vocal cords are wider apart and only their innermost margins are set into vibration by the current of air. The action of the vocal cords may be imitated by means of two pieces of thin sheet rubber stretched edge to edge across a tube. Alternatively, an artificial larynx may be made of a stretched rubber strip fixed edgewise across a flattened air passage. The pitch of the note emitted by such an arrangement is determined by the tension and length of the rubber strip and by the air pressure used to excite the vibrations.

Speech – Articulate speech is produced in the mouth and pharynx, if there is at the same time a production of voice in the larynx. If the latter is not called into play, the air 'breathed' through the resonant cavities of the mouth gives rise to whispered speech. Speech is characterised by rapid, but smooth, changes of pitch, whereas singing usually maintains the pitch for longer periods at a constant value. Speech sounds are divided into vowel sounds and consonants. The analysis and synthesis of these sounds was considerably advanced by the studies of Helmholtz,* who formed the view that the special characters of each vowel sound were due to a combination of harmonic components of the 'cord tones,' different for each vowel, reinforced by resonance in the oral cavities, *i.e.* in the throat, mouth, and nose. This has been called the *harmonic or steady state theory* of the vowel sounds. More recently the question has been examined by Paget. In a series of papers † he has described the results of his analysis and

* *Sensations of Tone.*

† Sir Richard Paget, *Nature*, March 1922; *Proc. Roy. Soc.*, A, **102**, p. 752, 1923; **106**, p. 150, 1924; **114**, p. 98, 1927.

synthesis of the more important vowels and consonant sounds. Observations on 'breathed' (*i.e.* without sounding the vocal cords) vowel sounds indicated that in *every* case the oral cavity as a whole, from larynx to lips, actually gives *two* simultaneous resonances for each vowel sound. In certain cases, more frequent in 'American accent,' a third resonance, due to the upper or nasal cavity, may be introduced. The two components are each more or less constant for each breathed vowel sound. They consist of an upper series of notes ranging—for a bass voice—from about 608 to 2579 p.p.s., and a lower series ranging from 304 to 912 p.p.s. By means of plasticine models, Paget successfully demonstrated that recognisable reproductions of vowel sounds could be made by passing a larynx note through two resonators in series, such that when joined they resonate respectively to the two characteristic resonances heard in the voice when the vowel is breathed. These resonances in speech are varied by means of the lips and tongue, which divide the oral cavity into two parts. The tongue varies the relative volumes of the two cavities and the size of the aperture or 'neck' of the inner cavity. The shape of the cavities is relatively unimportant. It was found that various consonant sounds could be produced by manipulating the plasticine 'vowel models' as had previously been done by Kratzenstein (1779), von Kempelen, Willis (1829), and Wheatstone (1837) * in the case of a single resonator. Using a conical reed pipe, Tyndall * produced sounds resembling human speech: "Holding the palm of my hand over the end of the pipe, so as to close it altogether, and then raising my hand twice in quick succession, the word 'mama' is heard as plainly as if it were uttered by an infant." As a result of his experiments Paget † concludes that the consonants are recognised by the nature and the changes of the resonances which are set up, and by their curves of approach or recession from the resonances of the vowel with which they are associated, *i.e.* all the consonant sounds are as essentially musical as the vowels. Willis (1829), Hermann, and Scripture have proposed what is now described as the *inharmonic theory* (in contrast with Helmholtz's harmonic theory). It is suggested that the characteristic frequencies of the vowel sounds are the natural vibrations of the oral cavities excited impulsively by the more or less periodic puffs of air from the glottis. On this view, there need be no harmonic relationship between the vowel frequencies and the 'cord tone.' The classical

* See Tyndall, *Sound*, p. 198 ; see also Rayleigh, *Sound*, **2**, p. 397, etc.
† *Loc. cit.*

Fourier analysis is not considered applicable to such vowel sounds. D. C. Millar,* on the other hand, is in entire agreement with the Helmholtz theory. He has carried out extensive investigations on the optical recording and Fourier harmonic analysis of speech sounds (see also p. 441). He concludes that all vowels may be divided into two classes: (*a*) Those having a single characteristic region of resonance (represented by f*a*ther, r*aw*, n*o*, and gl*oo*m, at frequencies 1050, 730, 460, and 326 p.p.s. respectively), and (*b*) those having two characteristic regions of resonance (represented by m*a*t, m*e*t, m*a*te, m*ee*t, at frequencies 800 and 1840, 691 and 1953, 488 and 2461, 308 and 3100 p.p.s. respectively). The second series are in general agreement with Paget's observations. J. Q. Stewart † has recently reproduced vowel sounds by combining transient electrical oscillations of different frequencies. As Stewart remarks, the difference between the Helmholtz theory and the Willis theory is not serious, as the disagreement concerns methods rather than facts. There is general agreement between the two sets of data. A comprehensive investigation of speech sounds has recently been made by I. B. Crandall,‡ who employed a Wente non-resonant condenser microphone and a 'distortionless' resistance-capacity valve amplifier in obtaining oscillograph records of the various vowel and consonant sounds (see p. 441). Analysis of the records gave the predominant frequencies characterising these sounds, and the relative amounts of energy present at each frequency. The results confirm and amplify the work of Millar and Paget and yield much fundamental information relative to speech sounds. Fig. 61, due to Crandall, shows the energy distribution in speech as a function of frequency. In the male voice the maximum energy occurs at a frequency of 120 p.p.s., the female voice having a maximum about an octave higher. Measurements

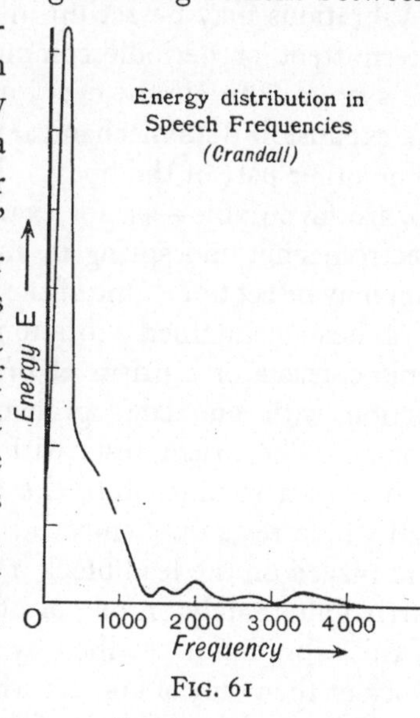

Fig. 61

* Millar, *Science of Musical Sounds*, p. 215, etc.
† *Nature*, Sept. 2, 1922.
‡ *Bell Syst. Techn. Journ.*, p. 586, 1925.

by Fletcher * on the interpretation of speech by articulation tests, indicate that it is the *high-pitched* sounds which are essential to intelligibility—in spite of the fact observed by Crandall that the energy of the voice is mostly of low pitch! The clearness of speech is unimpaired if all sounds up to 500 p.p.s. are filtered out, 60 per cent. of the total sound energy being thereby removed. It is important to observe in this connection that the ear has its maximum sensitivity in the region of frequency 1000 to 4000 p.p.s. Sabine † found that the rate of emission of sound energy in ordinary conversational speech was of the order 125 ergs/second, and for public speech 2500 ergs/second. Sacia ‡ obtained results of a similar nature, viz. 100 ergs/second for speech, including 50 per cent. silent intervals; and 'peak' values, in accented syllables, rising to 20,000 ergs/second.

Maintenance of Vibrations by Heat

Vibrations may be set up in a solid or a volume of gas by the intermittent or periodic communication of heat to some part of the system. In almost every case the heated body expands, and the expansion does mechanical work in compressing or displacing some other part of the body. If the phases of the forces thus set up are favourable—as, for example, in the analogous case of the electromagnet and spring in an electric bell—a continuous vibration may be set up. One of the earliest and most familiar examples of a heat-maintained vibration is known as *Trevelyan's rocker*. This consists of a prism of brass or copper almost triangular in section with one edge grooved to form two adjacent parallel ridges. The prism rests with this grooved edge on a block of lead with a rounded top, the end of the prism terminating in a ball which rests on a smooth surface. When the prism is heated and placed on the lead block it begins to vibrate, the weight being carried alternately on one or other of the two ridges. The cause of this vibration is ascribed by Leslie to the expansion of the cold block at the point of contact with the hot metal, the rocking being due to inequality of inertia of the portions of the rocker on opposite sides of the ridge. If both surfaces are quite clean rapid communication of heat is possible. The expansion of the lead under one of the ridges of the rocker produces a 'hump,' the force of

* *Phys. Rev.*, **19**, p. 221, March 1922; and *Bell Syst. Techn. Journ.*, p. 586, 1925.

† Sabine, *Phys. Rev.*, p. 21, 1923.

‡ *Bell Syst. Techn. Journ.*, pp. 382 and 627, 1925.

gravity throwing the rocker on to the other ridge where the operation is repeated. In order to account for the maintenance, however, it is necessary to suppose a slight lag between contact and expansion, otherwise the motion would be checked. It is important that the block on which the rocker rests should have a large temperature-coefficient of expansion. The rocker experiment was repeated later by Page * in a somewhat different manner, the local heating at the points of contact being obtained by means of an electric current passing between the block and the rocker. On the same principle Gore † succeeded in producing the continuous motion of a copper ball bridging a pair of concentric circular rails connected to the terminals of a battery. The 'gravity' theory of Leslie has recently been questioned by Chuckerbutti,‡ who concluded, as a result of a series of photographic observations of frequency and amplitude of 'rockers,' that the gravity theory was inadequate. He suggested that the *elastic* vibrations of the rocker and handle-bar play an important part, and eventually displace such simple gravity rocking movement as may occur in the earlier stages. Still more recently E. G. Richardson § and Bhargava and Ghosh ‖ have examined the question experimentally, and have obtained confirmation of Leslie's gravity theory, which requires that the product N^2a (N = frequency and a = amplitude) should be constant. No elastic vibrations, as required in Chuckerbutti's view, were detected.

Singing Flames – Under certain circumstances, a small gas flame introduced into a resonant chamber of air or other gas will emit a musical sound. The fact that the note emitted has practically the same pitch as the resonant note of the cavity was noted by Chladni (1802) in repeating the first experiments of Huggins (1777) with a hydrogen flame. De la Rive obtained similar effects by heating water in a thermometer bulb, and ascribed the singing in all cases to the intermittent condensation of water-vapour. Faraday (1818) disproved this explanation when he produced the sounds in tubes raised to a temperature above 100° C., and obtained similar results with flames of carbon monoxide.¶ Schaffgotch and Tyndall independently succeeded in generating very powerful sounds by burning ordinary gas flames in tubes

* *Silliman's Journal*, **9**, p. 105, 1850.
† *Phil. Mag.*, **15**, p. 519, 1858, and **18**, p. 94, 1859.
‡ *Ind. Assoc. for Cultivation of Science Proc.*, **6**, pts. 3 and 4, 1921.
§ *Phil. Mag.*, **45**, p. 946, 1923.
‖ *Phys. Rev.*, **22**, 517, 1923.
¶ See Tyndall, *Sound*, Lecture VI, on 'Singing Flames.'

containing air—incidentally showing how the flame could be used as a sensitive *detector* of vibrations in tune with the tube. Using a revolving mirror, Wheatstone proved that the flame was intermittent, and at one phase may withdraw itself entirely within the supply tube.* Sondhauss † showed that the singing was due to intermittent heating of the air near the jet, and that the length of the supply pipe was definitely related to the length of the singing tube. He found it was necessary that the vibrations of the gas in the supply tube and the air in the singing tube should be in step near the jet. A satisfactory explanation of all the phenomena was first supplied by Rayleigh,‡ who pointed out that the fundamental factor of importance is the phase of the heat supply relative to that of the vibration of the gas in the singing tube. He showed theoretically that the heat must be supplied at the instant of greatest condensation (or removed at the instant of greatest rarefaction) in order to maintain the vibrations, *i.e.* the driving force, the heat supply must be in phase with the condensation. It should be observed that this differs from the ordinary cases of forced vibration where the vibration is best encouraged if the impressed force is a maximum when the condensation is zero. The difference in the present case arises from the fact that the heat supply is intermittent, *not* sinusoidal. The maintenance of the vibration in the air column may be regarded in the following way. Sudden application of heat at the instant when the pressure-amplitude is a maximum 'a,' increases the pressure to $(a+\delta)$. Apart from loss of amplitude due to viscous forces and radiation of sound, the pressure-amplitude will therefore increase by δ at each complete vibration, and will ultimately reach a very large value. The equilibrium state is, of course, reached when the increments of amplitude due to the heat supply are just balanced by the losses due to friction and sound radiation. To produce the maximum effect it is therefore important that the flame must be placed at or near a point of maximum pressure-amplitude, *i.e.* near a node in the singing tube. In addition to this it is of course necessary that the oscillations of the flame must maintain the required phase relationship with the vibrations in the singing tube. The correct phase relation is made possible by the fact that there are stationary waves set up not only in the singing tube but also in the gas supply tube. Now the pressures

* See also Töpler, *Pogg. Ann.*, **128**, p. 126, 1866.

† *Pogg. Ann.*, **109**, pp. 1 and 426, 1860.

‡ *Theory of Sound*, **2**, p. 226.

in both tubes must always be equal at the jet. Consequently the pressures in both tubes near the jet must be increasing and decreasing simultaneously, *i.e.* the jet must be at a node in the supply tube to correspond to the node in the singing tube. The required condition is fulfilled if the length of the supply tube is an odd multiple of $\lambda/4$ (where λ is the wave-length, *in the gas*, of the note in the singing tube). The corresponding antinode will be at a point where the supply pipe opens into a larger gas pipe or reservoir.

If the gas jet is placed at an antinode in the 'singing tube,' where the air is at its normal density, there will be no tendency to maintain or discourage the vibrations, but the frequency of vibration of the tube will be altered. If the heat is supplied a quarter period before the maximum pressure occurs there will be an increase of frequency, whereas a decrease will be produced if the heat is supplied a quarter period after the occurrence of the maximum pressure. The various cases are illustrated graphically in fig. 62. Of these, in case (1) only will the vibrations be main-

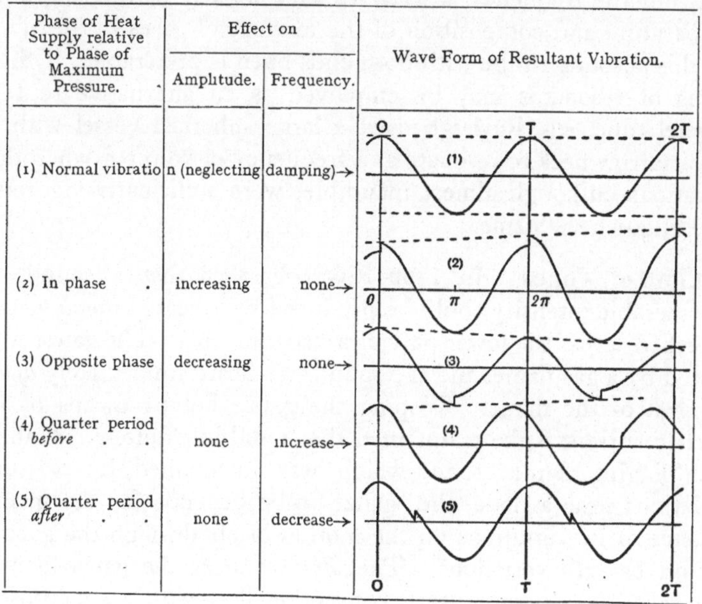

Phase of Heat Supply relative to Phase of Maximum Pressure.	Effect on Amplitude.	Effect on Frequency.	Wave Form of Resultant Vibration.
(1) Normal vibration (neglecting damping) →			
(2) In phase .	increasing	none →	
(3) Opposite phase	decreasing	none →	
(4) Quarter period *before* . .	none	increase →	
(5) Quarter period *after* . .	none	decrease →	

Fig. 62

tained; in case (2) they will be definitely discouraged; whilst in (3) and (4) they will die away, after excitation, due to the usual damping forces of friction and sound radiation. The vibrations which we have considered may or may not start spontaneously. In carrying out the experiment it is often necessary to initiate the vibration in some other way, *e.g.* by blowing across the open end of the tube or by exciting it with a source of sound of the same pitch. It will be understood also that the fundamental tone of the 'singing tube' will be modified by the general rise of temperature and by the products of combustion of the gas in the tube. E. G. Richardson * has carried out experiments on singing tubes which verify Rayleigh's theory. The flame of a König manometric capsule, attached to a singing tube on a level with the jet, was viewed, simultaneously with the singing flame, through a stroboscope. It was found that the two flames vibrate in the same phase, as the theory requires, heat being given to the air at each condensation. Stationary waves were formed both in the 'singing' tube and in the supply tube, but the lengths of supply tube favourable to singing were found to cover a much wider range than Rayleigh surmised.

The tone of a singing flame is almost as pure as that of a tuning-fork, but the frequency is liable to vary slightly with changes of temperature and composition of the mixture of gases in the tube. On this account a tube with both ends open is preferable. Other forms of resonator may be employed as an alternative to the parallel tube, *e.g.* Rayleigh used a large spherical vessel with a long, narrow neck to resonate at a frequency of 25 p.p.s. when the vibrations, although almost inaudible, were sufficiently vigorous to extinguish the flame.

'Gauze' Tones – In 1859 Rijke observed that a sound of considerable intensity could be produced by a heated metal gauze stretched across the lower part of a vertical tube. The gauze was heated by a gas flame, and the sound was heard immediately *after* removal of the flame. Keeping the gauze hot by means of an electric current Rijke maintained the sound indefinitely. Unlike the 'singing flame' tube, which was maintained by *pressure* variations near a node, the 'gauze' tube depends for the maintenance of its vibrations on the motion of air through the gauze, *i.e.* on *velocity* variations. The air-flow near the gauze is the resultant of a direct upward convection flow with a superposed alternating flow due to the vibrations. This results in a maximum temperature fluctuation in the direction of the upward part of each vibration; for cold air is drawn over the gauze at the upward

* *Proc. Phys. Soc.*, **35**, pp. 47–54, Feb. 1923.

movement, and air almost at the gauze temperature on the downward movement. When the gauze is in the lower half of the tube, *i.e. below* the node at the midpoint, this *upward* resultant assists the upward wave of compression, and as a consequence the vibrations are maintained. It will be evident that the vibrations would be discouraged if the gauze were situated *above* the node, *i.e.* in the upper half of the tube. The experiment is very effective if a large metal pipe, several feet long and about 6 inches in diameter is used, the vibrations set up may become very violent, but they only persist a few seconds, during the cooling of the gauze.*

Similar principles are involved in the explanation of the complementary phenomenon observed by Bosscha and Riess.† In this case the gauze is placed in the *upper* half of the tube, and a current of *hot* air is passed up the tube through the *cold* gauze. To maintain the vibrations it is necessary to use a special water-cooled gauze. As Rayleigh states : "In both Rijke's and Riess's experiments the variable transfer of heat depends upon the *motion* of vibration, while the effect of the transfer depends upon the variation of *pressure*. The gauze must therefore be placed where both effects are sensible, *i.e.* neither near a node nor near a loop."

Heating by Alternating Currents. The Thermophone, Singing Arc – If a conducting wire is heated by alternating current, its temperature rises and falls once in each half-cycle of the current, *i.e.* at twice the frequency of the A.C. supply. If the wire is sufficiently thin, this temperature fluctuation is accompanied by corresponding changes of length. A stretched horizontal wire will therefore be set into resonant vibration when heated by a current of frequency equal to that of the wire. Interesting observations relative to the production of sound by wires heated in this way are described in a paper by E. Ludin.‡

In addition to the change in length of the heated wire or strip another important effect may be observed, namely, *the heating of the air immediately surrounding the wire*. The air is alternately heated and cooled in each half cycle of the current and, if retained within a cover of small volume, the expansions and contractions are observed as sound. This effect is employed in a device invented by P. de Lange § and known as the *thermophone*. A theory

* See *Phil. Mag.*, **17**, p. 419, 1859, and **7**, p. 155, 1879.
† *Pogg. Ann.*, **107**, p. 342, 1859; **109**, p. 145, 1860; and **127**, p. 166, 1866.
‡ *Arch. des Sciences*, **4**, p. 383, Sept., Oct. 1922.
§ *Proc. Roy. Soc.*, **91**, p. 239, April 1915.

of the instrument, which can be used as a precision source of sound energy, has been developed by Arnold and Crandall,* who experimented with a simple form of thermophone using a platinum strip 7×10^{-5} cm. thick. When an alternating current $i \sin nt$ is passed through a fine wire of resistance R the heat developed is proportional to $Ri^2 \sin^2 nt = \frac{1}{2}Ri^2(1-\cos 2nt)$. The heat developed consequently varies between 0 and $Ri^2/2$ at a frequency $2n/2\pi$, *i.e.* at *double* the frequency $n/2\pi$ of the current. If, now, a sufficiently large initial direct current i_0 be passed through the wire, the double frequency term, in

$$R(i_0+i \sin nt)^2 = R(i_0^2+\tfrac{1}{2}i^2)+2Ri_0 i \sin nt - \tfrac{1}{2}Ri^2 \cos 2nt,$$

may be made negligible. The fluctuations of heating effect then vary with the frequency $n/2\pi$ of the current. The thermophone is therefore used with a moderate direct current on which the A.C. is superposed. The sound reproduction of the device is of good quality, but is somewhat feeble relative to the response of ordinary electromagnetic telephone receivers. Within certain limits of frequency the thermophone has possibilities as a sound standard of small intensity-range. Incidentally, Wente (*loc. cit.*) has indicated how the device may be used as a means of measuring the thermal conductivity of a gas, troublesome convection effects being entirely avoided.

The Singing Arc – Duddell † showed that a continuous current arc shunted by an inductance L and capacity C in series would, for certain values of these quantities, emit a musical note of frequency $N = 1/2\pi\sqrt{LC}$. This effect is due to the superposition of an oscillating current of frequency N on the continuous current, with corresponding heating effects in the gases in and around the arc. The singing property of the arc is made possible by its so-called *negative*-resistance (the slope of the potential current curve dV/di being always negative). The best effects are obtained when the impedance of the oscillating circuit, including the arc, is fairly small compared with the resistance in series with the main supply of current. The singing carbon arc in air may be used as a source of sound or of alternating current. If, however, the arc is formed in a vacuum or in an inert gas at reduced pressure, as in the case of a tungsten arc ('pointolite') there will be alternations of current but no sound. A carbon arc fed with

* *Phys. Rev.*, **10**, p. 22, July 1917; see also E. C. Wente, *Phys. Rev.*, **19**, p. 333, April 1922.

† *Electrician*, p. 46, 1900.

alternating current behaves as a source of sound. If a microphone be coupled to the arc through a transformer, it is possible to cause sounds to proceed from the arc similar to those falling on the microphone. Conversely, sound waves falling directly on an arc, produce fluctuations of temperature and consequently of resistance and current. The current variations after passing through a transformer may actuate a telephone receiver. It should be noted also that the frequency of alternation of the arc may have any desired value in or above the audible range (up to 10^5 p.p.s.). In this respect the singing arc is primarily of importance as a means of generating high-frequency current oscillations.* The arc is, however, very inefficient either as a source or a receiver of sound.

Spark Sound Waves – A powerful electric spark may be the source of a single intense pressure pulse or, if the electrical circuit is tuned, the spark discharged may be oscillatory with accompanying pulses of alternating pressure. The pressure effects are undoubtedly due to the large fluctuations of temperature and density of the medium in the path of the spark, the corresponding expansions and rarefactions spreading as spherical sound waves which may be sonic or supersonic according to the frequency of the electrical oscillations in the spark discharge. We shall refer later to the use of short wave-length spark-sounds or pulses to demonstrate various 'optical' phenomena.

Vortex Sounds

(*a*) Æolian Tones – When a steady flow of fluid passes an obstacle with appreciable velocity it is usually found that eddies are formed behind the obstacle. These eddies arise from the forces of viscous drag between the surface of the obstacle and the fluid. Such effects are familiar in the case of a stream of water 'swirling' past a pile or rock. Similar effects are also produced in a stream of air. Careful observation shows that a cylindrical obstacle, such as a rod or wire, in a stream of air sets up a double series of vortices,† as shown in fig. 63. These vortices arising on opposite sides of the wire revolve in opposite directions, eventually becoming detached and carried along with the stream. The stream of air past the wire is thus set into transverse vibration, as alternate left- and right-handed eddies are formed and

* See 'L'Arc Electrique,' M. Leblanc (*Les Presses Universitaires de France*); also Bragg, *World of Sound*, p. 108.

† Mallock, *Proc. Roy. Soc.*, **9**, p. 262, 1907.

detached. This process results in the production of alternating pressure-waves in the stationary air outside the stream. If the dimensions of the wire and the velocity of the air-flow are correctly adjusted the pressure fluctuations become audible. An experimental investigation of these vortex vibrations was first undertaken by Strouhal * and later by Kohlrausch, † who described the sounds as 'reibungstöne.' Strouhal stretched a wire vertically on a frame which could be revolved at various known speeds about an axis parallel to the wire. It was found that the frequency N of the tone excited in this way was independent of the length and

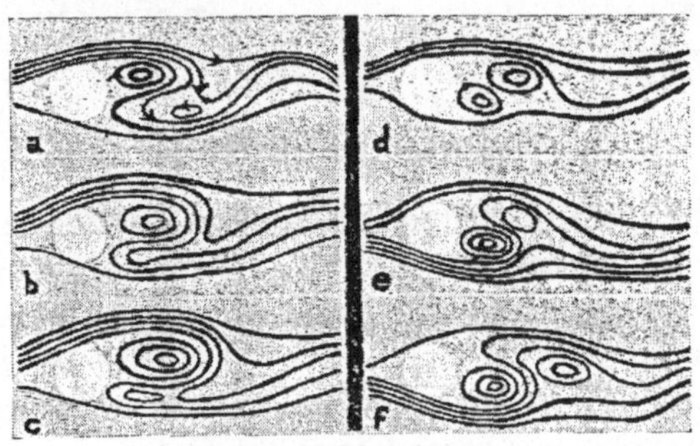

FIG. 63—Vortices
(Bragg's *World of Sound*)

tension of the wire, and varied only with the thickness d of the wire and the speed v of rotation. The results are expressed approximately by the relation

$$N = 0{\cdot}185 v/d \qquad (1)$$

The detachment of vortices and the wavering of the stream from side to side behind the wire is accompanied by corresponding alternating transverse forces of reaction on the wire itself. These forces tend to set the wire into vibration transversely to the stream.‡ When the velocity of flow is such that the value of the frequency N in the above expression coincides with one of the natural frequencies of the wire, $N' = \frac{s}{2l}\sqrt{\frac{T}{m}}$ (see equation (7*a*), p. 87), the latter is set into resonant vibration and the sound emission is greatly increased. The music of the Æolian harp is produced in this way. The harp consists of a number of wires of graduated thicknesses mounted on a sounding-board. The wires are all tuned to the same low fundamental note which consequently has

* *Wied. Ann.*, **5**, p. 216, 1878. † *Wied. Ann.*, **13**, p. 545, 1881.
‡ Rayleigh, *Phil. Mag.*, **7**, p. 161, 1879; and **29**, p. 433, 1915.

a wide range of overtones in the audible region. When the wind 'plays' the harp one or more of these wires will resonate and emit the note of the appropriate overtone.

Strouhal observed further that a rise of temperature was accompanied by a fall of pitch. Rayleigh * suggests that this effect must be associated with change of kinematic viscosity ν of the air. In such a case the frequency N must be of the form

$$N = f(\nu/vd) \,.\, v/d \qquad . \quad . \quad . \qquad (2)$$

where f is an arbitrary function; and there is dynamical similarity if $\nu \propto vd$. Strouhal's experimental results indicate that ν/vd is always very small.

Rayleigh (*loc. cit.*) gives the following formula as agreeing well with observation,

$$N = 0{\cdot}195\frac{v}{d}\left(1 - \frac{20{\cdot}1\nu}{vd}\right) \qquad . \quad . \quad . \qquad (3)$$

where, for air at 20° C., the value of $\nu = \mu/\rho = 0{\cdot}1505$ c.g.s., and for water at 15° C., $\nu = 0{\cdot}0115$. Rayleigh has also demonstrated the general behaviour of a wire in a stream of air by the analogous case of a pendulum dipping in a revolving tank of water. The pendulum is free to swing across the stream, along a radius of the tank, and its oscillations are sufficiently slow to be counted. E. G. Richardson † has obtained confirmatory results with wires stretched across a wind or a water channel. Observations were also made on the critical values of flow past objects of aerofoil section with the stream incident at various angles.

Examples of Æolian tones are numerous, the 'singing' of telegraph wires, the 'sighing' and 'roaring' of wind in trees, and the 'whistling' of wind through tall grasses, are a few of the more familiar.

(*b*) **Edge and Jet Tones** – The tones produced when a stream of gas passes a sharp edge or emerges from a jet are analogous to the Æolian tones to which we have just referred. Edge tones arise when a blade-shaped sheet of air or other gas streams out from a slit and meets a more or less sharp-edged wedge. A tone first occurs when the 'edge' is at a certain minimum distance from the jet, this critical distance being dependent on the velocity of the gas issuing from the jet.‡ The frequency of the tone falls off

* *Sound*, **2**, p. 413. † *Proc. Phys. Soc.*, **37**, p. 178, 1925.
‡ Schmidtke, *Ann. d. Physik*, **60**, 8, p. 715, 1919.

inversely as the distance x of the edge from the jet. At a certain value of x, however, the frequency 'jumps' to the octave higher, then again falls with increase of x as before. Air emerging from a slit-jet tends to form alternately spaced vortices on each side of the air stream. A photograph (by the 'Schlieren' method) of a high-velocity jet of air from a circular opening is shown in fig. 54, p. 177, where it is clearly seen that the emergent jet is divided into sections of nearly equal length. Kruger and Schmidtke * have shown that the frequency of the tones given out by air or other gases issuing from a small circular jet conforms with the relation $N = kv/D$, where v is the velocity of the jet, D the diameter of the opening, and k a constant equal to 0·045 approximately for various mixtures of air, CO_2, oxygen, and coal-gas. The regular vortex structure of the jet explains the peculiar observations on edge tones. The distance between the jet and the edge is critical, because it must approximate to a multiple of the 'vortex separation' in the jet.

Wachsmuth † carried out extensive experiments on the production of jet and edge tones. He has drawn attention to an important application in flue organ pipes. With normal blowing pressure, the frequency of the first edge tone should coincide with the fundamental frequency of the air column. In the high-frequency sound generator of Hartmann and Trolle, to which we have previously referred (fig. 55), a small resonating column of air is maintained in vibration by a high velocity jet having the regular vortex structure shown in fig. 54. The 'bird-call,' a very convenient source of sound of high frequency, probably depends for its action on vortex formation. A stream of air issues from a circular hole in a thin metal plate and impinges on a similar hole in a parallel plate held a short distance away in a telescopic tube. By varying the distance apart of the holes, or by varying the strength of the air blast, the note can be controlled over a wide range of frequency, its upper limit being quite inaudible.

The formation of a regular series of vortices with an alternating flow of air through an orifice has been beautifully demonstrated in a paper by Carrière.‡ The 'stationary stratification' of the air near the orifice of a resonator and the formation of eddies in an 'edge-tone' are also clearly demonstrated.

* *Ann. d. Physik*, **60,** 8, p. 701, 1919.
† *Ann. d. Physik*, **14**, p. 469, 1904.
‡ *Journal de Physique*, **9,** 6, p. 187, 1928.

Sirens

The earliest form of siren, due to Seebeck (1805–1849), consisted of a revolving disc perforated with a ring of equally spaced holes which interrupted a jet of air from a tube mounted opposite. The frequency of the successive puffs of air issuing through the holes is equal to the product of the number of holes and the revolutions per second of the disc. In the later form of siren invented by C. de la Tour, the holes in the revolving disc were drilled obliquely and the single air jet was replaced by a corresponding fixed ring of jets (oblique holes in a disc forming the cover of a 'wind-chest'). The blast of air emerging from the fixed jets impinged on the holes in the disc and, by its reaction, set the disc in rotation. In this form of siren the intensity of the note increases with frequency, both these quantities varying with the strength of the air blast. Improved forms of Seebeck's siren are now in common laboratory use. The disc of holes is now driven by an electric motor fitted with a speed indicator and speed control, the arrangement forming a very convenient means of determining the pitch of a note directly. It can be regarded as a point source of sound of which the amplitude is approximately proportional to the pressure of the air supply. The wave-form of the sound is dependent on the manner in which the effective area of the orifice (the area common to the tube and the hole) varies with time as the disc revolves. With equal circular holes spaced a hole-diameter apart and equal in diameter to the jet, the wave-form will approximate to that shown in fig. 64A, indicating that

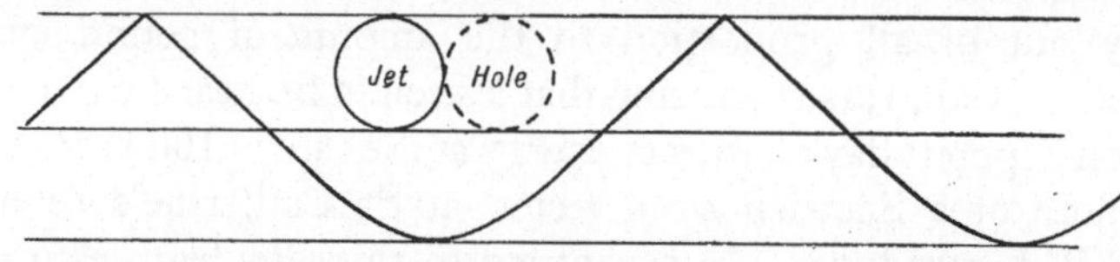

FIG. 64A—Calculated Wave-form (circular holes and jet)

a number of harmonics are present in addition to the fundamental. It would be an interesting exercise in Fourier analysis to discover the frequencies and relative intensities of the various harmonics in the wave-form, and to verify the result experimentally. Milne and Fowler * have recently devised a special form of siren of the Seebeck type for giving pure tones. The jet is rectangular in section and the holes are so shaped that the area of the jet exposed varies sinusoidally. Fig. 64B shows the shape of a hole in the

* *Proc. Roy. Soc.*, **98**, p. 414, 1921.

disc of such a siren. An examination of the sound emitted by the special siren indicates a remarkable freedom from harmonics in comparison with the sound from the ordinary siren with circular holes. Dove and Helmholtz produced sirens having several rings of holes provided with corresponding air-jets, so that one or more tones could be produced simultaneously.

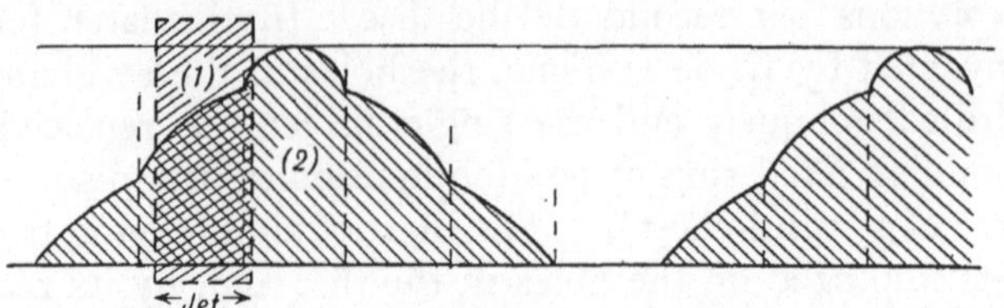

FIG. 64B—Rectangular Jet (1) and special-shaped Holes (2) to make Area of Opening vary sinusoidally

Sirens driven by compressed air or steam are in common use as sources of sound. Powerful sirens of the la Tour type, driven by a jet of water, have also been used by Sir Charles Parsons for signalling under the sea.*

During recent years much progress has been made in the design of *large-power sirens* for use in fog-signalling from lighthouses and lightships for the safety of ships at sea. In a paper "On the Production and Distribution of Sound," Rayleigh † refers to an investigation (at St Catherine's, Isle of Wight) of the relative merits of various forms of sound generators, and concludes in favour of a motor-driven siren operating at 180 p.p.s. with an air pressure 25 lb./sq. in. Such sirens, however, consume an amount of energy out of all proportion to the amount of sound energy emitted. It appears surprising that a siren to be heard eight miles away on a 'good' day consumes energy at the rate of 100 H.P. The high note of a Scottish siren tested at St Catherine's required 600 H.P. to produce. As a contrast to this, Rayleigh estimated the power to produce 'an unpleasantly loud blast' of a small horn blown by the lips to be 0·00027 H.P. The 7-in. disc siren used at St Catherine's, examined in a similar way, was shown to be extremely inefficient—the maximum pressure fluctuation (calculated) being less than 0·1 lb./sq. in. as compared with the pressure 25 lb./sq. in. actually used. L. V. King ‡ has carried out extensive

* *Engineering*, p. 517, April 18, 1929.

† *Phil. Mag.*, **6**, p. 289, 1903; *Scientific Papers*, **5**, p. 126.

‡ L. V. King, *Frank. Inst. Journ.*, **183**, p. 259, March 1917; *Phil. Trans.*, **218**, p. 227, 1919; Hart and Smith, *Principles of Sound Signalling* (Constable); E. T. Paris, *Journ. Sci. Instrs.*, **3**, p. 187, March 1926.

tests of sirens and other types of fog-signalling apparatus, using a Webster 'phonometer' (see p. 435) to measure the sound intensity at various distances from the source. We shall have to refer later to his observations in dealing with transmission of sound. Using a diaphone (see below), King concluded that of the 35 H.P. supplied as compressed air only 2·4 H.P. actually appeared as sound, and of this only 0·3 H.P. remained at a range of ten miles. Rayleigh has shown that degradation of sound energy is to be expected, and is most marked at large amplitudes. This is confirmed by King's experiments. M. D. Hart has determined the 'degradation coefficient' in certain cases, and has shown experimentally with a siren operated at 4 to 5 lb./sq. in. that 15·5 per cent. of the energy is lost between 40 and 100 cm. from the source, and a very much greater fraction between 0 and 40 cm. *For the efficient propagation of sound, a high intensity at any point must therefore be avoided.*

The Diaphone – This is a powerful long-range fog-signalling device invented by J. P. Northey in 1903. It is essentially a siren of the same class as those of Seebeck and la Tour, since it differs from them only in the fact that the opening and closing of the ports is effected by a reciprocating instead of a rotary motion. The air supply causes the piston to oscillate through a small amplitude, and also issues intermittently through the ports to produce the sound. A section of a diaphone by Chance Bros., Birmingham, is shown in fig. 65. The actual diaphone used by L. V. King was fitted with a $4\frac{1}{2}$-in. diameter piston having 10 ports and an amplitude of 0·15 in. The

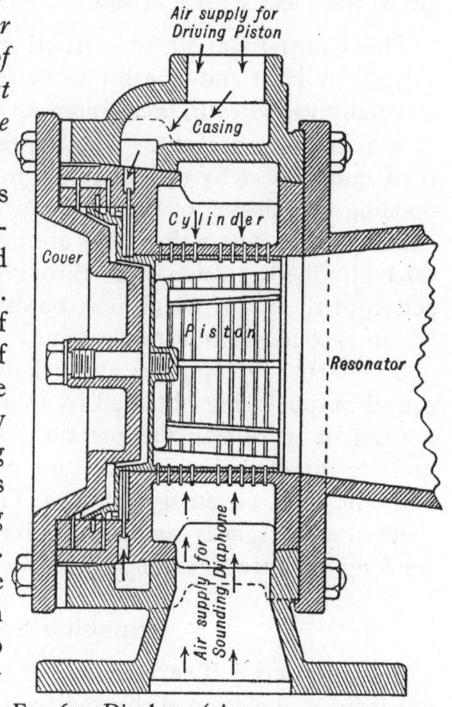

FIG. 65—Diaphone (piston 9 ports ·05 inch wide at intervals of 0·29 inch. At 30 lbs. pressure-stroke 0·47 inch, 102 oscillations per second, giving a sound frequency of 204 p.p.s.)

(By courtesy of Messrs. Chance Bros., Birmingham)

operating pressure was 25 lb./sq. in., giving 90 oscillations per second, *i.e.* a sound frequency of 180 p.p.s. The diaphone was fitted also with a conical trumpet of semi-vertical angle 6·5 degrees, diameter 4¼ to 16½ in. in a length of 4 ft. 10 in., having a resonance frequency of 180 p.p.s. at which the diaphone operated. W. S. Tucker * describes experiments with a diaphone horn near the Casquet Rocks, Channel Islands, in which the sound was received by a doubly resonated hot-wire microphone (see p. 406) in tune with the diaphone, and recorded photographically. "At a distance of 4½ miles the sound was still too intense to obtain the whole record on the photographic strip, using full sensitivity . . . it is believed that a range of at least 20 miles would have been recorded by the microphone. The superiority of the diaphone (over the ordinary siren) was clearly demonstrated."

The Centrifugal Siren – Another form of siren, which is considered by Hart and Smith † to be the most efficient, consists of a radially vaned cylindrical rotor revolving in a stationary casing in which ports are cut. The vanes are connected in pairs at their outer edges by cylindrical segments, so that rotation causes opening and closing of the fixed ports. Air is drawn through an axial aperture where the vanes are cut away. The sound is produced by the expulsion of air through the ports by the centrifugal action of the rotor, which may be driven by an electric motor or by an internal combustion engine. The siren is very powerful and efficient. It operates at a relatively low pressure, increased sound output being obtained by increasing the area of the ports instead of increasing the pressure. No resonator is employed, and its directional properties are good. The centrifugal siren combines the advantages of the Trinity House siren and the diaphone as regards signal characteristics, and is more efficient for long-distance signalling.

Explosion Sounds

Infra-Sound – The ejection of a shell from a gun and the subsequent explosion of the shell are both accompanied by a large-amplitude pressure wave which can be detected at long ranges. The explosion at the gun is due to the relatively slow burning of cordite as compared with the sudden detonation of T.N.T. in the shell. The pressure waves in the two cases are therefore distinguished by the difference in abruptness of the wave-front. The

* *Journn. Roy. Soc. Arts*, **71**, p. 121, Jan. 5, 1923. † *Loc. cit.*

'*ondes de bouche*' (from the mouth of the gun) consist principally of waves of very low frequency (of the order 1 p.p.s.), described by Esclangon * as *Infra-Sounds*. These are accompanied by sounds of audible frequency. An observer at an appreciable distance from the gun receives what appears to be a continuous train of waves, *i.e.* a 'reverberation,' due to the numerous reflections of the pulse from objects along its track. Such reverberations may last several seconds, the duration often depending upon meteorological conditions. Gun records, obtained by means of a hot-wire microphone and Einthoven string oscillograph (see p. 478 on Sound Ranging), show extended and complicated vibrations in the case of sounds coming 'down-wind,' the effect being due apparently to the downward curvature of the sound ray (see p. 290) and to the repeated reflection at the surface of the ground. With a flank wind, the reverberation is generally negligible and the simple character of the explosion 'sound' is retained. Reverberation effects are similarly observed under water on the explosion of a mine or a depth charge, the complications arising from wind, of course, being entirely absent.

A single explosion impulse is often sufficient to set a resonator into vibration, thereby producing a musical note or a noise by *shock excitation*. A sharp hammer blow on a resonant metal bar or a sudden withdrawal of a cork from an empty bottle ('air filled,' to be more precise) are simple illustrations of this. The ejection of a shell from a gun results in a large-amplitude heavily damped oscillation. The record of the gun wave, using the hot-wire microphone and Einthoven oscillograph, clearly reveals this effect. The longer the gun the lower its 'natural frequency' and the slower the oscillation.† The gun barrel may be regarded as a stopped pipe ($N = c/4l$), but the composition and temperature of the contained gas at the instant of ejection of the shell is uncertain. The record gives an approximate value of the frequency. Hot-wire microphones, for use in locating guns, were consequently fitted in 'resonators' of various possible gun-frequencies. A gun was recognised by its 'signature.'

A rapid succession of explosions at equal intervals of time may also result in a sound of a more or less musical character. The exhaust of a high-speed multi-cylinder gas or petrol engine, *e.g.*

* See E. Esclangon, *L'Acoustique des Projectiles et des Canons*.

† P. Villard, *Comptes rendus*, **179**, p. 617, Oct. 1924, ascribes the variation of frequency in these cases to the different weights of the explosive used ; see also Esclangon, *loc. cit.*

an aero engine, emits a powerful harsh sound, having a well-defined pitch, which varies with the speed. On account of the impulsive nature of the pressure fluctuations in this sound, a Fourier analysis of its wave-form would reveal a long train of harmonics in addition to the fundamental. A similar 'musical noise' of high frequency is produced in the operation of a Wehnelt electrolytic interrupter (used as a 'break' for large X-ray induction coils). This also is due to a regular succession of explosive impulses, arising from the sudden generation of gas under the liquid. The wave-form of such a series of impulses recorded on a cathode-ray oscillograph indicates the abruptness of each 'detonation.' The highly compressed spherical pulse from a powerful electric spark, and the '*onde de choc*' of a projectile (see p. 269) appear to an observer as a sharp 'crack.' The thickness of the pulse as revealed by spark-shadow photographs (see p. 340 and fig. 82) is extremely small, a fraction of a millimetre, indicating a very abrupt rise of pressure in the front of the pulse. We shall have to refer to the applications of such impulsive waves in the study of wave transmission.

Sounds from Rotating Propellers

The propeller, or air screw, of an aeroplane behaves as a powerful source of sound, which may be heard, under favourable conditions, at distances of several miles. The sound arises from the rotation of the source and sink system associated with pressure differences on the rotating blades. Using a Tucker resonant hot-wire microphone, with a continuously variable tuning, A. Fage * has analysed the sounds emitted from various types of air screw. He found that the *sound of rotation* consists of a large number of harmonics having as fundamental a note of frequency equal to the product of the number of blades and the rotational speed. In addition to these sounds, '*tearing sounds*' associated with the shedding of eddies from the blades (see p. 207 on Edge Tones) were also observed, but the frequencies were not determined. Sounds arising from the *flexural vibrations of the air-screw* blades and the shaft were found to have frequencies in agreement with calculation. The predominant frequency in the rotation sound is very intense and can readily be detected, under favourable conditions of wind, temperature gradients, etc., at very long ranges; particularly if a tuned resonator is used to receive the sound.

The greater part of the under-water sound from a moving ship

* *Proc. Roy. Soc.*, **107**, p. 451, 1925.

comes from the screw propeller. As it revolves, the pressure behind the blades is reduced and, above a certain speed, a partial vacuum and unstable vortex cavities are formed. When such 'cavitations' collapse, either on themselves or on the blades, a noise is produced. This noise has no predominant frequency, although it is to some extent characterised by the beat of the engines or the 'rush' of the turbines which drive the propeller. In considering the various possible causes of erosion in ships' propellers, Parsons and Cook * arrived at the conclusion that such cavitation effects are quite sufficient to account for the serious erosion observed. Calculation shows that the pressure of the blow, when a vortex cavity collapses upon a central nucleus one-twentieth the diameter of the original cavity, may reach 68 tons per sq. in., or 765 tons per sq. in. if the diameter of the nucleus is only 1 per cent. of that of the cavity initially. Such enormous forces acting on the metal blades account not only for the rapid disintegration but also for the considerable output of noise. Similar effects are observed with high-speed aeroplane propellers. If the screw of a ship revolves sufficiently slowly cavitations are not formed and the noise emitted is almost negligible. A submerged submarine, wishing to avoid detection, would therefore cruise at a 'silent speed.' At a moderate speed of say 10 or 15 knots, a surface ship could be detected at a range of several miles by the noise emitted from its propellers.

Directional Sources of Sound

A given amount of sound energy confined to a parallel beam (plane waves) or to a cone of small angle will be transmitted to a much greater distance than the same amount of energy spreading uniformly in all directions (spherical waves). Certain types of sound source, *e.g.* those of large area, double sources, etc., possess inherent directional properties.† Other types, *e.g.* small 'simple' sources, may become directional when used in conjunction with such devices as mirrors or trumpets. The latter, however, must conform with the condition that the radiating area is large. The combination of a small non-directional source and a *large* reflector forms a directional source. A line, a ring, or an area of equally spaced non-directional sources may also, under certain conditions, function as a directional source.

* *Engineering*, p. 515, April 18, 1919.

† With the exception of the 'breathing' or radially vibrating sphere (a mathematical fiction) which is non-directional, whatever its radius.

Sources of large Area – We have already dealt with the case of radiation from a 'piston source' of radius R emitting a sound of wave-length λ (see p. 149). It was shown, making use of the optical analogy of diffraction through a circular aperture, that the semi-angle of the primary beam is $\sin^{-1} 0{\cdot}61\lambda/R$, the solid angle of the cone approximating to $\Omega = 0{\cdot}37\pi\lambda^2/R^2$. This result indicates increased sharpness of direction with increased area of source. H. Stenzel (*loc. cit.*) has dealt with this case by more elaborate mathematical analysis, making use of Rayleigh's equations for the velocity potential in the neighbourhood of a circular piston source in an infinite wall.* The deductions are essentially the same as those already given (p. 147, etc.). The polar curves shown in fig. 66 are taken from Stenzel's paper. It

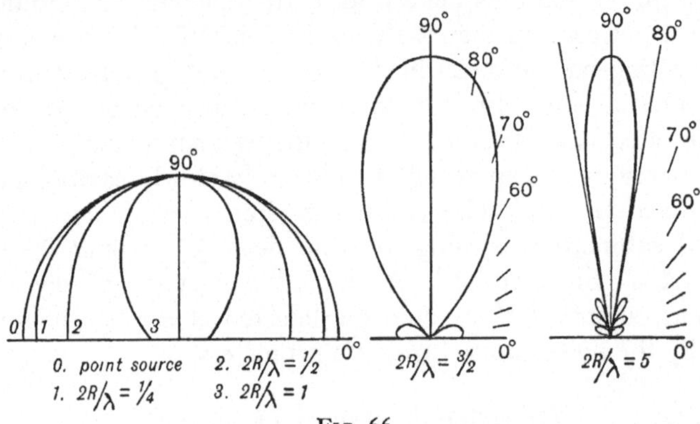

FIG. 66

will be seen that the primary beam gradually improves in directional properties as the radius of the disc R increases relatively to λ. When 2R is less than $\lambda/4$ the source is practically non-directional, *i.e.* it may be regarded as a 'point' source radiating spherical waves. When 2R is very great compared with λ the 'beam' of sound becomes approximately parallel, like a beam of light passing through a large aperture. The intensity distribution around sources of sound of varying area has recently been examined experimentally by F. W. Hehlgans.† Using a high-frequency quartz oscillator (68,000 p.p.s., $\lambda = 0{\cdot}5$ mm. in air) and varying the area exposed by means of an adjustable 'slit,' he obtained polar curves of intensity in good agreement with the theoretical curves shown in fig. 66 due to Stenzel. The paper describes also many other interesting properties of high-frequency piezo-electric sound sources and receivers. In practice it would

* *Sound*, **2**, p. 162 *et seq.* † *Ann. der Phys.*, **4**, 86, p. 587, 1928.

be impossible to attain such a parallel beam of sound within the audible range of frequencies, for the linear dimensions of the source (*e.g.* a mirror with a point source at the focus) would have to be many times the wave-length of the sound employed (in air λ varies from 11 ft. to 1·1 ft., over a frequency-range 100 to 1000 p.p.s.). Large concrete mirrors 20 ft. or so in diameter have been used to receive sounds of such wave-lengths * in a fixed direction, but there are many objections to the use of such large reflectors as a directional source of sound. The other alternative is to employ reflectors or other sources of moderate dimensions and work with sounds of very high frequency. The ticks of a watch placed at the focus of a concave mirror, say 2 ft. diameter, may be directed along a fairly definite line and received by another similar mirror. Rayleigh has verified the more important points in the theory by means of a speaking-trumpet. For sounds of moderate pitch the directional property is almost negligible, but as the pitch rises to a hiss the trumpet emits approximately plane waves, the sound intensity at any point appreciably off the axis of the trumpet being very small.

Owing to the enormous loss due to viscosity and heat conduction at high frequencies (see p. 321), it is impracticable to use sounds of very high frequency for long-distance transmission in air. It is an unfortunate fact that the range of transmission in air is a maximum in the neighbourhood of 200 p.p.s. where the wave-length of the sound is about 5 ft. A source of 'large area' operating at such a wave-length would be out of the question. Under water, however, Langevin, Boyle, and others † have used, with a certain amount of success, quartz 'piston' sources at very high frequencies 50,000 to 100,000 p.p.s. The wave-length in these cases is relatively short, of the order of 1 to 2 cm., consequently a piezo-electric source 30 or 40 cm. in diameter would emit a tolerably 'parallel' beam (see p. 149). The applications of such sound transmitters have already been considered.

Multiple Sources – The double-source, consisting of two adjacent simple sources vibrating in opposite phase, was shown (p. 66) to exhibit directional properties. A vibrating diaphragm supported in an annular ring, and freely exposed to the medium (air or water) on both sides, behaves as a directional source (pp. 156 and 416), the sound intensity being a maximum in front of, or behind, the diaphragm, and zero in the plane of the ring. The prong of a

* See *Ency. Brit.*, 13th ed. (supplementary vols.), article on 'Sound,' A. O. Rankine.

† *Loc. cit.*, pp. 146 and 471.

tuning-fork exhibits similar properties, due to the interaction between the two sets of waves which radiate, in opposite phase, from the two sides of the prong. The principle may be extended to a large number of simple sources, suitably spaced, which may be vibrating in the *same* phase. Suppose we have 'm' *sources spaced at equidistant intervals* 'd' *along a straight line* and vibrating with the same frequency N, amplitude, and phase (see fig. 67A). We require to know the polar distribution of amplitude in any plane passing through the line of sources. It simplifies matters if we arrange that the maximum resultant amplitude at a *distant*

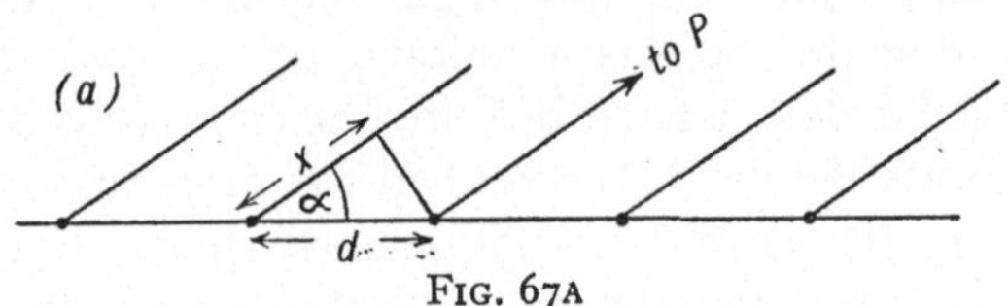

FIG. 67A

point P is equal to unity, each transmitter having an amplitude $1/m$. If the motion of each source is represented by $\frac{1}{m}\cos 2\pi Nt$, the resultant displacement at a point P oriented at an angle α with respect to the line of sources, will be

$$R' = \frac{1}{m}\left\{\cos 2\pi Nt + \cos 2\pi\left(Nt + \frac{x}{\lambda}\right) + \ldots + \cos 2\pi\left(Nt + \frac{(m-1)x}{\lambda}\right)\right\}$$

The summation is conveniently performed vectorially, since each source has the same amplitude $1/m$ and is displaced so as to make a phase angle $\beta = 2\pi x/\lambda$ with the previous one. The magnitude of the resultant vector (closing the polygon) represents the magnitude of the resultant amplitude $\frac{1}{m}\cdot\frac{\sin(m\beta/2)}{\sin(\beta/2)}$ (see p. 17) and its phase by $(2\pi Nt + (m-1)\beta/2)$. Expressing the orientation α by $\cos\alpha = x/d$, the resultant displacement at P therefore becomes

$$R' = \left[\frac{1}{m}\cdot\frac{\sin\left(\frac{m\pi d}{\lambda}\cdot\cos\alpha\right)}{\sin\left(\frac{\pi d}{\lambda}\cdot\cos\alpha\right)}\right]\cos 2\pi\left\{Nt + \frac{(m-1)}{2}\cdot\frac{d}{\lambda}\cdot\cos\alpha\right\}$$

the term in square brackets being the resultant amplitude R at P. Writing $\phi = \frac{\pi d}{\lambda}\cdot\cos\alpha$, we have $R = \frac{\sin m\phi}{m\sin\phi}$, which has maximum values when $\phi=0$ and 2π. Between these positions there are $(m-2)$ secondary maxima whose positions, since $dR/d\phi=0$, are given by the transcendental equation

$$m \tan \phi = \tan m\phi,$$

which may be solved graphically. Combining in this equation the value of R, we find

$$R^2 = 1/(m^2 \sin^2 \phi + \cos^2 m\phi),$$

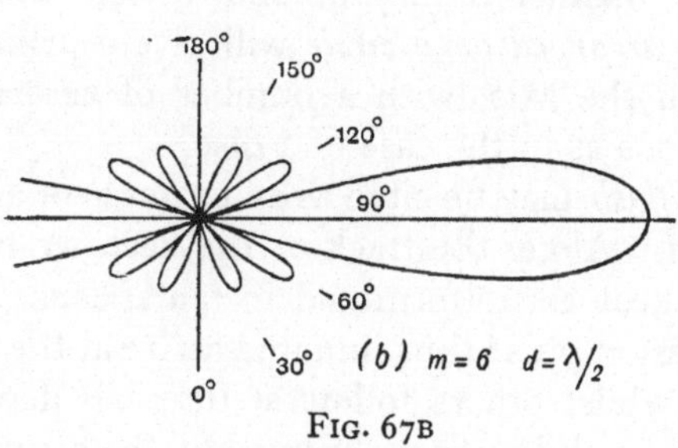

FIG. 67B

which is the equation of an ellipse of semi-axes 1 and $1/m$. The extreme values of R, *i.e.* the primary and secondary maxima, all lie on this ellipse. The greater the number of sources, the flatter the ellipse, *i.e.* the more sharply directional is the 'ellipsoidal disc' of sound. The *exact* positions of the secondary maxima may be obtained by solving the transcendental equation graphically. The positions are given approximately by $\sin \alpha = 3\lambda/2md$, $5\lambda/2md$, etc. Beyond $m=6$ the ratio of primary maximum to the largest secondary maximum hardly increases, the increase in the number of simple sources beyond this value therefore serves no useful purpose. By suitably choosing the ratio d/λ, many primary and secondary maxima may be obtained between $\alpha=0$ and $\alpha=\pi/2$, there being always $(m-2)$ secondaries between each pair of primaries. The case of $m=6$, $d=\lambda/2$ is shown in the polar curve, fig. 67B. The value of R is zero whenever $\sin(\pi/2 \,.\, m \cos \alpha)$ is zero, *i.e.* whenever $m/2 \,.\, \cos \alpha$ is an integer. In the example illustrated ($m=6$, $d=\lambda/2$), the directions of zero amplitude in the first quadrant are given by $\alpha=70° \, 48'$, $48° \, 42'$, and $0°$, with intermediate maxima as shown. The simple treatment of a line of multiple sources given above includes, of course, the case of the *double source* $m=2$, which has primary maxima at $\alpha=90°$ and $270°$ and no $((m-2)=0)$ intermediate secondaries. When $d=\lambda$ the primaries occur at $0°$ and $180°$. If d is greater than λ there are more than two primaries. For further examples the reader is referred to the original paper by H. Stenzel * which deals with the theory of numerous types of directional source. From the above, it will be seen that *a vertical line of equally spaced sources, $\lambda/2$ apart*, will give a definite concentration of energy in a horizontal plane at right angles to the line of sources. The importance of such an arrangement as a fog-signalling device, where a maximum energy-concentration is required in a horizontal plane surrounding a light vessel, will be realised.

In a similar manner it may be shown that *a number of equidistant sources arranged on a circle* will give a primary maximum of intensity on the axis, with a number of zero and secondary maxima positions as in the case of a disc.

[A *lightning-flash* may be cited as an example of an irregular *line source of sound*. Along the track of the flash we may regard the generation of heat as uniform and instantaneous. Certain parts of the compression pulse thus initiated arrive at the observer at the same instant, whilst others follow at intervals depending on the track of the flash. The effect observed by the ear is a more or less continuous rumble broken up by short periods of increased intensity. The duration of this irregular rumble depends, of course, on the distances of the various extremities of the flash from the observer.]

The directional beam of sound obtained by means of multiple sources may be rotated either by direct mechanical rotation of the line of sources, or alternatively by giving *artificial phase displacements* (*e.g.* electrically) to the individual sources. In the latter case, *the line of sources is fixed and the polar distribution is varied* in accordance with the *relative* phase changes imposed on the sources (see also binaural compensator, p. 413, where this principle is applied to directional reception of sound).

The directional and intensifying properties of trumpets, horns, and concave reflectors used with small point sources of sound have already been considered (see p. 193).

Some Methods of Observing the Motion of Vibrating Bodies

The blurred outline of a vibrating string or the prong of a tuning-fork is an indication that the string or the fork is in vibration with a double amplitude corresponding to the limits of blurring. Beyond this, however, very little is revealed to the eye. The slow to-and-fro motion of a pendulum can be followed fairly

* *Electrische Nachrichten Technik*, Bd. 4, Heft 6, p. 239, 1927, and p. 165, May 1929.

closely by the eye, but as the frequency increases the separate swings are eventually merged into a general 'blur.' In the case of a body vibrating with a small amplitude at high frequency it is necessary to adopt special means to record the vibrations, or to make them appear sufficiently 'slow' and sufficiently great in amplitude for inspection by the eye. This may involve (*a*) mechanical or optical magnification of the vibrations, (*b*) graphical or stroboscopic methods of viewing. The vibrations of gases and liquids have been studied indirectly by means of various forms of microphone (*e.g.* the hot-wire type), and more directly by means of manometric flames and membranes, and by methods of interferometry. In the following brief summary, however, we shall be more directly concerned with the motions of vibrating solid bodies, *e.g.* wires, rods, or diaphragms. The large-amplitude vibrations of the prong of a massive tuning-fork, or any similar vibrator, may be *mechanically recorded* on a smoked surface which is moving at right angles to the line of vibration. A stiff wire or bristle attached to the prong is allowed to touch the smoked surface as lightly as possible. The recording surface may, for example, be a flat plate of glass or metal falling past the vibrating fork under the action of gravity, in which case the wavy line traced by the bristle is drawn out as the plate accelerates. The period of vibration is then obtained from

$$T=\frac{1}{m}\sqrt{\frac{(l_2-l_1)}{g}},$$

where l_1 and l_2 are the *successive* lengths traversed on the plate in m oscillations of the vibrator. The tuning-fork records shown in fig. 37 were obtained by the 'dropping plate' method. As an alternative, the record may be made by tracing a wavy line on the smoked surface of a metal cylinder or disc rotated at a constant known speed. This principle is used in the case of a gramophone recorder, in which a vibrating diaphragm is connected by means of a lever to a cutting stylus which engraves the vibratory motion on the revolving disc. Such methods of direct recording are only possible where the mass and amplitude of the vibrator are both large and the friction and mass of the bristle are negligible. The method is unsuitable, for example, to record the vibrations of a thin string vibrating with a small amplitude. In this case, an optical method is preferable. The motion of the string may be viewed in a rotating mirror if a speck of chalk dust be attached to the string and brightly illuminated. A better method, which is employed in the Einthoven string oscillograph, is to project a magnified image of an element of the string on a revolving mirror or a photographic film. If the string is very thin, *e.g.* an Einthoven fibre 1μ or 2μ diameter, a microscope objective is used to focus light on it, whilst a second high-power objective and an eyepiece project a magnified image on the slit of the recording camera. Details of the arrangement may be found in electrical textbooks.*

The following *methods of obtaining large mechanical and optical magnification* of the displacement of the vibrating body are instructive. They have a wide application in the experimental study of vibrations.

(1) In certain cases, *e.g.* the transverse vibration of a diaphragm or a reed, some part of the vibrating surface tilts through a small angle relative to its equilibrium position. Provided the mass of the vibrator is not too small, a minute fragment of mirror (*e.g.* an oscillograph mirror, about 1 mm. square, of silvered microscope cover-glass) should be attached at a point where the angular motion is greatest, *e.g.* at a point about half the radius from the centre of a diaphragm. The linear oscillation of a spot of light reflected from the small mirror to a revolving mirror or photographic plate, at a distance d from the diaphragm, will be $2\theta d$, where θ is the angular oscillation of the small mirror.

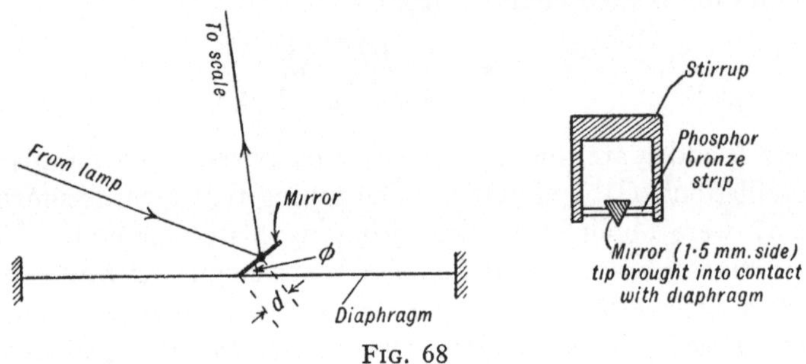

Fig. 68

(2) Method (1) requires a fairly large amplitude of vibration as the optical magnification is small. A great improvement in the method is due to A. E. Kennelly,† who made use of a high-frequency optical lever of very light construction to explore the small amplitude vibrations of a telephone diaphragm. In this method, a small triangular mirror (about $1\frac{1}{2}$ mm. side) is attached rigidly to a phosphor bronze strip (as used for galvanometer suspensions) which is stretched tightly on a metal stirrup. The

* *E.g.* see Irwin's *Oscillographs* (Pitman).

† *Electrical Vibration Instruments.*

tip of the mirror is arranged to press lightly on the part of the vibrator of which the motion is required, and is adjusted to make an angle of about 45° with the direction of vibration. The magnification factor M of the arrangement shown in fig. 68 is $2D/d \cos \phi$. For $2D=50$ cm., $d=0{\cdot}05$ cm., $\phi=45°$, $M=1400$ approx. It is necessary of course that the natural frequency of the small mirror on its phosphor bronze 'spring' is considerably higher than the frequency of the diaphragm under test. Also it is necessary that the inertia of the mirror system should not appreciably affect the vibrations of the diaphragm. This optical lever gives a large magnification of the diaphragm movements, a small linear displacement of the diaphragm producing a large angular displacement of the small mirror and a correspondingly large deflection of a spot of light reflected from it to the distant rotating mirror or film. A good optical arrangement for use with such small mirrors is shown in fig. 69. The method is metrical and may be used to determine the actual amplitude of vibration.

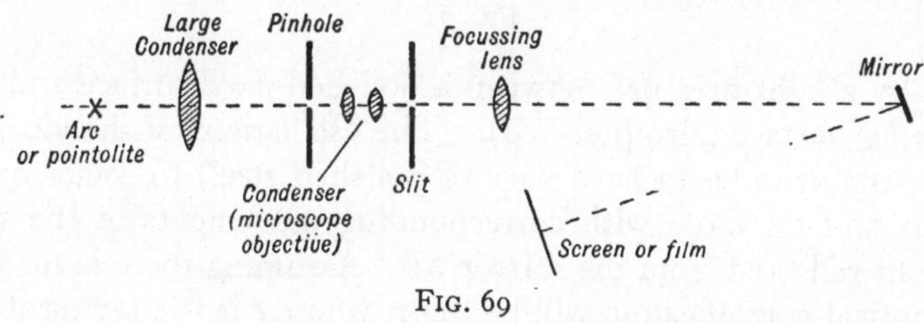

FIG. 69

A somewhat similar method has been employed by the writer in certain types of oscillograph.* Two forms of this arrangement are shown in fig. 70 adapted for longitudinal and transverse vibrations respectively. A small oscillograph mirror M (about 2 mm. long by 1 mm. wide) is mounted by means of rubber solution across two parallel knife edges as shown. The rubber

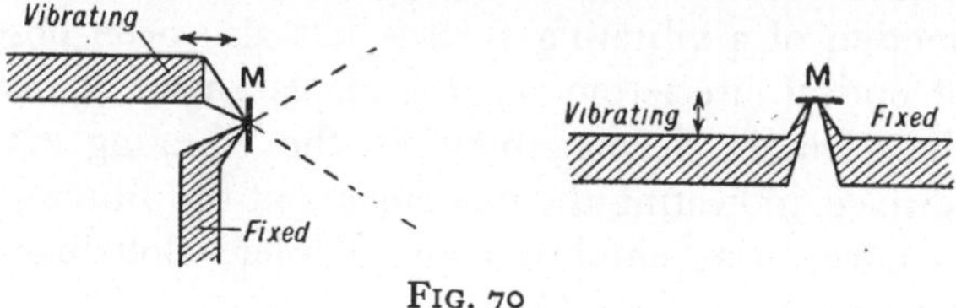

FIG. 70

solution, after setting, holds the small mirror tightly, as with an elastic band, down to the knife edges. Vibration of one of

* See A. B. Wood, 'Piezo-Electric Oscillographs,' *Phil. Mag.*, **1**, p. 631, Sept. 1925.

the latter causes the mirror to rock on the other as a pivot. If d is the effective distance apart of the knife edges and D the distance from the mirror to the film or rotating mirror, the optical magnification is $2D/d$. This may be as great as 10^4. The method illustrated in fig. 70 has a useful application in a simple form of oscillograph in which the vibrator is the reed of a Brown telephone earpiece. For moderately low frequencies this oscillograph is extremely convenient in a laboratory.

(3) Another method, involving an optical lever device, is illustrated in fig. 71, in which R is a polished cylinder of steel and M is a small mirror rigidly attached to it. The cylinder is

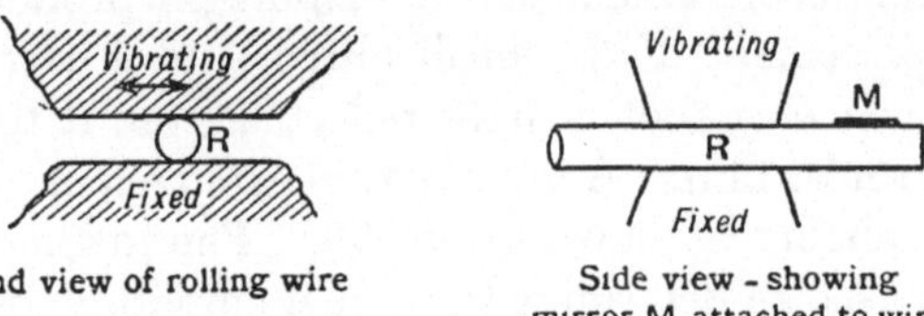

FIG. 71

held by a light pressure between a polished *fixed* surface and the *vibrating* surface (also polished). The oscillations of the vibrator cause the cylinder (a hard wire of polished steel) to rotate backwards and forwards with corresponding movements of the spot of light reflected from the mirror M. Assuming there is no slip, the optical magnification will be $2D/r$, where r is the radius of the wire and D the distance from the mirror to the point of observation of the spot of light. With a wire 1 mm. in diameter the amplitude of the spot of light at 1 metre distance will be 4000 times that of the vibrating surface in contact with the wire. The magnification may be still further increased by reducing the diameter of the wire, always ensuring that the wire with its attached mirror behaves as a rigid system.

The motion of a vibrating surface has also been observed by means of optical interferometry, the displacement of a system of interference fringes, formed between the vibrating surface and a fixed surface, indicating the movements of the vibrator. These methods have been employed by Töpler, Boltzmann, Raps, Webster, and others (*loc. cit.*).

Stroboscopic Methods* – If a body is in motion and is illuminated for a sufficiently short period of time it will appear to be at

* A comprehensive summary of stroboscopic methods is given by J. F. Crowley in *The Illuminating Engineer*, p. 189, Aug.–Sept. 1923.

rest. If the motion is periodic, then by suitable devices, the body may be exposed or illuminated at successive instants whilst it is passing through the same space interval in the same direction. The resultant effect of such successive momentary exposures is equivalent to a prolonged continuous exposure. Devices which give the appearance of rest or of slow motion to bodies in rapid motion are known as stroboscopes.* There are four possible ways in which the necessary interruption of vision for stroboscopic observation may be achieved, viz. by an interruption (1) of the light between the object and the eye, (2) between the illuminant and the object, (3) of light at the object, and (4) of the illuminant itself. *Methods* (1) and (3) involve the mechanical operation of some form of shutter at the correct frequency, such as a rotating disc with a radial slit (or slits) which periodically comes into the line of vision or allows the light to fall on the object. Plateau's disc with radial slits was first used to determine the speed of revolution of wheels and shafts. A disc with corresponding markings was placed on the rotating shaft and illuminated by light rendered intermittent by the slotted disc. Stroboscopic beams of light, interrupted by such a disc placed at a focus, are similarly used to view vibrating objects (wires, reeds, etc.). C. V. Drysdale introduced the use of a *stroboscopic tuning-fork*, of known frequency with shutters attached to the prongs, to produce the necessary interruptions in the light. He also employed a more elaborate disc with geometrical patterns of various shapes for the motor-shaft, and was thus able to measure speeds in steps. A somewhat similar apparatus, with prongs of adjustable frequency, is known as the "Crompton-Robertson" stroboscopic vibrator. In a more recent arrangement Drysdale uses the tuning-fork to drive a phonic motor (see p. 126) with a conical drum from which a slotted disc is driven by friction.† The speed of the disc is controlled by varying its position of contact with the conical drum, and a continuous adjustment of frequency is thus obtained. In a still later form the slits are replaced by electrical contacts which 'flash' a neon lamp at the required synchronising frequency (see Method (4) below). A good example of type (1) stroboscope is known as the *Ashdown 'rotoscope.'* It consists essentially of a viewing slit in front of which revolve a number of vanes which permit the vibrating or rotating object to be viewed inter-

* The first practical instrument of this kind, a slotted disc for viewing a vibrating tuning-fork, is ascribed to Plateau, 1801–1883.

† The Roller Stroboscope—manufactured by H. Tinsley & Co.

mittently. The shutters are spring-driven, and the speed is easily controlled and indicated on dials. The vibrations of a tuning-fork or a string are very effectively made to appear 'stationary' or 'in slow motion' by this apparatus. It is extremely convenient also for viewing the vibrating reeds of a frequency meter, as suggested on p. 41, to demonstrate phase effects near resonance. *Rotating and vibrating mirrors* are frequently used for stroboscopic work. König and Wheatstone both used a revolving set of four plane mirrors to observe the fluctuations in height of a vibrating flame. When the mirrors are rotated fast enough the images appear as a saw-like band across the mirrors. A similar application of rotating mirrors is used to observe the movements of a spot of light or the image of a fibre in the Duddell and Einthoven oscillographs. The appearance on the mirrors of the rapidly vibrating spot is a continuous curve which represents the instantaneous values of current or voltage applied to the oscillograph. The British Woollen Research Association has recently devised a similar mirror stroboscope for viewing high-speed parts (spindles, etc.) in spinning-machines. *Method* (3), viz. interruption of the light at the object itself, is exemplified in a device due to M'Leod and Clark (1870), who used a cylindrical drum rotating about a vertical axis and covered on its curved surface with black paper ruled with white lines parallel to the axis and spaced about an inch apart. A horizontal beam of light, controlled by a mirror on the vibrating body, is projected on this drum, its oscillations being parallel to the white lines. When rotating at the correct speed, a white curve is seen representing the wave-form of the vibrations. This method is applied in the Hilger audiometer (p. 437), in which an extremely thin collodion membrane with a 'sputtered' silver mirror attached reflects a beam of light on the striped drum. The black parts of the drum are used to intercept the light to the eye, whilst the white parts reflect. When the wave appears stationary the frequency of vibration must be the product of the number of white lines and the revolutions per second of the drum (or a multiple of this). In *Method* (4) the source of light itself is cut off periodically. The introduction of the neon lamp * (1911) has rendered this method practicable and extremely convenient. The lamp, filled with neon gas at reduced pressure, glows brightly with a reddish coloured light when 180 volts, or thereabouts, are applied to its electrodes. If the voltage is reduced to, say, 150 volts the discharge fails and the light disappears.

* Now supplied commercially as the 'Osglim' lamp.

The effect of varying the voltage outside these limits produces practically instantaneous response in the glow. It may, in fact, be used as an intermittent light-source up to a frequency of the order of 10,000 p.p.s. The flashing of the lamp may be controlled by the Anson-Pearson method.* Using a condenser C in parallel with the lamp and a resistance R in series, the flash frequency may be adjusted by varying C or R. More generally, however, the lamp is operated by a tuning-fork fitted with electrical contacts which interrupt the primary current of a small induction coil, the secondary of which is connected to a neon lamp. This gives flashes of a fixed frequency. Alternatively the frequency may be varied by means of a commutator revolving at any required speed.

All the above methods depend for their success on the physiological effect known as 'persistence of vision.' If the frequency of interruption falls appreciably below 25 p.p.s. a definite 'flicker' is observed and the image on the retina appears no longer continuous. The methods are only applicable, of course, where the amplitude of the motion is easily perceptible to the naked eye. For this reason it is often necessary to introduce the methods of optical and mechanical magnification (mentioned above) before viewing stroboscopically.

* *Proc. Phys. Soc.*, **34**, pp. 175 and 204, 1922.

SECTION III

TRANSMISSION OF SOUND

The motion of a vibrating body is communicated to an elastic medium with which it is in contact and longitudinal waves are produced. The velocity with which these waves travel through the medium is dependent, as we have seen, on the fundamental physical quantities, elasticity and density. In fact, the satisfactory agreement between the observed and calculated velocity of sound waves is sufficient evidence of the longitudinal nature of the vibrations. When sound is transmitted through an 'infinite' medium it may spread spherically, or may, theoretically at any rate, remain in the form of a parallel beam of plane waves. In either case there is a diminution of intensity or attenuation, as the distance from the source increases, due to losses of a frictional character, viz. viscosity and heat conduction. When the sound waves near the source are of large amplitude, attenuation (or degradation) becomes extremely serious and the wave-velocity is also affected. Again, an *infinite* medium for sound propagation is a mathematical fiction. All media capable of transmitting sound are limited in extent, and sooner or later the wave must stop, or change from one medium to another. Obstacles in the medium give rise to the phenomena of reflection, refraction, interference, diffraction, etc., all of which we shall have to consider.

VELOCITY OF SOUND WAVES OF SMALL AMPLITUDE

The velocity c of small-amplitude elastic waves in any extended medium is given by equation (17), p. 52, viz. $c=\sqrt{\kappa/\rho}$, where κ is the appropriate elastic constant and ρ the normal density of the medium. It is important to note that the variations of density $\delta\rho$ involved in the transmission of the wave are always *small* compared with ρ, otherwise the relation for c is no longer valid. With regard to the elastic constant κ, this is the bulk modulus of elasticity in the case of a fluid medium. Solids, however, change in shape as well as in volume, consequently the coefficient of rigidity μ must be introduced and κ replaced by $\kappa'+\frac{4}{3}\mu$.

(*A*) Velocity of Sound in Gases

Newton made the first theoretical estimate of the velocity of sound in air, assuming that the relation between pressure and density was that given by Boyle's law, i.e. $p/\rho = p_0/\rho_0 =$ constant. On this assumption the bulk modulus of elasticity κ (see p. 51) is

$$\frac{dp}{dv/v_0} \quad \text{or} \quad \rho_0 \frac{dp}{d\rho} = \rho_0 \frac{p_0}{\rho_0} = p_0.$$

The velocity c will therefore be

$$c = \sqrt{\kappa/\rho_0} = \sqrt{p_0/\rho_0} \quad . \quad . \quad . \quad (1)$$

for small variations of ρ_0. Taking $p_0 = 10^6$ dynes/cm.^{2} approximately and $\rho_0 = 0{\cdot}001293$ for air at 0° C., the velocity c would be $2{\cdot}78 \times 10^4$ cm./sec., which is about 16 per cent. lower than the observed value. This discrepancy was not satisfactorily explained till Laplace pointed out the error in deducing the elasticity from Boyle's law, which only holds at constant temperature. Now the compressions and rarefactions in sound waves take place so rapidly that the heat developed in compression, or the cooling in rarefaction, has not time to be transferred to the surrounding medium. That is, the changes are not isothermal, but adiabatic, and the relation between pressure and volume or density is given by

$$pv^\gamma = \text{const.} = p_0 v_0^\gamma \quad \text{or} \quad \frac{p}{\rho^\gamma} = \frac{p_0}{\rho_0^\gamma}, \quad \text{that is,} \quad p = \rho^\gamma \frac{p_0}{\rho_0^\gamma},$$

where γ is the ratio of specific heats. From this we obtain

$$\frac{dp}{d\rho} = \gamma \rho^{\gamma-1} \frac{p_0}{\rho_0^\gamma}, \quad \text{whence} \quad \kappa = \rho_0 \frac{dp}{d\rho} = \gamma p_0 \quad \text{and} \quad c = \sqrt{\gamma p_0/\rho_0} \quad (2)$$

This is as we should expect, since

(Adiabatic elasticity) $= \gamma$. (isothermal elasticity).

The value of γ for air is approximately 1·41, consequently the value of the velocity obtained from Newton's isothermal formula must be multiplied by $\sqrt{\gamma}$, *i.e.* by 1·187. We then find $c = 3{\cdot}3 \times 10^4$ cm./sec. at 0° C., which is in good agreement with observation. This agreement appears to justify Laplace's assumption that the changes taking place in the medium traversed by the sound wave are adiabatic. Stokes has examined the question theoretically and has shown that the elasticity must be either isothermal *or* adiabatic—any intermediate condition would involve rapid dissipation of energy and the sound waves would not travel an appreciable distance through the air, a result which is contrary to experience.

Since the ratio p/ρ is constant at constant temperature, *i.e.* ρ changes in proportion to p, the *velocity is independent of the pressure.* At very high pressures, however, where departures from Boyle's law become noticeable, this is not strictly accurate.*

Influence of Temperature – The density of a gas varies with temperature $\theta°$ C. in accordance with the relation $\rho_\theta=\rho_0/(1+\alpha\theta)$, where α is the temperature coefficient of volume-expansion at constant pressure. The velocity c_θ at temperature $\theta°$ is therefore

$$c_\theta=\sqrt{\gamma p_0(1+\alpha\theta)/\rho_0}=c_0\sqrt{(1+\alpha\theta)}.$$

Now for any gas $\alpha=1/273$ per degree Centigrade nearly, therefore

$$c_\theta/c_0=\sqrt{T/T_0} \quad . \quad . \quad . \quad (3)$$

where T and T_0 are the absolute temperatures corresponding to $\theta°$ and 0° C. That is, *the velocity varies directly as the square root of the absolute temperature.*

Equation (2) indicates that *the velocity is independent of frequency.* It is, however, dependent on the nature of the gas, since the ratio of specific heats γ and the density ρ are involved. Provided γ is the same, *the velocity varies inversely as the square root of the density.* For example, the velocity is four times as great in hydrogen as in oxygen. When the values of ρ, p, and c are known the value of γ may be determined. This has direct application to the determination of the ratio of specific heats of rare gases (see p. 240).

Influence of Moisture – The velocity is only slightly affected by the humidity of the air. The presence of water-vapour produces a slight lowering of mean density ρ, the value of the ratio of specific heats γ being practically the same for dry air or air saturated with water-vapour. The calculated velocity in air saturated with moisture at 10° C. is from 2 to 3 ft./sec. greater than in dry air.

Relation between Sound-Wave Velocity and Molecular Velocity in Gases – As we have seen, the to-and-fro vibrations of the medium are transmitted forward from layer to layer with

* Witkowski, *Bull. Acad. Cracow*, p. 138, 1899; and P. P. Koch, *Ann. d. Physik*, **26**, p. 551, 1908.

a definite velocity. Regarded from the standpoint of the kinetic theory of gases, this to-and-fro motion must be superposed on the ordinary random motions of the gas molecules. The velocity with which the 'sound wave' is propagated must depend primarily, for small-amplitude waves, on the velocity of the molecular movements. Disregarding at first the adiabatic changes of temperature which take place due to rapid alternations of pressure in the sound wave, it is a simple matter to deduce the velocity of sound in terms of the velocity of the molecular movement. According to kinetic theory, the pressure in a gas arises from the to-and-fro motions of the particles and is transmitted in all directions due to this cause. In determining the velocity of sound we have to consider the value of that *component* of molecular motion corresponding to the direction of the sound wave. Consequently it follows that the velocity of sound must be less than the mean velocity of the molecules which move at random in all directions. In confirmation of this we know, for example, that in air at 0° C. the velocity of sound is 331 metres/sec., whereas the molecular velocity is given by $\sqrt{\overline{u^2}}=485$ and $\bar{u}=447$ metres/sec. ($\sqrt{\overline{u^2}}$ being the root mean square velocity and $\bar{u}$ the mean velocity of the molecules). The kinetic theory of gases shows very simply that the pressure in a gas is given by

$$p=\tfrac{1}{3}\mathrm{N}m\overline{u^2}=\tfrac{1}{3}\rho\overline{u^2} \quad . \quad . \quad . \quad (4)$$

where N is the number of molecules per c.c., m the mass of a molecule, and ρ the density of the gas.

This relation expresses the molecular velocity in terms of the pressure and density of the gas, thus

$$\overline{u^2}=\frac{3p}{\rho} \quad . \quad . \quad . \quad . \quad (5)$$

Since the energy of the motion of the molecules in a given direction, *e.g.* the x axis, is one-third of the whole energy,* we may write

$$\overline{u_x^2}=\frac{p}{\rho} \quad . \quad . \quad . \quad . \quad (6)$$

where u_x now refers to the component of molecular velocity along the x axis. But $\sqrt{\overline{u_x^2}}$ is identical with the velocity c of sound waves through the molecular system. Thus we may write

$$c=\sqrt{\overline{u_x^2}}=\sqrt{\frac{p}{\rho}} \quad . \quad . \quad . \quad (7)$$

* See O. E. Meyer, *Kinetic Theory of Gases*, pp. 18 and 74.

which is Newton's 'isothermal velocity' of wave-propagation in the gas. If we employ the mean molecular velocity $\bar{u}$ instead of $\sqrt{\overline{u^2}}$ we must write (4) in the form

$$p=\frac{\pi}{8}\cdot\rho(\bar{u})^2 \quad \text{since} \quad \sqrt{\overline{u^2}}=\bar{u}\sqrt{\frac{3\pi}{8}} \quad . \quad . \quad (8)$$

which leads to

$$c=\bar{u}\sqrt{\frac{\pi}{8}}=\sqrt{\frac{p}{\rho}} \quad . \quad . \quad . \quad (9)$$

As we know, this relation for the velocity of sound is inaccurate since no allowance has been made for the local heating and cooling of the gas, *i.e.* the local increase and decrease in velocity of the molecules, as the sound wave passes through it. The ratio of pressure to density only remains constant at constant temperature. If the latter quantity is allowed to fluctuate we must introduce the factor γ, the ratio of specific heats of the gas. Thus equation (9) for the velocity becomes

$$c=\bar{u}\sqrt{\gamma\frac{\pi}{8}}=\sqrt{\frac{\gamma p}{\rho}} \quad . \quad . \quad . \quad (10)$$

We arrive therefore at the simple relation which exists between the velocity c of sound in a gas and the mean velocity $\bar{u}$ of its molecules.

This relation also involves the ratio of specific heats of the gas, a quantity which also can be derived from molecular data. It is shown in textbooks on heat * that

$$\gamma=1+\frac{2}{r},$$

where r is the number of degrees of freedom of the molecule.

For monatomic gases . .	$r=3$ and	$\gamma=1{\cdot}66$.
,, diatomic gases . . .	$r=5$,,	$\gamma=1{\cdot}40$.
,, triatomic gases . . .	$r=7$,,	$\gamma=1{\cdot}29$.

It will be evident that a knowledge of the velocity of sound in a gas, at known pressure and density, provides a powerful method of determining valuable molecular data, viz. the mean velocity of its molecules, the ratio of specific heats and the number of degrees

* See, for example, Preston's *Heat*, p. 289.

of freedom of the individual molecules. In this connection the following table is of interest:—

Gas.	Atoms per Molecule.	Observed Ratio of Specific Heats, γ.	Mean Molecular Velocity, cm./sec. $\bar{u}$.	Observed Velocity of Sound, cm./sec.	$\bar{u}\sqrt{\gamma\frac{\pi}{8}}$, cm./sec.
Argon .	1	1·67	$3{\cdot}8 \times 10^4$	$3{\cdot}08 \times 10^4$	$3{\cdot}08 \times 10^4$
Hydrogen	2	1·41	17·4 ,,	12·86 ,,	12·9 ,,
Air .	2	1·40	4·47 ,,	3·31 ,,	3·31 ,,
Carbon dioxide	3	1·30	3·6 ,,	2·57 ,,	2·56 ,,

Equation (10), which gives the velocity of sound waves through the gas in terms of the molecular velocities, implies a small-amplitude sound wave, in which the additional velocities imparted to the molecules in the direction of wave-transmission are not comparable with the ordinary molecular velocities. An example will serve to illustrate this point. A sound wave of frequency 2000 p.p.s. and displacement amplitude 10^{-8} cm. (a sound of moderate intensity) will impart to the molecules in air a 'particle' velocity of $2\pi \times 2000 \times 10^{-8}$ cm./sec., *i.e.* $1{\cdot}2 \times 10^{-4}$ cm./sec., which is quite negligible compared with the mean velocity of random motion of the molecules $4{\cdot}47 \times 10^4$ cm./sec. The displacement-amplitude of the molecule due to the sound wave is in this case comparable only with the diameter of the molecule, of the order 10^{-8} cm., and is very small compared with the mean free path of the molecules of the order 10^{-5} cm. at N.T.P. It is only in the case of large-amplitude waves, which we shall consider later, that the wave-velocity requires a correction for the particle-velocity.

(*a*) Experimental Determination of the Velocity of Sound in Free Air

The velocity of sound in free air is affected by the wind. If the component of the velocity of the wind in the direction of the sound is v, then the observed velocity of sound will be $c \pm v$ according as the wind is 'with' or 'against' the sound, provided that v is constant and not unusually great (v^2/c^2 negligible) at the time of making the observations. The temperature and humidity affect the velocity in the manner we have seen.

Earlier Determinations – The first recorded careful determination of the velocity of sound in the open air appears to have been made by certain members of the Paris Academy in 1738 by what is now known as the *method of reciprocal observation*, which eliminates the wind correction. Cannons were fired at half-hour intervals alternately at the two ends of a base line 18 miles long, and the intervals between the flash and the report were noted by observers at each end of the base. Reduction of their observations gave a value of 332 metres/sec. in dry air at 0° C. A second series of experiments was made at Paris in 1822 by the Bureau des Longitudes. The method of reciprocal firing was used at intervals of five minutes, and accurate chronometers replaced the pendulum clocks of the first series. At 15·9° C. the mean velocity was found to be 340·9 metres/sec., whence $c=331$ metres/sec. at 0° C. In 1864 Regnault improved the method of reciprocal observation by the use of electrical recording, by which the 'personal equation' of the observer was eliminated. Regnault found, however, that his receiver, a stretched membrane with an electrical contact, also had a 'personal equation' which varied with the intensity of the sound wave. This error was evaluated and a correction applied. The instant of firing was recorded by the breaking of a wire, forming part of the electrical recording circuit, stretched across the muzzle of the gun. The mean values of velocity at 0° C. obtained were

From 0 to 1280 metres	$c=331{\cdot}37$ m./sec.
,, 0 to 2445 ,, ...	$c=330{\cdot}71$ m./sec.
Whence from 1280 to 2445 metres . .	$c=329{\cdot}9$ m./sec.

The decrease of velocity with diminishing intensity is, as we shall see later, in accordance with theory.

Observations of the *velocity of sound at low temperatures* (−38° F. to +35° F.) were made in the Parry Arctic Expedition. The reduced observations give a value 331 metres/sec. at 0° C., and an increase with temperature of approximately 0·55 metres/sec. per °C. A. W. Greely,* experimenting at temperatures between −10° C. and −45° C., gives a value of velocity $333+0{\cdot}6\theta$ metres/sec., where θ is the temperature in degrees Centigrade.

In 1853 J. Bosscha † determined the velocity of sound by what is known as *the method of coincidences*. The method was simplified by König in Paris (1863). The velocity is determined as follows: The sound (*e.g.* from an electromagnetic 'hammer' or 'tapper') is transmitted, at regular accurately known time-intervals, simultaneously from two points at a known variable distance apart.

* *Phil. Mag.*, **30**, 507, 1890. † *Pogg. Ann.*, **92**, 485, 1854.

When the sources are close together the taps are heard together, but when one source is moved further away the taps are separated by an interval, at first increasing and then decreasing till they again coincide. Increasing the distance continuously such coincidences are observed at equal intervals of distance. Thus if N is the number of signals per second and d the mean distance apart of the points of coincidence, the velocity c will be Nd. In 1864 Kahl further simplified the method by using only one source of sound, the other being replaced by the 'image' of the first reflected from a wall. The observer with the tapper moves away from the wall, noting the points at which direct sound and echo coincide. The velocity is then $c=2\mathrm{N}d$, where d is the distance apart of the points of coincidence. In another modification of this method the distance apart of the two 'tappers' (or of one and its 'echo') is fixed, whilst the number N of taps per second is varied until the taps are heard simultaneously. The accuracy of timing in the coincidence method is, according to Szathnari (1877), about 1/400 sec.

More Recent Determinations – Probably the most accurate measurements of sound velocity in the open air were made during the war in connection with location of guns by sound-ranging methods (see p. 478). E. Esclangon,* employing a refinement of Regnault's recording method, made an extensive series of velocity observations during 1917 and 1918 at temperatures from 0° to 20° C. and under various conditions of wind and humidity. The sound waves from guns of various calibres were received by electrical detectors (*e.g.* the tuned hot-wire microphone, see p. 402) at distances of 1400 metres and 14,000 metres along the same line. The instants of arrival of the sound wave at the receivers were recorded on an Einthoven string galvanometer, the time difference being considered accurate to ±·002 second. The final result for the velocity of sound in dry still air at 15° C. was given as 339·8 metres/sec., as compared with Regnault's value (obtained in 1864) of 339·7 metres/sec. Angerer and Ladenburg,† also recording on an Einthoven string galvanometer, measured the time interval between firing a charge of powder, which broke a wire in the recording circuit, and the arrival of the impulse at a microphone a known distance away. In the immediate vicinity of the explosion (where the wave is definitely of *large* amplitude) the

* *Comptes rendus*, **168**, p. 165, Jan. 1919; and *L'Acoustique des Canons et des Projectiles*, p. 254, Gautier Villars, Paris, 1925.

† *Ann. d. Physik*, **66**, 5, pp. 293–322, Jan. 1922, and *Engineering*, **113**, p. 529, April 28, 1922.

velocity was found to reach 1150 metres/sec. At greater distances, of the order of 10 km., the mean velocity, reduced to dry air at 0° C., was found to be 330·8 ±0·1 metre/sec. T. Vautier * has measured the velocity of weak explosive sounds such as are produced by electric sparks or percussion caps of an unloaded pistol. He gives a mean value for the velocity of small-amplitude waves of 330·57 metres/sec. at 0° C. *The value generally accepted* for sound-ranging purposes in the British and French armies was *337·16 metres/sec. in dry air*, or 337·6 metres/sec. in air of average humidity, *at 10° C.*

T. C. Hebb † has rightly pointed out that the accuracy of sound velocity measurements over a long base (*e.g.* of several miles) is impaired by uncertainties regarding temperature, wind, and humidity at different points along the base line. He considers that such sources of error, combined with errors due to personal equation and different intensities of source, are sufficient to explain the variation between different results. To overcome such difficulties, Hebb, at the suggestion of A. Michelson, made a direct measurement in 'free air' of the wave-length of a sound emitted from a source of known frequency. In this method, two parabolic reflectors M_1 and M_2 (of plaster of Paris, 5 ft. diameter, 15 in. focal length) are placed coaxially so that sound waves sent out from the focus of M_1 are collected at the focus of M_2. At the focus of each mirror is placed a telephone transmitter (carbon granular microphone), the two transmitters T_1 and T_2 being each connected in series with a battery and one primary winding of a telephone transformer. The latter has two primaries and a common secondary winding which connects to a pair of telephones. In addition to the microphones, a source of sound, a high-pitched whistle (2376 p.p.s.), was placed at the focus of M_1. The sound heard by the telephones will be the *vector* sum of the effects at T_1 and T_2. By varying the distance apart of M_1 and M_2 the relative phase of these effects will vary, at certain points annulling and at intermediate points reinforcing each other. This affords a direct method of measuring the *wave-length* λ of the sound. If the mirror M_2 be moved a hundred wave-lengths, and if the minima can be located to a tenth of a wave, the value of λ thus obtained will be accurate to 1°/₀₀. The temperature being known at all points between the mirrors and the frequency of the sound of the whistle adjusted to that of an accurately calibrated tuning-

* *Comptes rendus*, **181**, p. 1055, Dec. 1925.

† *Phys. Rev.*, **20**, p. 89, 1905, and **14**, p. 74, 1919.

fork, the velocity is readily obtained. Hebb gives a mean value of velocity $331{\cdot}29 \pm {\cdot}04$ metre/sec. at 0° C., as compared with a mean for previous reliable determinations of 331·75 metres/sec. Assuming the value 1·405 for the ratio of specific heats γ, as determined by Rontgen, the calculated velocity is 331·80 metres/sec.

(*b*) Velocity of Sound in Gases contained in Tubes

The majority of determinations of the velocity of sound in air or in other gases have been made with the gas enclosed in a pipe or tube. The reasons for this are sufficiently obvious, for the experiment is not only brought within the limits of a small laboratory, but the methods, or at any rate some of them, are applicable in the case of gases which are obtainable only in small quantities. The tube method also makes it possible to study the effects of variations in the factors which influence velocity, such as temperature, density, γ, pressure, and humidity. As we shall see, when we have to deal with attenuation in the transmission of sound waves (p. 317), the velocity of sound may be considerably modified due to the effects of viscosity and heat conduction when the gas is contained in a tube. The viscous forces and the heat loss in the gas near the walls of the tube tend to reduce the velocity, the compressions and rarefactions being no longer perfectly adiabatic. In a very narrow tube, therefore, the velocity will tend towards Newton's isothermal value $\sqrt{p/\rho}$, Laplace's adiabatic velocity $\sqrt{\gamma p/\rho}$ being applicable to tubes of large diameter only. H. Helmholtz,* and later G. Kirchhoff,† calculated the change of velocity due to these causes. The velocity c' in the tube of radius a is given by

$$c' = c\left\{1 - \frac{1}{a}\sqrt{\frac{\nu'}{4\pi N}}\right\},$$

where c is the ordinary velocity in a large volume of the gas, N is the frequency of the sound, and ν' is a constant known as the 'kinemetic viscosity and heat conduction coefficient,' which is equal to $2{\cdot}5\ \mu/\rho$ (μ being the ordinary 'static' viscosity coefficient and ρ the density of the gas (see p. 319). Experimental observations of the velocity of sound in pipes may be divided into two classes, (*a*) those typified by Regnault's observations, in which a direct measurement is made of the time taken for a sound pulse (*e.g.* an explosion wave) to traverse a known length of pipe; and (*b*) those based on Kundt's method in which the gas contained in

* *Crelles Journ.*, **57**, p. 1, 1859. † *Pogg. Ann.*, **134**, p. 177, 1868.

the pipe is set into resonant vibration, the velocity being determined from observations of wave-length and frequency.

Direct Method – V. Regnault (1862–3),* using the electrical recording method, measured the time taken for the pressure wave emanating from a small explosion to travel through the air enclosed in tubes of various lengths up to 4900 metres, and diameters varying between 10 and 110 cms. In the longer tubes intermediate receivers were sometimes used, and in all cases a record was made of the successive reflected waves from the ends of the tube. He found that the velocity increased with the diameter of the tube, reaching the ordinary value in very wide tubes. Thus in a tube 10·8 cm. in diameter c=324·25 metres/sec., whilst in a tube 110 cm. diameter c=330·3 metres/sec. at 0° C., in agreement with calculation. Regnault also verified in these experiments that the velocity is independent of pressure between 247 mm. and 12,670 mm. of mercury. Subsequent analysis of Regnault's observations by Rink gives a mean value of the velocity, in a large diameter pipe (110 cm.), *for small-amplitude waves* of 330·5 metres/sec. at 0° C. We shall refer again to these experiments when dealing with waves of large amplitude. Further observations of a similar nature to those of Regnault were made later by Violle and Vautier † (c=331·1 metres/sec.) and by the Phys. Tech. Reichsanstalt, Berlin, 1907 (c=331·92 metres/sec. in dry air at 0° C. and free from CO_2).

Kundt's Dust Tube Method ‡ – Reference has already been made to this method as a means of observing the motion of air in pipes (see p. 178). Its main importance lies, however, in the application to the measurement of the velocity of sound in small quantities of gas under controlled conditions of temperature, pressure, purity, etc. In the simplest arrangement the gas is contained in a cylindrical glass tube (say 5 cm. diameter and about 80 cm. long) containing a sprinkling of fine, dry powder (lycopodium seed or cork dust), and is thrown into *resonant vibration* by any convenient means, *e.g.* (1) by the longitudinal vibrations of a rod (clamped at the mid-point), one end of which is inserted in the tube; or (2) by similar vibrations set up by stroking the glass containing-tube; or (3) by means of electromagnetic, piezo-electric, or other form of vibrator mounted at

* *Comptes rendus*, **66**, p. 209, 1868. † *Phil. Mag.*, **26**, p. 77, 1888.
‡ A. Kundt, *Pogg. Ann.*, **127**, p. 497, 1866, and **135**, pp. 337, 527, 1868.

one end of the tube. At resonance, the dust is thrown into vigorous vibration, forming heaps or discs at equidistant intervals along the wave-tube (see fig. 57, p. 181). The distance apart of these discs can be measured with considerable accuracy, and is equal to half a wave-length of the sound in the gas contained in the tube. The velocity $c=\sqrt{\gamma p/\rho}=\mathrm{N}\lambda=2\mathrm{N}d$, where N is the frequency, λ the wave-length, and d the distance apart of the dust heaps in the tube. In order that the experiment shall be successful it is necessary that (*a*) the vibrator (*e.g.* the end of the rod) at the end of the glass tube shall not be seriously damped by touching the walls of the tube, (*b*) the length of the tube shall be adjusted accurately for resonance, and (*c*) the interior of the tube and the dust shall be carefully dried—the tube should preferably be rotated so that the dust is on the point of slipping. If the vibrations are set up by electromagnetic means as from the diaphragm of a 'loud speaker' or telephone excited from a valve-oscillator, the length of the tube may be fixed and the frequency of excitation varied until resonance occurs. The experiment is capable of a number of variations having important applications:

(1) Absolute determination of velocity of sound in a gas, the frequency N of the sound being found by recognised methods

$$c=\mathrm{N}\lambda \quad (\lambda=2d).$$

(2) Comparison of velocities of sound in a gas and in a solid rod (the longitudinal vibrations of the rod setting the gas in the wave-tube into resonance),

$$c_{\text{gas}}=\mathrm{N}\lambda_{\text{gas}}=2\mathrm{N}d \quad \text{and} \quad c_{\text{rod}}=\mathrm{N}\lambda_{\text{rod}}=2\mathrm{N}l$$

$$(l=\text{length of rod}).$$

Therefore $$d/l=c_{\text{gas}}/c_{\text{rod}}.$$

(3) Comparison of velocities of sound in two different gases (double-tube method). The same rod is used to excite resonant vibrations in *two* wave-tubes, one at each end. Then if d_1 and d_2 are the respective nodal separations in the tubes, and c_1 and c_2 the corresponding velocities, $d_1/d_2=c_1/c_2$.

(4) Determination of the ratio of specific heats γ of a gas. The method is the same as in (1), $c=\sqrt{\gamma p/\rho}=2\mathrm{N}d$,

whence $$\gamma=4\mathrm{N}^2d^2\rho/p.$$

When gases are available in small quantities only (*e.g.* argon, helium, krypton, etc.) this is the only known experimental method of determining γ. It is also the most convenient method for the determination of γ for vapours of volatile liquids.

(5) Measurements of the variation of velocity with temperature, pressure, humidity, or with the nature of the gas. The velocity of sound in mixtures of gases and in vapours has been measured by this method.

The value of the velocity determined by Kundt's tube is, of course, subject to the Helmholtz-Kirchhoff correction for the radius of the tube (see p. 238). The correction may be eliminated experimentally in the following manner:—

Writing $\sqrt{\nu'/4\pi N}=k$ (see p. 238), the velocity in a tube of radius a_1 is $c_1=c(1-k/a_1)$, and in a tube of radius a_2 the velocity $c_2=c(1-k/a_2)$, for the same gas and frequency of excitation. Thus on eliminating k, the velocity c in a large volume of gas is given by $c=(a_2c_2-a_1c_1)/(a_2-a_1)$. The relative values of c_1 and c_2 may be found by direct experiment by the double-tube method (3) above, enclosing the same kind of gas in tubes of different radii, and using the same source to excite the vibrations. The two prongs of a tuning-fork may alternatively be used to excite the two tubes.

By the dust-tube resonance method, Kundt showed for a given gas that the velocity increased with the diameter of the tube and with the wave-length (as Kirchhoff's correction indicates). He showed also that the velocity was independent of the pressure and proportional to the square root of the absolute temperature. Values of the velocities in different gases were also obtained in terms of the velocity in air. No tube correction is required in this comparison if tubes of equal diameter are employed for the gases to be compared.

Kundt and Warburg * determined the ratio of specific heats of mercury-vapour by the double-tube method. One tube contained air and the other mercury-vapour at high temperature, fine quartz dust being used to indicate the nodal points. The value of γ obtained was 1·66, indicating the monatomic nature of the vapour. Rayleigh and Ramsay † determined γ for argon and helium by Kundt's tube method, obtaining a value of 1·66 in each case. Behn and Geiger ‡ have introduced a modification of the double-tube method for use in such cases where the gas is chemically pure and must necessarily be enclosed. The gas

* *Pogg. Ann.*, **157**, p. 353, 1876.

† *Phil. Trans.*, **186**, p. 187, 1895, and *Proc. Roy. Soc.*, **8**, p. 86, 1895.

‡ *Verh. d. Deuts. Phys. Ges.*, **9**, p. 657, 1907.

is led into a closed wave-tube by a T-piece at the midpoint, which is clamped. The wave-tube is excited into vibration by stroking or by electromagnetic means, and *its frequency is tuned to that of the enclosed gas* by loading the ends. One end is then inserted into the air-wave tube for comparison of velocities, the vibrations of the closed wave-tube serving to excite both 'gas' and air columns into resonance.

The method of excitation of the dust tube employed by Andrade and Lewer (see p. 179) permits of considerable accuracy in the measurement of the antinodal distance ($\lambda/2$) between the 'dust discs' and of the frequency (N) of resonant vibration—data which yield an accurate value of velocity ($N\lambda$). The method is extremely simple in practice, and many interesting phenomena relating to the amplitude and direction of vibration of the air particles in different parts of the tube are readily observed.

Kundt's method of determining the ratio of specific heats of gases lends itself to a study of the number of atoms per molecule in gases. Einstein * has suggested that it might be applied to cases of partial dissociation, *e.g.* $N_2O_4 \rightleftarrows 2NO_2$.

Velocity by Resonant Gas Column - In Kundt's method special means are employed for indicating resonance in a column of gas and for measuring the length of the stationary waves set up in the tube. In discussing the modes of vibration of a column of gas in a wide tube closed at one end (p. 171) it was shown that the fundamental frequency N is given by $c/4l$ (where l is the effective length of the tube), and the frequencies of the overtones were 3, 5, 7, etc. times that of the fundamental. Consequently if N and the effective length l are known the value of c for the gas contained in the tube may be determined. The end correction δ at the open end of the tube may be eliminated by finding two successive lengths, l_1 and l_2, of the tube which resonate to the same frequency N, *the difference between these lengths* (l_2-l_1) *must be equal to half a wave-length of the sound in the tube* (see p. 171); the successive resonating lengths will be ($\lambda/4-\delta$), ($3\lambda/4-\delta$), ($5\lambda/4-\delta$), and so on, each differing by $\lambda/2$ from the one on either side of it. This provides a useful laboratory method of measuring the velocity of sound ($c=N\lambda$) in air, a tuning-fork of known frequency N being used as the source of sound. The successive resonant lengths, adjusted aurally, may be obtained by a gradual variation in the height of a water column closing the lower end of a vertical tube. D. J. Blaikley † carried out an extensive

* *Berlin Ber.*, **3**, 380, 1920. † *Proc. Phys. Soc.*, **6**, p. 228, 1884–5.

series of measurements of a similar nature to this. He used a tube with a pear-shaped cavity near the mouth of a cylindrical portion which was closed at the other end by a sliding piston. The pipe was excited by blowing, and the object of the pear-shaped mouth was to render the natural overtones of the pipe inharmonic, and therefore would not affect the fundamental vibration in any way. The tube was made to 'speak' with a certain position of the piston, then the latter was moved outwards to a second position which gave the same note. The distance between the two positions of the piston is equal to $\lambda/2$. Experimenting with tubes of various diameters, Blaikley obtained good agreement with the Kirchhoff velocity formula given on p. 238. His final value for the velocity of small-amplitude sound waves in free air at 0° C. was 331·68 metres/sec., and the ratio of the specific heats 1·4036. G. W. Pierce * has recently used a modification of this method as an instrument of precision. His apparatus is shown in fig. 72. A brass resonance tube 120 cm.

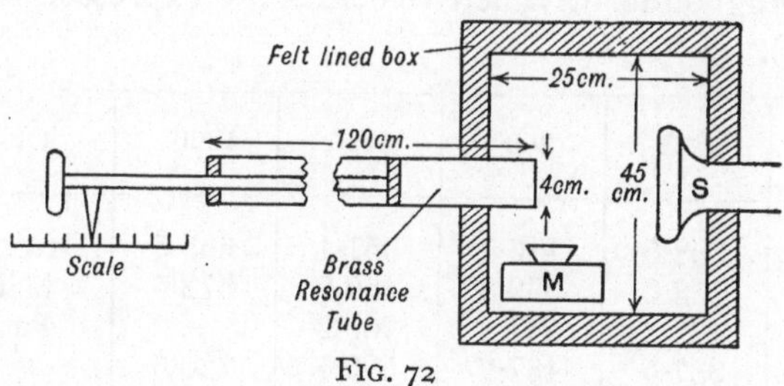

FIG. 72

long and 4 cm. internal diameter, provided with a piston and scale, was mounted in a felt-lined box containing also a telephone earpiece S and a microphone M as shown. The telephone was excited electrically from a tuning-fork of known frequency (995·88 p.p.s.). The distances between S, M, and the tube were so adjusted that the sound direct from S to M just neutralised the sound from the tube to M when the sound emitted by the tube was a maximum. This gave a fiducial point of *silence* in the middle of what would otherwise be a maximum, thus permitting of very accurate setting and measurement of the half wave-length displacements of the piston. The microphone M communicated through a transformer with a telephone receiver on the head of the observer. By means of an electric filter in this circuit the fundamental frequency of the sound (995·88 p.p.s.) could be cut

* *Proc. Amer. Acad.*, **60**, p. 284, 1925. See also G. D. West, *Journ. Sci. Instrs.*, **6**, p. 254, Aug. 1929.

out, and the harmonic $3 \times 995{\cdot}88$ ($=2987{\cdot}8$) p.p.s. used instead. Pierce found for the velocity of sound in air $331{\cdot}94 \pm 0{\cdot}07$ m./sec. at 0° C., uncorrected for the effects of the tube.

Velocity of Sound in Gases and Vapours at Different Temperatures – The resonance tube methods of determining the velocity of sound in gases and vapours have been employed by numerous investigators. E. H. Stevens * determined the velocity in air and various organic vapours at temperatures between 0° and 185° C. A. Kalahne,† using a somewhat different arrangement, a telephone diaphragm exciting the tube into resonance, worked up to temperatures of 900° C. In both cases the velocity was found to vary as the square root of the absolute temperature. Dixon, Campbell, and Parker ‡ made observations of velocity in argon, nitrogen, CO_2, and methane at temperatures from 0° to 1000° C., whilst S. R. Cook § measured the velocity in oxygen and air at various temperatures down to −183° C. (see following table in which velocities are expressed in metres per second).

	0° C.	300°	600°.	1000°.	Observer.
Argon .	308·5	446·5	551·1	665·5	Dixon, Campbell, and Parker.
CO_2 .	258·3	369·3	450·9	572·5	
Methane	429·2	587·3	709·2	..	
Nitrogen	337·5	487·2	599·4	720·6	

	+21° C.	−28·4°.	−66·5°.	−137·5°.	−183°.	
Oxygen .	328·5	282·4	264·3	210·1	173·9	S. R. Cook.

Cook's observations indicate a fall of velocity with temperature somewhat greater than theory predicts on the assumption of constant density and ratio of specific heats.

Measurements of the velocity in water vapour by various observers ‖ give values increasing from 401 metres/sec. at 0° C. to 424 metres/sec. at 130° C.

* *Ann. d. Phys.*, **7**, p. 285, 1902.

† *Ann. d. Phys.*, **11**, p. 225, 1903, and **20**, p. 398, 1906.

‡ *Proc. Roy. Soc.*, **100**, p. 1, 1921.

§ *Phys. Rev.*, **23**, p. 212, 1906.

‖ W. Jaeger, *Wied. Ann.*, **36**, p. 165, 1889; W. Trietz, *Dissert. Bonn.*, 1903; and A. Lechner, *Wiener. Ber.*, **118**, p. 1035, 1909.

(*B*) Velocity of Sound in Liquids

The relation $c=\sqrt{\kappa/\rho}$ for the velocity of sound in gases is equally applicable to liquids in bulk. Strictly speaking, we must also use the adiabatic coefficient of volume elasticity $\gamma\kappa$, *not* the isothermal value found by static measurements. It is true that in certain cases the difference between isothermal and adiabatic volume elasticity is small, thus for water at 8° C. $\gamma=1{\cdot}001$, affecting the velocity only 1 part in 2000, but in others, *e.g.* ethyl ether at 18° C. $\gamma=1{\cdot}32$, it may become very important. In the case of sea-water (35°/₀₀ salinity) γ varies from 1·0004 at 0° C. to 1·0207 at 30° C.; and for the same water at 0° C. the ratio increases from 1·0009 at 100 bars pressure to 1·0126 at 1000 bars (1 bar=0·987 atmosphere).* For accurate calculation of velocity, therefore, it is necessary to know γ the ratio of specific heats. It is shown in textbooks of thermodynamics † that

$$\gamma=\frac{1}{1-\alpha^2\kappa_\theta \mathrm{VT}/c_p},$$

where α =coefficient of cubical expansion,
κ_θ= ,, ,, isothermal elasticity,
V =volume of unit mass,
T =absolute temperature,
c_p =specific heat at constant pressure (expressed in ergs).

The velocity of sound in a liquid is therefore given by $c=\sqrt{\gamma\kappa_\theta/\rho}$ where γ has the above value.

(a) *Liquids in Bulk* – Water, either fresh or salt, is the only liquid which permits of direct measurements of velocity over appreciable distances. The first recorded observations in this medium are those of Colladon and Sturm ‡ who made a direct measurement of the velocity of sound in Lake Geneva. A submerged bell was struck by a hammer which simultaneously 'flashed' a quantity of powder, thereby indicating to an observer at the other side of the lake the instant of striking the bell. A 'quarter-second' stopwatch was used to measure the time-interval between the flash and the arrival of the sound signal. Consider-

* See D. J. Mathews, 'Physical Oceanography,' *Dictionary of Applied Physics*, **3**, pp. 665–692; and *Tables of Velocity of Sound in Pure Water and in Sea Water for use in Echo-Sounding and Sound-Ranging*, Hydrographic Dept., Admiralty, H.M. Stationery Office Publication, 1927.

† See, *e.g.*, Poynting and Thomson's *Heat*, p. 289.

‡ *Ann. d. Chim. et Phys.*, **36**, pp. 113 and 225, 1827.

ing the circumstances of the experiment, the final value of the velocity, 1435 metres/sec. at 8·1° C., is in surprisingly good agreement with the calculated value 1436·4 metres/sec.

Practically all recent determinations of the velocity of sound in water in bulk have been made by firing explosive charges in the sea. These determinations were rendered imperative by the rapid progress in methods of measurement of depths and distances in the sea by means of sound signals. With such objects in view experiments were organised by the French, British, and American navies. The German hydrographic department also studied the theoretical aspect of sound velocity in sea-water with special reference to depth sounding.* The theoretical values of velocity in sea-water at different temperatures, salinities, and pressures (depths) have been calculated by D. J. Mathews (*loc. cit.*) using Knudson's values of salinity and density and Ekman's † values of elasticity. As will be seen, these calculated values are in good agreement with direct experimental determinations.

In 1919 M. Marti ‡ made a careful determination of velocity in the sea in Cherbourg roadstead at a depth of 13 metres. Three accurately surveyed hydrophones (microphones on diaphragms) were laid on a straight line at intervals of 900 metres approximately. The sound wave was produced by an explosion of a charge of dry guncotton on an extension of the line of hydrophones, first on one side then on the other, about 1200 metres from the nearest hydrophone. The passage of the explosion wave over the microphones was recorded automatically on a smoked drum chronograph, a tuning-fork of frequency 500 p.p.s. traced a 'time mark' alongside the explosion records. The mean value of the velocity reduced to a temperature of 15° C. in sea-water of density 1·026 at a pressure of one atmosphere, was found to be 1504·15 metres/sec. The agreement with the value calculated by Mathews 1500·9 is not so good.

During the years 1920–22 an extensive series of velocity measurements in the sea was made at St Margaret's at Cliffe, Dover, for the British Admiralty by the writer and H. E. Browne, assisted in the later stages by C. Cochrane.§ Four microphone

* H. Maurer, *Ann. d. Hydrogr.*, **52**, p. 75, 1924; and A. Schumacker, **52**, p. 780, 1924. See also F. Aigner, *Unterwasserschalltechnik*, Berlin, 1922; and Special Publication No. 4 (March 1925) of International Hydrographic Bureau, Monaco.

† *Public. des Circonstances de Conseil Permanent pour l'Exploration de la Mer*, No. 43.

‡ *Annales Hydrographiques*, 1919–1920.

§ *Proc. Roy. Soc.*, **103**, p. 284, 1923.

receivers (hydrophones) were laid at intervals of about 4 miles, on a line approximately N. and S., to the eastward of the Goodwin Sands. The positions of these hydrophones were accurately surveyed, from Ordnance Survey points in Kent, by the Hydrographical Department of the Admiralty, the positions being known to ± 7 ft., *i.e.* 1 in 10,000 of the total base line. The time intervals of passage of the explosion wave between the various pairs of hydrophones were recorded on a film by a six-stringed Einthoven galvanometer with photographic recorder.* The four hydrophones were connected by cables to the corresponding galvanometer strings, whilst a fifth string was operated (through a microphone circuit) by the half-second ticks of a chronometer whose daily rate was accurately known by comparison with Eiffel Tower W/T time-signals. The sixth string of the galvanometer was used to record a 'wireless' signal sent from a destroyer at the instant of firing a charge. The film was traversed every tenth and hundredth of a second by the time marks of a phonic motor controlled by a tuning-fork, these time marks being checked by the ticks of the accurate chronometer. Time-intervals could be estimated to ±·001 second, which represents a high degree of accuracy when it is considered that the sound wave took about fourteen seconds to traverse the whole base line. The final accuracy, however, depends on a knowledge of the sea conditions. During the velocity measurements, continuous observation of sea temperatures and salinities were made from the destroyer and at the North and East Goodwin Lightships, using deep-sea reversing thermometers (with N.P.L. calibrations). An electrical recording thermometer laid near the base line gave a *continuous* record of sea temperature. A correction was at first applied to measured velocities on account of tidal flow (of the order 1 or 2 ft./sec.), but later experiments were made at neap tide slack water, when the tidal stream ceases to flow for upwards of an hour, and no tidal correction is required. Preliminary observations showed that there was no difference in velocity when the explosive charge was increased from 2¼ to 300 lb.—the distance of the charge from the nearest hydrophone being about twelve miles. No certain variation of velocity with depth of charge, 10 to 108 ft., could be detected. A considerable number of observations of velocity were made by the *Radio-acoustic Method* (see p. 478), *i.e.* the time-interval was measured between the instant of firing the charge, indicated by the wireless signal, and the arrival of the sound at the

* A standard pattern made by the Cambridge Scientific Instr. Co.

hydrophones. On account of slight uncertainties (such as the position of the charge, the *exact* moment of firing, the temperature and salinity of the sea at points off the base line) these results were not relied on in estimating the velocity (although they agreed within a few feet per second with the more accurate measurements).

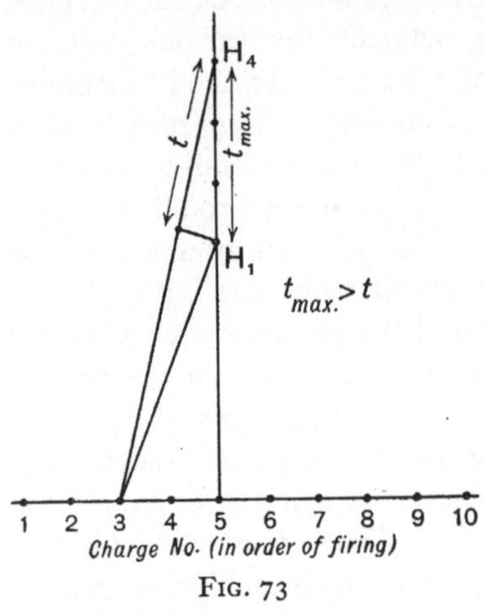

FIG. 73

In the more accurate method, however, a serious error may arise if the position of the charge is uncertain in an east and west direction (the base line running north and south), the 'effective' length d' of the base line then being less $(=d \cos \theta)$ than the actual length d. To eliminate this error *a new method, known as the 'Multiple-charge Method,'* was devised. In this method about ten small charges 1, 2, 3, etc., were fired, at approximately equal intervals of time and distance, along a line at right angles to an extension of the base line of hydrophones, H_1, H_2, H_3, etc. (see fig. 73). It will be evident that the recorded time-interval between a pair of hydrophones will be a maximum for that charge which is nearest 'in line.' On either side of the base line the effective distance and consequently the recorded time-interval will be less. To make this clear, the results of such a series of actual observations are shown in fig. 74, where the maximum time-interval t_{max} is clearly shown near charge No. 5. In this particular case the distance apart of the hydrophones was 70,245±7 ft.

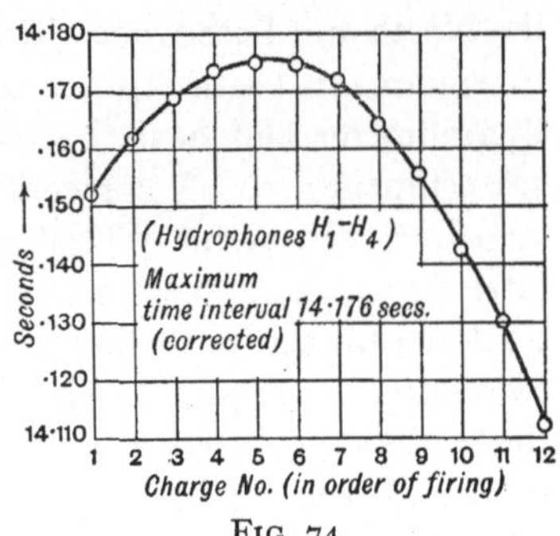

FIG. 74

and $t_{max}=14{\cdot}176\pm{\cdot}001$ sec., whence $c=4955{\cdot}4\pm1$ ft./sec., the mean temperature and salinity on the base line being 16·95° C. and 35·02°/₀₀ respectively.

Experiments by this method at different seasons of the year gave the relation between velocity and temperature. Comparative measurements of velocity in Gareloch (Clyde) and at Dover indicated a salinity coefficient of velocity of +3·73 ft./sec. per 1°/₀₀ increase of salinity. The results of the complete investigation are summarised in the following expression for velocity—

$$c=4626+13{\cdot}8t-0{\cdot}12t^2+3{\cdot}73\ \mathrm{S\ feet/sec.}$$

or

$$c=1410+4{\cdot}21t-0{\cdot}037t^2+1{\cdot}14\ \mathrm{S\ metres/sec.},$$

where t is the temperature °C. (within the range 6° to 17° C.) and S is the salinity in parts per thousand (*e.g.* 35°/₀₀ at Dover). The observed values of velocity indicate a value $\gamma=1{\cdot}0094\pm0{\cdot}0005$ for the ratio of specific heats of sea-water, of 35°/₀₀ salinity at 17° C., as compared with $\gamma=1{\cdot}0090$ calculated from thermodynamic data.

Measurements of velocity in the sea by the radio-acoustic method have also been made by E. B. Stephenson,* who obtained a value 1453·3±2 metres/sec. at 0·3° C. and S=33·5°/₀₀, as a mean of four observations. Similar measurements have been made by E. A. Eckhardt.† For the reasons given above, however, it is considered that velocity measurements made over long distances by this method are open to serious criticism.

A question of some importance in oceanic depth-sounding by echo methods (see p. 467) relates to the variation of velocity with depth. In certain oceans, for example the North Atlantic, depths exceeding 4000 fathoms (about 4½ miles) are sometimes encountered. The pressure near the bottom in such cases becomes very great. The elasticity and density of the water vary with pressure and consequently with depth, the elasticity increasing more rapidly than the density with increasing pressure. Values of elasticity, density, and velocity have been calculated by D. J. Mathews ‡ at depths from 0 to 30,000 ft. (5000 fathoms), at which depth the pressure reaches 1000 atmospheres (approx. 67 tons/sq. in.), and the velocity has increased by about 160 metres/sec. above the 'surface' value. Heck and Service § have also published a list of tables giving velocity in terms of pressure and depth. In the same paper they describe an experimental investigation in *U.S.S. Guide* to determine the variation of velocity with depth by means of an echo method, using the 'Sonic depth finder' developed by H. C. Hayes.‖ The ship, equipped also with standard apparatus for taking 'wire' soundings, temperatures,

* *Phys. Rev.*, p. 181, Feb. 1923. † *Phys. Rev.*, **24**, p. 452, 1924.

‡ *Loc. cit.* See also Ekman (*Pub. de Circ.*, No. 49), Maurer and Schumacker, *Ann. d. Hydr.*, pp. 86 and 93, 1924.

§ U.S. Coast and Geodetic Survey, Publication No. 108, 1924.

‖ *Journ. of Frank. Inst.*, **197**, p. 323, 1924.

water samples, etc., cruised in the North Atlantic Ocean, the Caribbean Sea, and the Pacific Ocean, measurements of velocity being made in depths from a few fathoms to 4500 fathoms. The 'sonic' apparatus measured the time-interval for a sound impulse to travel from the surface to the bottom and back again, it being assumed that the echo returned from a point vertically below the ship (the accuracy of this assumption depending on the slope of the bottom). The results are rendered very uncertain due to (1) the unreliability of 'wire' soundings at great depths on account of the drift of the ship, and (2) to the weakness of the echoes at such depths. Considering the difficult nature of the observations, the agreement between observed and calculated velocities was fairly good.

(b) *Liquids in Tubes* – In discussing the production of stationary waves in air or gas-filled tubes, it was assumed that the walls of the tube were rigid. The same theory would apply also to liquid-filled tubes, were it not for the fact that the assumption regarding the walls of the containing tube is no longer valid. At the same frequency and intensity, the displacement-amplitude of a sound wave in air is 60 times greater than in water, the pressure amplitude being correspondingly less. If stationary waves are formed in a thin-walled tube filled with a liquid such as water, the pressure variations at the nodes may be very great and the tube will yield radially at those points, causing the pressure to drop. The actual pressure-fluctuation at the nodes in the liquid will therefore be less than in the case of a *rigid* tube. The yielding of the wall of a tube containing a resonating liquid column can readily be detected, for audible frequencies, with a stethoscope applied to the outside of the tube. The outer surface of the tube will appear to have antinodes and nodes corresponding respectively to the nodes and antinodes in the liquid. The yielding of the tube near the nodes results in a lowering of the wave-velocity in the liquid. This was first demonstrated experimentally by Wertheim (1847), who was successful in exciting liquid-filled organ pipes into resonance by 'blowing.' The pipes were fitted with suitable mouthpieces immersed in the liquid, and were 'blown' with a jet of the same liquid. In this manner Wertheim obtained, from the frequency of the note and the length of the pipe, a value 1173 metres/sec. for the velocity in a water-filled pipe. The correct explanation of this low velocity is, however, due to Helmholtz (1848), who predicted that the velocity would increase with increase in the thickness of the walls and with decrease in the diameter of the tube. H. Lamb * has dealt with the matter theoretically, assuming that the deformation of the tube has a statical value corresponding to the instantaneous distribution of pressure in the liquid. This assumption is justified by the fact that the wave-velocities in solids (*e.g.* glass or steel) are considerably greater than in liquids. If c_0 be the velocity of sound in the liquid 'in bulk,' and c the actual velocity in the tube of radius a and thickness h (assumed small), then

$$c_0 = c\sqrt{1+2\kappa a/h\mathrm{E}},$$

where κ is the volume elasticity of the liquid and E Young's modulus for the material of the tube. When the walls are *very thick* it is found that

$$c_0 = c\sqrt{(\kappa+\mu)/\mu},$$

where μ is the rigidity. Kundt and Lehmann applied the dust-tube method (see pp. 174, 178, and 239) to the determination of velocities of sound in liquids. Fine iron filings or powdered sand (free from air) were sprinkled along the tube. On exciting the liquid into resonance the powder collected at the nodes in the usual manner. The results confirmed Helmholtz's predictions, and application of Lamb's formula to their experimental values gives $c_0 = 1436$ metres/sec. at 19° C. for the velocity of sound in open water (fresh). K. Dorsing † has measured in this way the velocities of sound in air-free distilled water, salt solutions, alcohol, ether, etc. A. Cisman ‡ has recently measured, by a method similar to Regnault's for gases, the velocity of sound in various liquids in twelve different tubes. The liquids were contained in U-tubes, one end being closed by an iron disc which, on attraction by an electro-magnet, generated a wave of rarefaction. The time of passage of the wave was recorded photographically. The values obtained, after applying the necessary 'tube correction,' agreed in all cases, within ±2 per cent., with the velocities in the free medium. The liquids used were: distilled water, salt solutions of different strengths, alcohol, ether, benzine, toluol, carbon disulphide, and carbon tetrachloride.

It should be mentioned that resonance may easily be set up in a liquid column by means of a steel diaphragm closing one end

* *Dynamical Theory of Sound*, p. 175.

† *Ann. d. Phys.*, **25**, p. 228, 1908.

‡ *Journ. de Phys.*, **7**, pp. 73 and 345, 1926; Jonescu, *Journ. de Phys.*, **5**, p. 377, 1924; Veenekamp, *Dissert.*, Gottingen, 1922; and H. G. Green, *Phil. Mag.*, **45**, p. 907, 1923.

of the tube and excited electromagnetically, using current of variable frequency from a valve oscillator. At frequencies within the audible range, the resonance within the liquid can be observed by ear, and the yielding of the walls of the tube examined by a stethoscope. This provides a very convenient laboratory method of determining the velocity of sound in liquids.

The velocity of sound in fluid mixtures, *e.g.* air-water, is considered on p. 326.

(*C*) Velocity of Sound in Solids

The transmission of waves through fluids is relatively simple. Apart from surface waves, depending on surface tension s and gravitational forces * (velocity of ripples $=\sqrt{2\pi s/\rho\lambda}$ and of gravity waves $=\sqrt{g\lambda/2\pi}$), only longitudinal waves need be considered in fluid media. A perfect fluid will not transmit shearing forces, so that transverse waves are not possible. In solids, however, which readily transmit both compressional and shearing forces, longitudinal, torsional, and transverse waves may be set up. We have already referred to the more important cases, viz.:

(1) Transverse waves in wires, $c=\sqrt{\mathrm{T}/m}$, in which the elastic properties of the material may be disregarded (see p. 85).

(2) Transverse waves in bars, $c \propto \frac{1}{\lambda}\sqrt{\frac{\mathrm{E}}{\rho}}$ (see p. 109), the velocity depending on wave-length λ, as well as on the elasticity E (Young's modulus) and the density ρ of the material.

(3) Longitudinal waves in wires and bars, $c=\sqrt{\mathrm{E}/\rho}$ (see p. 136).

(4) Torsional vibrations in wires or bars, $c=\sqrt{\mu/\rho}$ (see p. 152).

In all these cases the solid is restricted in certain dimensions, *e.g.* the diameter relative to the length.

(a) *Solids in Bulk* – Any local disturbance in an unlimited solid medium breaks up into two types of wave, which ultimately assume transverse and longitudinal characters. The longitudinal or 'dilatational' wave, in which the displacement is wholly in the direction of propagation, has a velocity $c=\sqrt{(k+\frac{4}{3}\mu)/\rho}$ or $\sqrt{\{(1-\sigma)\mathrm{E}/(1-\sigma-2\sigma^2)\rho\}}$, where k is the volume elasticity, μ the rigidity, E Young's modulus, and σ Poisson's ratio. In the second type, the transverse or 'distortional' wave, the displacement is everywhere at right angles to the direction of propagation and the

* Rayleigh, *Sound*, 2, p. 344.

velocity becomes $c=\sqrt{\mu/\rho}$. It was pointed out in dealing with the longitudinal vibrations of a bar that transverse motion also takes place, this motion becoming more and more important as the cross-section of the bar increases relatively to the length. The vibrations of a *limited* solid may therefore become very complex. Even in an extended solid the circumstances of wave-propagation may be seriously modified by the existence of a free surface. This is exemplified in the case of earthquake disturbances. Seismograph records at considerable distances from the source or 'focus' indicate three distinct sets of waves, (1) preliminary tremors, dilatational waves, which travel *direct* through the earth from source to recorder with a velocity about 10 km./sec., (2) distortional waves, also direct, with a velocity about 6 km./sec., and (3) large-amplitude waves, known as 'Rayleigh waves,' which Rayleigh (1885) ascribed to surface waves, analogous to large water waves, penetrating only to a small depth and propagated by elastic forces in the earth's crust. On account of the longer, circumferential, path of the Rayleigh waves they are delayed, relatively to types (1) and (2), more than in proportion to the wave-velocities. Love * has attempted to deduce the volume-elasticity k and rigidity μ of the interior of the earth from the various wave-velocities indicated in seismograph records.

Apart from such records, experimental information relating to the velocity of transmission of sound waves in extended solids is very scanty. This is, of course, mainly due to the fact that the only available solid of very large bulk is the earth's crust, which consists of a heterogeneous mixture of minerals having different values of elasticity and density. During recent years the velocity of sound in strata of different compositions (mineral ore, 600 metres/sec.; soft earth, 300 metres/sec.) has been determined, and applied to the location of mineral strata by echo methods † (see p. 474).

(b) *Limited Solids*. *Bars* – A direct determination of the velocity of sound in a long solid bar, or in a tube, was made by Biot. A bell mounted at the end of a long chain of cast-iron pipes (950 metres) was struck with a hammer. At the other end of the pipe two sounds were received: the first *via* the iron, the second *via* the air in the pipe. The time interval t between these two sounds gives a measure of the velocity of sound in iron c_i in terms of the known velocity in air c_a. Thus, if l is the length of the pipes,

* *Phil. Trans.*, A, **207**, p. 215, 1908.
† See Berger, *Schalltechnik*, p. 87, 1926.

$c_i \doteq lc_a/(l-tc_a)$. The value thus obtained for cast iron was 3500 metres/sec. Provided the diameter of a bar or wire is small compared with its length, the velocity $\sqrt{E/\rho}$ of longitudinal waves in it may be determined most simply by observing its fundamental frequency N of longitudinal vibration. The various methods of exciting such vibrations have already been dealt with on p. 138 *et seq*. At audible frequencies, the most desirable point of clamping the bar is at the middle (see Case II, p. 136) when the restraining action of the clamp has a negligible influence on the frequency of vibration. When the bar sounds its fundamental note, the frequency, which may be determined directly by means of a siren or a calibrated monochord, is given by $N=c/2l$, where l is the length of the bar and c $(=\sqrt{E/\rho})$ is the longitudinal wave-velocity required. When the bar is used to excite an air column into resonant vibration, as in Kundt's tube experiment (see p. 240), the velocity c_{bar} may be compared directly with the velocity c_{air} of longitudinal waves in air. In this case

$$c_{bar}/c_{air}=l_{bar}/l_{air},$$

where l_{air} refers to the half wave-length separation of the dust heaps in the tube.

The method is sometimes used, alternatively, to determine Young's modulus E for a bar in which the velocity of sound is assumed. If the bar is obtainable only in a short length, its fundamental frequency may be so high that it is not possible to estimate its pitch, or it may be above the audible limit. In such a case Kundt's tube is invaluable, for the dust heaps at the nodes in the wave-tube give at once the wave-length of the sound in air, and consequently the frequency N. The velocity in the bar is then $c_{bar}=2Nl$ where l is its length. Such methods are, of course, equally applicable to a wire which may be mounted on a sounding-board and excited into longitudinal vibration by drawing a resined leather along the wire. For the comparison of the velocities in two wires or bars the lengths may be varied until they emit a sound of the same frequency, then $c_1/c_2=l_1/l_2$. Wertheim measured the velocity of sound in a series of bars of different metals, in terms of the frequency and length, obtaining values in good agreement with calculation $(c=\sqrt{E/\rho})$. The *velocities of sound in wax and other soft solids*, which are not easily excited into resonance by recognised methods, were determined by Stefan. He cemented the 'soft' bar on the end of a resonant metal bar in which the velocity was known. This resonant bar was used to excite vibrations in the soft bar, the velocity being deduced by a somewhat involved calculation. In wax at 17° C. Stefan obtained $c=800$ metres/sec., -40 metres/sec. per 1° C. rise of temperature. Warburg determined the velocities of sound in soft materials (stearine, wax, paraffin, tallow) by coupling a bar of the material with a bar of glass and setting the combination into *transverse* vibration. The nodal distances were obtained by means of sand, whence the longitudinal velocities could be calculated (see p. 112). Warburg's value for the velocity in wax, 880 metres/sec. at 16° C., agreed fairly well with Stefan's estimate. More recently, Herbolsheimer * has determined the velocity of a pressure-pulse in gelatine in bulk and in the form of strips or bars. Polarised light was passed through two points, a known distance apart, and after passing through an analyser (Biot's method †) was received on a photographic film with time marks. The passage of the pressure pulse produced a sudden illumination of the film, due to stress in the gelatine rotating the plane of polarisation, and the time-interval was recorded. The velocities in *gelatine strips* (containing varying percentages of gelatine and sugar in a solution of glycerine and water) varied between 5·90 and 10·80 metres per second, the velocity increasing slowly with time. For *gelatine in bulk*, however, the velocities were found to approximate to 1600 metres/sec., *i.e.* in the neighbourhood of the velocity of sound in water. Kellog, measuring the elasticity E and velocity of sound in indiarubber, found that E varied from $2{\cdot}7\times10^7$ at 0·1 p.p.s. to $6{\cdot}4\times10^7$ at 200 to 600 p.p.s., corresponding to velocities of 52 and 80 metres/sec. respectively.

In this connection it is interesting to observe that the ratio of longitudinal wave-velocities for solids in bulk c_0 and in the form of a bar c_l is given by

$$\frac{c_0}{c_l}=\sqrt{\frac{(1-\sigma)E}{(1-\sigma-2\sigma^2)\rho}}\Bigg/\sqrt{\frac{E}{\rho}}=\left\{\frac{(1-\sigma)}{(1-\sigma-2\sigma^2)}\right\}^{\frac{1}{2}} \quad \text{(see p. 252),}$$

where σ is Poisson's ratio. Taking $\sigma=0{\cdot}49$ for indiarubber, $c_0/c_l=17$. If σ were equal to 0·5 the ratio would be infinite, that is, the longitudinal wave would *spread* infinitely rapidly. Herbolsheimer's observations with gelatine strips and in bulk therefore indicate a value of σ approximately 0·5. For a substance like cork, $\sigma=0$ approximately, whence the longitudinal wave-velocity in a rod will be the same as in bulk.

* *Zeits. f. Physik*, **3**, p. 182, 1920. † See Tyndall, *Sound*, p. 168.

Velocity in Crystalline Solids – The elasticity of a crystalline substance varies along different axes, and as a consequence the velocity of elastic waves has a different value according to the direction of propagation. In most crystals the variation has no practical interest, but in the case of quartz, which is now extensively used as a standard of high-frequency vibrations, the question may become important. P. W. Bridgeman * has measured the compressibility of quartz rods, cut along different axes, and has found at 30° C. that the ratio of compressibilities parallel and at right angles to the optic axis is

$$\frac{\text{parallel}}{\text{perpendicular}}=\frac{0{\cdot}753\times10^{-6}}{1{\cdot}072\times10^{-6}}=\frac{1}{1{\cdot}42},$$

whence the ratio of velocities (proportional to $\sqrt{1/\text{compressibility}}$) is 1·19 to 1.

The Velocity of Torsional Waves in Bars may be determined in a similar manner to that described for measuring the velocity of longitudinal waves. The bar, clamped at the midpoint, is excited into torsional vibration of frequency N, whence

$$c=\text{N}\lambda=2\text{N}l,$$

as before. In this case, we have already seen (p. 152) $c=\sqrt{\mu/\rho}$, which can be calculated from the physical constants, rigidity μ and density ρ, of the material. The ratio of longitudinal to torsional velocity is $\sqrt{\text{E}}/\sqrt{\mu}$ or $\sqrt{2(\sigma+1)}$. In the case of steel, taking $\sigma=0{\cdot}3$, the ratio becomes 1·6. Since σ can never be greater than 0·5, the ratio of longitudinal to torsional velocities can never exceed 1·73 in any material. The ratio of velocities, equal to the ratio of frequencies, is easily obtained experimentally and provides a very simple means of determining Poisson's ratio σ for the material of the bar.

VELOCITY OF SOUNDS OF HIGH FREQUENCY

When the frequency of the sound is above the audible limit, or too high to be estimated by ear, the velocity is most conveniently measured by a method due to Rayleigh, viz. *the stationary wave method.* In Rayleigh's original experiment † the source of sound was a bird call (see p. 208) placed about two metres from a plane

* *Proc. Amer. Acad.*, **58,** p. 163, 1923. See also R. B. Sosman, *Properties of Silica* (Chem. Catalogue Co. Inc., U.S.A.).

† *Sound*, **2,** p. 403.

reflector, a sheet of metal or glass. Between the source and the reflector a sensitive flame (see p. 371) could be moved to and fro to locate the positions of the nodes and antinodes in the stationary waves formed by superposition of the direct and reflected waves. The flame was least disturbed at the nodes, these positions being well defined by the *sudden* cessation of flaring. The nodal separation being $\lambda/2$, the velocity may be obtained from the relation $c=N\lambda$ *provided the frequency is known.* In the bird-call experiment the frequency in general is *not* known, the method being more suitable to determine the *frequency* in terms of an assumed velocity. Recent developments in the production of supersonic vibrations of *accurately known frequency* by piezo-electric crystals and magnetostrictive rods (nickel, etc.) have rendered Rayleigh's stationary wave principle a method of precision in the determination of velocities in gases and liquids. In this connection the work of Pierce in the cases of gases, Boyle, Wood, Hubbard and Loomis for liquids, is noteworthy.

(a) *In Gases* – In 1907 W. Altberg * measured the wave-length of the sounds emitted (of frequency $N=340{,}000$) from an electric spark, and found that the product $N\lambda$ was in moderately good agreement with the recognised value for the velocity of sound in air. A. Campbell and D. W. Dye † obtained a similar result at 900,000 p.p.s., using a high-frequency spark as source and a Kundt's dust tube to indicate the wave-length. A series of measurements of the velocity of sound waves of very high frequency was made in 1923 by K. Palaiologos.‡ Supersonic vibrations were produced in air by imposing high-frequency alternating currents on the direct current in an arc. The arc current was thus made to fluctuate at a frequency equal to that of the A.C., and thereby became a source of sound (see p. 204). The wave-length of the sound emitted was measured by means of diffraction gratings (see p. 309); a concave mirror with the high-frequency (short wave-length) sound source (the arc) at its focus produced a parallel beam, which, after passing through the grating, was focussed by a second concave mirror on the detector. The wave-length was calculated from the usual grating formulæ. Palaiologos found, for frequencies 2×10^5 to 2×10^6 p.p.s. ($\lambda=1{\cdot}7$ mm. to $0{\cdot}17$ mm.), that the velocity of sound was constant and equal to 335 metres/sec. at 0° C.

The oscillations of a quartz piezo-electric resonator have been

* *Ann. d. Physik*, 23, 1907. † *Electrician*, 66, p. 862, March 1911.
‡ *Zeits. f. Physik*, **12**, p. 375, 1923.

applied by G. W. Pierce * to the precision measurement of the velocity of high-frequency sounds. A crystal of quartz, cut in the manner described on p. 141, is mounted in front of a smooth reflecting surface as shown in fig. 75 (*a*). The crystal is main-

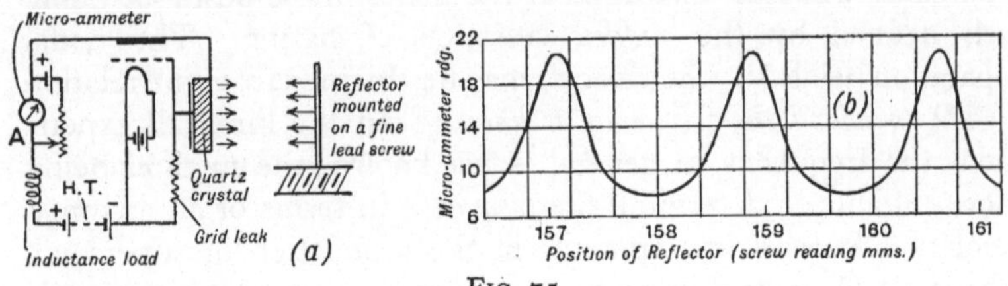

FIG. 75

tained in continuous oscillation at its own natural frequency by means of the circuit shown. The sound waves thus produced are reflected back to the crystal and form a system of stationary waves. The reflector may be moved to and fro on a fine 'lead' screw to explore these waves and to measure the wave-length. No additional apparatus for detecting the sound is necessary, for the reflected sound wave, falling on the emitting face of the crystal vibrator, even when the reflector is in some cases at a distance of 300 half-waves away, reacts on the crystal with sufficient force to cause readings of the milli-ammeter A to fluctuate visibly in accordance with the phase of arrival of the reflected sound wave. In order to render the fluctuations more evident for precision measurements, a sensitive galvanometer (or micro-ammeter) is used at A shunted by a potentiometer as shown, to neutralise the main current through the instrument. A set of typical observations given by Pierce is shown in fig. 75 (*b*). The maxima could be located to 0·05 mm. and a hundred or more successive half-waves may be counted. The accuracy of wave-length measurement is of the order of 1 in 3000. The frequency of the vibration of the crystal was determined by accurate wave-meter methods as in standard radiofrequency measurements.† The ultimate time standard in such measurements is a master clock, and the connecting links in the measurement, a standard tuning-fork and a multivibrator (see p. 132). It was proved by Pierce that the reaction of the reflected waves on the crystal vibrator had a negligible effect on its frequency. As a result of such measurements Pierce came to the following conclusions:—

* *Proc. Amer. Acad.*, **60**, p. 271, 1925. See also Hubbard and Loomis, *Phil. Mag.*, **5**, p. 1177, 1928.

† See also *Dictionary of Applied Physics*, **2**, Article on 'Radio Frequency Measurements,' p. 627.

(1) The velocity of sound in free air for a range of frequencies between 40,000 and 1,500,000 p.p.s. varies with frequency in a peculiar manner (see fig. 13 of Pierce's paper). At 0° C. the values were 331·94, 332·47 (max.), and 331·64 metres/sec. at frequencies 1000, 50,000, and $1{\cdot}5 \times 10^6$ p.p.s. respectively. The effect of humidity was negligible; at 80 per cent. humidity the velocity differed by less than 0·02 per cent. from the velocity in dry air.

(2) In CO_2 the velocity at 0° C. was found to be 258·52 metres/sec. at a frequency of 42,000 p.p.s., increasing to 260·15 metres/sec. at 200,000 p.p.s. At still higher frequencies CO_2 was found to become *opaque to sound waves*. The theoretical significance of these results is not yet clear. They appear to suggest, however, that the pressure variations of sound waves are not quite adiabatic for frequencies below 50,000 p.p.s. in air, or below 200,000 p.p.s. in CO_2. We shall refer later (p. 321) to Pierce's observations on the absorption of high-frequency sounds in air and CO_2. Pierce proposes to make velocity measurements in gases using magnetostrictive rods * (nickel-iron alloys, etc.) to replace quartz. The new method is particularly suitable at frequencies below the range of the quartz oscillator, *i.e.* below 40,000 p.p.s. The above method of measuring velocity by means of quartz oscillators has recently been applied by E. Griffiths † as a means of determining the percentage of CO_2 (or other gases) mixed with air. The distance apart of the nodes is an indication of the composition of the gaseous mixture, assuming it to be composed of gases which do not react chemically.

(b) *In Liquids* – The method has also been employed, in a somewhat different manner, by R. W. Boyle and J. F. Lehmann ‡ to demonstrate the existence of stationary waves under water when a beam of plane waves of high frequency is opposed by a similar beam in the opposite direction. Two quartz transmitters in a tank of water were arranged 60 cm. apart and face to face. They were connected in parallel to the same valve oscillator so that each radiated energy of the same frequency. Whilst both were emitting sound waves of 96,200 p.p.s. particles of coke dust were sprinkled in the track of the beam and eventually settled on a whitened plate suspended just below the centre of the energy field. The particles of dust were forced by radiation-pressure

* *Amer. Acad. Proc.*, 63, p. 1, 1928.
† *Proc. Phys. Soc.*, **39**, p. 300, Aug. 1927.
‡ *Trans. Roy. Soc. Canada*, **19**, p. 159, 1925.

into the planes of minimum intensity of vibration, and ultimately settled in lines at the base of each nodal surface, the distance apart of these lines being $\lambda/2$. In a particular experiment there were 44 half wave-lengths in 34·86 cm., which gives a value $1{\cdot}52 \pm {\cdot}03 \times 10^5$ cm./sec. for the velocity in water at 15° C. (waves of 96,200 p.p.s.).

R. W. Boyle and G. B. Taylor,* using a single quartz oscillator and a reflector to produce the stationary waves, attempted to measure the variation of velocity of sound in water at different frequencies from 43,000 to 600,000 p.p.s. Their earlier results indicate a falling off in velocity from $1{\cdot}51 \times 10^5$ to $1{\cdot}42 \times 10^5$ cm./sec. at these frequencies respectively. This reduction of velocity with increase of frequency is not explicable on the assumption that viscosity is the cause (see p. 318). Using a very viscous oil at high frequencies (6×10^5), they found the velocity to be the same as at the low frequency. In a later paper,† Boyle and Taylor ascribe the apparent decrease in water to a possible progressive error in the wave-meter used to determine the frequency. A continuation of the experiments under more exact experimental conditions indicated *no detectable* change in velocity from 29,000 to 527,000 p.p.s. The mean value of the velocity in water over this range of frequencies was found to be $1{\cdot}48 \times 10^5$ cm./sec. ± 1 per cent. at 18·5° C. A similar set of observations of the velocity of ultrasonic waves in castor oil of viscosity 0·50 at 20° C. gave a constant value $1{\cdot}43 \times 10^5$ cm./sec. over the whole range of frequency. R. W. Wood, A. L. Loomis, and J. C. Hubbard ‡ have made measurements of the wave-length of high-frequency sounds in various liquids by a method analogous to that of Pierce in gases. A quartz piezo-electric oscillator 10 cm. in diameter, vibrating at frequencies 200,000 to 400,000 p.p.s., produced practically plane waves in a small dish of liquid placed upon it. These waves were reflected from a plane surface immersed in the liquid, a fine micrometer screw adjusting the distance of this surface from the face of the quartz disc with which it was parallel. As the reflector passed through each nodal point of the stationary waves in the liquid, a neon lamp loosely coupled to the oscillating circuit was extinguished due to the reaction of the system of stationary waves upon the crystal. It was thus possible to measure 20 or more half wave-lengths with a precision

* *Trans. Roy. Soc. Canada*, **19**, p. 197, 1925.
† *Ibid.*, **21**, p. 79, 1927.
‡ *Phil. Mag.*, **4**, p. 417, 1927, and *Nature*, Aug. 6, 1927.

of 0·01 mm. Typical results are included in the following table:—

VELOCITY OF SOUND IN METRES/SECOND

Liquid.	Temperature °C.			
	5°.	15°.	25°	35°.
Distilled Water . .	1439	1477	1509	1534
1 per cent. NaCl Solution .	..	1487	1520	1542
2·5 " " "	..	1510	1539	1561
5·1 " " "	..	1540	1569	1589
Mercury	..	..	1469	1468
Carbon Disulphide . .	1215	1184	..	..
Chloroform . . .	1066	1027	..	..

No variation in velocity with frequency could be detected. The velocity was also independent of the material and dimensions of the containing vessel, a result which would be anticipated with a source (large compared with $\lambda=3$ to 8 mm.) of plane waves. It was found that small quantities of dissolved air materially affect the velocity.

(c) *In Solids* – The velocity of sound c in a solid bar may at once be deduced from a knowledge of its length l and the frequency N of its fundamental tone when in longitudinal vibration ($c=2Nl$). When the bar is very short, however, the frequency cannot be determined by aural methods and special means for its measurement are necessary. Kundt's dust tube is perhaps the simplest of such methods. On striking one end of a short metal bar, the other end of which is inserted in the 'dust tube,' the stationary wave dust pattern is very easily produced, and the wave-length in air determined, whence the frequency N of the sound can be calculated, assuming a knowledge of the velocity in the air contained in the dust tube. The velocity in the bar is then equal to $2Nl$, where l is the length of the rod. In this way R. J. Lang * determined the velocity of longitudinal waves in bars of steel and brass, varying in length from 200 to 5 cm., and in frequency from 1280 to 50,150 p.p.s. The ratio of the velocities of the sound waves in air and in the metal bars was found to remain constant (within experimental error) over the whole range of frequencies.

* *Trans. Roy. Soc. Canada*, **16**, p. 163, 1922.

The absolute values of velocities in steel and brass were found to agree with the values calculated from the static elasticity E and density ρ—that is, from $c=\sqrt{E/\rho}$. The ratio of adiabatic to isothermal elasticities (that is γ) of a metal may be calculated by the method indicated on p. 245 for liquids. For iron, the ratio is found to be 1·0026; whence it might be anticipated that the velocity determined by direct experiment would be 1·3°/$_{oo}$ higher than the value calculated from the isothermal or static elasticity. This difference is, however, within the limits of error of Lang's measurements. In a paper on "Magnetostriction Oscillators," G. W. Pierce * describes a study of the velocity of sound, and its variation with temperature, in a number of alloys exhibiting magnetostrictive properties. A graduated series of 'stoic' metal bars (about 36 per cent. Ni and 64 per cent. Fe) from 20·8 to 6·9 cm. in length give a mean value of (length $\times$ frequency) of 2079·6$\pm$2, and consequently a mean velocity of 4160 metres/sec. at 20° C. No definite change in velocity was observed within the frequency-range 10,000 to 30,000 p.p.s. of these rods. It was found that the velocity in the Ni-Fe alloys has a minimum and the temperature coefficient a maximum value at 36 per cent. to 40 per cent. nickel —that is, about the composition of Invar and Stoic metal.

The velocity of high-frequency vibrations in a piezo-electric material like quartz is best determined from a knowledge of its linear dimensions and its natural frequencies. The latter may be determined accurately by the methods indicated on pp. 132 and 142, or approximately by means of a Cady or a Pierce oscillator circuit (see p. 145) combined with a simple form of electrical wave-meter (a known variable inductance or capacity and an indicator of resonance such as a neon lamp or an electrostatic volt-meter). On p. 144 is given a table of frequencies of the partials of two quartz bars of lengths 47·0 and 62·2 mm. The fundamental frequencies are 58,675 and 44,120 p.p.s. respectively, corresponding to an average velocity $c=2Nl=5{\cdot}50\times10^5$ cm./sec. in quartz. There is no indication in the table (p. 144) that there is any progressive variation of velocity with frequency up to 3 or 4×10^5 p.p.s.

The velocity of sound in a bar may conveniently be determined if the bar is excited into resonant vibration by electromagnetic, piezo-electric, or by magnetostrictive methods. The source of electrical excitation is a valve-maintained oscillator of continuously variable tuning and the frequency of resonance is determined

* *Amer. Acad. Proc.*, **63**, 1, p. 1, 1928.

from the constants of the electrical oscillating circuit or preferably by means of a wave-meter. R. W. Boyle * has pointed out that cautious judgment must be exercised when determining a resonant frequency, particularly in regard to the overtones. His observations show that the vibratory modes in high-frequency bars may be very complex—longitudinal, torsional, and flexural vibrations may take place simultaneously. At certain resonant frequencies, dust particles sprinkled on the *end* of a cylindrical bar in vibration arranged themselves in patterns similar to some of Chladni's figures, *e.g.* four-, six-, eight-, and twelve-pointed stars could be obtained.

VELOCITY OF SOUND WAVES OF FINITE AMPLITUDE

We have assumed hitherto that the displacement amplitude and the condensation are always small, in which case the sound waves travel through the medium with velocity $c=\sqrt{\kappa/\rho}$ without change of type. In the majority of cases this assumption is quite justifiable. When the condensation s becomes large, however, as in an explosion wave, the velocity may be considerably modified. The curve connecting pressure p and density ρ in the medium cannot be regarded as a straight line for large fluctuations about the normal condition. The question has been examined mathematically by Poisson, Stokes, Earnshaw, Riemann, Rankine, Rayleigh, and others,† who show that a progressive wave of finite condensation cannot be propagated without change of type, the velocity of propagation in any part of the wave being dependent on the value of the condensation s at that point. Referring to the equations for the propagation of plane waves (see pp. 51 and 52) we have

$$\frac{d^2\xi}{dt^2}=-\frac{1}{\rho_0}\frac{dp}{dx} \quad . \quad . \quad . \quad (1)$$

Now $(p/p_0)=(\rho/\rho_0)^\gamma$ for adiabatic changes, and we may write

$$\rho/\rho_0=(1+s)=(1+\Delta)^{-1}=(1+d\xi/dx)^{-1}.$$

Consequently

$$p=p_0\left(1+\frac{d\xi}{dx}\right)^{-\gamma} \text{ and } \frac{dp}{dx}=-\gamma p_0\frac{d^2\xi}{dx^2}\left(1+\frac{d\xi}{dx}\right)^{-(\gamma+1)},$$

* *Nature*, p. 13, Jan. 5, 1929.

† See Rayleigh, *Sound*, **2**, p. 32 *et seq.*; Lamb, *Sound*, p. 174 *et seq.*, and *Hydrodynamics*, p. 470.

whence

$$\frac{d^2\xi}{dt^2}=\frac{\gamma p_0}{\rho_0}\left(1+\frac{d\xi}{dx}\right)^{-(\gamma+1)}\cdot\frac{d^2\xi}{dx^2} \quad . \quad . \quad (2)$$

It will be seen that for small values of $(d\xi/dx)$ this equation approximates to equation (18), p. 52, for small-amplitude waves travelling with a velocity $c_0=\sqrt{\gamma p_0/\rho_0}$. The value of the velocity c for large-amplitude waves relative to the undisturbed medium, as given by equation (2), may therefore be written

$$c=c_0\left(1+\frac{d\xi}{dx}\right)^{\frac{-(\gamma+1)}{2}} \quad . \quad . \quad . \quad (3)$$

or

$$c=c_0(1+s)^{\frac{\gamma+1}{2}} \quad . \quad . \quad . \quad . \quad (4)$$

In these expressions for the velocity of a large-amplitude wave we may write $s=\delta\rho/\rho$. The velocity of propagation, therefore, increases with increase of density or condensation. This results in a change of type in the wave as it advances, for the velocity in the more highly compressed parts of the wave is greater than in the less compressed or rarefied portions. Consequently the parts of the wave of greater density gain continuously on the parts of smaller density. The wave becomes steeper in front and more gradual behind as indicated in fig. 76. Such a process would ultimately lead to a perpendicular wave-front beyond which the analysis has no meaning. As Rayleigh points out, this tendency is held in check by the divergence of the wave and the influence of attenuation (viscosity and heat conduction), both of which tend to reduce the condensation and consequently the velocity to the 'normal' or small-amplitude values. Change of type in a progressive wave may be illustrated by sea-waves approaching a shelving beach. On account of the forward velocity of the water near the crests being greater than that near the troughs (*i.e.* near the sea-bed), the crests gain on the troughs and the wave-fronts become steeper and steeper until they curl over and break. Rayleigh's explanation * of the change of type in a pressure wave as it advances will help to make it clear. From the ordinary theory

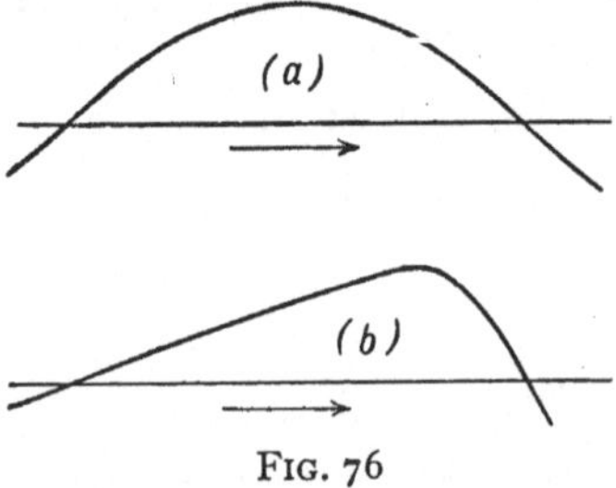

FIG. 76

* *Sound*, 2, p. 33.

of propagation of small disturbances the wave advances with a velocity c_0 relative to the undisturbed medium. Imagine now a wave so long that the fluctuations of particle-velocity and density are insensible for a moderate distance along it, and at a place where the particle-velocity (u) is finite we shall superpose a small secondary wave. The velocity with which this wave travels through the medium is c_0, but on account of the local motion of the medium itself, the velocity of advance will be (c_0+u) and will depend on the part of the long wave in which it is placed. This applies to the parts of the long wave itself; thus after a time t a point of particle-velocity u will be in advance of its original position by a distance $(c_0+u)t$; that is, the particle-velocity u will be transmitted with velocity (c_0+u). Symbolically $u=f\{x-(c_0+u)t\}$, a relation first obtained by Poisson (1808).

Experimental Observations – Measurements of the increase of velocity of waves of finite condensation are numerous. Regnault (1862–3) found that explosion waves in air (contained in pipes, see p. 239) increased in velocity with size of charge, and diminished in velocity with increasing distance. He found, as theory anticipates, that the limiting velocity at large distances (*i.e.* small condensation) was the same in all cases. In all such methods of velocity measurements, which include the region of large condensation near the source, the velocity will be abnormally high. Reference has already been made to the observations of Angerer and Ladenberg (p. 236), which indicated a velocity of 1150 metres/sec. in the vicinity of an explosion, falling to 330 metres/sec. at a distance of 10 kilometres. The velocity at any point distant x from such a source has been calculated by Riemann (1859), and may be stated in the form

$$c=\frac{dx}{dt}=c_0\left(1+\frac{k}{x^2}\right)^{\frac{1}{2}},$$

in which k is a quantity depending on the thickness of the highly condensed region of the pressure pulse. It will be seen that the velocity of the pulse of finite amplitude approximates to the normal 'small condensation' velocity c_0 when the distance x from the origin is large. The rigorous mathematical treatment of waves of large condensation is very complex even when it is limited to the case of plane waves. Actually, we have to deal with spherically divergent waves in viscous media, the condensation s diminishing as the distance from the source increases. This case has received a good deal of attention experimentally, and a number of ingenious photographic methods have been devised for studying the progress of a highly compressed pulse. The first of these to which we shall refer is *Töpler's "Schlieren Method"* for the observation of atmospheric striæ (described on p. 342). In 1867 Töpler * showed by an original optical method that the spark accompanying an impulsive electric discharge, as a Leyden jar, produced a highly compressed wave or pulse in the surrounding atmosphere. These experiments were continued by his son.† Mach and Gruss,‡ extending Töpler's experiment, demonstrated that this pulse had a velocity greater than that of sound and that the velocity increased with increase of spark intensity. The Schlieren method has been employed more recently by W. Payman, H. Robinson, and W. C. F. Shepherd,§ who have obtained very complete data relative to the velocity and form of the pressure pulse emerging from the mouth of a tube, due to the sudden release of compressed air or the explosion of a gaseous mixture or a detonator. Not only did they obtain 'instantaneous' spark photographs of the pressure pulse at different instants after leaving the tube, but they also obtained a continuous record indicating the progress of the wave-front with time and distance from the origin. From this record it is simple matter to deduce

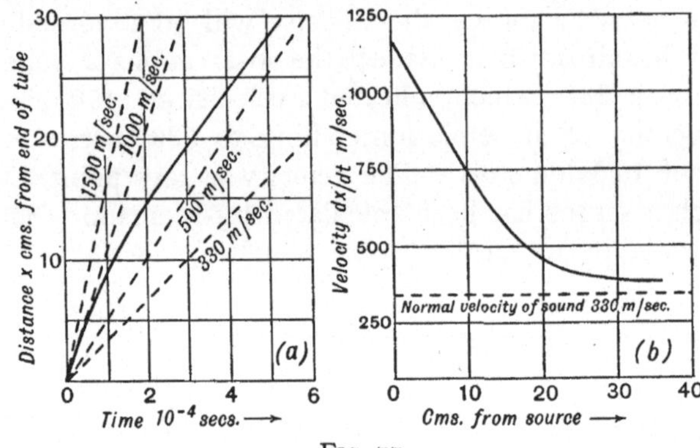

FIG. 77

the variation of velocity with distance from the source (the open end of the tube). A tracing from such a record is shown in fig. 77 (*a*), which indicates the distance x traversed by the pressure pulse in any time t. The slope of the curve dx/dt at any point x

* *Ann. der Physik*, **131**, p. 33, 1867.
† *Ibid.*, **27**, pp. 1043, 1051, 1908.
‡ *Wien Ber.*, **78**, 2, p. 467, 1879.
§ Safety of Mines Research Board, Papers Nos. 18 and 29, 1926.

from the origin indicates the instantaneous velocity at that point. The derived curve, fig. 77 (*b*), in which dx/dt is plotted as a function of x, therefore indicates the variation of velocity with distance from the source.* It will be seen that the velocity of the compressed spherical pulse falls from 1150 metres/sec. at the source (the mouth of the tube) to about 380 metres/sec. at 30 cm. distance. The spherical wave was still observed 6 metres away, or about 0·03 second after firing the charge, its speed, 360 metres/sec., being then slightly greater than the normal velocity of sound, although the gaseous products of the detonation ceased to move forward beyond a distance of 1 metre. 'Following-waves' due to the reflected pulse from the opposite end of the detonating tube were also recorded, the velocities being dependent on the composition of the burnt gases. Waves produced by solid particles travelling at high speed (600 to 800 metres/sec.) and breaking through the main 'sound wave' are clearly shown in some of the photographs.

The second photographic method of observing the progress of high-velocity impulses is known as *Dvorak's Shadow Method* † (described on p. 340). This method is more direct and has been more frequently used than the Schlieren method. C. V. Boys ‡ in 1892 obtained photographs of bullets in flight, and the accompanying pressure waves in the air, by the shadow method. Similar photographs have been obtained more recently by Quayle § and by Cranz.|| Foley ¶ used the method to photograph the pressure pulse sent out from an intense electric spark. The shadow of the high-pressure region forming the envelope of the pulse was photographed at known short time-intervals after the instant of production of the spark, and the variation of velocity with distance deduced therefrom. It was found that the velocity varied from 660 metres/sec. at a distance of 3·2 mm. from the spark to 380 metres/sec. at 18 mm. distance.

In all the experimental determinations to which we have referred the wave of finite amplitude consisted of a single pulse, produced by sudden release of compressed gas or by some form of explosive impulse. Little or no information is available relative to the variation of velocity of *continuous waves* with

* See also E. Jones, *Proc. Roy. Soc.*, **120**, p. 603, 1928.

† *Ann. d. Physik*, **9**, p. 502, 1880. ‡ *Nature*, **47**, pp. 415, 440, 1893.

§ *Journ. Franklin Inst.*, **193**, p. 627, 1923, and *Nature*, **115**, p. 765, 1925.

|| *Lehrbuch der Ballistik*, **2**, Berlin, 1926.

¶ *Phys. Rev.*, **35**, p. 373, 1912, **16**, p. 449, 1920, and *Nat. Acad. Sci. Proc.*, **6**, p. 310, 1920.

condensation, although it is well known that the change of wave-form as a wave of finite amplitude advances has an important bearing on the efficiency of powerful sound generators used in long-range signalling. In this connection it may be well to *emphasise the distinction between a wave of large 'amplitude' and a wave of large 'condensation.'* A wave of frequency $N=100$ p.p.s. and amplitude $a=10^{-4}$ cm. would generally be regarded as having a *small* amplitude, yet the *same* amplitude in a wave of frequency $N=10^6$ p.p.s. would be considered *large.* The important distinction lies, however, in the particle-velocity u (or the condensation s). In the former case the maximum value of $u=2\pi Na=0{\cdot}0628$ cm./sec., whereas in the latter $u=628$ cm./sec.; the corresponding values of maximum condensation $s=2\pi Na/c$ being $1{\cdot}9\times10^{-6}$ and $0{\cdot}019$ respectively, if air is the medium of transmission ($c=3{\cdot}3\times10^4$ cm./sec.). At the low frequency, therefore, the condensation is very small and the wave-velocity will have the normal value for small-amplitude waves, whereas at the high frequency the condensation must be regarded as 'finite' and the wave-velocity greater than normal. We find from equation (4), p. 264, when $s=0{\cdot}019$ that $c=c_0(1+0{\cdot}019)^{1{\cdot}2}=1{\cdot}023\ c_0$; that is, the velocity is about 2 per cent. above the ordinary velocity of sound. It seems not improbable that the variations of velocity (of the order of several parts per thousand) observed by Pierce (see p. 259) with highly resonant quartz oscillators of high frequency may be to some extent, if not entirely, due to variation of condensation ($\propto Na$) rather than to variation of frequency only, as Pierce suggests.

If we assume an amplitude 2×10^{-4} cm. for the quartz oscillator of 50,000 frequency, the velocity will be approximately $2°/_{oo}$ above normal, which accounts for the whole of the difference observed. Unfortunately, Pierce does not refer to observations of amplitude of the various quartz crystals used in the measurements of velocity, consequently we are not able to check this possibility. Pierce's method, however, would lend itself admirably to an investigation of the variation of velocity with *condensation*, provided simultaneous observations were made of amplitude, frequency, and wave-velocity.

Information which is available regarding the *propagation of waves of finite amplitude in water* is very scanty and unreliable. Threlfall and Adair * made observations of the velocity of explosion waves from various charges of guncotton (9 to 64 oz.), and

* *Proc. Roy. Soc.*, **46**, p. 496, 1889.

obtained large increases with increasing size of charge. There is no doubt, however, that some error in the method of observation led to a value of 2013 metres/sec. for a 64-oz. charge, as compared with the velocity of 1500 metres/sec. for small-amplitude waves. More recently D. A. Keys * has employed a method suggested by J. J. Thomson † to record the pressure wave produced by an explosion under water. In this method the pressure affects a piezo-electric disc of tourmaline crystals and generates a proportional electrical charge. The charge is recorded instantaneously, and measured, by means of a cathode ray oscillograph ‡ with photographic recording arrangements. Measurements with two tourmaline discs at a known distance apart, receiving an explosion wave from $2\frac{1}{4}$ lb. guncotton at a distance of about 50 ft., give a value of velocity slightly higher than that of sound. A similar result was obtained using a single receiver subjected to the direct pressure pulse and subsequently to the pulse reflected from the sea-bed, at a known depth below the receiver. The method is of interest, but the results can only be regarded as qualitative in character.

Apart from the complicated nature of the evidence revealed by earthquake waves, there is no available information relative to the propagation of waves of 'finite' amplitude in solids.

Sounds from High-speed Projectiles. 'Onde de Choc' (or '*Onde balistique*') – One of the best examples of the propagation of a pressure pulse with a velocity greater than normal, is provided by a high-speed bullet in its flight through the air. A bullet travelling at low speeds (less than c_0) compresses the air in front of it and a spherical *sound* wave of normal velocity is transmitted. If, however, the speed of the bullet exceeds the velocity of sound, the condensation at the nose can be transmitted laterally but not forwards. Photographs of bullets in flight, taken by Boys, Cranz, Quayle, and others § by spark photography, reveal some very interesting features (see fig. 78, photo by Cranz). The photograph shows clearly the existence of two wave-fronts somewhat conical in shape, one from the nose and the other from the base of the bullet. The wave from the base must also be a wave of condensation, for Rayleigh has shown that the only kind of wave of finite amplitude which can be maintained is one of condensation. The detailed structure of the envelope of the wave, as

* *Phil. Mag.*, **42**, p. 473, 1921. † *Engineering*, **107**, p. 543, 1919.
‡ See A. B. Wood, *Proc. Phys. Soc.*, **35**, p. 109, Feb. 1923, and *Inst. Elect. Eng.*, **63**, p. 1046, Nov. 1925. § *Loc. cit.*, p. 267.

revealed by the photograph, confirms this conclusion. At present, however, we are mainly concerned with the 'bow' wave from the nose of the projectile. This can be simply explained by Huyghens' principle (see p. 67). It is clearly the envelope of spherical waves which originate at the nose of the bullet at successive instants in its flight. If these were 'small-amplitude' waves—that is, if the condensation at the nose of the bullet could be regarded as small—

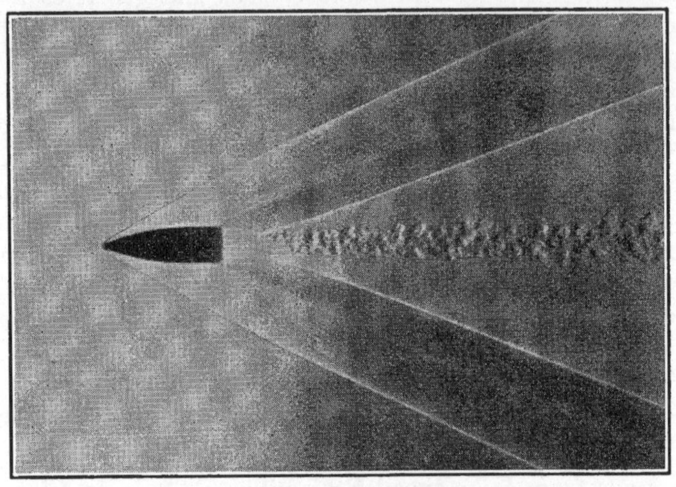

FIG. 78—Photograph of a high-speed Bullet in Flight
(By courtesy of Prof. W. Cranz)

the envelope would be a cone of semi-angle θ given by $\sin\theta = c/v$, where c is the velocity of sound and v the velocity of the bullet. When v is less than c, the angle θ is imaginary and the spherical waves have no envelope—that is, no wave-front is formed. Now the condensation at the nose of the bullet is finite, and as a consequence the velocity of propagation is greater than the normal velocity of sound. The nearer to the nose of the bullet, the greater this effect will be—that is, the angle θ of the 'cone' will increase towards the nose of the bullet. The actual wave-front under such conditions will therefore be a blunted cone, and this is what is actually observed (see fig. 78). The 'blunting' effect is increased the flatter the nose of the bullet, for the condensation is then much greater than that due to a sharp-nosed bullet. (The vortex eddies in the wake of the bullet are of interest, see p. 205.)

When a bullet or shell, travelling with a velocity greater than sound, passes an observer it makes a sound like the 'crack' of an explosion. This is described by the French as *Onde de choc* or *Onde balistique.** A. Mallock † has shown that it is possible to deduce the position of a gun by observing the apparent direction of the *onde de choc* at different points along the track of the shell. As the shell slows down in its flight and v approaches c, the angle θ of the cone gradually increases, the observed direction of the sound varying accordingly and therefore becoming less and less inclined to the direction of flight of the shell.

CHANGE OF MEDIUM

(*A*) Reflection

When a wave of sound meets the bounding surface between two different media it is partially reflected, and a wave travels in the negative direction through the incident medium with the same velocity as it approached the boundary. The geometrical laws of reflection of sound waves are the same as apply to light waves, the angles of incidence and reflection being equal and in the same plane. In many cases, however, the length of the sound waves is comparable with the linear dimensions of reflecting objects, when the geometrical laws cease to apply and the phenomena must be regarded essentially as diffraction. For the present, however, we shall assume the wave-length small compared with the dimensions of reflectors, the ordinary laws of geometrical optics being applicable.

Reflection of Plane Waves at the Boundary of Two Extended Media – The geometrical laws of reflection and refraction follow directly from the facts that the velocity in each medium is independent of the direction of the wave-front and that the traces of all wave-fronts on the plane of separation have equal velocities. Consequently, if θ_1, θ_r, and θ_2 are the angles of incidence, reflection, and refraction respectively (see fig. 79), we must have

$$\left.\begin{aligned} &\frac{\sin\theta_1}{c_1}=\frac{\sin\theta_r}{c_1}=\frac{\sin\theta_2}{c_2}, \quad \text{or} \quad \theta_1=\theta_r, \textit{ the law of reflection} \\ \text{and} \quad &\frac{\sin\theta_1}{\sin\theta_2}=\frac{c_1}{c_2}, \textit{ the law of refraction.} \end{aligned}\right\} \quad (1)$$

At the boundary of the two media it is necessary (*a*) that the component of particle velocity $\dot{\xi}$ normal to the bounding surface must

* See E. Esclangon, *L'Acoustique des canons et des projectiles* (Gautier-Villars, Paris).

† *Proc. Roy. Soc.*, p. 115, 1908.

be continuous, and (*b*) that the pressure-variation δp must also be continuous. If suffixes *i*, *r*, and *t* refer respectively to the incident, reflected, and transmitted waves, and 1 and 2 refer to the incident and transmitting media, we may write for condition (*a*),

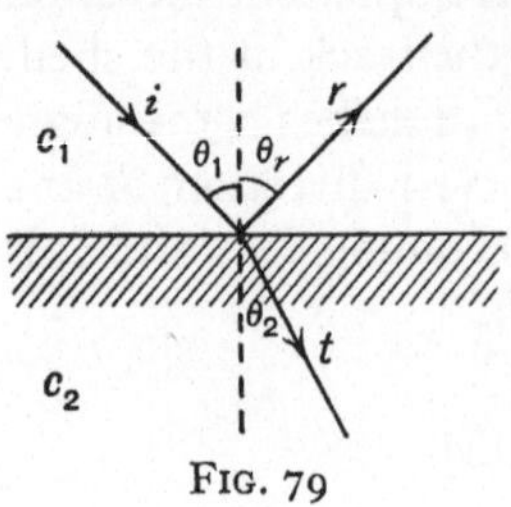

FIG. 79

$$\xi_i \cos\theta_1 + \xi_r \cos\theta_1 = \xi_t \cos\theta_2 \qquad (2)$$

and for condition (*b*),

$$\delta p_i + \delta p_r = \delta p_t \quad \text{or since} \quad \delta p = \pm c\rho\xi$$

(see equation (25), p. 55)

$$c_1\rho_1\xi_i - c_1\rho_1\xi_r = c_2\rho_2\xi_t \quad . \quad . \quad . \quad (3)$$

the negative sign in the second term of this equation indicating the reversed direction of propagation of the reflected wave. Eliminating ξ_t from (2) and (3) we obtain

$$\frac{\xi_r}{\xi_i} = -\frac{c_2\rho_2 \cos\theta_1 - c_1\rho_1 \cos\theta_2}{c_2\rho_2 \cos\theta_1 + c_1\rho_1 \cos\theta_2} = r \quad . \quad . \quad (4)$$

Similarly, by elimination of ξ_r we find

$$\frac{\xi_t}{\xi_i} = \frac{2c_1\rho_1 \cos\theta_1}{(c_2\rho_2 \cos\theta_1 + c_1\rho_1 \cos\theta_2)} = t_{12} \quad . \quad . \quad (5)$$

These expressions not only represent the condition at the boundary, but also at corresponding points in the respective waves. They are obtained without assumptions regarding wave-length and phase and are therefore applicable to waves of any type whatever.

At Normal Incidence $\theta_1 = \theta_2 = 0$ – If we denote the radiation resistance ρc by R the expressions for the reflected and transmitted waves become respectively

$$\frac{\xi_r}{\xi_i} = \frac{R_1 - R_2}{R_1 + R_2} = r \quad . \quad . \quad . \quad (6a)$$

and

$$\frac{\xi_t}{\xi_i} = \frac{2R_1}{R_1 + R_2} = t_{12} \quad . \quad . \quad . \quad (6b)$$

According to the *principle of reversibility*,* if all velocities in a dynamical system are reversed and there is no dissipation of energy,

* See Schuster's *Optics*, p. 45.

the whole previous motion is reversed. Consequently if the sound wave be now transmitted from medium (2) to medium (1) we have

$$t_{21}=\frac{2R_2}{R_1+R_2} \quad \text{and} \quad t_{12}t_{21}+r^2=1 \quad . \quad . \quad (7)$$

which is independent of any assumption as to change of phase at reflection. If I_a, I_r, I_t represent the energy in the incident, reflected, and transmitted waves respectively, then $I_a=I_r+I_t$. It would be wrong to conclude from this that $\xi_i^2=\xi_r^2+\xi_t^2$ because the squares of amplitudes only express relative intensities *in media of the same radiation resistance.* The energy in a wave is proportional to $c\rho\xi^2$ (see equation (27), p. 55), consequently we may write

$$c_1\rho_1\xi_i^2=c_1\rho_1\xi_r^2+c_2\rho_2\xi_t^2,$$

that is,

$$\xi_i^2=\xi_r^2+\frac{R_2}{R_1}\xi_t^2 \quad \text{or} \quad r^2+\frac{R_2}{R_1}t_{12}^2=1 \quad . \quad . \quad (8)$$

Defining the normal *reflection coefficient* as r^2 the ratio of reflected to incident *energy*, we must regard $t_{12}^2(R_2/R_1)$ as the normal *transmission coefficient.*

Phase Changes on Reflection – Referring to equation (6*a*), it is important to observe (a) *when* $R_2=R_1$ there is *no reflected wave*, the transmission to the second medium being complete; (b) *when* $R_2>R_1$ the radiation resistance in the first medium being less than in the second (*e.g.* air to water), the *particle-velocity is negative*, i.e. *reversed in phase. The phase of the condensation* $s_r=\dot{\xi}_r/c$, however, *remains unchanged* because both $\dot{\xi}$ and c are reversed in sign in the reflected wave, (c) *when* $R_2<R_1$ (*e.g.* water to air), the phase of the particle-velocity $\dot{\xi}_r$ remains unchanged, but *the phase of the condensation is reversed.* Equation (6) may therefore be written

$$\frac{s_r}{s_i}=\frac{R_2-R_1}{R_1+R_2} \quad . \quad . \quad . \quad . \quad (9a)$$

and

$$\frac{s_t}{s_i}=\frac{2R_1}{R_1+R_2} \quad . \quad . \quad . \quad . \quad (9b)$$

The phase of the transmitted wave is unaffected in all cases.

When the sound wave travels from a medium of low velocity c_1

to a medium of higher velocity c_2 (i.e. $c_2>c_1$), we see from equation (1) that there will be a *critical angle* of incidence θ_1 when θ_2 becomes 90°—that is, when $\sin\theta_1=c_1/c_2$, beyond which there is *total reflection.* On account of the greater velocity of sound in water, total reflection may occur when the waves are incident from air on water, but not from water to air.

To illustrate the application of the above equations, take the case of transmission from air to water. At normal incidence equations (6*a*) and (9*a*) are applicable. For air $R_1=\rho_1c_1=40$, and for water $R_2=\rho_2c_2=1{\cdot}5\times10^5$, whence the ratio of reflected to incident amplitudes is 0·9994, indicating almost total reflection. The phase of the condensation is unchanged (see equation (9*a*)), whilst that of the particle-velocity is reversed (the nodal condition). From water to air the reflection coefficient is again almost unity, but the condensation is now reversed in phase, whilst that of the particle-velocity is unchanged (the antinodal condition). In practically all cases of sound transmission from a gaseous to a solid or a liquid medium (or *vice versa*), i.e. *whenever the radiation resistances are widely different, there is almost complete reflection.* The difficulty of transmitting sounds from water (*e.g.* the noise of a ship's propeller in the sea) to the air-filled ear-cavity of an observer will be apparent. Transmission from a liquid to a solid, however, is much more favourable, for now R_1 becomes comparable with R_2. Take, for example, sound waves passing from water to steel where $R_1=1{\cdot}5\times10^5$ and $R_2=7{\cdot}8\times5\times10^5=3{\cdot}9\times10^6$ respectively. Here we find $r=0{\cdot}925$, representing a reflection coefficient r^2 of 85 per cent., and a transmission of 15 per cent. of the incident energy. Similarly from water to oak, where $R_2=3{\cdot}4\times10^5$ we find $r=0{\cdot}39$, $r^2=0{\cdot}15$, and 85 per cent. of the energy is transmitted. From water to steel $c_1/c_2=0{\cdot}33$, whence the *critical angle* $\sin^{-1}0{\cdot}33$ is $19\frac{1}{2}°$; from air to water $c_1/c_2=0{\cdot}22$ and $\sin^{-1}0{\cdot}22$ is 13°. In the latter case, as we have already seen, the reflection is almost total, even at *normal* incidence. Such calculations have an important bearing on the possibility of detection if objects under water, such as wrecks or icebergs, by echo methods * (see pp. 150 and 471). The density of ice is about 0·92 and the velocity of sound in it has been estimated at $2{\cdot}1\times10^5$, giving a value of $R_2=1{\cdot}9\times10^5$, whence the reflection coefficient r^2 from sea-water to ice is 0·027. The possibility of detecting pure ice under water by sound reflection methods would, on this estimate, be negligibly

* See Papers by Boyle and Reed, and Boyle and Taylor, *Trans. Roy. Soc. Canada*, 3, pp. 233 and 245, 1926.

small. Actually, however, an iceberg contains large inclusions of rock and air-cavities, both of which are much more effective than ice as reflectors of sound in water. It is these inclusions which make it possible to detect icebergs by echo methods.*

Method of Images – A convenient method of regarding problems in the reflection of sound employs the optical analogue of 'images.' A source of sound on one side of the plane boundary between two extended media may be regarded as having an image at an equivalent distance on the other side of the boundary. The resultant effect at any point in the first medium will then be made up of the sum of the effects due to the source and its image. It is important to attach the correct phase to the vibrations of the image with respect to the actual source. Thus if a sound wave in air is reflected from a solid wall or a water surface (*i.e.* if $R_1<R_2$), the image will be exactly similar to the source and at a distant point, almost equidistant from the source and the image, there will be reinforcement of the sound received, due to the presence of the reflector. If, however, the sound source is under water and its image is in air ($R_1>R_2$), the phase of the image will be opposite to that of the source. At a distant point, therefore, when the paths of the direct and reflected sounds are nearly equal, the resultant effect would be approximately zero. This result has an important bearing on the long-range transmission of sound through the sea. At smaller distances from the source interference phenomena (see pp. 296 and 299) are observed analogous to Lloyd's interference bands in Optics.

Reflection of Plane Waves from a Plate of Finite Thickness – This is an extension of the case we have just considered, the thickness of the second medium now being limited. The case of greatest interest and practical importance is that in which a slab of solid material is immersed in an 'infinite' medium. The mathematical treatment is analogous to the optical case of multiple reflection in a thick slab of glass, the boundary conditions, pressure and particle-velocity relations, being similar to those just considered. Rayleigh † has derived the ratio of reflected to incident amplitudes at normal incidence, and more recently R. W. Boyle and W. F. Rawlinson ‡ have extended the analysis to any angle of incidence. We shall be content with a quotation of Rayleigh's result,

* See Papers by Boyle and Reed, and Boyle and Taylor, *Trans. Roy. Soc. Canada*, 3, pp. 233 and 245, 1926.

† *Sound*, **2**, pp. 87, 88.

‡ *Trans. Roy. Soc. Canada*, **22**, 3, p. 55, 1928.

making the necessary changes in nomenclature. Thus for normal incidence:

$$\frac{\xi_r}{\xi_i}=r=\left(\frac{R_1}{R_2}-\frac{R_2}{R_1}\right)\Bigg/\left\{4\cot^2\frac{2\pi l}{\lambda}+\left(\frac{R_1}{R_2}+\frac{R_2}{R_1}\right)^2\right\}^{\frac{1}{2}} \quad . \quad (10)$$

where ξ_r and ξ_i refer as before to the reflected and incident amplitudes, R_1 and R_2 ($=\rho_1 c_1$ and $\rho_2 c_2$) are the acoustic resistances of the medium (1) and the plate (2) respectively, l is the thickness, and λ the wave-length of the sound in the material of the plate. The transmitted energy may readily be deduced from equations (8) and (10), if we assume there is no loss of energy due to absorption. It will be seen that the reflected amplitude varies between zero and a maximum amplitude as the thickness of the plate varies. The velocity of sound in most solids is greater than in liquids or gases, consequently for a solid plate in a fluid medium we may in general write $c_2>c_1$. In such a case equation (10) shows that $\cot^2(2\pi l/\lambda)$ is infinite, and consequently the reflected amplitude is zero, whenever l is zero or a multiple of $\lambda/2$. A *half-wave plate* therefore reflects none and *transmits all* the incident energy. When l is a multiple of $\lambda/4$ the value of $\cot^2(2\pi l/\lambda)$ is zero, and the reflected amplitude is a maximum. A *quarter-wave* plate therefore reflects a maximum and transmits a minimum of the incident sound energy. In this case equation (10) reduces to

$$r=\frac{\xi_r}{\xi_i}=\frac{R_1^2-R_2^2}{R_1^2+R_2^2} \qquad \text{(when } l=m\lambda/4) \quad . \quad (11)$$

The square of this ratio (r^2) is the energy reflection-coefficient. These deductions are exhibited graphically in fig. 80, taken from Boyle and Rawlinson's paper. The curves refer to large plates of duralumin immersed in water. The thickness l is expressed as a ratio of the wave-length λ of the sound in the metal. Assuming $c_1=1{\cdot}5\times10^5$, $\rho_1=1{\cdot}0$ for water, and $c_2=6{\cdot}5\times10^5$, $\rho_2=2{\cdot}8$ for duralumin, then $R_1=1{\cdot}5\times10^5$, $R_2=18{\cdot}2\times10^5$. At a frequency of 10^5 p.p.s., therefore, a plate of duralumin about 1·8 cm. thick would reflect over 97 per cent. of the

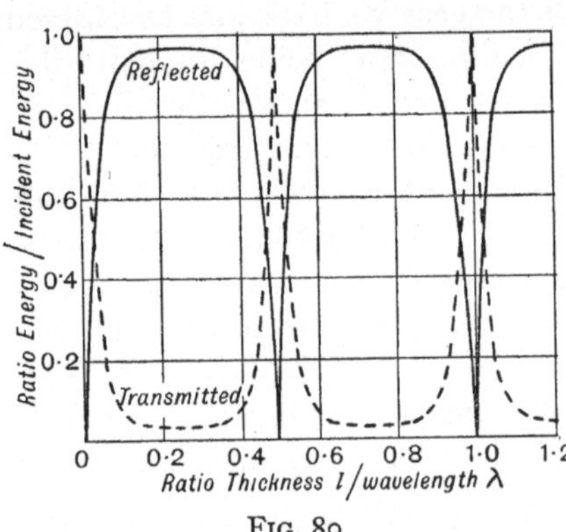

FIG. 80

sound energy incident normally upon it. These relations for the reflection of plane waves from flat plates have been verified experimentally by Boyle and Lehmann.* They used a quartz oscillator (see pp. 146 to 151) as the source of sound under water, at a frequency of 135,000 p.p.s., and measured the reflected energy by the torsion pendulum method (see p. 432). A number of torsion vanes were made of lead of different thicknesses, the energy reflected from the vane at first increasing rapidly with increasing thickness, remaining practically constant (from $l=0{\cdot}1$ to $0{\cdot}4\lambda$) at approximately 100 per cent. reflection until the half wave-length thickness (0·765 cm.) was approached. The critical angle of total reflection was also determined. The results were found to be in general agreement with the theory. Interesting practical examples of partitions transmitting *more* energy with *increase* of thickness are mentioned in experiments by Watson † on sound-proof partitions. Slabs of cork, each $\frac{3}{4}$ in. thick, closed the doorway of a room in which an organ pipe was emitting sound. A thickness of *two slabs* of this board was found to *reflect less and transmit more* sound than either one or three slabs of the same material. Such examples, however, should be regarded with caution, for deductions are rendered uncertain by absorption in the material and by the possibility of excitation of transverse vibrations, whereas the theory assumes only longitudinal vibration.

Important practical applications of the theory of transmission of sound through plates are described in a paper by H. Brillie.‡ Calculations are made on the transmission of sound through the hulls of ships, and on the choice of a material as intermediary between water outside, and air inside the ship. Brillie considers that a *thick* rubber slab or tube receiver is much more efficient in this respect than a metal receiver (a steel diaphragm, for example). This opinion is based on the calculation that the acoustic (radiation) resistance of rubber is of the same order as the geometric mean ($2{\cdot}4\times10^3$) of the values 40 for air and $1{\cdot}5\times10^5$ for water.

An interesting and important application of equation (10) relates to *the reflection of sound waves from a thin film of air* enclosed in a medium of much larger acoustic resistance, such as water or rock. A film of air 0·01 cm. thick ($R_2=40$) enclosed between *thin* metal plates under water ($R_1=1{\cdot}5\times10^5$) would

* *Trans. Roy. Soc. Canada*, **21**, 3, p. 115, 1927.
† *University of Illinois*, *Bull.* 127.
‡ *Le Génie Civil*, Aug. and Sept. 1919.

according to (10) reflect about 93 per cent. of the incident energy at a frequency of 1000 p.p.s. ($\lambda_{air}=33$ cm.). The percentage would be still greater if the medium enclosing the air film were solid rock. Assuming R_1 to be of the order $1{\cdot}5\times10^6$ for rock, 99·93 per cent. of the energy would be reflected from the thin air film. The significance of such a result and its bearing on geological prospecting by echo methods (see p. 474) will be appreciated.

Reflection from a Succession of Thin Parallel Laminæ – Tyndall * demonstrated by means of a high-pitched sound and a sensitive flame that a material like muslin, consisting of a network of fine threads, transmitted the greater part of the sound incident upon it, only a small proportion being reflected. If, however, the muslin be wetted, so as to close the meshes of the net, the amount reflected is considerably increased. In explaining the phenomena of iridescence in certain crystals Rayleigh † utilised the partial reflection of sound from *dry* muslin to illustrate an important principle. "If a pure tone of high (inaudible) pitch be reflected from a single sheet so as to impinge upon a sensitive flame, the intensity will probably be insufficient to produce a visible effect. If, however, a moderate number of such sheets be placed parallel to one another and at such equal distances apart that the partial reflections agree in phase, then the flame will be powerfully affected. The parallelism and equidistance of the sheets may be maintained mechanically by a lazy-tongs arrangement, which nevertheless allows the common distance to be varied. It is then easy to trace the dependence of the action upon the accommodation of the interval to the wave-length of the sound. Thus if the incidence were perpendicular, the flame would be most powerfully influenced when the interval between adjacent sheets was equal to the *half* wave-length; and although the exigencies of experiment make it necessary to introduce obliquity, allowance for this is readily made." The experiment has been more recently utilised by W. H. Bragg ‡ to illustrate the reflection of X-rays from successive parallel planes of atoms in crystals. As a means of measuring the wave-length and frequency of inaudible sounds in air the method should be quite as effective as the more generally known stationary-wave method. In fact, it should be more instructive, for the parallel planes serve as a

* *Sound.*

† *Sound*, **2**, p. 311, *Proc. Roy. Inst.*, April 1889; *Phil. Mag.*, **24**, p. 145, 1887, and **26**, p. 256, 1888.

‡ R.I. Lectures.

reflection grating (of the 'crystal' type) for analysing a complex high-frequency sound, the different wave-lengths having their maxima at different angles of reflection, or corresponding to different spacing distances of the reflecting laminæ. In the proposed arrangement shown in fig. 81 (*a*) S is a source of high-frequency sound (a Galton's whistle or a Pierce's oscillating quartz crystal) at the focus of a concave mirror M_1 which renders the sound beam practically parallel. This beam falls on the 'muslin' sheets L_1L_1, L_2L_2, etc., and, after reflection, is collected by a second concave mirror M_2 and focussed on a sensitive flame F. If θ is the angle of incidence and d the 'spacing distance' of the planes, then the difference of path between a ray reflected at the first and second surfaces is CB+BD in fig. 81 (*b*).

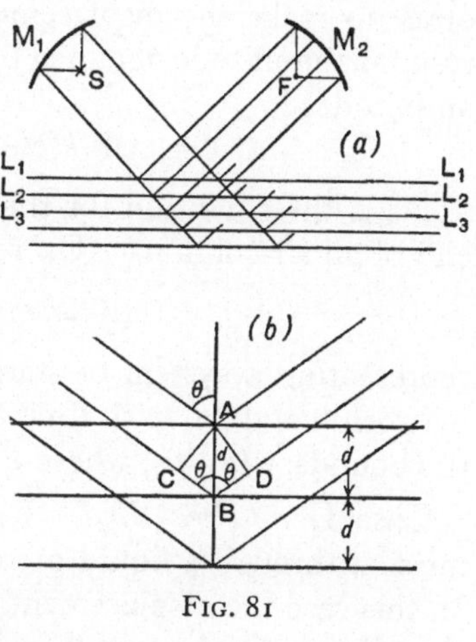

FIG. 81

From the construction shown it is evident that CB=BD, and the path difference is $2d \cos \theta$. The maximum effect is obtained at F therefore when this quantity is equal to an integral number m of wave-lengths λ—that is,

$$m\lambda = 2d \cos \theta.$$

The value of λ may be found by varying the angle θ as in the spectrometer method, or by varying the distance apart of the reflecting planes with the lazy-tongs arrangement.

Stationary Waves – When plane waves are reflected at normal incidence from the interface formed by two different media, a reflected wave travels back into the incident medium and 'interferes' with the oncoming incident wave. We have made frequent reference to such an effect in relation to the transverse vibrations of strings (p. 86) and to the longitudinal vibrations of rods (p. 136) and of gas columns in tubes (p. 170). The same mathematical treatment is also applicable to extended media. Consider a wave of displacement $\xi = a \cos k(ct - x)$, where $k = 2\pi/\lambda = 2\pi N/c$, advancing in the incident medium towards the surface of separation ($x=0$). A similar wave in the opposite direction would be represented by $\xi = a \cos k(ct+x)$. We shall now refer to two important cases in both of which *perfect reflection* is assumed, *i.e.* the acoustic resistances R_1 and R_2 are widely different.

Case I – '*Fixed*' *End.* R_2 much greater than R_1 (*e.g.* transmission through a gas to a liquid or a solid reflector). In order to make the displacement $\xi=0$ at the interface $x=0$, the resultant amplitude obtained by addition of the direct and reflected amplitudes is

$$\xi = a \cos k(ct-x) - a \cos k(ct+x).$$

(This is the condition for the fixed end of a string or the closed end of an air column.) On reduction this becomes

$$\xi = 2a \sin (kct) \sin (kx) \quad . \quad . \quad (1)$$

representing a system of stationary waves, of amplitude varying between 0 and $2a$, with time t and with distance x measured from the boundary ($x=0$), where ξ is always zero.

Case II – '*Free*' *End.* R_1 much greater than R_2 (*e.g.* transmission through a liquid or solid medium to a gaseous reflector). In this case the displacement ξ is a maximum (and the condensation $d\xi/dx=0$) when $x=0$. Consequently the resultant amplitude is given by

$$\xi = a \cos k(ct-x) + a \cos k(ct+x).$$

(This condition is approximately fulfilled at the open end of a pipe, or the free end of a rod.) The expression for ξ may be written

$$\xi = 2a \cos (kct) \cos (kx) \quad . \quad . \quad . \quad (2)$$

representing a system of stationary waves with an antinode at the boundary ($x=0$).

A third case, of considerable practical importance, arises when the reflection from the bounding surface of the two media is *not* perfect, the reflected amplitude being less than the incident amplitude.

Case III – *Imperfect Reflection.* (Fixed or Free End.) We shall regard the bounding surface to be in the nature of a fixed end $R_2 > R_1$, allowing a proportion of the incident sound energy to be transmitted or absorbed by the reflecting medium. The procedure for a free end reflector is similar. Denoting the incident and reflected amplitudes by a and r respectively, the resultant displacement will be given by

$$\xi = a \cos k(ct-x) - r \cos k(ct+x),$$

which becomes

$$\xi=(a+r)\sin kct \sin kx+(a-r)\cos kct \cos kx \quad . \quad (3)$$

The motion may thus be regarded as being due to two superposed stationary waves, of amplitude $(a+r)$ and $(a-r)$, the nodes and antinodes of one being a quarter of a wave-length (*or* $\pi/2$) distant from the nodes and antinodes of the other. This results in a series of positions of maximum $(a+r)$ and minimum $(a-r)$ amplitude ('*pseudo*' nodes and antinodes) a quarter of a wave-length apart. In the case of perfect reflection $r=a$, the expression (3) for ξ becomes identical with (1). If we write $\alpha=(a+r)$ and $\beta=(a-r)$, then the reflection coefficient (see p. 273) will be

$$\left(\frac{r}{a}\right)^2=\left(\frac{\alpha-\beta}{\alpha+\beta}\right)^2=\left(\frac{1-\beta/\alpha}{1+\beta/\alpha}\right)^2 \quad . \quad . \quad (4)$$

an expression in terms of the ratio (β/α) of minimum/maximum amplitude in the stationary wave. This serves as a basis for experimental methods of measuring the reflecting properties of materials, and their absorption or transmission properties also. We shall refer to such measurements later (see pp. 324 and 488).

The demonstration of stationary waves in air or other gases by means of high-frequency sounds reflected normally from a plane surface has already been mentioned (see Rayleigh's 'bird call' experiment, p. 208, and Pierce's piezo-electric observations, p. 258). We have also referred to Boyle and Lehmann's demonstration of stationary waves in water (p. 260) and the measurements of Wood and Loomis with stationary waves in various other liquids * (p. 260). See also fig. 87.

Echoes – The direct reflection of a sound of short duration from a surface of large area such as the wall of a building or a cliff is generally described as an echo.† The echo is only appreciated by the ear, however, if the time-interval separating the direct and reflected sounds is of the order of one-tenth of a second or more. In such cases the best echoes are obtained when the dimensions of the reflector are large compared with the wave-length of the incident sound. Thus a sharp crack or a sound of an impulsive nature gives a loud and clear echo, whereas with a low-pitched sound of long wave-length the echo may be scarcely perceptible. Simple echoes from surfaces of large area have been

* See also F. L. Hopwood, *Journ. Sci. Instrs.*, Feb. 1929.

† Many interesting examples of 'echo' are given in Tyndall's *Sound*, p. 17.

utilised under certain circumstances as a convenient means of measuring a distance in terms of the known velocity of sound c in the medium and the time-interval t for the sound to 'go and return' to the observer, *i.e.* the distance d from the reflector is $ct/2$. The height of an airship above the ground or the depth of the sea-bed beneath a ship may be found in this way (see p. 466). Under certain circumstances, however, the echo is different in character from the incident wave, the character of the reflection depending on the nature and position of the reflecting surface in its relation to the wave-form of the sound falling upon it.

Reverberation in large halls or auditoriums is due to multiple reflections of the speaker or singer's voice from the walls, floors, and ceiling. We shall consider the practical aspect of such reverberation when dealing with the acoustic properties of buildings (see p. 484).

Scattering of Sound. Harmonic Echoes – As the wave-length of the sound diminishes the effectiveness of a small reflector improves. If therefore the incident sound is complex, consisting, for example, of a fundamental tone plus a series of high harmonics, the component tones will be reflected in increasing proportion towards the higher frequencies, *i.e.* towards the shorter wave-lengths. It has sometimes been noticed, for example, that the pitch of the echoes returning from a group of trees appears to be raised one or more octaves above the pitch of the fundamental incident sound. Rayleigh has dealt with the question mathematically,* referring to the reflections as 'secondary waves,' and the phenomenon as *scattering* of sound. He shows that the *amplitude* of the secondary waves varies directly as the volume of the 'scatterer,' and inversely as the *square* of the wave-length λ of the incident sound. That is, *the intensity* of the sound returned to the observer *varies inversely as the fourth power of the wave-length.* As Rayleigh points out, this result might have been foreseen without calculation. The ratio of the 'scattered' to the 'incident' amplitude must necessarily vary directly as the volume V and inversely as the distance r, and in order that the result, the ratio of amplitudes a_s/a_i must have the dimensions of a pure number we must divide by λ^2, since λ is the only other linear magnitude involved, *i.e.*

$$\frac{a_s}{a_i} \propto \frac{V}{\lambda^2 r} \quad \text{or} \quad \frac{I_s}{I_i} \propto \frac{V^2}{\lambda^4 r^2}.$$

* *Sound,* **2,** p. 152.

The law of scattering, $I_s \propto 1/\lambda^4$ is also well known in optics, where light waves are scattered by particles whose dimensions are small compared with the wave-length of the light falling upon them. The blueness of the sky is attributed to such a cause. Returning to the case of sound waves, it will be seen therefore that the higher components of a complex note are reflected back from small scattering objects in a much greater proportion than the fundamental tone. The octave of a tone will be sixteen times more intense in the scattered than in the incident sound. Echoes of such a character have been called 'Harmonic Echoes.' A complementary phenomenon is of familiar occurrence. The sound received on the remote side of a 'harmonic reflector' must also differ from the incident complex sound. In this case, however, the change of quality is due to a deficiency in the higher-frequency components which were present in the original sound. The shorter wave-length components having been scattered forwards in a much greater proportion than the fundamental, or long wave-length component, the sound behind the obstacle appears to be 'purified.' The total effect of any obstacle in scattering or reflecting sound waves may be calculated by integration of the effects due to its elementary parts. In this way it is possible to trace the transition from a small obstacle, in which the *volume* but not the *surface* is important, to a reflector of large area which will ultimately reflect optically.

Musical Echo from Palings. Echelon Reflection (Gratings) – An observer walking on a hard footpath near a row of regularly spaced, or stepped, palings, may notice that each footstep is followed by an echo having a musical ring. Any 'sharp' or impulsive sound is reflected from a stepped structure in this way. The successive elements or strips each reflect the impulse or, as we have just seen, its higher harmonics, and the observer receives a regular succession of reflected pulses which, if sufficiently rapid, blend into a musical note. The frequency of the note will depend on the spacing d of the steps and on the direction θ from which the reflections proceed. The time between the successive pulses, *i.e.* the *period* of the musical echo, will be $2d \cos \theta/c$, and the frequency $c/2d \cos \theta$. The earlier reflections, from the strips nearest the observer, will usually be masked by the direct sound, and as a consequence only those reflections will be heard which come from more distant strips, *i.e.* when θ is small. For example, from a row of palings of spacing 10 cm. the frequency of the note heard by the observer will be $3{\cdot}3 \times 10^5/20 = 1650$ p.p.s. or some multiple of this frequency.

R. W. Wood * has demonstrated the actual existence of a succession of reflected pulses from an echelon or stepped reflector by the spark method of shadow photography (see p. 340). An echelon reflector may serve as a diffraction grating for sound-wave analysis just as in the optical case (see p 309).

Reflection from Curved Surfaces. Mirrors. Whispering Galleries – The well-known optical phenomena of reflection at spherical or parabolic surfaces have their analogue in sound, provided the dimensions of the reflector are considerably greater than a wave-length of the incident sound. Reference has already been made (pp. 193 and 215) to the directional or 'focussing' properties of spherical reflectors used as sound transmitters with a small source at the focus. The semi-angle of the cone of the diverging primary beam was shown to be $\sin^{-1} 0{\cdot}61\lambda/r$, which indicates the reflected beam does not become 'parallel' until the wave-length λ is very small compared with the radius r of the aperture. Conversely, the reflector acting as a collector of a parallel beam of sound does not form an 'optical image' unless r is very great compared with λ. The image of a point source of light was shown by Airy † to consist of a bright disc surrounded by bright rings separated by circles at zero intensity. The radii of these circles are given by $a = mf\lambda/2r$ (where f is the focal length of the reflector and m has successive values 1·220, 2·233, 3·238, 4·241, etc.; 0·839 of the total energy falling on the central disc, 0·071 on the first bright ring, and so on). The optical deductions are equally applicable to sound waves. Thus a *perfect* parabolic concave reflector, 1 metre in diameter ($r=50$ cm.) and focal length 50 cm., would form a diffraction image with a central 'bright' spot 2·0 cm. diameter, if the source were a distant point emitting sound of frequency 10,000 p.p.s. in air ($c=3{\cdot}3\times10^4$ cm./sec.), the wavelength λ being 3·3 cm. The 'resolving power' of the mirror, which indicates the limiting angle θ which can be recognised between two distant point sources, may be regarded as the angle subtended by the radius of the central disc at the pole of the mirror, *i.e.* $\theta = a/f = 1{\cdot}22\ \lambda/2r = 0{\cdot}61\ \lambda/r$. In the example just mentioned this angle is 2·3 degrees. The Yerkes telescope, with an aperture of 100 cm., resolves two distant stars subtending an angle of 0·125 second of arc. A concave sound reflector cannot be regarded, therefore, as a means of forming a sound image, in the optical sense. Such reflectors, however, are very efficient in

* *Phil. Mag.*, **48,** p. 218, 1899.
† *Trans. Camb. Phil. Soc.*, **5,** p. 283, 1834.

indicating the direction, within a few degrees, of a high-frequency source. To demonstrate the focussing of sound waves in the laboratory, Rayleigh's experiment with two concave reflectors, a bird-call (or other high-pitched source), and a sensitive flame are all that is required (see also p. 371).

The reflecting properties of concave surfaces are beautifully demonstrated by the method of spark photography (see p. 340). In fig. 71 a comparison is made between the reflection of the sound pulse from a spark at the focus of a parabolic and a spherical mirror respectively. It will be seen that the parabolic reflector gives a tolerably plane wave (parallel beam), whereas the aberration produced by the spherical reflector is clearly shown.

The well-known *whispering gallery* of St Paul's Cathedral owes its peculiar properties to the reflection of sound by the curved walls of the gallery at the base of the hemispherical dome. A person whispering along the wall on one side can be heard clearly by a listener close to the wall at any other part of the gallery. A complete explanation of all the phenomena is still to be found, but Rayleigh * has given a theory which, at any rate, agrees with the principal facts. He pointed out that the sound tends to creep around inside of a curved wall without ever getting far from it. "A whisper seems to creep round the gallery horizontally, not necessarily along the shorter arc, but rather along that arc towards which the whisperer faces. This is a consequence of the very unequal audibility of a whisper in front of and behind the speaker. The abnormal loudness with which a whisper is heard is not confined to the position diametrically opposite to that occupied by the whisperer, and therefore, it would appear, does not depend materially upon the symmetry of the dome." Moreover, whispered speech, which contains a large proportion of high-pitched sounds, is heard more distinctly than ordinary-voiced speech, especially if the speaker looks along the gallery towards the listener. Rayleigh demonstrated these effects on a small scale by means of a sheet of iron (12 ft. × 2 ft.) bent into the arc of a circle, with a bird-call at one end and a sensitive flame at the other. A narrow screen interposed radially at any point inside the curved surface *and near it* prevented the sound from reaching the flame. Rayleigh's theory shows (1) that the thickness of the narrow belt skirting the wall decreases with diminishing wavelength, (2) that the intensity is a maximum near the wall and gradually diminishes away from it, and (3) that the intensity does

* *Sound*, **2**, p. 126, and *Scientific Papers*, **5**, p. 617.

not fluctuate circumferentially, *i.e.* parallel to the wall. Experiments by C. V. Raman and G. A. Sutherland * with a high-pitched source and sensitive flame confirm Rayleigh's conclusion (1), but disagree with (2) and (3). A further investigation by Raman † confirms a view held by Sabine ‡ that the inward slope of the wall is important for the best effect. Circumferential and radial 'nodal' lines were observed in five different whispering galleries, the sound intensity being a maximum at opposite ends of a diameter. It was shown, further, that an impulsive sound would travel several times round the gallery, a succession of sounds separated by time intervals equal to the circumference of the gallery divided by the velocity of sound (within ± 1 per cent.) being observed. F. L. Hopwood § has demonstrated the existence of radial and circumferential nodal lines by means of sound waves of very high frequency (750,000 p.p.s.) in oil. The plane waves of very short wave-length emitted by a quartz oscillator entered a slit tangentially and passed round a model whispering gallery. The stationary sound distribution inside the model was revealed by sprinkling fine coke dust (Boyle's method) in the oil. B. Ray ‖ has studied the whispering gallery effect experimentally by analogous optical methods. He finds, in confirmation of Raman's views, that a *succession* of belts of maximum intensity are found near to the curved surface. These maxima appear in the form of a series of interference bands at one end of the curved strip of mirror which has a source of light on its surface at the other end.

Rayleigh has suggested that in the propagation of earthquake disturbances the surface of the earth behaves somewhat like a whispering gallery. It is not unlikely also that sounds which travel long distances in the sea are dependent on such an effect, and on repeated reflection at the surface and the sea-bed. In a sea of average depth, 150 ft. (the North Sea, for example), the maximum possible range of *direct* transmission will be about thirty miles, whereas ranges two or three times as great as this have frequently been observed. The possibility of a curved path due to temperature gradients, however, must be taken into account in any theory of long-range transmission of sound.

Huyghens' Construction for Reflected Waves – The more important features of reflection of sound waves are readily obtained

* *Nature*, **108**, p. 42, Sept. 8, 1921.
† *Indian Assoc. Cult. Sci.*, **7**, p. 159, 1922.
‡ *Collected Papers*, p. 255.
§ *Journ. Sci. Instrs.*, Feb. 1929.
‖ *Calcutta Math. Soc., Bull.* **12**, 4, p. 225, 1922.

by the direct application of Huyghens' principle of secondary waves (see p. 67). This principle states that the wave-front of a disturbance may at any instant be obtained as the envelope of the secondary wave proceeding from all points of the wave-front at some preceding instant. We shall consider a few simple examples. (1) *Reflection of plane waves from a plane surface.* Spherical secondary waves leave each point of the surface as soon as the incident wave-front reaches it. Thus in fig. 82 I. (*a*), as the wave-front WW′ impinges on the surface WR, each point between W and R becomes in turn a source of secondary waves. By the time W′ in the incident wave has reached R, the secondary hemicylindrical wave from W has spread over a radius WR′ = W′R. A plane RR′ drawn through R and tangential to this hemicylindrical surface will therefore also be tangential to the corresponding surfaces emanating from the respective secondary sources between W and R. Consequently RR′ is the new plane wave-front. The direction of propagation is normal to the wave-front so that the incident beam AB now becomes BC. The figure is symmetrical about the normal, so that the angle of incidence is equal to the angle of reflection. Fig. 82 I. (*b*) is an instantaneous spark photograph by Cranz * showing a number of reflections of the head waves from a bullet passing between two parallel plates. (2) *Spherical waves reflected from a plane surface* are represented graphically on Huyghens' construction in fig. 82 II. (*a*). When the spherical waves diverging from S meet the plane surface, the reflected wave-front formed by the secondary waves is represented by the mirror image of that portion of the incident wave, which in the absence of the plane would have spread beyond it. The reflected spherical wave appears to diverge from a point S′ which is the mirror image of S in the plane reflector. Spark photographs by Foley and Souder,† fig. 82 II. (*b*), show the reflection of the spherical pulse from a plane surface. (3) *Spherical waves reflected from parabolic and spherical mirrors.* The Huyghens' construction for spherical waves diverging from the focus is indicated graphically in fig. 82 II. and III. The comparative spark photographs by Foley and Souder † and by A. H. Davis ‡ for the parabolic and spherical reflections respectively, are shown in fig. 82 II. and III. (*b*). The resemblance between the graphical construction and the photograph is very striking. In both cases the aberration of the

* *Lehrbuch der Ballistik*, **2**, p. 452, 1926

† *Phys. Rev.*, **35**, p. 373, Nov. 1912.

‡ *Proc. Phys. Soc.*, **38**, p. 224, April 1926.

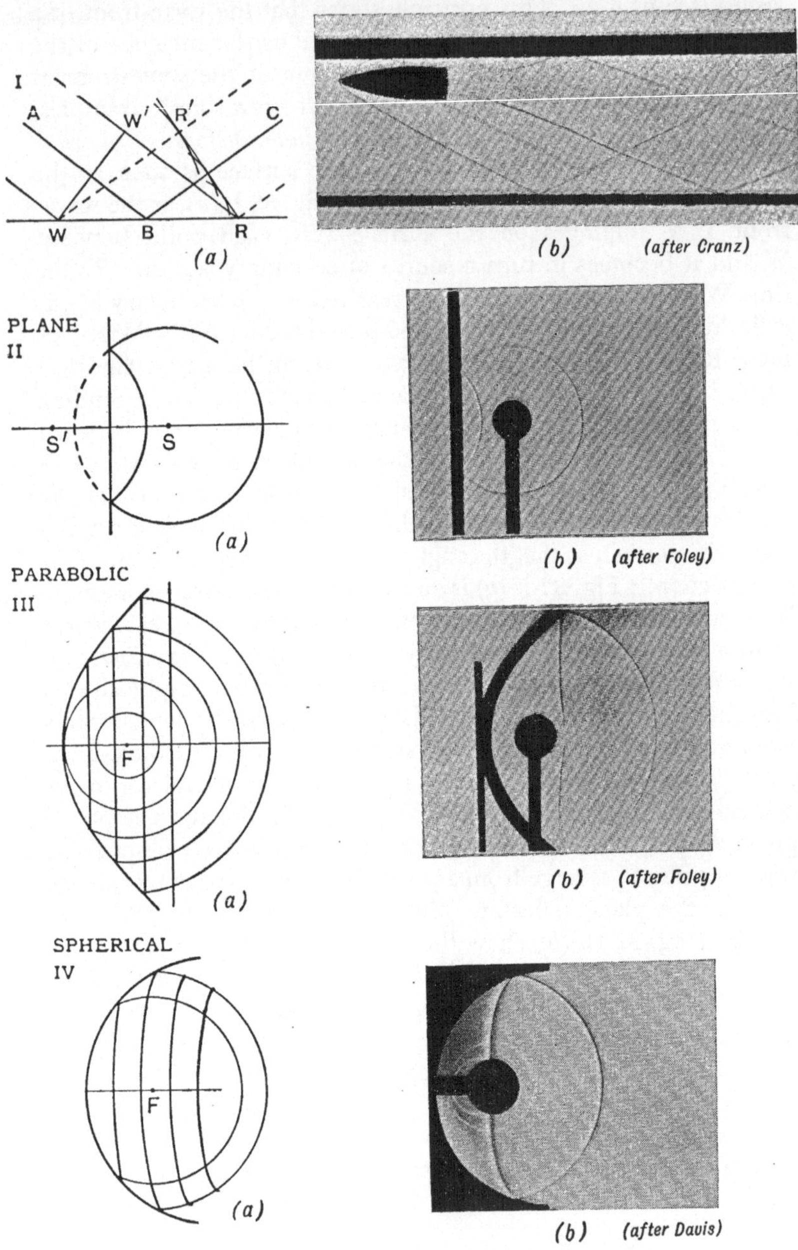
I
A
W′
R′
C
W
B
R
(a)
(b)
(after Cranz)
PLANE
II
S′
S
(a)
(b)
(after Foley)
PARABOLIC
III
F
(a)
(b)
(after Foley)
SPHERICAL
IV
F
(a)
(b)
(after Davis)

FIG. 82 (*a*) and (*b*)

spherical reflector and the superiority of the parabolic reflector as a means of producing plane waves is clearly shown.

It will be observed that in all cases the spark photographs exhibit practically true optical reflection. This is only possible, as we have seen, when the dimensions of the reflector are large compared with the wave-length. Now a single pulse, strictly speaking, has no wave-length. The basis of comparison in such a case is the *thickness* of the pulse which, as the photographs show clearly, is small compared with the diameters of the mirrors.

(*B*) Refraction

The change in direction in sound waves on reaching the boundary between two different media is realised most simply from a consideration of Huyghens' secondary waves. When the incident plane wave-front WW′ (see fig. 83) reaches the surface at W a secondary wave starts from that point with a velocity of sound c_2 in the second medium. At a time t afterwards the secondary waves from W′ and W will have reached R and R′ respectively. Each point of WW′ will similarly have reached a corresponding point of RR′, which is therefore the new wave-front in medium 2. Now $W'R/c_1 = WR'/c_2 = t$, that is,

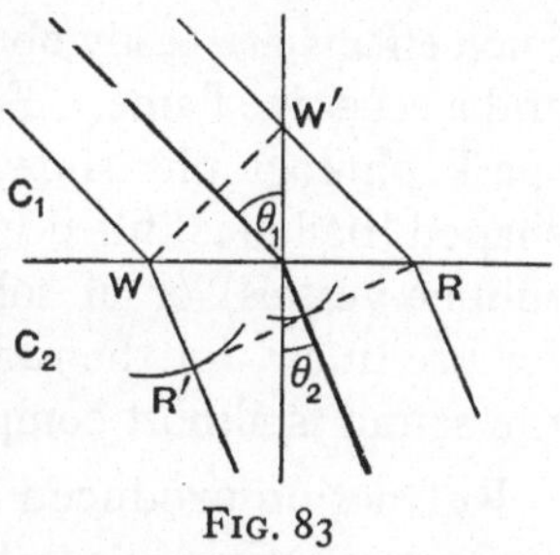

FIG. 83

$$WR \sin \theta_1/c_1 = WR \sin \theta_2/c_2$$

or

$$\frac{\sin \theta_1}{\sin \theta_2} = \frac{c_1}{c_2},$$

which is the 'law of sines' to which we have already referred (see equation (1), p. 271). Hence the normal to a wave-front in passing from one medium to another is deviated towards or away from the normal to the surface, according as the velocity of the wave in the first medium is greater or less than in the second medium. The *critical angle*, when $\theta_2 = 90°$, is given by $\sin \theta_1 = c_1/c_2$, and total reflection may take place when the sound travels from a medium of low to one of higher wave-velocity. The particle-velocity in the transmitted wave relative to that in the incident wave is given by equations (5) and (6*b*) on p. 272. Equation (8), p. 273, shows that the energy in the transmitted wave is $t_{12}^2(R_2/R_1)$, where R_1 and R_2 are the acoustic resistances ($\rho_1 c_1$ and $\rho_2 c_2$) and t_{12}

the ratio of the particle-velocities of the incident and transmitting media respectively. At normal incidence

$$t_{12}=2R_1/(R_1+R_2)$$

and the corresponding energy-ratio for the transmitted and incident waves is

$$\text{normal transmission coefficient}=\frac{4R_1R_2}{(R_1+R_2)^2},$$

which reduces to $4R_1/R_2$ when R_2 is large compared with R_1.

Sondhauss in 1852 demonstrated the refraction of sound waves through prisms containing various gases, and determined the *refractive index* $\mu(=c_1/c_2=\sin\theta_1/\sin\theta_2)$ relative to air. He was also successful in demonstrating the focussing action of a convex lens of carbon dioxide enclosed in a thin envelope of collodion. Such effects are easily obtained by means of a high-pitched source and a sensitive flame. Foley and Souder * have recently obtained spark photographs showing the focussing action of small lens-shaped 'balloons' filled with different gases. Lenses of liquid (in rubber vessels) or of solid material are obviously unsatisfactory for use in air, for the ratio R_2/R_1 in all such cases is so great that the sound is almost completely reflected at the front surface.

Refraction produced by Wind and Temperature Gradients – Refraction takes place whenever the wave reaches a point at which the wave-velocity changes. This change of wave-velocity may be brought about not only by complete change of medium, but also by a change of properties in the *same* medium (if it may still be so regarded). Thus the velocity of sound waves in air is affected by the wind and by temperature. If these factors are not constant over the whole wave-front and the whole track of the sound wave, refraction will take place. It is a familiar observation that sound travels better with the wind than against it. This is due to the fact that *the velocity of the wind increases from the earth's surface upwards.* The effective velocity of sound is consequently equal to its normal velocity plus or minus the velocity of the medium. Imagine a plane wave of sound travelling parallel to the earth's surface, *i.e.* with its wave-front perpendicular to the ground. In still air this condition will continue. If, however, a steady wind is blowing in the *same* direction the upper part of the wave-front where the wind-velocity is greater will travel faster than that part of the wave-front near the ground. The result is that the wave-front tends to bend downwards towards the

* *Loc. cit.*

ground. An observer therefore hears the sound by a 'ray' which leaves the source with a slightly upward inclination. Similarly the wave-front travelling in the *opposite* direction to the wind is bent upwards and, at a moderate distance from the source, passes high above the head of an observer.* Similar effects may be noticed when there is a uniform temperature gradient from the earth's surface upwards. The warmer the air the greater the velocity of sound ($c \propto \sqrt{\text{abs. temp.}}$). When the temperature gradient is positive upwards, the wave-front will bend towards the ground, and conversely when the temperature gradient is negative upwards the sound will be deflected upwards and lost. Temperature or wind refraction in the atmosphere is analogous to the optical phenomenon of *mirage*. Such wind and temperature gradients may therefore have a very important influence on the range of transmission of sounds in the atmosphere. The method of calculation of the curvature of a 'sound ray' is equally applicable to the cases of wind and temperature refraction, provided the correct sign is introduced for the velocity-gradient upwards. In fig. 84, AB and A′B′ represent portions of two adjacent 'rays' in the same vertical plane. WF and W′F′ are two consecutive wave-fronts intersecting at C and cutting off lengths δs_1 and δs_2 from the rays AB and A′B′ respectively. From the construction shown we must have $\frac{\delta s_1}{r} = \frac{\delta s_2}{(r - \delta r)}$, where r is the radius of curvature of the element δs_1 and δr is the separation of δs_1 and δs_2 measured along the radius. Since the lengths δs_1 and δs_2 are traversed by the wave-front in the same time interval,

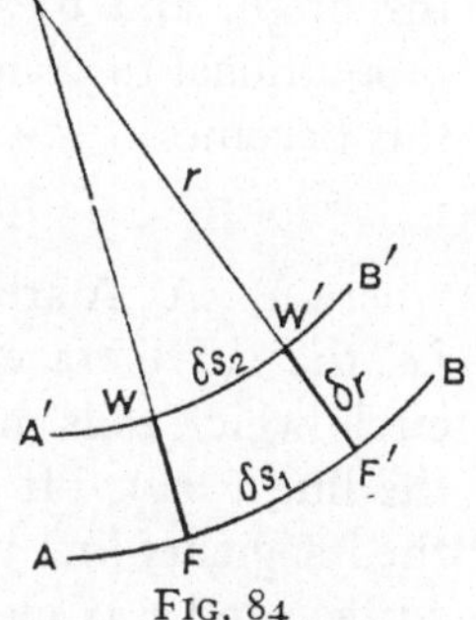

FIG. 84

$$\frac{\delta s_1}{c} = \frac{\delta s_2}{c + \delta c}.$$

Eliminating δs_1 and δs_2 from these relations we obtain

$$\frac{1}{r} = -\frac{1}{c} \cdot \frac{dc}{dr} \quad . \quad . \quad . \quad . \quad (1)$$

* W. S. Tucker observes, however (*Journ. Roy. Soc. Arts*, **71**, p. 132, Jan. 1923), that this rule is frequently broken. If a following wind has a velocity which *increases* with height it is favourable for sound transmission, the same condition obtaining with an adverse wind which diminishes in strength with height.

If the inclination of δs to the vertical is ϕ, then

$$\frac{dc}{dr}=\frac{dc}{dy}\sin\phi \quad \text{and} \quad \frac{1}{r}=-\frac{1}{c}\frac{dc}{dy}\sin\phi \quad . \quad . \qquad (2)$$

y being the vertical co-ordinate. This gives the curvature $(1/r)$ of the ray at all points, and depends on the particular assumption in regard to (dc/dy).

In considering any problem of continuous refraction it is important to remember *the sine law of refraction is valid over the whole track of a ray*, i.e. *the ratio* c/*sin* ϕ *is constant* at all points of the ray. It is possible to deduce equations (1) and (2), starting with this fact as a basis. Assuming a uniform temperature gradient (suppose negative) upwards, at a certain height the temperature would fall to the absolute zero. If we take this level as the origin and measure y downwards, the temperature will be proportional to y and the velocity c to $\sqrt{y}$. Writing $c=A''\sin\phi$ this becomes

$$y=A'\sin^2\phi=A(1-\cos 2\phi) \quad . \quad . \qquad (3)$$

where A'', A', A are constants. The relation represents a cycloid, *i.e.* the curve traced out by a point on the circumference of a circle which rolls in contact with a fixed straight line (in this case the line $y=0$). If the temperature gradient is positive upwards, the imaginary line $y=0$ of zero temperature is 'below' the earth's surface and y is measured upwards. In this case the generating circle of the cycloid rolls above the line $y=0$ and the cycloids have their vertices downwards. Only the relatively flat portions of the cycloids, which may be regarded as circular arcs, will, of course, be of interest in practice. For the reasons given above it seems not improbable that seasonal variations will be observed, apart from rapid variations due to numerous causes, in the range of transmission of sounds above the surface of the sea, *e.g.* from light-vessels. The temperature of the sea varies slowly, and lags behind the mean seasonal variation of the temperature of the atmosphere above it. In the Straits of Dover, for example, a continuously recording electrical thermometer indicated a maximum sea temperature in September and a minimum in March, representing a lag of about two months or so behind the corresponding temperatures of the atmosphere. From March to August, therefore, when the sea temperature is definitely lower than the air temperature and the gradient upwards is positive, a good range of transmission should be anticipated; whereas from

September to February, when the sea is warmer than the atmosphere and the gradient is negative, the sound transmission should be poor. It is interesting to refer in this connection to experiments by E. S. Player.* Observing, from North Foreland, the sound of a siren on a light-vessel several miles away, it is noted: From May to September inclusive the siren was audible on six occasions, inaudible on two; from October to March it was audible on two occasions, inaudible on nine.

Temperature Gradients in the Sea – The cycloidal form of track assumed by a sound ray in air due to wind or temperature gradients may also be found in the sea. H. Lichte † has observed large seasonal variations in the range of sound signals under the sea, and ascribes the various effects to the existence of horizontal temperature gradients in the sea, these gradients changing according to the season of the year. In this case the cycloidal pattern is repeated more frequently than in the case we have just considered, for a sound ray meeting the surface of the sea is completely reflected down again, only to rise once more and so on, the number of times it meets the surface in a given distance depending on the depth of the source of sound. A reverse temperature gradient results in corresponding effects at the sea-bed. These surface and bottom reflections play an important part in long range sound transmission under the sea. With a particular sound source emitting a measured amount of power, Lichte and Barkhausen ‡ observed a change in range from 10 km. in summer to 20 km. in winter in the Baltic Sea.

Interference

One of the most important theorems applicable to sound waves is Huyghens' *Principle of Superposition* (see p. 66). On this principle the resultant displacement of a particle of the medium through which two or more trains of waves are passing is obtained by the vector addition of the separate displacements due to each wave-train independently. This principle is also applicable to velocities and accelerations, but *not* to the squares of these quantities. Thus, two periodic vibrations of the same frequency, of amplitudes a and b and phase difference ϵ, combine to form a periodic vibration of amplitude $(a^2+b^2+2ab\cos\epsilon)^{\frac{1}{2}}$. If the amplitudes are equal ($a=b$) and the phases the same $\epsilon=0$, super-

* *Journ. Roy. Met. Soc.*, **52**, p. 354, 1926. (The observations relate to rainy days only.)

† *Phys. Zeits.*, **20**, p. 385, 1919.

‡ *Ann. d. Physik*, **62**, July 1920.

position gives a vibration of double amplitude $2a$; but if the phases are opposed $\epsilon=180°$ the resultant amplitude is zero. In the more general case, the amplitude may vary between $(a+b)$ and $(a-b)$ according to the phase difference ϵ. Such superposition of vibrations is termed *interference*. The term is more frequently applied to the special case in which the resultant amplitude varies between double the ordinary amplitude and zero. Schuster's definition * of interference in light waves is the following: "If the *observed* illumination of a surface by two or more pencils of light is not equal to the sum of the illuminations of the separate pencils, we may say that the pencils have interfered with each other, and class the phenomenon as one of interference." This definition, with the appropriate modification, is also applicable to sound waves.

The phenomena of interference may be observed on a smooth water surface which is disturbed simultaneously at any two points. J. H. Vincent † has obtained very beautiful photographs of such effects on the surface of mercury, the simultaneous disturbances being produced by means of a double 'dipper' attached to the prong of a vibrating tuning-fork. Such a photograph, taken from one of Vincent's papers, is shown in fig. 85. It is clear that in certain directions the crests and troughs reinforce each other, whilst in intermediate directions they neutralise. The result is a definite 'interference pattern' superposed on the ordinary wave systems. Analogous effects are easily demonstrated with sound waves proceeding from the tuning-fork. Each prong, as it vibrates,

FIG. 85—Interference pattern on Mercury Surface
(Dr. J. H. Vincent)

* *Optics*, p. 57.

† *Phil. Mag.*, p. 411, June 1897, p. 191, Feb. 1898, p. 290, Sept. 1898, p. 338, Oct. 1899.

is the source of two sets of waves in opposite phase, one side of the prong producing a compression pulse at the same time as the other side is producing a pulse of rarefaction. These two simultaneous pulses, in opposite phase, will neutralise in certain directions and assist in others. A similar interaction goes on, of course, between the waves from the two prongs. The result is a region of disturbance around the fork having symmetrical positions of comparative silence. As a vibrating fork is rotated in a vertical position near the ear, the sound intensity rises and falls four times per revolution. Any 'double' or 'multiple' source of sound shows these effects, the principle being used specifically to produce *directional* beams of sound. In such cases the sound energy is concentrated, on the 'superposition or interference' principle, in certain directions whilst in other directions the energy transmission is zero. A more direct method of demonstrating interference between two trains of sound waves makes use of the 'trombone' tube * shown in fig. 86. The source of sound S should preferably emit a fairly pure note of moderately high pitch, *e.g.* a tuning-fork of frequency 1000 p.p.s. or more, a musical buzzer or a loud speaker electrically maintained. A stethoscope attached at L to the fixed tube is used to convey to the ears the sound which arrives at L *via* the paths STL and ST′L. Provided these paths are equal or differ by a whole number of wavelengths, the two sets of sound waves arrive in the same phase and reinforce each other at the ears of the listener. When the lengths differ by a multiple of $\lambda/2$, however, there is complete neutralisation (silence) at L. The method is very effective in demonstrating interference, and provides a good laboratory method of measuring the velocity of sound in a tube, the frequency of the source being known. An important application of this method relates to the directional reception of sound (see 'binaural compensator,' p. 413). One of the best examples of interference of sound is that in which *stationary waves* are formed between the wave-trains incident on and reflected from a plane reflecting surface (see pp. 279 to 281). The formation of nodes and loops in vibrating strings, rods, and

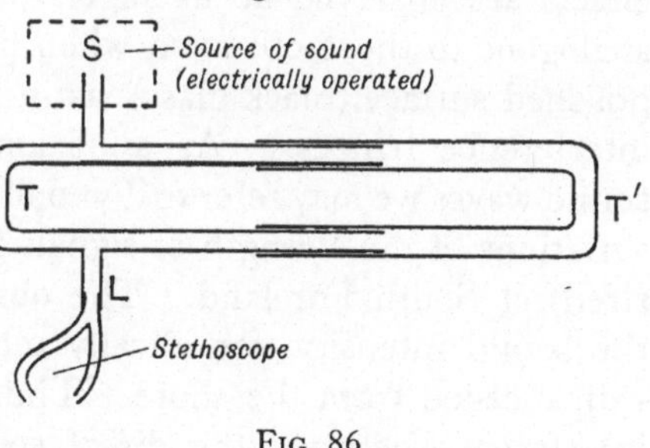

FIG. 86

* See Tyndall, *Sound*, p. 261.

air columns are also due to interfering wave-trains moving in opposite directions, reinforcing and neutralising at intervals of $\lambda/4$.

The phenomenon of beats (p. 19) may possibly be regarded as an example of interference. Two vibrating tuning-forks of *nearly* equal pitch produce a resultant sound, which increases and decreases in intensity with time, the frequency of this intensity-variation being equal to the difference of frequencies of the forks. This 'beating' effect is due to the fact that at certain equal *time* intervals the wave-trains agree in phase and reinforce each other, whilst at intermediate periods they are opposite in phase and tend to neutralise each other.

As in the optical analogy, interference may take place between sound waves from a source and its image (see p. 275), or from two images of the same source (as in Young's double-slit and Fresnel's biprism and inclined mirrors). The former is exemplified at normal incidence by stationary waves. Interference effects are observed at all angles of incidence, the effects being analogous to those observed when light is reflected from a single polished surface (black glass), as in Lloyd's method of producing interference fringes.* As an example of such interference in sound waves we may refer to Tyndall's observations of the apparent variations of the strength of signals from a source of sound (a fog siren) at South Foreland. The observer on a ship noticed that the sound intensity periodically increased and decreased as the ship receded from the shore. The effect was clearly due to the interference between the direct sound and the sound reflected from the surface of the sea. At greater distances, when the path difference between the direct and reflected waves is always less than $\lambda/2$, there can be no fluctuation of intensity, and the reflected wave (in almost the same phase) will assist the direct wave at all ranges. Experiments by F. B. Young and the writer † have demonstrated the existence of interference of sound waves *under* water also. A small source of sound of a definite frequency (580 p.p.s.) was supported at a known depth below the surface of a large reservoir. An exploration of the sound distribution in a vertical plane passing through the source was made by means of a small hydrophone 'pressure explorer' (a diaphragm with microphone attached, mounted on a watertight case). As the hydrophone was lowered from the surface to the bottom the sound received passed through maxima and minima. The

* See textbooks on *Optics*. See also p. 299.

† *Proc. Roy. Soc.*, **100**, p. 261, 1921.

positions of minima at different distances from the source were plotted and found to lie on a festoon type of curve near the source and then at greater distances to become indistinct. The first section of the 'festoon' was found to agree with the theoretical position of the interference zone due to sound from the source and its *negative image* in the surface of the water. The succeeding zones were explained by the multiple reflections from bottom and surface. The pressure amplitude of the sound near the surface was consistently very low, as would be expected since all points in the surface are equidistant from a positive source and its negative image (that is to say, there is no pressure fluctuation at a free end, water to air). This cancellation of the direct ray by the surface-reflected ray has an important bearing on the long-range transmission of sounds under water. Imagine a ship of 20-feet draught signalling with a Fessenden oscillator (see p. 390) of frequency 500 p.p.s., and a second ship 5000 feet away receiving the signals at the same depth 20 feet. The direct ray from ship to ship has to travel 5000 feet, and the ray reflected from the surface in opposite phase (*i.e.* $\lambda/2$ equivalent path-difference, by reflection at a free surface) has to travel $(5000^2+40^2)^{\frac{1}{2}}=5000{\cdot}16$ feet. Taking the velocity of sound in the sea as approximately 5000 ft./sec., the wave-length of the sound is 10 feet. The phase difference of the two 'rays' on arrival at the second ship is therefore $(0{\cdot}5-0{\cdot}016)\lambda$; at a range of 10 miles (say 50,000 feet) the phase difference would be $(0{\cdot}5-0{\cdot}0016)\lambda$, and so on. The surface-reflected wave at any appreciable range, therefore, practically neutralises the direct wave. In spite of this fact, signalling ranges of 20 to 30 miles are not unusual. The calculation assumes (1) that the surface of the sea is smooth, which is rarely the case, and (2) that the sea as a medium for sound transmission is homogeneous; temperature gradients may, however, produce considerable deviations in the path of the sound wave. Another important factor which has been neglected in the calculation is the reflection of sound from the sea-bed. In most cases this reflection is very efficient, and may in certain cases account almost entirely for the sound received at long ranges. The interference between two independent trains of high-frequency waves in water has been demonstrated by R. W. Boyle and J. F. Lehmann.* They mounted two piezo-electric quartz transmitters (see p. 147), vibrating with the same frequency (96,000 p.p.s.) and amplitude, side by side in a tank of water in such a way that their primary

* *Loc. cit.*, p. 150.

beams overlap. Fine coke dust sprinkled in the water settled in the nodal planes, and was ultimately caught on a whitened surface below the sound beam, the hyperbolic interference pattern being clearly shown. A photograph of such a pattern is produced in fig. 87 (*a*). In another experiment, to which we have already

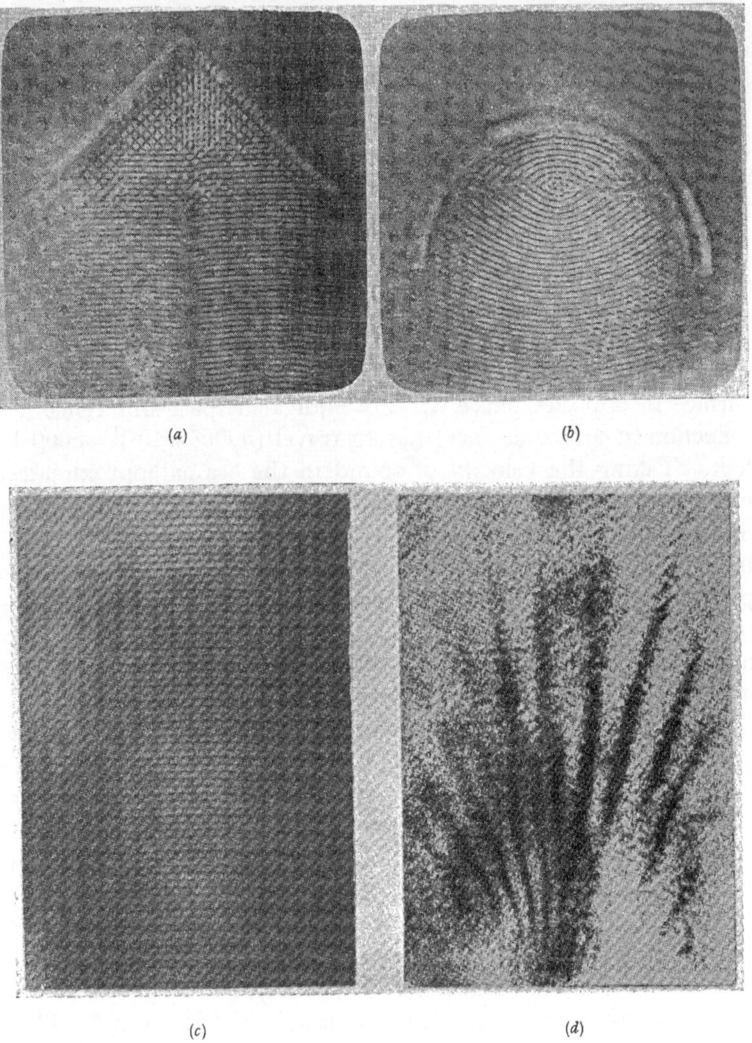

(*a*) (*b*)

(*c*) (*d*)

Fig. 87—Interference between High-Frequency Sound Waves

(*a*) Right-angle reflector
(*b*) Concave reflector
(*c*) Plane reflector (showing stationary waves)
} Frequency, 750,000 p.p.s. in oil (by courtesy of Dr. F. Ll. Hopwood)

(*d*) Interference between two high-frequency beams. Frequency, 96,000 (by courtesy of Dr. R. W. Boyle)

referred (p. 260), the two transmitters were placed face to face, and stationary waves, indicated by the dust pattern, were formed between them. F. L. Hopwood * has recently demonstrated by this method the interference effects produced by reflection of high-frequency sound waves (750,000 p.p.s.) from various surfaces (*e.g.* inclined planes, spherical mirrors, and a 'whispering gallery') (see fig. 87 (*b*) and (*c*)).

S. R. Humby,† using a high-frequency sound source in air, has demonstrated the existence of Lloyd's interference bands in front of a plane reflector, the observed hyperboloid 'fringes' agreeing accurately with the theoretical positions. Similar observations were made with two separate sources operated electrically from the same circuit, the form of the interference bands being, of course, the same as in the case of a single source and its 'image source' in the plane reflector.

Diffraction

It is a familiar observation that sound waves bend round corners whilst the propagation of light is sensibly linear. A carefully planned experiment is necessary to show that light also 'bends round a corner,' whereas the most casual observation reveals this effect in the case of sound. The sound-shadow of an obstacle is therefore not sharply defined, and appears to be much smaller than the corresponding optical or 'geometrical' shadow. The explanation lies in the fact that the length of the waves of sound is of the order of a million times the wave-length of light, and, as we shall see, the extent of the encroachment of the waves inside the geometrical shadow increases with increase of wave-length. To obtain sharp sound-shadows with obstacles of moderate size it is therefore necessary to use very short waves, *i.e.* sounds of high frequency. All such phenomena, involving the encroachment of waves (of any type) inside the geometrical shadow of an obstacle, are described as *diffraction*. The study of diffraction phenomena is often, inappropriately, left to text-books on optics. In almost every branch of sound, however, diffraction effects predominate. In the treatment of directional emission and reception, reflection, refraction, and scattering of sound it is generally important to specify the relationship between the dimensions of the bodies concerned and the wave-length of the sound—the importance of diffraction being thereby implied. Such

* *Journ. Sci. Instrs.*, Feb. 1929.

† *Proc. Phys. Soc.*, **39**, p. 435, 1927. See also p. 311.

phenomena are conveniently studied with the aid of Huyghens' principle of secondary waves (see p. 67). According to this principle every point on a wave-front at any instant may be regarded as the origin of a fresh supply of spherical waves, and the resultant effect at a distant point P ahead of the wave will be the sum of the effects of these secondary spherical waves. The new wave-front as it passes through P will be the envelope of such spherical waves. Thus if a *plane wave* of sound falls upon a hole of small diameter (compared with a wave-length) in a rigid wall, the hole becomes a source of *spherical waves*, and there can be no shadow of the boundary of the hole. If the hole be large the wave passing through it will be sensibly plane, but the diffraction of waves near the edge of the geometrical shadow will ultimately give the wave a slight divergence, depending on the relative dimensions of the aperture and the wave. In order to obtain quantitative data regarding the intensity fluctuations in diffracted waves it is necessary to employ methods due to Fresnel.

Annular Half-wave Zones – In fig. 88 let WF represent a section through a plane wave-front. We require to know the resultant effect at a point P distant r from the pole O of the wave-front (OP being the normal to WF). With P as centre, describe a series of spheres of radii r, $r+\lambda/2$, $r+\lambda$, $r+3\lambda/2$, . . . and so on, cutting WF in AA_1, BB_1, CC_1, etc. The plane wave-front will consequently be divided into a series of circles of radii $\rho_1=OA$, $\rho_2=OB$, $\rho_3=OC$, each circle being distant from P by $\lambda/2$ more than its smaller neighbour. The radius of the mth circle will clearly be

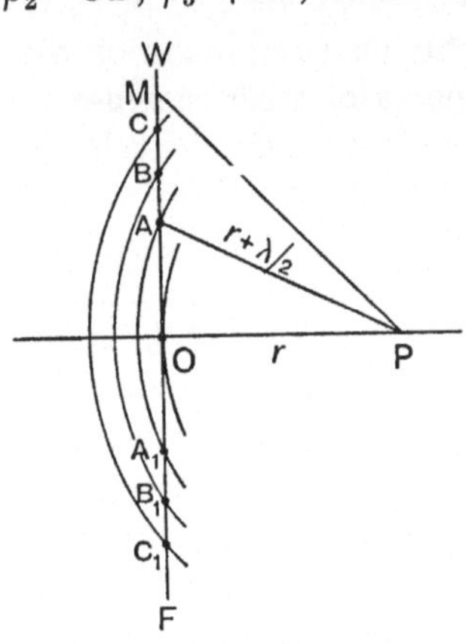

$$\rho_m=\{(r+m\lambda/2)^2-r^2\}^{\frac{1}{2}}=\sqrt{m\lambda r}$$

approximately. The area of this circle is therefore $\pi m\lambda r$, and the area of the annular zone between the mth and the $(m+1)^{th}$ circle is $\pi\lambda r$. That is, the *Fresnel annular zones* are all approximately of equal area $\pi\lambda r$. The contributions of alternate zones to the resultant at P will therefore be equal in magnitude but opposite in phase (each zone being $\lambda/2$, or π farther from P than its neighbour). The resultant effect at P will be the vector sum of the contributions of each element of the wave-

Fig. 88

front. It has been shown on p. 18 that the effect at P is represented by a vector polygon in the form of a spiral curve, each half-wave zone corresponding to a half-turn of the spiral. The resultant R at P is given by

$$\mathrm{R}=\frac{\mathrm{R}_1}{2}\pm\frac{\mathrm{R}_\mathrm{L}}{2},$$

and the phase $\theta=\pi/2$ behind that of the vibration at O, R_1 and R_L being the resultant effects at P due to the first and last zones respectively. When the last zone is at a considerable distance from P or is reduced by an obstacle to a negligible area, *the effect at P is then equal to half the effect of the first zone* (see p. 18). This is otherwise evident, for any zone after the first is neutralised by half the sum of those immediately before and after it, the remaining half of the first zone outstanding. The resultant effect of one zone will of course be less than if its elements were all in the same phase. When the latter is progressively increasing it is easily shown (see p. 17) that the resultant amplitude at P is $m\delta s\,.\,\sin\theta\,.\,/\theta$, whereas it would have been $m\delta s$ had all the elementary effects arrived in phase, *i.e.* the mean effect of a zone ($2\theta=\pi$) is $2/\pi$, and of the whole infinite wave-front is $1/\pi$, times the area $\pi\lambda r$ of the zone. As regards its effect at P, therefore, *the whole wave-front may be replaced by a small area $\pi\lambda r$ of effective amplitude a/π* (a being the amplitude at O).

It is possible to check this deduction by using it to calculate the resultant amplitude at P, in the case of a plane wave, since we know it must be the same as that at O, viz. a. Writing $\delta s=\pi\lambda r$ and $a'=a/\pi$, we obtain from equation (33), p. 62, for the particle-velocity $\dot{\xi}$ of spherical waves in a *semi*-infinite medium

$$\dot{\xi}=-\frac{k\mathrm{A}'}{2\pi r}\sin(nt-kr)\quad\text{where}\quad \mathrm{A}'=a'n\delta s\quad\text{and}\quad k=\frac{2\pi}{\lambda}$$

and

$$\xi=\frac{a'\delta s}{\lambda r}\cos(nt-kr),$$

whence the amplitude at a distance r from the source as $a'\delta s/\lambda r$, which is equal to a, the amplitude in the original plane wave. The procedure adopted for a plane wave-front is also applicable to a spherical or a cylindrical wave-front, provided the magnitude of the first zone be correctly estimated. Many of the simpler cases of diffraction may be treated by this elementary method of dividing the surface into half-wave zones. If the manner in which the zones diminish in amplitude with increasing distance is known, the spiral method of determining the resultant effect is invaluable.

Laminar Half-Wave Zones. Fresnel's Integrals.* Cornu's Spiral – In certain cases, *e.g.* the diffraction at a straight edge of an obstacle, it is more convenient to deal with zones or strips parallel to the edge, regarding the incident wave as plane or cylindrical with its axis parallel to the edge. In such cases we need only consider the effects in a plane *section* through the wave, the edge, and the point P. This method is due to Fresnel, who restricted his calculations of diffraction phenomena to obstacles bounded by straight lines of infinite length. In fig 89 let WF represent a section through a wave-front advancing towards a plane SS. The wave may be regarded as cylindrical or plane. We require to find the resultant effect at a point P on SS. If the displacement in the wave at O, the 'pole' of P, be represented by cos nt the contribution to P of an element δs situated at Q, distant s from O and $(r+\delta)$ from P, will be $\cos\{nt-k(r+\delta)\}\,.\,\delta s$. The displacement at P due to all the elements of the arc WF will be proportional to

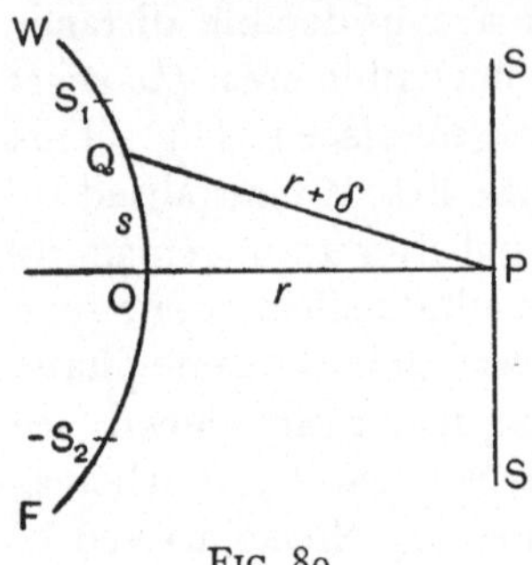

FIG. 89

$$\int \cos\{(nt-kr)-k\delta\}ds.$$

This may be written

$$\cos(nt-kr)\int\cos k\delta\,.\,ds+\sin(nt-kr)\int\sin k\delta\,.\,ds.$$

Now

$$k=\frac{2\pi}{\lambda}\quad\text{and}\quad\delta=\frac{s^2}{2r}\text{ app.,}\quad\text{that is}\quad k\delta=\frac{\pi s^2}{\lambda r}.$$

If we write

$$\text{A}\cos\alpha=\int\cos\frac{\pi s^2}{\lambda r}\,.\,ds\quad\text{and}\quad\text{A}\sin\alpha=\int\sin\frac{\pi s^2}{\lambda r}\,.\,ds$$

then the expression proportional to the displacement at P becomes

* A rigorous treatment of Fresnel's Diffraction Theory is given in Drude's *Optics*, chap. iv.

A cos $(nt-kr-\alpha)$. The *intensity* at P is proportional to A^2, that is to

$$\left(\int \cos\frac{\pi s^2}{\lambda r}\,.\,ds\right)^2+\left(\int \sin\frac{\pi s^2}{\lambda r}\,.\,ds\right)^2;$$

or if we write $s^2=v^2\,.\,\lambda r/2$ *the intensity at P is proportional to*

$$\left(\int \cos\frac{\pi v^2}{2}\,.\,dv\right)^2+\left(\int \sin\frac{\pi v^2}{2}\,.\,dv\right)^2 \quad . \quad . \quad (1)$$

The expressions in brackets are known as *Fresnel's Integrals*. Integrating between certain limits of v, we obtain the resultant of the secondary disturbances arriving at P and arising from that portion of the wave-front lying between the limits of s. The values of these integrals have been determined by Fresnel and tabulated. As we gradually increase the upper limit, the integrals pass through maxima and minima values, approaching $\frac{1}{2}$ as a limit. By a simple geometrical method Cornu has indicated the importance of these integrals in their application to diffraction problems. Writing

$$x=\int_0^v \cos\frac{\pi v^2}{2}\,.\,dv \quad \text{and} \quad y=\int_0^v \sin\frac{\pi v^2}{2}\,.\,dv \quad . \quad (2)$$

and plotting x and y for various values of v, it is found that a

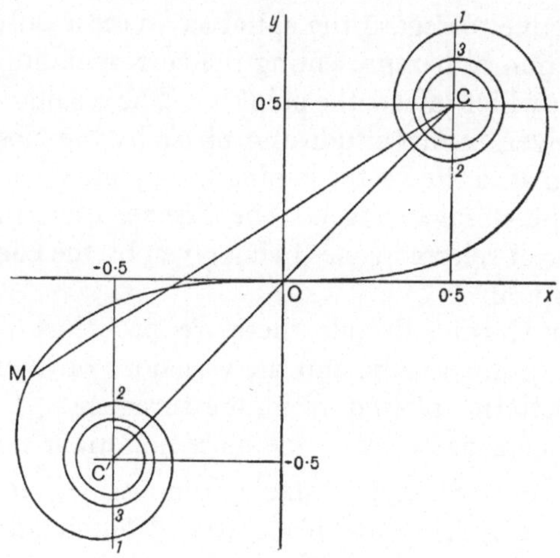

FIG. 90—Cornu's Spiral

spiral curve, known as Cornu's Spiral, results. This curve is shown in fig. 90. When $v=0$, $x=y=0$ also, and the curve passes through the origin. When v changes to $-v$ the expression under the integral is unaffected, but the upper limit of integration and consequently x and y change sign. The curve is symmetrical about the origin, and approaches asymptotically, to the points $C(\frac{1}{2} . \frac{1}{2})$ and $C'(-\frac{1}{2} . -\frac{1}{2})$, for values of v equal to $\pm\infty$. From equations (1) and (2) it is obvious that the resultant intensity at P due to a wave-front extending from s_1 to s_2 (corresponding to v_1 and v_2), is given by

$$I \propto (x^2+y^2)_{v_1}^{v_2} \quad . \quad . \quad . \quad (3)$$

When $v_1=0$, that is when we are dealing with one-half of the wave-front only, the resultant amplitude is the radius vector from O to the point v_2 in question on the spiral, the phase being the angle between this radius vector and the OX axis. When v_2 is infinite and $v_1=0$ the resultant is OC. In fact, the resultant amplitude between the limits v_1 and v_2 is represented by the vector joining these points on the spiral. Thus MC in fig. 90 represents the resultant from $-v$ to $+\infty$. The amplitude due to the *whole* wave-front, from $+\infty$ to $-\infty$ is CC′, and to one-half the wave-front OC=OC′. It will be observed that the phase lag in the case we are considering, viz. laminar zones, is 45° ($\pi/4$), whereas we found that the lag in the case of circular zones was 90° ($\pi/2$). In this respect the spirals shown in fig. 90 and fig. 6 should be compared.

It is instructive to regard the spiral as a vector polygon, as on p. 18, any section of it representing the corresponding portion of the wave-front which affects the point P. The resultant amplitude at P is then given in magnitude and phase by the closing side of the polygon (that is the vector joining the points v_1 to v_2). Each half-turn of the spiral represents one Fresnel zone, the resultant of any number of selected zones being given by the corresponding chord of the spiral.

By means of Cornu's Spirals, therefore, problems of diffraction can be solved geometrically, and the variations of intensity in the diffraction 'pattern' plotted from measurements of the spiral. We shall now consider a few of the more important examples.

Diffraction at a Straight Edge - Suppose WF in fig. 91 represents the cylindrical or plane wave-front incident on the semi-infinite rigid plane ES. We require to know the distribution of amplitude or intensity at various points P in the plane P_1NP_2. At N, the edge of the geometrical shadow, a complete half-wave is exposed. Consequently $v_1=0$ and $v_2=\infty$, and the amplitude at N, from the spiral (fig. 90), is the radius OC, *i.e.* half the amplitude, or one quarter the intensity, in the unrestricted wave. As the point P recedes from N into the 'shadow' the value of v_1 gradually increases from zero to infinity, that is, the point v travels from O along the spiral towards C. The amplitude thus steadily diminishes from OC=0·5 at the edge to zero when $v_1 = v_2 = +\infty$. The vector is a line joining C with the point v_1, which travels from O to C as the point P recedes farther and farther into the shadow. This line rapidly shortens without maxima or minima. *Outside* the geometrical shadow P is exposed to a complete half-wave plus a variable number of zones OE. The amplitude is therefore given by the vector-sum of OC and OM (see fig. 90), *i.e.* CM. Consequently as P moves outwards from N the point M ($-v_1$) passes from O round the lower spiral and eventually reaches C′ when P is an appreciable distance from N. As the distance P from N increases, the amplitude consequently passes through maxima and minima, gradually approaching the normal amplitude and occurring closer and closer together. As an example, consider a plane wave of sound of frequency 5500 p.p.s. ($\lambda = 0{\cdot}2$ ft. in air) incident on the edge. At a distance of 20 ft. behind the screen the intensity variations shown in fig. 92 would be observed. The widths of the successive zones are given by $\sqrt{m\lambda r}$, *i.e.* 2, 2·82, 3·46, etc., ft., and the relative amplitudes (and consequently intensities) are read off the Cornu's spiral for successive phase changes of $\pi(\lambda/2)$. Thus at 5500 p.p.s. the intensity falls off to about 3·6°/₀₀ at a distance of 6·3 ft. inside the geometrical shadow. At a frequency of 550 p.p.s. the same intensity would be observed at a distance of about 20 ft. inside the geometrical shadow. Considering that the ear is sensitive over a range of intensities of the order 10^{12} (see p. 351) a residual intensity of 3·6°/₀₀ would constitute a sound of moderate loudness if the intensity of the sound incident on the edge were fairly great. The 'shadow' of the edge is thus far from perfect, even

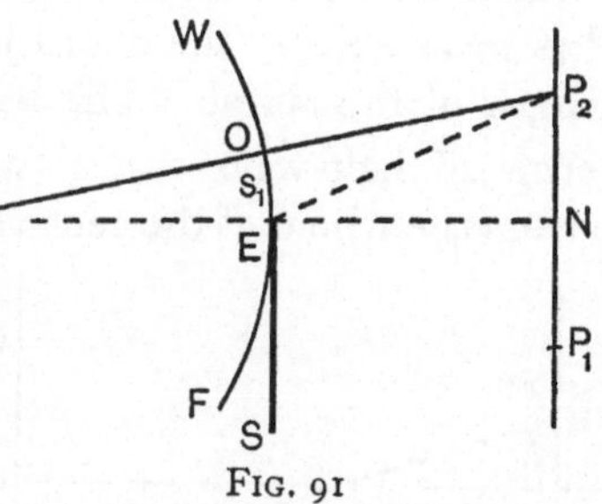

FIG. 91

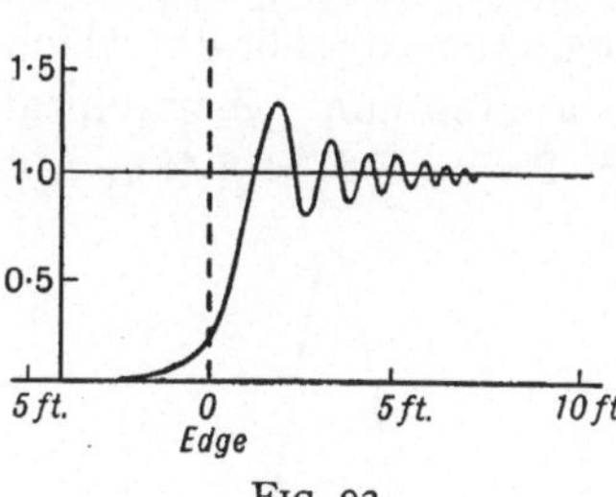

FIG. 92

at a frequency as high as 5500 p.p.s. At 55,000 p.p.s. the range inside the shadow would be reduced to a tenth and the shadow would be more nearly 'optical.'

Parallel Aperture – In this case the amplitude of vibration at P is measured by the chord joining the extremities of a constant length of the spiral. The width of the aperture is measured in terms of half-wave zones (width $\sqrt{m\lambda r}$). If the aperture just covers the whole of the central zone the amplitude at the midpoint will be a maximum, whilst it will be a minimum if the aperture covers two zones. The variations of amplitude on a plane whose distance r from the aperture is large compared with the width e of the opening are easily obtained directly by an approximate method (see fig. 93). Plane waves fall on the aperture AB and spread to the plane CPC.

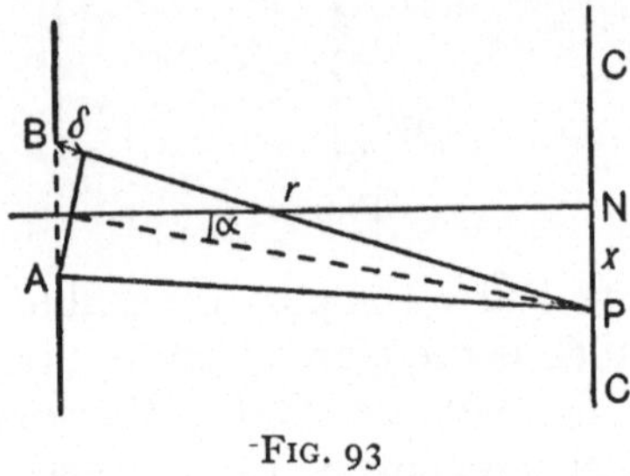

FIG. 93

When r is large we may regard BP and AP practically parallel and making an angle α with the axis. The resultant at P is then the sum of the effects due to elementary strips of the wave-front in the aperture. As we have seen on p. 17, this is given by

$$R=\frac{s\,.\,\sin\theta}{\theta} \quad \text{where} \quad \theta=\frac{2\pi}{\lambda}\cdot\frac{\delta}{2}=\frac{\pi e\sin\alpha}{\lambda},$$

e being the width of the aperture, δ the path difference between the extreme 'rays,' and s the 'effective' amplitude in the aperture. If we write $dR/d\theta=0$, *i.e.* $\theta=\tan\theta$, we obtain the values of θ at which the amplitude on CC is a maximum or a minimum. A graphical solution of this relation gives 0, $1{\cdot}43\pi$, $2{\cdot}46\pi$, $3{\cdot}47\pi$, $5{\cdot}48\pi$, and further values approximating to $13\cdot\pi/2, 15\cdot\pi/2, 17\cdot\pi/2$ and so on. The maxima of intensity R^2, which cannot be negative, occur at all these points, with intermediate positions of zero intensity (at $\theta=m\pi$ (where m is an integer)). For $\theta=0$ the intensity is s^2. Writing $s^2=1$ the magnitudes of the secondary maxima are 1/21, 1/61, 1/120, etc., successively. The intensity fluctuations are shown in

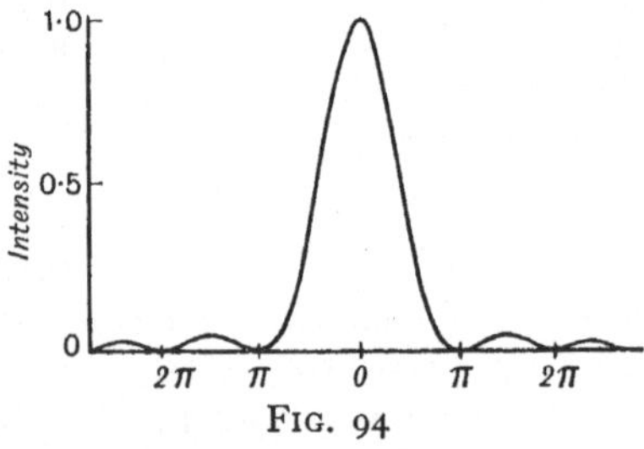

FIG. 94

fig. 94 which indicates that the greater part of the sound energy, over 95 per cent., is confined to the primary maximum extending from $\theta=0$ to π on either side of N. When e is large compared with a wave-length, practically the whole of the energy is confined to directions for which θ is small. Writing $NP=x$, $\sin\alpha=x/r$, and $\theta=\pi ex/\lambda r$. The amount of spreading of waves after passing through an opening depends entirely on the relation between the wave-length and the width of the opening. This explains the apparent discrepancy between sound and light, which delayed so long the general adoption of the wave theory of light.

Circular Aperture – When the aperture is such that the wave-front emerging from it can be divided into circular zones the resultant amplitude is easily determined for points on the axis. If the radius P of the aperture is such that an even number of zones is included, the amplitude at P is zero, whilst an odd number of zones gives an amplitude double that in the unobstructed wave. The condition for maximum and minimum amplitude is clearly $\delta=\rho^2/2r=m\lambda/2$ approximately, where r is the distance of P from O and m is an integer, the intensity being a maximum or a minimum according as m is odd or even. As in the case of a parallel aperture just considered the resultant amplitude is given by $A\sin\theta/\theta$. Off the axis the intensity cannot be calculated by simple methods. The problem has been treated mathematically by Verdet,* who shows that the intensity at a distance of many wave-lengths from the opening is given by

$$I=(\pi R^2)^2\left[1-\frac{1}{2}\left(\frac{m}{1}\right)^2+\frac{1}{3}\left(\frac{m^2}{2}\right)^2-\frac{1}{4}\left(\frac{m^3}{2\times3}\right)^2+\frac{1}{5}\left(\frac{m^4}{2\times3\times4}\right)^2\right]^2,$$

where $m=\frac{\pi R}{\lambda}\,.\,\sin\alpha$, R being the radius of the aperture and α the angle between the axis and the line joining P to the centre of the aperture. As α increases m will vary as $\sin\alpha$ and the expression in brackets becomes alternately positive and negative with intermediate zero values. The energy from the aperture will therefore be distributed in co-axial zones of maximum intensity separated by zones of minimum intensity. The positions and relative intensities of these maxima and minima are given in the following table:

---

* *Leçons d'optique physique*, t. 1, p. 301. See also Airy, *Math. Tracts Camb.*, p. 259, 305, 1842.

	m/π.	Intensity.
Central Maximum	0	1·00
First Minimum	0·61	0
Second Maximum	0·82	0·0175
,, Minimum	0·116	0
Third Maximum	1·333	0·00415
,, Minimum	1·619	0
Fourth Maximum	1·847	0·00165
,, Minimum	2·120	0

The intensities of the secondary maxima are almost negligible compared with that of the primary or central maximum. Therefore most of the energy is confined to the central beam, of which the angle of divergence is determined from the value of m at the first minimum. Denoting the semi-angle of the beam by ϕ,

$$0{\cdot}61=\frac{R}{\lambda}\sin\phi \quad \text{or} \quad \sin\phi=0{\cdot}61\frac{\lambda}{R}.$$

It will be seen that the central beam will be narrow when the radius of the aperture is large compared with the wave-length λ. When λ is greater than R, the value of $\sin\phi$ is imaginary and the beam is 'spherical,' *i.e.* the aperture may be regarded as a 'point' source of Huyghens' secondary waves.

The diffraction of plane waves of sound through an aperture is analogous to the radiation of sound from a piston source which we have already considered on p. 147. The applications of the theory to the directional transmission of sound by means of quartz oscillators has been investigated by P. Langevin * and by R. W. Boyle * (see p. 150). The late Lord Rayleigh * applied the theory to the directional transmission and reception of sound by means of trumpets, horns, and reflectors (see p. 192). Further important applications of this nature are dealt with by I. B. Crandall.*

Circular Disc. Zone Plate – The experimental demonstration by Arago that the optical shadow of a disc had a bright spot at its centre was regarded as final proof of the wave theory of light. Such a result for sound waves also follows immediately from the Huyghens-Fresnel construction. The resultant intensity on the axis of the disc is due to the action of the whole wave-front with the exception of the zone or zones covered by the disc. The effect is therefore approximately equal to one-half that of the zone

* *Loc. cit.*

bordering the disc. If the latter covers only one or two zones the intensity on the axis will be the same as if the disc were removed. The rings of secondary maxima separated by circles of zero intensity are similar to those already mentioned for a circular aperture. As we shall see (p. 420) a diffraction disc with a sensitive sound detector on its axis constitutes an important type of directional sound receiver, the effect on the 'detector' being a maximum when the source is on the axis of the disc.

An interesting application of the principle of Fresnel's half-wave zones is known as the *Zone Plate.* If a circular grating be cut from a sheet of metal so that alternate zones of an advancing wave-front are blocked out (reflected or absorbed) the remaining active zones will arrive in phase at a point P on the axis.

Let a be the distance of a point source S on the axis and b the distance of a small detector at P on the opposite side. Then the radius ρ of the mth zone may be derived from

$$\sqrt{(a^2+\rho^2)}+\sqrt{(b^2+\rho^2)}-(a+b)=m\lambda/2,$$

that is,

$$\frac{m\lambda}{\rho^2}=\frac{ab}{a+b}$$

approximately. Such a circular grating or zone plate acts like a condensing lens, the intensity at P being many times greater than in the absence of the grating. On account of increasing obliquity, there is little to be gained by employing more than a few zones near the axis. The equivalent focal length f of the zone plate is given by the above relation for

$$\frac{1}{f}=\frac{1}{a}+\frac{1}{b}=\frac{m\lambda}{\rho^2}, \quad \text{whence} \quad f=\rho^2/m\lambda.$$

A zone plate with a detector fixed on its axis at the focus not only serves to increase the intensity but also gives accurate indications of the *direction* of the source of sound S.

Diffraction Gratings. Reflection from Stepped or Corrugated Surfaces – The diffraction grating so familiar in optics, has its counterpart in sound. When sound waves are reflected from a regular periodic structure such as a row of palings or a corrugated surface, the reflected waves reinforce or neutralise in certain directions, depending on the wave-length of the sound and the spacing of the reflectors. Similarly, with the waves transmitted through equidistant apertures. A plane wave incident normally on such a grating will be diffracted in accordance with the simple relation

$$\sin\theta = \pm m\lambda/d,$$

where m is the 'order' of the spectrum and has the values 1, 2, 3, etc., θ is the direction of a maximum, λ the wave-length, and d the grating constant (space+strip). When d is smaller than λ there are no lateral maxima and the incident beam is reflected in the ordinary way. Thus a row of palings or a rough wall reflects sound of moderate pitch like a perfectly smooth surface, little or no sound being returned towards the source except at normal incidence. When the frequency is high, however, and the wave-length λ is short (less than d), the sound is diffracted into the spectra of various orders. A regular row of palings may therefore serve as a 'reflection' or 'transmission' grating for the analysis of complex, high-pitched sounds, the various component wave-lengths reinforcing in corresponding directions. In order to obtain satisfactory results it is, of course, necessary, as in the optical case, to use focussing devices to render the incident beam parallel and to converge the reflected or transmitted beams on a small detector.

Experimental – As a means of demonstrating the phenomena of diffraction, sound waves have not received the attention which they deserve. It is not until comparatively recent years that the importance of such phenomena in the directional transmission and reception of sound has been fully realised. All the so-called optical diffraction effects are easily demonstrated with high-frequency sound waves, and experiments on a laboratory scale may be made with very simple apparatus. Using a bird-call or a Galton's whistle as a source of sound of short wave-length (about 1 cm. in air) and a sensitive flame as a detector, Rayleigh * demonstrated the more obvious diffraction effects. Thus the flame flares when it is placed at the centre of the sound shadow of a *circular disc*, proving there is sound of appreciable intensity at such a point, although a small distance off the axis there is silence and the flame is quiescent. Tucker and Paris,† using an electrically maintained tuning-fork as a source of sound, obtained the diffraction pattern behind a disc of wood 10 ft. in diameter and 1 in. thick. The fork was placed 30 ft. in front and on the axis of the disc, and a hot-wire microphone connected to a tuned

* *Sound*, **2**, pp. 142 and 143.
† *Phil. Trans. Roy. Soc.*, **221**, p. 389, 1921.

vibration galvanometer was used to explore the sound distribution behind the disc. The diffraction pattern consisted of a central maximum equal in intensity to the unobstructed sound surrounded by a ring of zero intensity and a further secondary maximum. The disc, with the microphone *fixed* behind it near the centre, served as an accurate means of locating the direction of the incident sound. Rayleigh also demonstrated effectively the antagonism between the parts of a wave corresponding to the first and second Fresnel's zones. The distances 'a' and 'b' of the bird-call and sensitive flame respectively from a screen with a hole 20 cm.diameter were adjusted so that the hole corresponded to the first two zones, the flame is then unaffected. On reducing the aperture to 14 cm. diameter corresponding to the first zone only, the flame flares violently. A similar effect is produced if the central zone is blocked off, leaving the outer zone to transmit the sound to the flame. A metal *zone plate* of eight zones and focal length $f=\rho^2/m\lambda=48$ cm. was constructed and found to give results in accordance with the simple theory. More recently S. R. Humby * has demonstrated quantitatively the various interference and diffraction effects of high-frequency sound waves, using as detectors both sensitive flames and the ear. A telephone excited by alternating currents from a valve oscillator (see p. 79) was used as a steady source of sound of a known high frequency. This form of source has distinct advantages over the older high-pitched sources such as Rayleigh's bird-call or Galton's whistle. The experiments on interference gave results analogous to the optical interference experiments of Lloyd (see p. 260), Fresnel, and Newton, and very close agreement was obtained between measured and calculated positions of points on the hyperboloid fringe systems. An experiment was made in which results analogous to certain 'Heaviside layer effects' were obtained. The ordinary diffraction effects could be demonstrated easily over a large frequency range using such telephone 'whistles.' Diffraction effects could be detected at the *straight edge* of a large wooden screen. The gradual and progressive weakening of the effect on the flame as it passes into the geometrical shadow and the presence of maxima and minima outside the shadow were clearly observed. With a *rectangular aperture* 20 cm. wide by 43 cm. high and a source 40 cm. away, a central maximum and several secondary, or lateral, maxima were measured at 40 cm. beyond the screen, using a wave-length of 4·4 cm. The bright spot at the centre of a

* *Proc. Phys. Soc.*, **39**, p. 435, 1927.

circular disc 33 cm. diameter was observed and two other zones inside the geometrical shadow were located. *Zone plates* were found to have focal lengths in accordance with theory. A diffraction grating made of cardboard strips fixed across an opening in a large screen proved effective. The grating had nine openings, each 3·4 cm., with strips 1·7 cm. wide. The source was placed at the focus of a metal mirror 80 cm. from the grating, the frequency being 11,600 (λ=2·9 cm.). With the sensitive flame 70 cm. behind the grating a number of maxima were observed 9·2 cm. apart. *Diffraction gratings* were used in 1907 by W. Altberg * to measure the wave-length of high-frequency sounds produced by an electric spark. The grating consisted of a series of parallel glass rods (1 to 6 mm. diameter) 1 cm. apart and the spark was placed at the focus of a concave reflector which thereby directed plane waves on the grating. A similar mirror received the diffracted sound and brought it to a focus on a sensitive detector. The sound spectrum was obtained by rotation of the grating with respect to the source and the receiver. Wave-lengths of the order of a millimetre, corresponding to 330,000 p.p.s., were measured in this way. More recently K. Palaiologos † has used reflection gratings of spacing 0·6 mm. to measure wave-lengths as short as 0·17 mm., the sound waves being detected and their intensity measured by means of a Rayleigh disc (see p. 426). In fig. 95 is reproduced a spark shadow photograph by Foley and Souder, showing clearly the Huyghens' spherical wavelets transmitted through and reflected from a small grating having four parallel slits. No interference is, of course, possible with a single

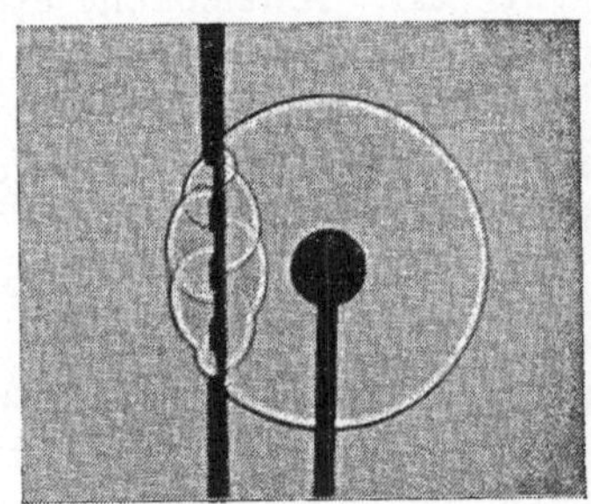
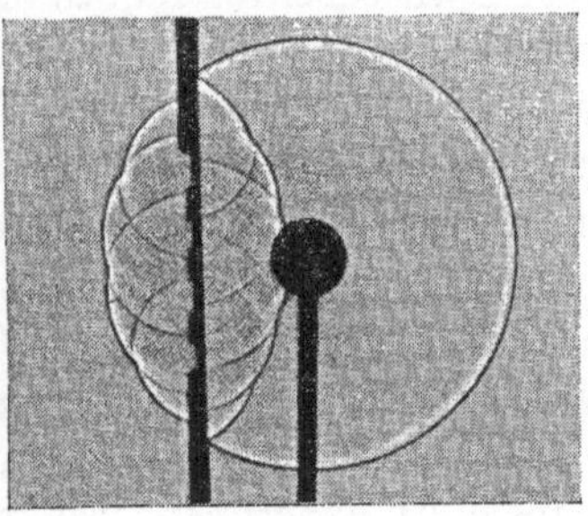

FIG. 95
(By courtesy of A. L. Foley)

* *Ann. der Physik*, **23**, p. 267, 1907. See also Dieckmann, *Ann. der Phys.*, **27**, p. 1066, 1908; and Neklepajow, *Ann. der Phys.*, **35**, p. 175, 1911.
† *Zeits. f. Physik*, **12**, 6, 375, 1923.

pulse of this character, consequently the reinforcement in certain directions is not shown. For this purpose ripple photographs similar to those obtained by J. H. Vincent are necessary. In these photographs a continuous train of ripples passes through the grating and the various orders of spectra are clearly shown.

Apart altogether from laboratory demonstrations of the above character, diffraction phenomena play an important part in the directional transmission and reception of sound. With regard to the former, reference has been made to the increasing directional efficiency of horns and concave reflectors as the wave-length diminishes, and to the sound beams of small divergence emitted from the Langevin quartz disc transmitter. In such cases, the mouth of the horn or reflector and the face of the quartz disc behave like an aperture through which plane waves are diffracted. The smaller the ratio λ/R the smaller the angle of divergence of the beam. Remembering that sound waves are of the order 10^5 or 10^6 times the length of light waves, the relative difficulty of producing a parallel beam or an image of the source in the two cases will be appreciated. With regard to reception, the 'bright sound-spot' at the centre of the shadow of a disc has been employed as an accurate means of locating a sound source even at low frequencies.* A sensitive detector on the axis of the disc gives a maximum response when the source, on the opposite side of the disc, is also on the axis; a few degrees on either side, and the response is zero.

Disturbance produced in a Sound Wave by a small Obstacle – We have already referred to the scattering of sound waves by a small obstacle (see p. 282), the amplitude in the scattered wave varying inversely as the square of the wave-length. Let us now consider the nature of the disturbance produced by a rigid obstacle of small dimensions compared with a wave-length. In the first place the obstacle behaves like a *simple source* due to its great rigidity. Relatively to the medium which it displaces the body virtually expands and contracts as the adjacent medium contracts and expands, thus giving rise to secondary waves in the same phase as the incident wave. A second wave-system is set up, due to the greater density of the obstacle which, on this account, fails to swing to and fro with the same amplitude as the medium which it displaces. The secondary waves therefore correspond to virtual oscillations of the obstacle exactly equal and opposite to

* See *Ency. Brit.*, 13th ed., Supplementary Vols., 'Sound,' A. O. Rankine.

that of the medium, tending to give rise at the surface facing the source to vibrations in the same pressure-phase as the primary waves, and at the remote surface to waves in the opposite pressure-phase. It thus behaves also as a *double source.* In this case, however, if the dimensions of the obstacle are small, the motion may result in 'local flow' from the positive to the negative side of the obstacle. F. B. Young and the writer * have made careful quantitative measurements of the acoustic disturbances produced by small bodies in plane waves transmitted through water. Observations were made of the sound distribution around various small discs (10 in. diameter) placed in the path of plane sound waves (580 p.p.s., $\lambda=8\frac{1}{2}$ ft.) in a large reservoir. The following measurements were made at a large number of points around each disc: (*a*) the *direction* of oscillation of the water particles, (*b*) relative amplitudes of displacement of the water particles, and (*c*) relative amplitude of the pressure-oscillations. For the purposes of (*a*) and (*b*) a small 'light-body' hydrophone (see p. 419) was employed. This consisted essentially of a hollow ebonite sphere (1 in. diameter) containing a 'button' carbon granule microphone (see fig. 96). The arrangement responds readily to sound waves which travel in a direction parallel to the axis of the microphone, but is insensitive to those arriving 'edge on' or parallel to the microphone diaphragm. The device acts as a directional hydrophone with an accuracy of about 5° at the 'minimum' position, the response at the 'maximum' position being proportional to the displacement or velocity of the water particles. Relative pressure - amplitudes were measured by means of a small button microphone mounted on a thin diaphragm forming the cover of a small cylindrical brass box (1 in. diameter). These 'velocity' and 'pressure' explorers could be traversed under water in a horizontal plane passing through the centre of the disc,

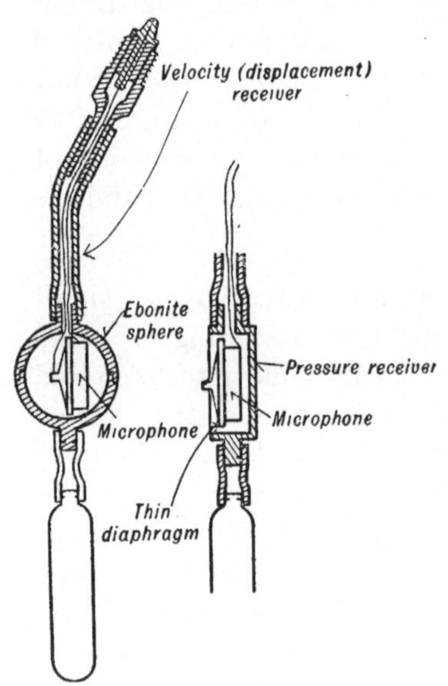

Fig. 96

* *Proc. Roy. Soc.*, **100**, p. 261, 1921.

and measurements of direction, displacement, and pressure-amplitudes made at known points relative to the disc. The principal phenomena observed were of two distinct types according as the 'sound field' is produced around (1) a solid body, or (2) a hollow (air-filled) body.

(1) *Solid Bodies* – The charted sound field around a lead disc, ¾ in. thick, in water (see fig. 97) is typical of solid bodies, which exhibit the following characteristics: (*a*) The lines of direction bend round the edges of the disc. At the same time a considerable portion of the sound energy is transmitted directly through it. The direction is clearly defined at every point, (*b*) There is a general diminution of displacement amplitude near either face of the disc, and (*c*) no variation of pressure amplitude (greater than 10 per cent.) is observable.

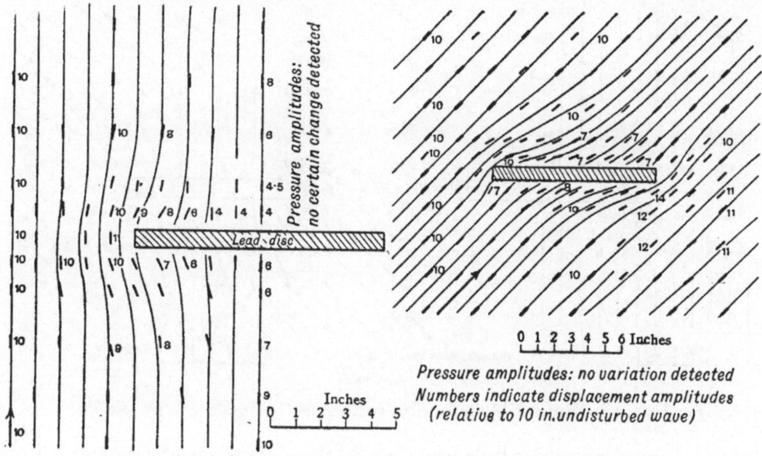

FIG. 97—Sound Distribution round a Lead Disc in Water

The observations in all respects confirm the view that the obstacle acts as a simple and a double source as outlined above. The form of the 'stream lines' in the charts is explained if they are regarded as the resultant of the primary oscillations and the secondary lateral oscillations. The absence of any appreciable effect due to the rigidity of the disc is to be expected from its small thickness, ¾ in., relative to a wave-length (about 8 ft. in water or in lead). Discs of wood, rubber, and paraffin-wax produced no definite disturbance, but appeared completely transparent to the sound waves in the water.

(2) *Hollow* (*air-filled*) *Bodies* – The oiled silk disc (shown in fig. 98) may be regarded as the simplest type of this class.

It consists of two sheets of oiled silk stretched over a light wooden ring, and is therefore a 'disc of air' in the water. The sound distribution shows the following characteristics: (*a*) The lines of direction converge on, or diverge from, the faces of the disc, (*b*) marked increases are produced in the displacement-amplitude near either face, (*c*) marked diminution of pressure-amplitude is produced near the faces, (*d*) regions of confused direction are produced in which the oscillations of the water particles are no longer rectilinear, and (*e*) marked changes of quality are produced if the incident sound is not pure. If the bounding wall of the disc consists of material having appreciable rigidity, such as tinplate, the surface appears to vibrate in sections, producing very sharply marked and localised disturbances with corresponding changes of quality. The 'hollow disc' in water

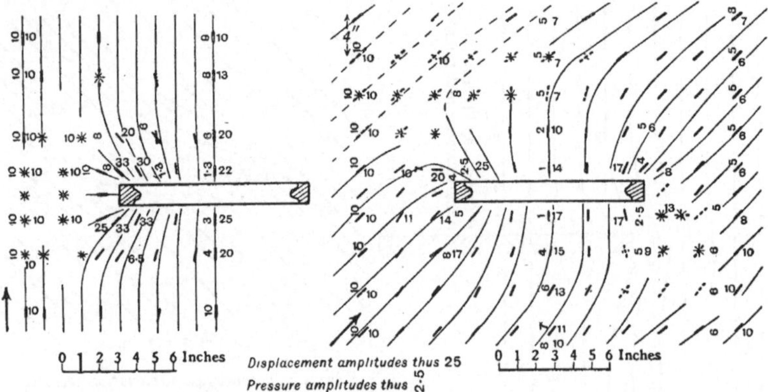

FIG. 98—Sound Distribution round an " Air-Disc " in Water

may in some respects be regarded as the converse of the solid disc. The 'air disc' shown in fig. 98 is more compressible than water, and will therefore tend to act as a simple source. The density of the disc being also less than that of water, the amplitude of the disc as a whole will exceed that of water particles in the undisturbed field. The disc will therefore behave as a double source. In the case of the air disc the phases of the vibrations of the simple and double sources will be opposite to the corresponding vibrations in the case of the solid lead disc. Various compromises between solid and hollow body characteristics may be produced by suitably constructed discs, the more pronounced hollow body effects tending to predominate. The above method of investigation has been applied to examine the sound distribution round more complicated structures in water, *e.g.* the single-plate directional hydrophone and its baffle plate (see fig. 125). In such cases the method has thrown a good deal of light on complex problems of sound distribution which are not amenable to mathematical treatment.

Attenuation of Sound Waves

Energy Loss due to Viscosity and Heat Conduction – We have hitherto assumed that spherical sound waves spread in an infinite homogeneous medium without loss of energy, in accordance with the law of inverse squares. It is found, however, that the rate of diminution of sound intensity with distance in extended media is much more rapid than this simple law of spreading would indicate. It is true that in certain cases, where the sound waves are confined between reflecting surfaces, the rate of decrease of intensity is reduced. Thus in the sea, sounds travel in a two-dimensional rather than a three-dimensional medium, with a correspondingly increased range. The increase of range obtained by means of speaking-tubes, confining the sound waves to *one* dimension only, is well known. Were it not for certain frictional effects, which we shall now consider, sound energy might be conveyed without loss to very great distances.

Apart from other considerations, energy loss must take place whenever there is relative motion between the various particles comprising the medium, such loss being due to ordinary viscous forces which tend to degrade the sound energy into heat. A *perfect* fluid is one in which tangential stresses cannot exist. In *actual* fluids, however, whether liquid or gas, such stresses do exist in varying degrees depending on the factor which is termed viscosity. It is this viscosity which tends to damp out relative motion between the various parts of a fluid. In the case of a fluid flowing over a plane solid surface, the layer in immediate contact with the solid must be at rest, the velocity of successive strata of the fluid increasing with increase of distance from the solid surface. The tangential stress between any two successive layers will tend to accelerate the lower or more slowly moving layer, and to retard the upper more rapidly moving layer. The tangential stress between the layers is proportional to the velocity gradient, *i.e.* Stress $\propto dv/dy$ or Stress$=\mu dv/dy$, where μ is the coefficient of viscosity. The viscous effect when relative motion of layers of a fluid takes place may be explained on the Kinetic Theory as being due to an interchange of molecules between the different layers, there being a loss of momentum by the higher velocity layers to those of lower velocity. This loss of momentum is the cause of the 'drag' or viscous force tending to prevent relative motion. Similar forces of a frictional character are present in solids as well as in liquids and gases. The vibrations of a cast-iron bar die away much more rapidly than those of a similar steel bar, the difference being due to the greater frictional forces in the cast iron. Returning to the case of gases, both theory and experiment indicate that the viscosity is, within wide limits, independent of the density or pressure of the gas, but increases with rise of temperature. It will be seen that *the viscosity* μ in a liquid is equivalent to the *rigidity*, or resistance to *shearing*, in a solid body. The equation of motion of plane waves in a medium of density ρ and elasticity κ is

$$\rho\frac{d^2\xi}{dt^2}=\frac{dp}{dx},$$

which, according to Rayleigh,* in the case of a medium of viscosity μ becomes

$$\frac{d^2\xi}{dt^2}=\frac{\kappa}{\rho}\frac{d^2\xi}{dx^2}+\frac{4}{3}\frac{\mu}{\rho}\cdot\frac{d^2\xi}{dxdt} \quad . \quad . \quad . \quad (1)$$

The solution of this equation indicates a wave advancing with the usual velocity $c=\sqrt{\kappa/\rho}$, diminished by a small amount due to viscosity. It can be shown that this velocity is given by

$$c^2=\frac{\kappa}{\rho}-\left(\frac{4}{3}\frac{\mu}{\rho}\right)^2\frac{k^2}{4}\left(\text{where } k=\frac{2\pi}{\lambda}=\frac{2\pi N}{c}\right) \quad . \quad (2)$$

In most cases the second term is negligible.

As Maxwell pointed out, the effect of viscosity in modifying the motion is dependent on the ratio of μ/ρ rather than on μ alone. Consequently he called this ratio $\mu/\rho=\nu_1$ the 'kinematic' coefficient of viscosity. Commencing with a simple harmonic wave of the form $\xi=\cos nt$, the solution of (1) indicates, at a distance x from the origin of the wave,

$$\left.\begin{aligned}&\xi=Ae^{-\alpha x}\cos(nt-\beta x)\\ \text{in which}\quad &\\ &\beta=n/c \quad\text{and}\quad \alpha=2\nu_1 n^2/3c^3\end{aligned}\right\} \quad . \quad . \quad (3)$$

The wave advances therefore with diminishing amplitude, falling to $1/e$ of its initial value when $x=1/\alpha$, *i.e.* when

$$x=\frac{3c^3}{2\nu_1 n^2} \quad\text{or}\quad \frac{3c^3}{8\pi^2\nu_1 N^2} \quad . \quad . \quad . \quad (4)$$

* *Sound*, 2, p. 315 ; and Lamb, *Sound*, p. 185.

This deduction has an important bearing on the range of transmission of sounds of various frequencies and in different media. At constant frequency the range will vary directly as the cube of the wave-velocity in the medium and inversely as the kinematic viscosity. In the same medium the range of transmission will vary inversely as the square of the frequency. The values of ν_1 for air and water are 0·132 and 0·013 respectively at 10° C., whilst c air$=0{\cdot}331\times10^5$ cm./sec. and c water$=1{\cdot}5\times10^5$ cm./sec. Before considering practical applications, however, we must refer to another equally important factor which influences the range of transmission of sound waves. It was first pointed out by Kirchhoff that the loss of energy due to heat conduction in the medium cannot be neglected. To include such heat losses Maxwell multiplied the value of ν_1 given above by 2·5. Thus the kinematic 'viscosity' coefficient $\nu=2{\cdot}5\ \mu/\rho$ when it includes losses of this nature. Stokes (1851), referring to the loss of energy due to heat radiation, remarked that vibrations of *very low* frequency should travel isothermally (with equalisation of temperature in the wave), whereas vibrations *of very high* frequency should travel adiabatically (with no heat transfer). In a gas therefore waves of very low frequency would have the Newtonian velocity $\sqrt{P/\rho}$, and waves of very high frequency the Laplacian velocity $\sqrt{\gamma P/\rho}$. For intermediate frequencies we should therefore anticipate intermediate values of velocity. In the extreme cases the sound wave should be propagated without attenuation due to heat transfer, whereas in the intermediate case it should be rapidly stifled. W. F. Herzfeld and F. O. Rice * have recently remarked, however, that Stokes's view of the matter appears to be incorrect. They say, "An erroneous opinion on the influence of the frequency on the completeness of the adiabatic state seems to be often held. It is believed that while ordinary sound frequencies are sufficiently high to guarantee the adiabatic condition, we should expect Newton's velocity for very low frequencies. But a close scrutiny shows that the *adiabatic* state is best guaranteed for the *low* frequencies, while for the higher frequencies the influence of heat conduction is larger. . . . The result comes about in the following way: while it is true that with decreasing frequency the time allowed for heat conduction increases, the amount of heat conducted at a given moment decreases much more rapidly. The latter is proportional to d^2T/dx^2 (T =temperature), and for a given amplitude this expression is inversely

* *Phys. Rev.*, **31**, p. 691, 1928.

proportional to λ^2. Therefore *the heat conducted during one period increases proportionally with frequency because of the increasing steepness of the temperature gradient.*"

Referring now to equation (4), we can calculate the distance x plane waves may travel before the amplitude is reduced to $1/e$th. In air $\nu = 2{\cdot}5 \times 0{\cdot}132 = 0{\cdot}33$, whence $x = 4{\cdot}1 \times 10^{10}/N^2$ metres. For low frequencies (*e.g.* $N=200$, $x=1000$ km.) the value of x is large and attenuation of this nature is negligible. At high frequencies, however, the effects of viscosity and heat conduction are more serious; thus, when $N=10^4$ or 10^5, $x=410$ metres and 4 metres respectively. There is consequently an upper limit of frequency for the propagation of vibrations to appreciable distances. In water $\nu=0{\cdot}033$ and $c=1{\cdot}5\times10^5$ cm./sec., whence $x=3{\cdot}84\times10^{13}/N^2$ metres. At a frequency of 1000 p.p.s., therefore, $x=3{\cdot}84\times10^4$ km., and at 10^4 p.p.s. the value is 384 km. *In water, therefore, the effects of viscosity are unimportant relatively to those occurring at the same frequency* in air. This partly explains the relatively large ranges of transmission of sound under the sea as compared with those obtained in air. A small charge of guncotton (9 oz.) exploded under water can be detected and recorded with certainty under the sea at a distance of 40 miles. It is very unlikely that such a range would be obtained in air. Strictly speaking, the comparison should be made with continuous sound waves from a source with a constant energy output in the two media. There is ample evidence to show, however, that bell sounds and other noises under water are attenuated far less than in air. Even in air, the influence of frictional forces (viscosity and heat losses) are not likely to be serious unless the frequency of the waves approaches or exceeds the upper limit of audibility. Rayleigh refers to the 'mellowing' influence of sounds by distance, as observed in mountainous countries, and suggests that the effects may be due to friction by which the higher and harsher components of the sound are gradually eliminated.

On account of attenuation due to viscosity, the intensity of a sound wave spreading in an infinite medium therefore falls off more rapidly than the inverse square law of distance indicates. This effect is still more marked when the waves are of large amplitude. In such cases—as, for example, the pressure pulse from a gun or the wave from a powerful sound-signalling source—it is probable that the large temperature fluctuations in each cycle of pressure may be such as to involve serious heat losses due to conduction and radiation. This heat 'loss' is actually a transfer of heat from the regions of high pressure to those of low pressure, the tendency being to produce equalisation of pressure, *i.e.* to suppress the wave. Such effects become more serious at large amplitudes—that is, near the source In this connection experiments by M. D. Hart, to which we have already referred (p. 211), are of considerable importance. He found that the sound intensity near a siren source diminished extremely rapidly at first, then more slowly as the distance increased. On such grounds he concluded that *sources of small intensity and large area are likely to be more efficient than sources of great intensity and small area.* This result has an important bearing on the design of sound-signalling devices for use in air, and indicates that high-pressure intensities must be avoided if attenuation is to be negligible.

Observations at very High Frequencies – In measuring the variation of velocity with frequency, G. W. Pierce * observed that carbon dioxide absorbed sound energy more readily than air. At frequencies of 10^5 and 2×10^5 the absorption in CO_2 was respectively four times and more than eighty times the absorption in air. At a frequency of a million p.p.s. CO_2 was found to be *opaque*. Pierce's experiments have been extended by T. P. Abello,† who measured the absorption coefficient for high frequency waves in mixtures of air with CO_2, N_2O, H_2, and He. A beam of sound waves of frequency $6{\cdot}1 \times 10^5$ p.p.s. from an oscillating quartz crystal was passed through various known mixtures of the gas and air contained in the tube. In the first series of experiments the sound intensity at the distant end of the tube was measured by means of a torsion vane pendulum (see p. 432) with quartz fibre suspension. Later, the emergent beam was allowed to fall on another tuned quartz crystal connected to a vacuum tube voltmeter.‡ A logarithmic decrease in the emergent intensity was observed as the percentage of CO_2, H_2, etc., was increased. For example, the intensity was reduced to half-value by the addition of 8 per cent. CO_2 or of 13 per cent. hydrogen. In argon no absorption was observed. As we have already remarked on p. 259, Pierce noticed also a progressive small *increase* of velocity in CO_2, combined with increased absorption

* *Loc. cit.*, p. 259. See also N. Neklapajev and P. Lebedew, *Ann. d. Physik*, **35**, p. 171, 1911.

† (1) *Nat. Acad. Sci. Proc.*, **13**, p. 699, 1927, and (2) *Phys. Rev.*, **31**, p. 1083, 1928. See also K. F. Herzfeld and F. O. Rice, *Phys. Rev.*, **31**, p. 691, 1928.

‡ See, for example, E. B. Moullin, *Journ. I.E.E.*, **61**, p. 295, 1923.

at very high frequencies. From equation (2) (p. 318), it is possible to calculate the diminution of velocity with increase of frequency. Thus for CO_2, $\nu=2{\cdot}5$, $\mu/\rho=0{\cdot}185$, and $c=2{\cdot}59\times10^4$ cm./sec., whence $\delta c=-9\times10^{-8}N^2$. At a frequency 10^5 p.p.s. there should, on account of viscosity and heat conduction, be a *decrease* of 9 m./sec. in velocity, whereas Pierce observed an *increase* of 1·4 m./sec. A possible explanation of Pierce's anomalous results is offered on p. 268. Boyle and Taylor * have looked for such a change of velocity in water in which $\nu=0{\cdot}013$ and $c=1{\cdot}5\times10^5$ cm./sec. The value of δc, calculated from equation (2), is 0·028 cm./sec. at 6×10^5 p.p.s., a change which is undetectable. Even in a very viscous oil at this frequency Boyle and Taylor could detect no certain diminution of velocity. S. L. Quimby † has determined the viscosity of glass and metals under the rapidly alternating strains due to longitudinal waves. At frequencies of the order 40,000 p.p.s. he found viscosity coefficients of the order 10^3 for glass, aluminium, and copper, whereas the constants determined by static methods are of the order 10^8. Oscillograph records, showing the decay of longitudinal oscillations in steel bars supported on threads, indicated that the fractional loss of energy per cycle due to internal friction in solid media is a constant of the material independent of the speed of performance of the cycle. The internal damping in solids may also be determined from observations of the sharpness of resonance. The difference in resonant properties between a steel and a copper bar are sufficiently obvious; a similar large difference in internal damping is noted by W. G. Cady between quartz and steel resonators (see p. 143). Viscosity in metals is also discussed in a paper by R. W. Boyle,‡ who observed the decay of amplitude in short steel bars by means of a high frequency granular microphone electrically coupled to a valve oscillator generating alternating currents of nearly the same frequency as the bar. The heterodyne or beat-note could be heard clearly even at 4×10^5 p.p.s., its duration giving an indication of the rate of decay of the oscillations, and consequently the damping or viscosity factor.

Sound Absorption in Narrow Tubes and Cavities. Porous Bodies – The viscous and heat conduction losses which

* *Loc. cit.*, p. 260.

† *Phys. Rev.*, **25**, p. 558, 1925, and **31**, p. 1113, 1928.

‡ *Trans. Roy. Soc. Canada*, **16**, p. 293, 1922. See also Ronda and Konno (*Phil. Mag.*, **42**, July 1921).

we have just been considering have an important bearing on the explanation of the absorption of sound by porous bodies. Certain types of sound-absorbing media may be regarded as a mixture of gas-solid or gas-liquid. Thus cork and felt are examples of the gas-solid 'mixture' and frothy liquids as a gas-liquid mixture. It is always found that sound waves are greatly attenuated in their passage through such materials. The effects of viscosity and heat conduction in degrading the sound energy are greatly increased when a gaseous medium (air, for example) is brought into contact with a large surface area of solid or liquid. The viscous forces are increased because the tangential motion of the gas layers is hindered by the proximity of the solid wall. Lamb * shows that the linear magnitude $h=(4\pi\nu_1/\mathrm{N})^{1/2}$ (where ν_1 =kinematic viscosity and N=frequency) may be regarded as a measure of the extent to which the viscous 'dragging' effect penetrates into the gas. At this distance the velocity is within $2°/_{\circ\circ}$ of the normal value at an infinite distance from a solid surface. In air, in which $\nu_1=0{\cdot}13$, we find $h=1{\cdot}29/\mathrm{N}^{\frac{1}{2}}$ cm. At a frequency of 1000 p.p.s. the viscous drag therefore extends 0·04 cm. from a solid surface. When the diameter of a tube is small, the conduction of heat from the centre of the air column to the wall becomes more and more rapid. Ultimately in a *very* narrow tube the temperature of the gas *remains* throughout the same as that of the solid conducting wall; the rarefactions and compressions being isothermal, as if no heat were developed at all. The velocity of sound in such a tube must be Newton's isothermal value $\sqrt{\kappa/\rho}$, and heat conduction will not appear in the viscosity factor. In the case of a narrow tube of circular section (radius r), Rayleigh † shows that the form of the wave at a distance x along the tube is represented by

$$e^{\alpha x} \cos nt \quad \text{where} \quad \alpha=2\sqrt{\nu_1\gamma n}/cr.$$

The attenuation constant α therefore varies directly as the square root of the product of kinematic viscosity ν_1, the ratio of specific heats γ and the frequency $\mathrm{N}(=n/2\pi)$, and varies inversely as the product of the velocity of sound c and the radius of the tube r. In the case of air $\nu_1=0{\cdot}13$, $\gamma=1{\cdot}41$, $c=3{\cdot}3\times10^4$, whence

$$\alpha=0{\cdot}65\times10^{-4}\sqrt{\mathrm{N}}/r.$$

At a frequency of 1000 p.p.s. the attenuation constant in a tube of radius 0·001 cm. is therefore $\alpha=2{\cdot}05$, the amplitude falling to $1/e$ of its initial value in 0·5 cm. approximately. When the diameter

* *Sound*, p. 192.

† *Sound*, 2, p. 331.

of the tube is so great that the thermal and viscous effects extend only a relatively small distance from the wall we arrive at the case already considered on p. 238, which refers to the velocity of sound in tubes as a function of the radius of the tube. Rayleigh's treatment of the absorption of sound in very narrow tubes is of course applicable in principle to any form of air cavity in which the gas is everywhere in close proximity to a solid wall (*i.e.* all distances are less than 'h'). The smaller the radius of the cavity the greater are the viscous forces and the more rapidly is the sound energy absorbed as it passes along the cavity. The absorbent properties of hangings, carpets, cushions, and such-like air-permeated bodies have long been appreciated as a means of suppressing reverberation. In large halls and public buildings where such commodities are not always available special sound absorbent materials are now frequently used as a remedy for excessive reverberation. We shall have to consider this question in greater detail later. It is convenient, however, to refer here to a laboratory method of measuring the absorption coefficients of porous materials such as felt, cork, wool, and various forms of 'acoustic plasters and tiles.'

Stationary-Wave Method of Measuring Absorption Coefficients of Porous Materials – We shall assume that the sound energy which is incident on the surface of the porous substance is either reflected or absorbed, none being transmitted. This condition is obtained in practice by backing the porous material with a *perfect* reflector. On p. 281 it has been shown that the *reflection coefficient* of an imperfect reflector, on which plane waves are incident normally, is given by

$$\left(\frac{r}{a}\right)^2 = \left(\frac{1-\beta/\alpha}{1+\beta/\alpha}\right)^2,$$

where a is the incident and r the reflected amplitude, and β/α is the ratio of minimum and maximum amplitudes observed in the stationary wave formed by interference between the direct and reflected waves. The *coefficient of absorption* is, by definition, the fraction of the incident sound energy lost at reflection, *i.e.*

$$\frac{(a^2-r^2)}{a^2} \text{ or } \frac{4}{(2+\alpha/\beta+\beta/\alpha)}.$$

This expression is due to H. Taylor,* but the stationary-wave method of determining absorption coefficients was first proposed

* *Phys. Rev.*, **2**, p. 270, 1913.

by J. Tuma,* and was followed by the experimental observations of F. Weisbach.† More recently E. T. Paris ‡ has improved the technique of the method and has obtained much valuable information relating to the reflection and absorption properties of materials. The arrangement which he used (see fig. 99) consisted of a

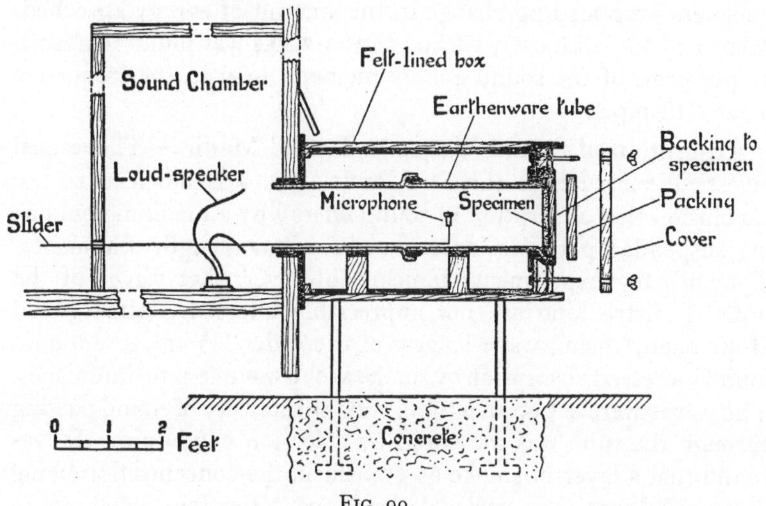

FIG. 99

(From Davis and Kaye, *Acoustics of Buildings*)

large earthenware pipe (30 cm. diameter, 226 cm. long) closed at one end by the specimen of porous material, backed by a good reflecting material. At the other end, a large 'loud speaker' actuated from a valve oscillator, supplied a sound of constant frequency and intensity. By means of a calibrated resonant hot-wire microphone (see p. 402) the amplitude along the axis of the tube was measured, the ratio of the maximum and minimum values, α and β, being then utilised in the above relation to calculate the absorption coefficient. The method is simple, direct, and accurate, and only small quantities of absorbing material are required. It is used in a slightly modified form at the National Physical Laboratory, and at the Bureau of Standards (Washington). Paris determined in this way the absorption coefficients of various kinds of acoustic tiles and plasters, values from 0·10 to 0·36 being observed. As the theory of absorption indicates, the absorption coefficients were

* *Sitz. K. Akad. d. Wiss.*, **3**, p. 402, 1902.

† *Ann. d. Physik*, **33**, p. 763, 1910. Also Eckhardt and Chrisler, Bureau of Standards, No. 256, 1926.

‡ *Proc. Phys. Soc.*, **39**, p. 269, 1927; *Proc. Roy. Soc.*, **115**, p. 407, 1927, and *Phil. Mag.*, **4**, p. 907, 1927.

found to increase with the frequency of the incident sound; in a particular specimen the successive values were 0·13, 0·26, and 0·31 at frequencies 380, 512, and 650 p.p.s. The coefficient of absorption was found to increase also with increasing thickness of porous material, approaching an upper limit when further increase of thickness produced no change in the amount of energy absorbed. A layer of 4 in. of loosely packed cotton-waste was found to absorb 91 per cent. of the sound energy incident upon it, the frequency being 512 p.p.s.

Absorption of Sound Waves in Turbid Media – Theoretical investigations such as those of Rayleigh and Lamb lead to the conclusion that absorption of sound energy by a medium containing suspended particles increases as the wave-length diminishes. Tyndall's fog experiments indicate, however, that waves of the order 1 metre long are not appreciably affected. Altberg and Holtzmann,* using wave-lengths of the order 1·5 cm. to 6·5 cm., found increased absorption by smoke as the wave-length diminishes. The waves were set up in a tube, and the intensity of sound passing through the tube was measured by a torsion pendulum. It was found that a layer of smoke 56 cm. thick, the concentration being $7{\cdot}3 \times 10^{-6}$ grm./c.c., varied from 24 per cent. for $\lambda=5{\cdot}6$ cm. to 70 per cent. for $\lambda=1{\cdot}5$ cm.

Absorption of Sound by Gas-Bubbles in Liquids. Velocity of Sound in Fluid Mixtures – The phenomena of sound absorption in porous solids are exemplified also in 'porous liquids,' if it is permissible to use the term. A tumbler full of water is resonant and rings clearly when struck with a hard object, but emits a dull, non-resonant sound when the water is frothy or full of gas-bubbles. The effect is simply demonstrated with a spoonful of effervescent salt in half a tumbler of water. As the bubbles disappear the resonant quality returns. The air-bubbles, therefore, cause a marked increase of damping. The phenomenon has been examined theoretically by A. Mallock,† who shows that the damping is due to increase of distortion of the liquid separating the bubbles. The pressure variations act almost entirely on the volume of gas and scarcely at all on the relatively incompressible fluid. Bubbles in a solid cannot behave in this way since their volume is fixed by the rigidity of the surrounding medium. A material like spongy rubber, however, can behave like bubbly water, by turning forward compression into sideways expansion.

* *Phys. Zeits.*, **26**, p. 149, 1925.

† *Proc. Roy. Soc.*, **84**, p. 391, 1910.

To determine the velocity of sound in a mixture of two fluid media (*e.g.* small non-resonant air-bubbles in water) we shall assume that the velocity is the same as that in a homogeneous fluid of the same mean density ρ and mean elasticity E as in the mixture. Let $\rho_1 E_1$ and $\rho_2 E_2$ represent the density and elasticity of the constituents 1 and 2 respectively. Then let x=the proportion of the first constituent by volume and $(1-x)$=the proportion of the second constituent by volume. The *mean density* ρ is therefore

$$\rho = x\rho_1 + (1-x)\rho_2 \qquad (1)$$

We must have also

$$\frac{1}{E} = \frac{x}{E_1} + \frac{(1-x)}{E_2},$$

whence the *mean elasticity*

$$E = \frac{E_1 E_2}{xE_2 + (1-x)E_1} \qquad (2)$$

and the *mean velocity*

$$c = \sqrt{\frac{E}{\rho}} = \sqrt{\frac{E_1 E_2}{\{xE_2 + (1-x)E_1\}\{x\rho_1 + (1-x)\rho_2\}}} \qquad (3)$$

These expressions are applicable to a mixture of any two fluid media, which do not react chemically. We shall apply equation (3) to determine the velocity of sound in a mixture of non-resonant air-bubbles and water. In this case we have

for air $\rho_1 = 0{\cdot}0012$ $\quad E_1 = 1{\cdot}2 \times 10^6$,
for water $\rho_2 = 1{\cdot}0$ $\quad E_2 = 2{\cdot}25 \times 10^{10}$.

Employing these values in equation (3) for the mean velocity and substituting various values of x (the fraction of air by volume) we obtain the curve shown in fig. 100. Here it is seen that a proportion of 1 part air in 10,000 parts water lowers the velocity of sound in the water by about 40 per cent. The curve also reveals that the velocity reaches a minimum value when the volume of air exceeds one-tenth that of the water. This minimum velocity is only about one-tenth (36 metres/sec.) of the velocity of sound in air alone. Under such conditions the mixture may be regarded as 'froth,' the water merely serving to load the air-bubbles (the springs) and consequently to lower the velocity of transmission, On these grounds a small lowering of velocity might be anticipated in the case of sound transmitted through dense mist or fog—that is, in air loaded with fine particles of water. So far as the writer is aware, there is no experimental information bearing on this point. It will be seen from the curve, as is otherwise obvious, that the velocity becomes equal to the velocity in water or air respectively as x is equal to zero or unity.

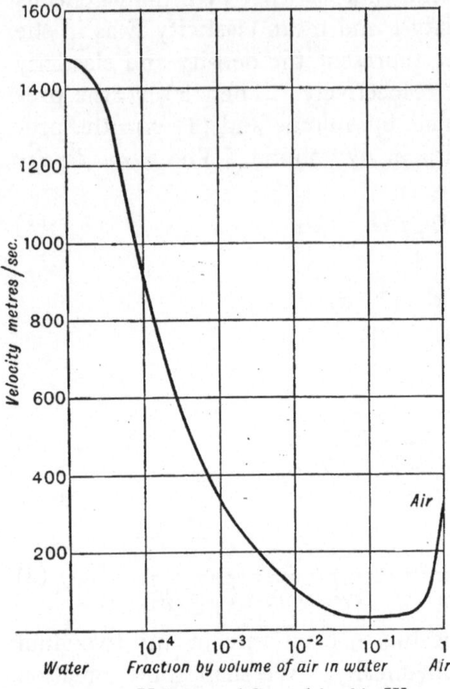

FIG. 100—Velocity of Sound in Air-Water Mixtures

The damping being due to the large volume variations in the liquid surrounding each bubble, it may be calculated in the case of a spherical bubble surrounded by a sphere of viscous (but incompressible) fluid subjected to a simple harmonic variation of radius. The efficiency of the bubbles in damping vibrations increases rapidly both as their diameter and distance apart diminishes. This may be verified experimentally.

From expressions (1) and (3) above we may deduce the reflection coefficient of a mass of air-bubbles in water. The acoustic resistance (ρc) of the mixture is clearly

$$\rho c = \sqrt{E\rho} = \sqrt{\frac{E_1E_2\{x\rho_1+(1-x)\rho_2\}}{xE_2+(1-x)E_1}} \quad . \quad . \quad (4)$$

The reflection coefficient of a "thick' (semi-infinite) layer is from equation (6a), p. 272.

$$r^2 = \left(\frac{\rho c - \rho_2 c_2}{\rho c + \rho_2 c_2}\right)^2 \quad . \quad . \quad . \quad (5)$$

where ρc refers to the mixture of air and water (see equation (4)) and $\rho_2 c_2$ refers to the clear water. The reflection from a layer of bubbly water of relatively small thickness may similarly be calculated using equation (10), p. 276. The percentage energy reflected (r^2) is plotted as a function of x (the ratio by volume of air to water) in fig. 101.

This curve illustrates the serious reduction of intensity which might occur when a sound wave encounters a mass of air-bubbles in the sea. The noise of a ship's propeller is seriously reduced by the bubbly water in the wake. In such cases, as the foregoing remarks will show, the incident energy is partially reflected and partially absorbed, the loss increasing rapidly as the proportion of air to water increases.

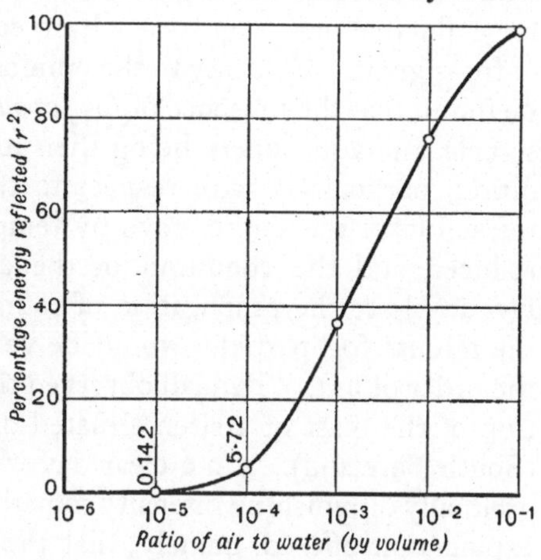

FIG. 101—Reflection of Sound from Air-Water Mixtures

Transmission of Sound through the Atmosphere. Audibility of Fog-Signals—The various factors affecting sound transmission, which we have hitherto considered independently, all combine in varying degrees to influence the range of propagation of sounds in air. Wind, temperature gradients, reflection, scattering, refraction, interference, diffraction, and viscosity all have an important influence on the character of the wave-front as it advances from a source of sound through a heterogeneous medium such as the atmosphere. The influence of the meteorological conditions on sound transmission through the atmosphere has long been recognised. The first experimental study of such influences was made by Tyndall* in connection with the audibility of fog-signals sent out from Trinity House sirens mounted at South Foreland Lighthouse and on various light-vessels. The importance of such investigations in relation to the navigation of ships in foggy weather, when lights cannot be seen, need not be stressed here. There is not only a large annual loss of life due to fog, but in addition to the loss of ships by collision the pecuniary loss is also very great when large numbers of vessels are prevented from entering harbour on account of fog. The systematic use of sound signals for marine protection is of comparatively recent date in spite of the fact that Tyndall and, later, Rayleigh had long ago indicated its possibilities and devoted special attention to its

* *Phil. Trans. Roy. Soc.*, 1874.

development. Tyndall came to the conclusion that 'temperature refraction' and a flocculent condition of the atmosphere, arising from the unequal heating and humidity, were responsible for the large fluctuations frequently observed in the range of audibility of fog-signals. Contrary to the opinion generally held at that time, he found that the presence of *fog favoured the transmission* of sound signals, the atmosphere being then in a more homogeneous condition, particularly with respect to temperature gradients. The refraction of the sound wave by temperature gradients would be reduced and the condition of the atmosphere would be more favourable to the propagation of sound. The scattering effect of the minute fog particles would be very small for wave-lengths of the order of 5 ft. Tyndall observed the duration of the reverberation of the blast of a siren situated on a cliff overlooking the sea (South Foreland). On a clear day with a smooth sea the echo of gradually diminishing intensity sometimes lasted 15 seconds. The explanation offered, namely, that the echoes were returned from invisible 'acoustic clouds,' seems not improbable. On this hypothesis Rayleigh offers an explanation of certain anomalous results obtained by Tyndall. The latter found that different kinds of signals did not always preserve the same order of effectiveness, a siren (giving a continuous blast) being sometimes better and sometimes worse, as regards range of audibility, than a gun giving a single impulse. Rayleigh points out, on the grounds that energy cannot be lost by reflection or refraction, that the intensity of radiation at a given distance from a continuous source of sound is not altered by the acoustic 'cloud,' the loss due to the intervening parts of the cloud being compensated by reflection from those which lie beyond the source. "If the sound is of short duration, however, the intensity at a distance may be very much diminished by the cloud on account of the different distances of the reflecting parts and the consequent 'drawing out' of the sound, although the whole intensity, as measured by the time integral, may be the same as if there had been no cloud at all."

The work of Tyndall and Rayleigh has in recent years been extended by a number of workers, notably G. I. Taylor, L. V. King, E. A. Milne, W. S. Tucker, E. T. Paris, M. D. Hart, E. S. Player, and others.* Such experimenters have dealt chiefly with two types of phenomena: (*a*) *scattering* effects due to acoustic clouds,

* E. A. Milne, 'Sound Waves in the Atmosphere' (*Phil. Mag.*, July 1921); W. S. Tucker, *Journ. Roy. Soc. Arts*, **71**, No. 3659, Jan. 1923; Hart and Smith, *Principles of Sound Signalling* (1925).

fog banks, air vortices, and other atmospheric discontinuities, and (*b*) reflection and refraction effects due to more or less continuous variations in such factors as wind, temperature, and humidity, which may influence the velocity of the wave-front. W. S. Tucker and his staff have made a study of 'meteorological acoustics' at a station situated near North Foreland, selected on account of its favourable situation relative to a number of light-vessels fitted with sirens. As an instance of the remarkable fluctuations of intensity of the received sound on a typical English summer day when the sky is clear or possibly flecked with white cumulus cloud, Tucker refers to a record of the sound emitted at one-minute intervals from the North Goodwin light-vessel siren (seven miles away). The record shows irregular variations of the order 30 to 1 in intensity within the space of a few minutes. On certain occasions the sound is 'scattered' by atmospheric 'patches' of abnormal acoustic resistance, and it sometimes becomes impossible to determine the direction of the source of sound although it is clearly audible. This diffusion or scattering of the sound is analogous to the effect of light penetrating a fog. A large collection of data of this nature correlated with weather, temperature, and humidity conditions has thrown a good deal of light on this complicated problem. The apparatus used by Tucker and Paris for measuring the strength of fog-signals has recently been described by Paris.* It consists essentially of a hot-wire resonated microphone tuned to the frequency of the sound and connected to a sensitive indicating or recording galvanometer (*e.g.* the Einthoven string galvanometer). With this apparatus an interesting acoustic survey † was made of the sound emitted in different directions from a diaphone fitted in the Caskets Lighthouse. A vessel carrying the tuned microphone and recorder circled round the lighthouse at a constant distance of two miles and the amplitude of the received sound signal was plotted as a function of the bearing of the ship with respect to the diaphone. The horn of the latter pointed upwards, so there could be no directional effect at the source. A south-easterly wind of 2 miles per hour was blowing and the atmosphere was clear. The result of such a survey is shown in fig. 102. The normal temperature gradient was such as to refract the rays of sound upwards from the surface of the sea. Thus to the south-east the tendency of the wind would be to enhance such an effect, whilst to the north-east the tendency

* *Journ. Sci. Instrs.*, **3**, 6, p. 187, 1926.

† See also L. V. King, *Phil. Trans.*, **218**, p. 219, 1919.

would be towards neutralisation. This is clearly shown in the polar curve of amplitudes. The fluctuations of observed amplitudes from blast to blast are of the kind generally encountered in this class of work and are not unusually large. They are no doubt due to atmospheric conditions, possibly to moving eddies. Tucker refers to records of a sound arriving by two distinct paths on a clear day with a strong following wind whose velocity alternately increased and decreased with increase of height. The effect heard by the observer was the same as if every blast of the siren were doubled. E. S. Player * considers that the only factor, apart from wind and temperature gradients, in which the changes are at all comparable with the large and sudden variations in the range of audible transmission is humidity. Comparative charts of humidity and audibility (loudness) of a siren indicate a close correlation. Player refers also to an exceptional example of temperature effect. With an off-shore wind with a bad gradient (12 m.p.h. at the ground level, rising to 20 m.p.h. at 1000 ft.) the listening conditions were distinctly adverse, yet the galvanometer gave abnormally large deflections which were confirmed aurally. The explanation was found in the fact that the wind had passed over many miles of hot ground (during a summer drought), the lower layers being cooled as they passed over the sea, with the result that a 'favourable' temperature gradient was produced, sufficient to neutralise the adverse effects of the wind and to set up good conditions for sound transmission. Sufficient has been said to indicate that sound is conveyed in a very capricious way through the atmosphere. Apart from the large variations of intensity to which we have referred, it is not exceptional to encounter large areas of silence in different directions and at different distances from the sound source. Such inconsistencies tend to make signalling through the atmosphere somewhat unreliable.

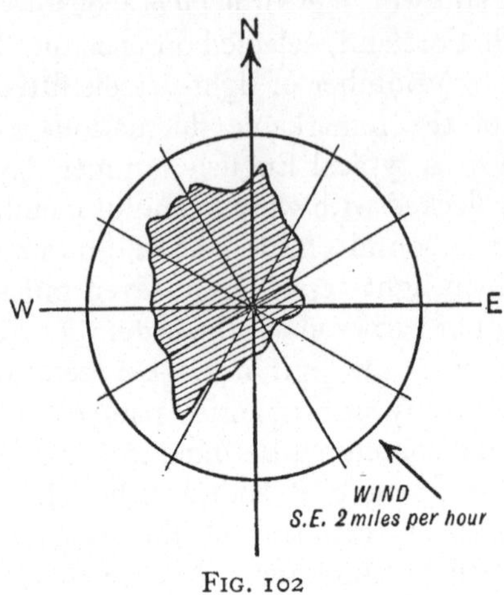

FIG. 102

* *Quart. Journ. Roy. Met. Soc.*, **52**, p. 351, 1926.

Explosive Sounds. Zones of Silence – The firing of a gun or the detonation of an explosive charge is accompanied by a pressure wave of very large amplitude which, in some cases, may be detected at very great distances (of the order of a hundred miles) from the source. Such effects are comparable in magnitude with 'natural' explosions accompanying a flash of lightning. Waves of such enormous intensity reveal abnormalities which are not noticed with ordinary sounds of small amplitude. The high velocity near the source where the amplitude is large, and the change of wave-form as the wave progresses are two of these abnormalities which have already been mentioned. Wherever a sound wave from a large explosion can be detected it is always accompanied by a succession of pressure pulses arriving by different paths after reflection or refraction in the atmosphere. A phenomenon common to all large explosions in air is the occurrence of regions where the sound is not audible (or otherwise detectable), whilst at *greater* distances the sound is quite distinct. Tyndall refers to such zones of silence experienced with ordinary fog-signals, a signal being audible, for example, from 0 to 2 and from 4 to 6 miles, but quite inaudible between 2 and 4 miles. The explosion effects to which we now refer are similar, but the scale is much greater. When a large charge is exploded, the resulting noise may, for example, be heard for a distance of 60 miles or so surrounding the charge, and again beyond 100 miles, but not within the range 60 to 100 miles. Within the inner zone of audibility the sound travels at first with abnormally high velocity and subsequently with the normal velocity of sound. The direct wave is steadily attenuated in its passage over the earth's surface, and at an average distance of about 60 miles becomes inaudible. The sound waves reaching the outer zone of audibility appear to have travelled with an abnormally low velocity, but this is not the case. Actually, the sound heard in this zone has travelled by an abnormally long *path*, *via* the higher atmosphere and ultimately down to the ground again. There is some divergence of opinion as to the actual mechanism of this abnormal wave, whether it is bent downwards into the outer audible zone by refraction (due to air currents and temperature gradients) or whether it is reflected from a layer of the stratosphere 30 km. or more from the ground, at which level the velocity of sound increases with height. The effects doubtless arise as a result of the peculiar meteorological conditions at the time of the explosion. As an example we may refer to the explosion of a large ammunition dump at Oldebroek, Holland, in January 1923, which was recorded at various distances up to 500 miles, but was not detected between 60 and 100 miles from the origin. Similar observations were made of the Oppau and La Courtine (May 1924) explosions. Numerous observations and explanatory theories were immediately forthcoming, in connection with which we may mention the names of Dufour, Deslandres, Esclangon, Villard, Maurain, Collignon, and others.* Dufour and Villard point out that *records* of explosions at long ranges (in the outer 'audible' zone) show no trace of vibrations of audible frequency which would constitute the *noise* of the explosion, the principal effect of the explosion at La Courtine being a pressure fluctuation in Paris (230 miles away) of frequency about one period per second. Villard considers that such slow oscillations are the primary effect, whereas the noise is an accessory phenomenon of negligible importance.

An interesting example of a large explosion wave being heard in two zones separated by a zone of silence has been recently described by F. J. W. Whipple,† the source of sound being *a detonating meteor* (September 6, 1926). From numerous observations it was deduced that this meteor exploded at a height of 30 to 40 km. (20 to 25 miles approximately). The sound of the explosion was heard (like thunder) at distances up to 60 or 70 km., then a zone of silence followed by a second audible zone. The hypothetical track of the sound rays from such an 'explosion' is shown in fig. 103, taken from Whipple's paper. It is not outside

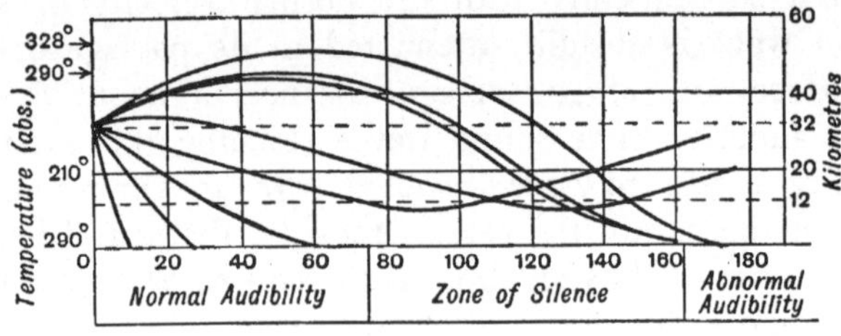

FIG. 103

the bounds of possibility that the explosion sounds of a 'detonating' meteor are produced by the '*onde de choc*' as the meteor passes through the air at high speed (estimated in this case to be 30 km. per second).

* See *Comptes rendus*, **178** and **179**, 1924.
† *Monthly Notices R.A.S.*, p. 89, Oct. 1928.

Transmission of Sound through the Sea – The sea is a much more homogeneous medium for sound transmission than the atmosphere. As we have seen, large fluctuations of intensity and changes of apparent direction of sounds in air are not infrequent and are liable to occur in a short space of time. In the sea the conditions are much more consistent. The effects of tidal currents, analogous to wind, are practically negligible, a 5-knot tide (about 9 ft./sec.) affecting the velocity of sound (5000 ft./sec.) only slightly. Temperature gradients in the sea, although less serious than in the atmosphere, under certain circumstances may become important. Thus H. Lichte * has observed seasonal variations in the range of transmission of under-water signals, which he ascribes to the refraction produced by temperature gradients in horizontal layers of the sea. As we have seen (p. 293) such gradients may cause a sound-ray to curve upwards or downwards (according to the sign of the temperature gradient), the wave being continued to greater distances by successive reflection at the surface or the bottom. On account of the smaller absorption of sound waves in water and the limit to the vertical spreading, considerably greater ranges are obtainable in water than in air for the expenditure of a given amount of power. Thus a Fessenden oscillator (see p. 390) having a 50 per cent. efficiency and consuming about 1 kw. can be heard at 20 or 30 miles under the sea, whereas a siren (of 10 per cent. efficiency) consuming 100 kw. in air can be heard at a range of about 10 miles only. A 9-oz. charge of guncotton exploded under water can be heard, and recorded, at a distance of 30 or 40 miles. The sounds from the propellers of a large vessel at speed may sometimes he heard at 8 or 10 miles, but the range of detection of such sounds (of a non-musical character) is liable to vary greatly according to the state of the sea or, more precisely, according to the intensity of extraneous 'water noises' (due to breaking waves, etc.). Systematic observations of the ranges of transmission of sounds of various frequencies have been made by Barkhausen and Lichte * in the Baltic Sea. Their results indicate that the decrease in amplitude of the sound wave with distance is exponential, energy *dissipation* due to various causes being more important than the effect of spreading. With a Fessenden oscillator, at a frequency of 517 p.p.s., they observed an attenuation equivalent to a fall of amplitude to half value every 2 to 3 km. Such a large attenuation cannot be explained on the basis of viscosity and heat losses in the water

* See *Unterwasserschalltechnik*, F. Aigner (Berlin, 1922).

—it was shown on p. 320 that a sound of frequency 1000 p.p.s. would travel about 4×10^4 km. before its amplitude fell to $1/e$th due to viscosity. A possible explanation may be found by assuming the presence of large masses of suspended air-bubbles in the sea, which increase the damping as indicated on p. 328.

Variation of Pitch with Motion. Doppler Effect

The pitch of a sound of constant frequency is dependent on the motion of the source and the observer. It is obvious that an observer approaching a source of sound will encounter more sound waves per second than if he remained stationary, and, similarly, if the source is approaching the observer. The two cases differ, however, in certain respects.

Source Stationary, Observer in Motion – In fig. 104 (a) let S be a stationary sound source and P an observer moving with velocity v along a line AOB distant OS$=d$ from the source.

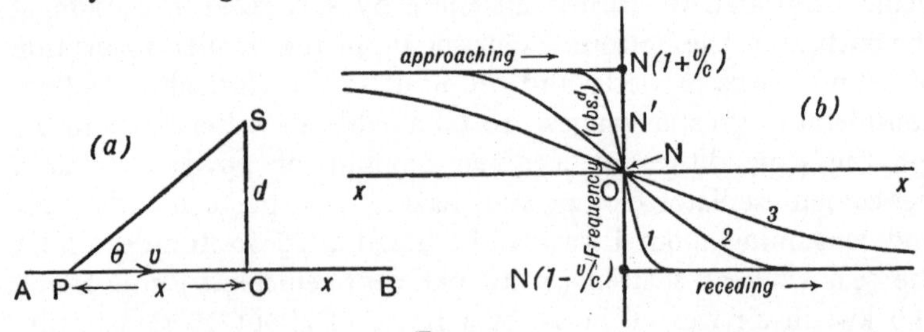

FIG. 104

The observer approaches S with velocity $v \cos \theta$, and the apparent velocity of the sound waves arriving at P is $(c+v \cos \theta)$. Consequently, if N is the actual frequency of the source, N′ the frequency of arrival of waves at P (*i.e.* the *apparent* frequency of the source), and λ the wave-length of the sound, we must have

$$\frac{c}{\mathrm{N}}=\frac{c+v \cos \theta}{\mathrm{N}'}=\lambda, \text{ that is } \mathrm{N}'=\mathrm{N}\left(1+\frac{v}{c} \cos \theta\right), \quad . \quad (1)$$

or writing

$$\cos \theta=\frac{x}{\sqrt{x^2+d^2}} \qquad \mathrm{N}'=\mathrm{N}\left(1+\frac{v.x}{c\sqrt{x^2+d^2}}\right). \quad . \quad (2)$$

Plotting N′ as a function of x for various values of d, we obtain a series of curves of the form shown in fig. 88 (b), curves (1), (2), and (3), representing increasing values of d. In all cases when x is large compared with d,

$$(\mathrm{N}'-\mathrm{N})=\pm\frac{\mathrm{N}v}{c} \text{ and } \frac{d\mathrm{N}'}{dx}=0 \quad . \quad . \quad (3)$$

approximately. As the observer passes through O, however, the value of $(N'-N)$ changes sign, and dN'/dx reaches a maximum value (when $x=0$) of Nv/cd.

Observer Stationary, Source in Motion – In the case we have just considered, viz. the observer in motion and the source stationary, the change of frequency is due to the change of rate at which the sound waves pass the observer—that is, to an apparent change of velocity of propagation, the wave-length remaining constant. If in fig. 104 (*a*) we now regard P as a moving source and S a stationary observer, we obtain a somewhat different result. There is in this case an apparent change of wave-length due to the motion of the source towards the observer, the velocity of sound propagation remaining the same.* The observer, instead of receiving waves spaced c/N apart will now receive them at distance intervals of $(c-v \cos \theta)/N$. That is to say,

$$\lambda'=(c-v \cos \theta)/N$$

and

$$N'=\frac{c}{\lambda'}=N\left(\frac{c}{c-v \cos \theta}\right) \quad . \quad . \quad . \quad (4)$$

or

$$(N'-N)=\frac{Nv \cos \theta}{(c-v \cos \theta)} \quad . \quad . \quad . \quad (5)$$

the maximum value of which is

$$(N'-N)_{max}=\pm\frac{Nv}{c-v} \quad . \quad . \quad . \quad (6)$$

This value is attained when θ is zero or π—that is, when x is large compared with d.

If we substitute in equation (4) the ratio $x/\sqrt{x^2+d^2}$ for $\cos \theta$ and differentiate with respect to x, we obtain an expression for the *rate of change of frequency* dN'/dx. This expression indicates

(1) When x is large compared with d

$$\frac{dN'}{dx}=0,$$

$(N'-N)$ having the maximum value $\pm Nv/(c-v)$.

* As Rayleigh points out (*Sound*, 2, p. 156), the mathematical treatment of a moving source is difficult. In the above simple procedure we have neglected any disturbance of the medium produced by the *motion* of the source or the observer.

(2) When x is zero

$$\frac{dN'}{dx}=\frac{Nv}{cd} \quad . \quad . \quad . \quad . \quad (7)$$

the frequency difference $(N'-N)$ changing sign.

It will be seen that for small values of d, the *source passing close to the observer*, the apparent frequency at first remains at a constant high value and *suddenly* changes to a constant low value. When d *is large* the change is more gradual and is not so easily noticed unless v is large also. It should be noted that the rate of *change* of frequency is directly proportional to the *actual* frequency and inversely proportional to d. The first of these cases is illustrated by the familiar example of the express train whistling whilst rushing at high speed through a station. A *sudden* drop of pitch in the note as the engine passes *close* to the observer cannot fail to be noticed. The second case is illustrated by a high-speed aeroplane passing overhead. The change of pitch is more gradual on account of the large value of d, but it is still quite easy to detect because v is large also. When an aeroplane passes an observer at high speed (v large) *and* at a short distance (d small) the effect is somewhat sensational, since $dN'/dx=Nv/cd$ is then very great.

Observer, Source, and Medium in Motion – In the cases just considered the medium and either the source or the observer were at rest. Let v now represent the velocity of the source *relative to the medium* (v indicating that component of velocity in a direction towards the observer). Also let u represent the velocity with which the observer is moving (relatively to the medium) away from the source, *i.e.* in the same direction positive as v. Then the wave-length will be

$$\lambda'=\frac{c-v}{N}$$

and the apparent velocity $c'=(c-u)$. Consequently the observed frequency will be

$$N'=\frac{c'}{\lambda'}=N\frac{c-u}{c-v} \quad . \quad . \quad . \quad . \quad (8)$$

(*Note.*—The minus sign in the numerator of the expression on the right arises from the fact that the velocity of the observer is in a direction *away* from the source.) It is more convenient perhaps to express (8) in terms of velocities relative to some 'fixed' body—for example, the ground. We must therefore substitute for u and v the values $(u'-w)$ and $(v'-w)$ where v', u', and w are the velocities of the source, observer, and medium relative to the ground, c being the velocity of sound relative to the medium. Thus the observed frequency

$$N''=N\frac{c-u'+w}{c-v'+w} \quad . \quad . \quad . \quad (9)$$

When u' and v' are the same in magnitude and direction the frequency is unchanged whatever the velocity of the medium. Any relative motion between observer and source, giving u' and v' different values, will result in a frequency change which is affected also by the motion of the medium. The effect produced by the latter is equivalent to a change in the velocity of sound. A wind in the same direction as the sound lessens the change of pitch due to the motion of the observer and source.

The principle of change of apparent frequency by relative motion is due to Doppler, who first applied it to the changing colour of certain stars as they approached or receded in the line of sight. The *Doppler effect*, as it is called, is not unfamiliar in these days of high-speed traffic. Interesting hypothetical cases of sources and observer moving with velocities greater than that of sound will be left to the ingenuity of the reader. Rayleigh (*Sound*, **2**, p. 154) instances a case in which a high-speed observer ($u=2c$) listening to a musical piece hears it in correct time and tune *but backwards*. The *onde de choc* and the shriek of high-speed bullets serve as further examples for consideration.

The Doppler effect has been demonstrated, and the principle verified in a number of ways. In the earliest recorded experiments Buijs Ballot * and Scott Russell observed the changes of pitch of musical instruments carried on locomotives. A laboratory method devised by Mach † employed a whirling tube with a whistle at one end, air being supplied to the whistle along the axis of the tube. An observer in the plane of rotation could hear a rise and fall of pitch as the whistle approached and receded. König ‡ used two tuning-forks, giving a few beats per second when stationary. If one of the forks were made to approach towards or recede from the ear whilst the other remained at rest, the number of beats per second was changed in accordance with Doppler's principle. A similar instance of this was recently

* *Pogg. Ann.*, **66**, p. 321.

† *Pogg. Ann.*, **112**, p. 66, 1881, and **116**, p. 333, 1862. ‡ *Loc. cit.*

demonstrated by S. R. Humby * at A. O. Rankine's suggestion. Two telephone earpieces were actuated simultaneously at a high frequency from the same oscillating valve circuit. A sensitive flame was used to detect the high-frequency vibrations. When one of the telephones was moved towards or away from the flame, the 'beat' frequency was shown clearly by the fluctuating response of the flame. In carrying out such an experiment, care must be taken to avoid complications due to stationary waves or interference effects.

Experimental Study of Wave Transmission

Many of the so-called 'optical' characteristics of sound to which we have referred may be studied on a laboratory scale provided the wave-lengths employed are sufficiently small. It is unnecessary to refer again to the many beautiful experiments of this nature which may be performed by means of high-frequency sound waves in air and in water (see reference to R. W. Boyle and others †). For many purposes, the methods which we shall now consider are convenient and instructive, and are applicable to complex as well as to simple 'optical' cases. With the object of studying the reflection of waves from typical mirrors, model sections of buildings, etc., much use has been made during recent years of two relatively old methods. The 'spark pulse' and the 'ripple' methods have both proved extremely convenient in the laboratory as a means of following the progress of a wave from its initiation to any subsequent moment.

Spark-Pulse Photography – When a small intense electric spark is produced in air, a thin spherical pulse in which the medium is highly compressed spreads outwards from it. The high-pressure layer forming the wave-front has a density and a refractive index considerably greater than that of the undisturbed medium at normal pressure. Two different processes have been devised to photograph the wave-front, both of which make use of this change of refractive index and employ a second spark to illuminate the pulse produced by the first. These are known as the 'Schlieren' and the 'Shadow' methods. As the latter method is simpler, it will be described first.

(a) *The Shadow Method* of Dvorak ‡ is illustrated in principle in fig. 105. Actually this represents diagrammatically the appar-

* *Loc. cit.*

† See, for example, F. L. Hopwood, *Journ. Sci. Instrs.*, Feb. 1929.

‡ *Ann. der Phys.*, **9**, p. 502, 1880.

atus used by A. H. Davis at the National Physical Laboratory,* which is an improvement on that described by Foley and Souder.† The principle of the method is briefly as follows. The sound pulse is produced by an intense electric spark, the '*sound spark*,' followed by a second spark, the '*light spark*,' at a known short interval of time. The highly compressed envelope of the sound pulse casts a shadow, when illuminated by the light spark, on a ground-glass screen or a photographic plate. As shown in the figure, an electrical machine capable of generating 100,000 volts charges two sets of leyden jars connected to *trigger* spark gaps.

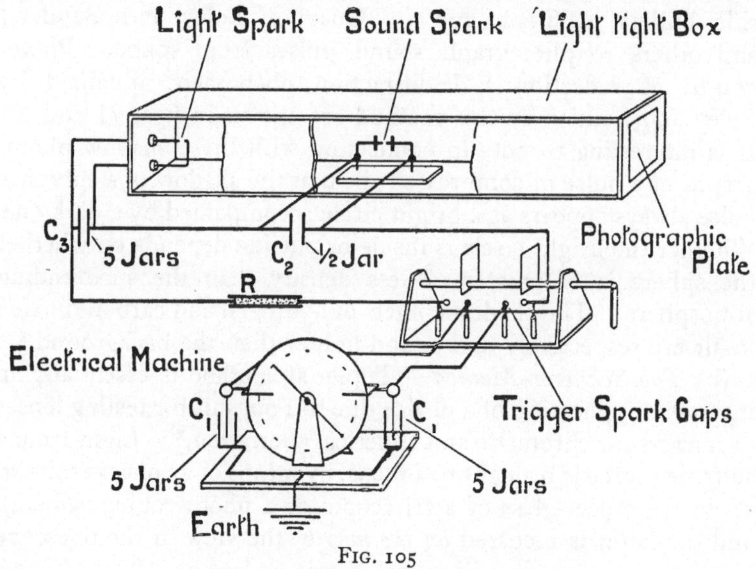

FIG. 105
(From Davis and Kaye, *Acoustics of Buildings*)

When the glass plates, shown in fig. 105, are inserted in or removed from the gap, the condensers are discharged through two spark gaps marked 'sound spark' and 'light spark.' These are arranged in a long blackened box as shown, in line with a viewing screen or the photographic plate. The slight delay which is required between the sound spark and the light spark is controlled by means of the condenser C_3 across the gap of the light spark. The latter is formed between two magnesium wire electrodes within a small capillary tube, a point source giving good definition being thereby obtained. The magnesium electrodes give a very brilliant and photographically active light spark. A liquid

* See Davis and Kaye, *Acoustics of Buildings*, p. 32 (Bell), and *Journ. Sci. Instrs.*, **3**, p. 393, 1926. † *Loc. cit.*

rheostat R is used to control the longer time-intervals, although the sparks are reduced in intensity. The sound spark-gap is arranged at a suitable point inside the reflecting surface which it is desired to study. Light from the sound spark is prevented from reaching the photographic plate by projecting discs mounted on the electrodes. To obtain photographs which show the progress of the sound pulse, a succession of exposures is made with a graduated series of time-intervals, obtained by varying the condenser C_3 and the rheostat R. The principle of the shadow method has been used by C. V. Boys * and others to photograph rifle bullets in flight, and by Quayle,† Foley and Souder,‡ and others to photograph sound pulses from sparks. Photographs of reflection and diffraction phenomena obtained by Foley and Souder by this method are shown in figs. 71 and 82. It is interesting to note in connection with these shadow photographs of a pulse of compressed air, that the shadow of a spherical pulse always appears as a bright circle accompanied by a dark one. Whether the bright circle is inside or outside depends on whether the sphere is of greater or less density than the surrounding atmosphere. The shadows of jets of hydrogen and carbon dioxide in air are respectively darker and lighter than the background.§

(b) *The Schlieren Method* – Töpler's‖ method is essentially an ingenious application of a device due to Foucault for testing lenses or mirrors for chromatic and spherical aberration.¶ Light from a suitable source is brought to a focus, by means of a concave mirror, on to the object-glass of a telescope (or a photographic camera), and the latter is focussed *on the mirror*, the view in the telescope being the surface of an illuminated mirror (see fig. 106). If now an opaque screen be drawn gradually across the object-glass of the telescope until it enters the focus of the light from the mirror, the image becomes darker and darker as more and more light is cut off, and ultimately a critical position will be reached when the mirror appears quite dark. In this position the screen covers a little more than half of the object-glass. Actually the screen is set a little in advance of this position so that about half of the object-glass is covered. It is important that the mirror should be of good optical quality, otherwise its im-

* *Nature*, **47**, pp. 415, 440, 1893.

† *Journ. Frank. Inst.*, **193**, p. 627, 1922.

‡ *Phys. Rev.*, **35**, p. 373, 1912 ; **14**, p. 143, 1919 ; **20**, p. 505, 1922.

§ See Payman and Robinson, Safety of Mines Research Board, Paper No. 18.

‖ *Pogg. Ann.*, **131**, p. 33, 1867. See also note above.

¶ See Edser, *Light*, pp. 98 and 99.

perfections will be revealed in the telescope or the camera, as Foucault originally intended. Since each point of the mirror has its corresponding position in the image, if the path of a ray be deflected at any particular point, the screen will be in the critical position for all points except that one. Provided the deflection is in the right direction, the ray of light will pass beyond the edge of the screen *s* into the lens and a bright image will be formed in its true position. It will be seen therefore that a pressure pulse passing in front of the mirror in the direction AB (see fig. 106) will result in a corresponding band of light across the image which will therefore represent the shape of the advancing pressure pulse. If the latter passes some distance in front of the mirror, as at AB, the telescope must be focussed on that plane whilst pointing directly at the mirror. Using a lens instead of the mirror, Töpler obtained excellent photographs of the pressure pulse from an electric spark, similar to those obtained by Dvorak's shadow method. These experiments were continued by his son* and extended by Mach† and others. R. W. Wood‡ used the method to illustrate the focussing of a beam of light by a concave mirror. Allen § and Cranz || used the Schlieren method for the photography of flames, whilst the latter also photographed the pressure waves emanating from bullets in flight (see fig. 78).

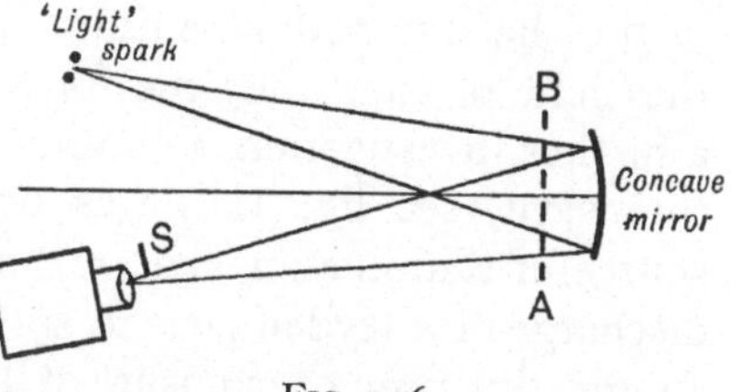

FIG. 106

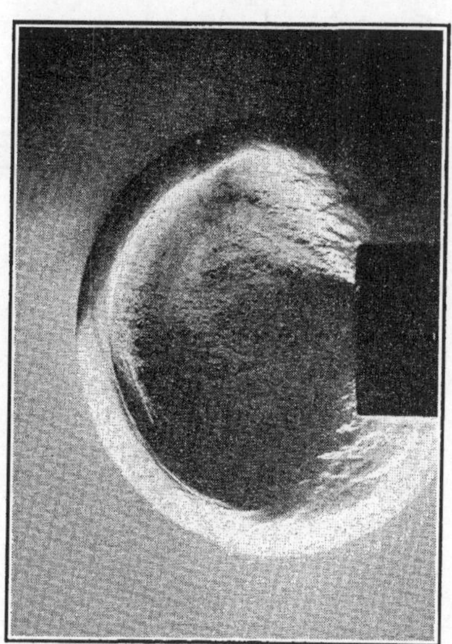
FIG. 107—Schlieren Photograph of Explosion Pulse emerging from Tube (Payman and Robinson)
(By courtesy of Safety in Mines Research Board)

* *Ann. der Physik*, **27**, pp. 1043, 1051, 1908.
† *Ann. der Physik*, **159**, 330, 1876 ; **41**, p. 140, 1890.
‡ *Phil. Mag.*, **48**, 218, 1899 ; **50**, p. 148, 1900.
§ *Proc. Roy. Phil. Soc. Glasgow*, **33**, p. 71, 1902.
|| *Lehrbuch der Ballistik*, **2**, Berlin, 1926.

More recently W. Payman and W. C. F. Shepherd * have made excellent use of the method in photographing the explosion impulse emerging from the mouth of a steel tube when various gas mixtures and detonators were fired in it (see fig. 107). As in the shadow method, a timed light spark was used to illuminate the field of view after the formation of the pressure pulse. In a further investigation a continuous record of the movement of the wave (see fig. 108) was obtained by using a continuous source of illumination (an arc) instead of an instantaneous spark discharge of a leyden jar. A special 'wave-speed' camera, giving an effective time of exposure of 10^{-5} second, was designed for the purpose. The photographs provided direct information relative to the velocity of the wave-front as it proceeded from the mouth of the tube (see p. 266). Specimen photographs obtained by the Schlieren method are shown in figs. 107 and 108.

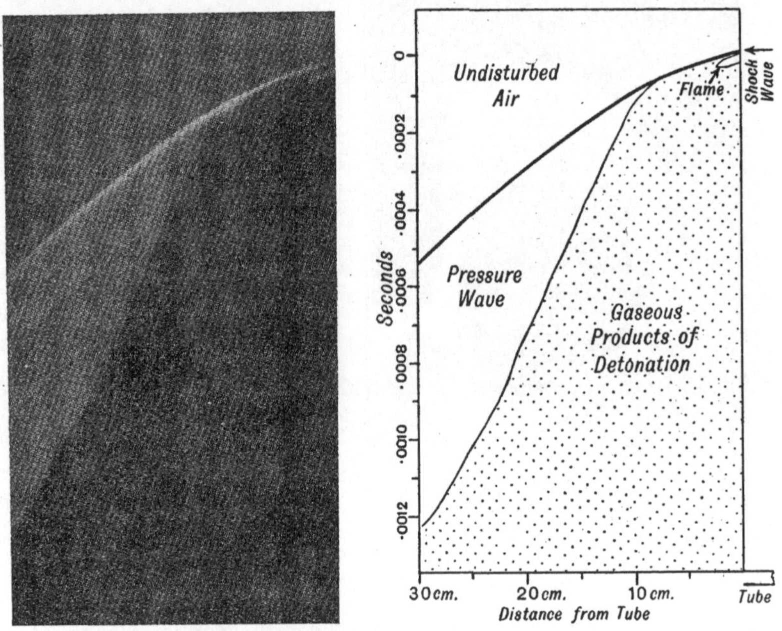

FIG. 108—Schlieren Photo of Pressure Wave (Payman, Shepherd and Robinson)

(By courtesy of Safety in Mines Research Board)

Ripple Tank Method – Results of a similar character may be obtained by means of the ripple tank. Tyndall made use of the analogy of ripples on the surface of water and mercury to illus-

* Safety of Mines Research Board, Papers 18 and 29, 1926.

trate the principle of superposition and the interference and reflection of sound waves.* Later, J. H. Vincent † obtained excellent photographs of ripples in a small wooden trough containing mercury, and demonstrated the various 'optical' phenomena of wave motion—reflection, refraction, interference, and diffraction. Kelvin ‡ has shown that the velocity V of waves on the surface of an inviscid, incompressible liquid under the action of gravity g and surface tension T is given by

$$V^2=\frac{g\lambda}{2\pi}+\frac{2\pi T}{\rho\lambda} \quad . \quad . \quad . \quad . \quad (1)$$

where ρ is the density of the liquid and λ the wave-length. When λ is large, the waves are controlled mainly by gravity and the velocity approximates to $\sqrt{g\lambda/2\pi}$. On the other hand, when λ is small, the influence of surface tension predominates and V becomes $\sqrt{2\pi T/\rho\lambda}$. Solving (1) for λ in terms of V, it is found that for no wave-length can V be less than $V_0=(4Tg/\rho)^{\frac{1}{4}}$. The value of λ corresponding to the minimum velocity is $\lambda_0=2\pi(T/g\rho)^{\frac{1}{2}}$, and the critical frequency N_0 is consequently $\frac{1}{2\pi}\left(\frac{4g^3\rho}{T}\right)^{\frac{1}{4}}$. Kelvin defines a ripple as a wave whose length is less than that of a wave which is propagated with the minimum velocity. In order to obtain true ripples, therefore, it is necessary to use vibration frequencies above a certain value. For ordinary mercury (T being 300 to 400 C.G.S. units) waves less than 1·3 cm. long are ripples, a frequency of about 15 p.p.s. producing the longest waves which may be regarded as ripples. Frequencies of 200 p.p.s. and upwards therefore give rise to true capillary ripples controlled by surface tension only. In water ($\rho=1$, $T=76$) we find the limiting velocity $V_0=23{\cdot}1$ cm./sec., $\lambda_0=1{\cdot}71$ cm., and $N_0=13{\cdot}6$. Lamb § has shown mathematically that the equations of motion of gravity waves on the surface of a liquid are closely analogous to those for cylindrical waves of sound. More recently A. H. Davis ‖ has pointed out, in the case of ripples in a small tank, that the analogy loses its generality, for the velocity of sound is independent of wave-length, whereas in the case of ripples there is a definite dependence upon wave-length. Provided, however, the depth

* *Sound*, Lecture 7. He refers also to Weber.

† *Phil. Mag.*, **43**, p. 17, 1897 ; **45**, p. 191, 1898, etc. (*loc. cit.*).

‡ *Phil. Mag.*, **42**, p. 375, 1871.

§ *Hydrodynamics. Arts.* 172 and 189.

‖ A. H. Davis, *Phys. Soc. Proc.*, **50**, p. 75, 1925.

of liquid is greater than one-half the wave-length of the small, simple, harmonic ripples, the analogy is still perfect. Even impulsive ripples, in spite of the accompanying subsidiary wavelets, bear a striking resemblance to impulsive sound waves. The comparative photographs of a ripple and a sound pulse reflected from a hemicylindrical mirror (see fig. 109 (Davis)) are

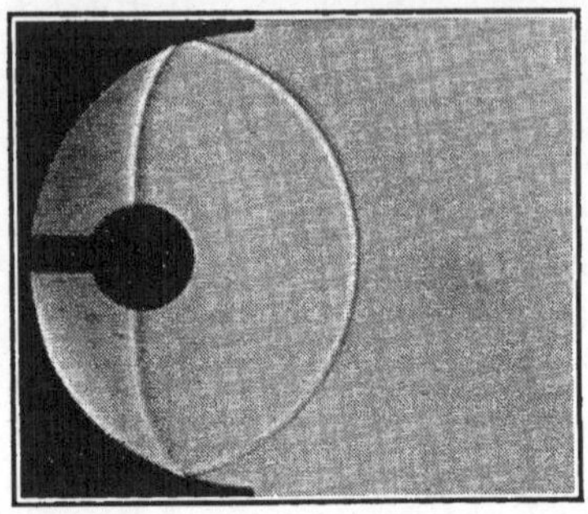

(*a*) Spark

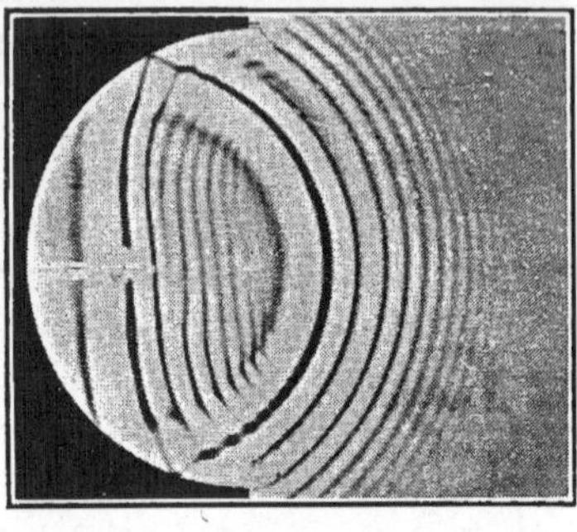

(*b*) Ripple

FIG. 109—Comparison of Spark and Ripple Reflections from concave Surface (Dr. A. H. Davis)

sufficient proof of this. The ripple pulse is produced by a round-headed dipper which is suddenly withdrawn from the water by means of an electromagnet. To prevent confusion due to reflections from the ends of the ripple tank, a 'shelving beach' is fitted which rapidly damps out the vibrations of the incident waves. This property of a beach was pointed out by J. H. Vincent. When continuous ripples are required, a style attached to one prong of an electrically maintained tuning-fork (say 50 p.p.s.) is used and the ripples are viewed stroboscopically,* *e.g.* by using a neon lamp illumination of the same frequency as that of the tuning-fork—preferably excited through an electrical contact on the other prong. Davis uses a ripple tank, 3 ft. × 5 ft., containing 2 in. of water, the bottom of the tank being of plate glass. Intermittent illumination is obtained by means of an arc and stroboscopic sectored disc driven by a phonic motor controlled by the tuning-fork. The light from the arc 10 ft. distant is passed vertically upwards through the glass bottom of the tank, and casts a shadow of the ripples on a horizontal screen suitably mounted about 5 ft. above. With the stroboscopic arrangement the ripples appear stationary and are easily photographed. A 'slow motion' may be given to the appearance of

* See Rayleigh, *Sound*, **2**, p. 345, and *Collected Papers*, **3**, p. 383.

the ripples by driving the sectored disc independently at a speed slightly different from that of the fork. The ripple tank method has the advantage over the spark methods of demonstrating visually the actual progress of waves. It is easily and quickly adaptable to illustrate a wide variety of wave phenomena.* Refraction effects are obtained by varying the depth of the water in the tank (*i.e.* by varying the wave-velocity); reflection and diffraction by introducing suitably shaped obstacles in the path of the waves. The method has been extensively applied to a study of reflections from complex structures such as model sections of auditoriums, but we shall have to refer to this question again in dealing with the acoustics of buildings (see p. 484). Davis has used a kinematograph camera to photograph the complete progress of a wave. He has also illustrated the transmission of sound waves through Quincke filters (see p. 190), curved conduits (such as loud-speaker horns), and vibrating partitions.† In the case of the curved horn, the ripple method revealed a pronounced stationary vibration (transverse to the axis) at a certain bend in the horn, indicating that the bend was behaving as a wave-filter over a moderate range of frequency.‡ At other frequencies the waves emerged freely from the flare of the horn.

Bullet Photography – A method which has occasionally been used to demonstrate wave phenomena (reflection and diffraction) employs the highly compressed 'bow-wave' from a flying bullet. A spark shadow photograph is taken of the bullet, or rather its accompanying wave, in its flight past the object to be studied. Thus an early photograph by C. V. Boys shows an excellent reflection of the pulse from a plane obstacle in its path. The photographs in fig. 82 by Cranz, showing the reflection of bullet waves from two parallel planes, require no additional explanation. As a method of studying wave transmission, however, it is, of course, far less convenient than the other two methods just described.

* See Davis and Kaye, *Acoustics of Buildings*; and Bragg, *World of Sound* (Bell).

† *Phys. Soc. Proc.*, **40**, p. 90, 1928.

‡ See also A. L. Foley, *Phys. Rev.*, **20**, p. 503, Dec. 1922.

SECTION IV

RECEPTION, TRANSFORMATION, AND MEASUREMENT OF SOUND ENERGY

THE manner in which sound energy is abstracted from the medium through which it is passing is dependent on a wide variety of circumstances. The choice of a 'receiver' of the wave energy will depend, for example, on (*a*) the character of the sound wave itself, involving considerations of wave-form, frequency, and amplitude; (*b*) the nature of the medium which transmits the wave, questions of radiation resistance, density, and velocity being involved; and (*c*) the ultimate object for which the sound energy is required, this aspect of the question involving the transformation of the energy of sound waves into a more convenient form. A comprehensive classification of sound receivers on these lines is practically impossible, for in many instances a number of conditions must be simultaneously fulfilled, and the most suitable receiver cannot be classified in one particular group. Before dealing in detail with particular types therefore a few remarks on receivers in general may not be out of place. It is clear that a sound receiver which is sensitive in air may be quite unsuitable for use in another medium, such as water or earth. Sounds differing widely in frequency, say 50 to 50,000, require entirely different treatment. If the sound is to be employed in long-range signalling it is important that the receiver should be sensitive to very weak signals at extreme ranges. This raises the question of tuning the receiver to the frequency of the sound wave used to transmit the signals. In such cases a knowledge of the resonant frequency and the degree of damping is important, whilst distortion of wave-form is a secondary consideration. In certain circumstances, however, when a faithful record or reproduction of the sound is required, energy and sensitiveness are subordinated to faithfulness in the reproduction of wave-form. A resonant receiver would be useless, or at any rate very undesirable, for such a purpose. We have therefore to distinguish between *resonant and non-resonant receivers* where sensitiveness and faithfulness are respectively of primary importance. Similarly we may require to distinguish between *pressure and displacement receivers* according as the sensitive element (*e.g.* some form of 'microphone') is placed at a point of maximum pressure fluctuation and minimum displacement or *vice versa.*

With certain reservations, it may be said that all forms of sound receiver require the introduction of an obstacle in the path of the sound waves. The receiver must either partake of, or otherwise influence, the motion of the particles of the medium or must respond in some way to the pressure variations on its surface. The receiver, regarded as a rigid obstacle, must reflect some of the incident energy in the ordinary way, and, regarded as a vibrator, it must re-radiate a certain proportion of the energy which excites it into vibration. If, however, the receiver is to fulfil its purpose efficiently, a reasonable proportion of the incident sound energy must be absorbed and converted into another form of energy most convenient for the purpose in view. This conversion of energy may, for example, be a direct mechanical transformation of longitudinal vibrations into transverse vibrations, as when sound waves in air fall on a diaphragm. Such a transformation may be followed by a conversion of mechanical energy into electrical or thermal energy. In sound reception we have therefore to consider (*a*) the transformation of one form of mechanical motion into another form, and (*b*) the transformation of mechanical into electrical or thermal energy. The simplest example is the ordinary telephone 'transmitter.'* Longitudinal vibrations from the mouth of a speaker are converted into transverse oscillation of the diaphragm. These in turn affect a microphone and set up corresponding electrical oscillations. The converse process takes place at the 'receiver' end of the telephone line.

A distinction is sometimes made between electrical sound receivers which convert sound energy *directly* into electrical energy (analogous to a dynamo) and those in which the conversion is *indirect* (analogous to a relay); these are exemplified respectively by the electromagnetic telephone and the carbon granular microphone.

Another type of mechanical transformation is sometimes utilised when it is required to increase or reduce the displacement or pressure amplitude of a vibrator. This is a true transformer action analogous to the electrical transformation of current and voltage. As we shall see, numerous examples of the application

* The term transmitter in this case is used to imply electrical transmitter—actually the transmitter is a *receiver* of sound waves, and *vice versa*.

of this principle are to be found in sound producers and receivers, *e.g.* in the gramophone sound-box and in various signalling devices. Mechanical 'amplifiers' (or reducers), such as mirrors, trumpets, or resonators, or electrical amplifiers (microphones and valves) are also available for similar purposes.

Metrical forms of sound receiver such as the Rayleigh disc and the sound pressure radiometer form another class of 'receivers,' which might also include the various 'phonometers,' hot-wire microphones, and piezo-electric devices. When sound vibrations are to be recorded as directly as possible, various optical devices may be utilised. The vibrations of a diaphragm on which the sound falls may be recorded optically by means of a minute tilting mirror. Vibrations of the medium may be recorded still more directly by means of such devices as optical interferometry, sound-shadow photography, or by means of sensitive flames.

The above remarks will be sufficient to indicate the difficulty of attempting a broad classification of sound receivers.

THE EAR

In order of importance the human ear is the foremost of all receivers of sound. This 'organ of hearing,' as it is sometimes called, has a marvellous range of frequency and sensitivity; it can appreciate vibrations of frequency varying from 20 to 20,000 periods per second, and can distinguish changes of intensity over a range of 10^{12} to 1. The ear with its associated nerve-endings has also remarkable powers of analysis, being able to resolve qualitatively a complex note into its Fourier components, and to discriminate between a number of different sounds at the same time. Before proceeding to deal with the structure of the ear and the mechanism of hearing, it will be well to refer to a number of experimental observations which must be explained on any theory of audition.

Sensitivity of the Ear to Changes of Intensity and Frequency. Weber-Fechner Law – The physical stimulus, or intensity, which produces the sensation of loudness is, as we have seen on p. 54, proportional to the square of the product of amplitude and frequency, when the vibration is simple harmonic. If the frequency is constant, then the intensity is proportional to the square of the amplitude of vibration. The measure of the sensation of loudness, however, is not so easy to define. The difficulty may be circumvented in a manner indicated by Weber's law:

"The increase of stimulus to produce *the minimum perceptible increase of sensation* is proportional to the pre-existing stimulus." That is, commencing with a certain sound intensity the increase of intensity δE which produces a *noticeable* change of sensation δS may be measured. From Weber's law Fechner derived the relation $\delta S = k\delta E/E$ or $S = k \log E$, where S is the magnitude of the sensation, E the intensity of the stimulus, and k a constant. Although it is not possible to measure S directly, it is not difficult to determine the ratio $\delta E/E$ as a function of intensity E. In certain experiments by F. B. Young and the writer,* in which sounds of moderate intensity were compared, it was found that the value of $\delta E/E$ at a frequency of 580 p.p.s. was approximately 0·10 under favourable conditions, *i.e.* in the absence of extraneous noises. V. O. Knudsen,† in an experimental investigation of the Fechner-Weber law, determined the sensitiveness of the ear to changes of intensity and frequency N. For the purpose, a telephone receiver actuated from a valve oscillator, of wide range of power and frequency, was used as a source of sound. The intensity or frequency could be varied periodically, about once a second, by changing automatically the resistance or capacity in the oscillating circuit. The method of observation was to change δE or δN continuously until the ear could only just distinguish the variation. Independent experiments with the telephone receiver showed that the sound energy emitted was proportional to the electrical input. The latter therefore served to measure the relative sound intensities at a fixed frequency. The frequency scale was determined by calibration. The experimental curves obtained for various values of E and N, indicated an agreement with a modified form of the Weber-Fechner law, as proposed by Nutting ‡ for the eye, namely,

$$\frac{\delta E}{E} = \kappa + (1-\kappa)\left(\frac{E_0}{E}\right)^m,$$

where E_0 is the 'threshold' intensity and $\kappa = 0{\cdot}10$ approx. The exponent m varies somewhat with frequency, being 1·65 for 200 p.p.s. and 1·05 for 1000 p.p.s., but at the same loudness level, *e.g.* 10,000 E_0, the ratio $\delta E/E$ is nearly independent of frequency, showing only a 10 per cent. variation from 100 to 3200 p.p.s. For moderate and large intensities (E much greater than E_0), the

* *Proc. Roy. Soc.*, **100**, pp. 264 and 266, 1921.

† *Phys. Rev.*, **19**, p. 261, 1922, and **21**, p. 84, 1923.

‡ *Bureau of Standards Bull.*, **3**, No. 1, 1907.

expression reduces to $\delta E/E = \kappa = 0{\cdot}10$ at any frequency, a result which agrees with that of Young and the writer. At a frequency near 1000 p.p.s. it was found that the normal ear could distinguish about 400 gradations or 'steps' of loudness between the threshold and a painful intensity 10^{12} times as great.

As regards frequency sensitivity, at the same intensity (or loudness level), Knudsen found that the ratio $\delta N/N$ diminished from 0·01 at 55 p.p.s. to 0·003 at 600 p.p.s., remaining constant at this value up to a frequency of 3200 p.p.s. This result refers to discrimination between two sounds of slightly different frequency (δN) sounded independently. The frequency sensitivity of the ear is of course much better than this if the two sounds are excited simultaneously, a frequency difference of 1 in 20,000 being detectable by 'beats.' Observations of a similar nature to those of V. O. Knudsen have been made by H. Fletcher and R. L. Wegel * using a valve-actuated air-damped telephone receiver over a frequency-range 60 to 6000 p.p.s. By means of a specially constructed *attenuator* the current entering the telephone could be varied about three millionfold by turning a dial switch. The arrangement was carefully calibrated so as to give alternating air pressure on the ear cavity in terms of the indications of the dial switch. A curve representing an average of 93 normal ears, which gives the pressure fluctuation δp required to produce a perceptible change of sensation at different frequencies, shows that $\delta p = 0{\cdot}1$ dyne/cm.^{2} at 60 p.p.s., falling to a minimum $\delta p = 0{\cdot}0007$ dyne/cm.^{2} at a frequency of about 2000 p.p.s. Careful observations made more recently by R. R. Reisz † are in general agreement. He finds that Weber's law $\delta E/E$ = constant holds true for moderate intensities, and that as a function of frequency it reaches a minimum value at about 2500 p.p.s., the minimum being less sharply defined at high intensities than it is at low. The maximum sensitivity of the ear to detect slight differences of intensity therefore occurs at a frequency near 2500 p.p.s. Reisz finds that the ear can detect the greatest number of gradations of loudness, viz. 370, at a frequency of 1300 p.p.s. A very instructive diagram, see fig. 110 (*a*), has been obtained by Wegel,‡ indicating the range of the average human ear with respect to intensity and frequency. The upper curve in the diagram represents a more or less arbitrary

* *Nat. Acad. Sci. Proc.*, **8**, p. 5, 1922.

† *Phys. Rev.*, **31**, p. 867, May 1928. See also Steinberg, *Phys. Rev.*, **26**, p. 507, 1926 ; F. W. Kranz, *Phys. Rev.*, **22**, p. 66, 1925 ; and Kingsbury, *Phys. Rev.*, **29**, p. 588, 1927.

‡ *Bell System Techn. Journ.*, **1**, Nov. 1922.

upper limit to the range of auditory sensation. It may be regarded, in fact, as the 'threshold of *feeling*' at which the loudness is so great that the effect is somewhat painful, and further increases of amplitude are not perceived as increases of loudness. The lower curve represents the other extreme, or the lower limit of amplitude at which a sound can be detected. The area enclosed between the two curves, from the threshold of audibility to the threshold of feeling, has been described as the 'auditory sensation area.' * Outside this area sounds are not 'heard' whatever the intensity. This does not mean that they are always unperceived, for large amplitude vibrations may be *felt* at very low frequencies, the ear in these circumstances behaving like a simple pressure indicator. The *infrasonic* waves to which Esclangon † refers, in connection with the pressure-wave from the mouth of a gun ('*onde de bouche*') may have a sufficiently large amplitude to affect the ear, in spite of the fact that the frequency is of the order of 1 p.p.s. only. At the other extreme also, powerful *supersonic* waves of sufficiently great intensity may be felt rather than heard.

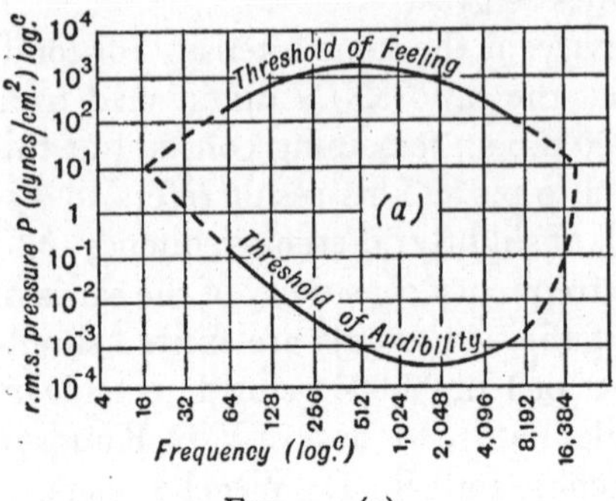

FIG. 110 (*a*)

The sensitiveness of the ear is shown graphically in fig. 110 (*b*) as a function of frequency. The ordinates, expressing sensitivity, are the reciprocals of the minimum amount of energy (in ergs/cm.² sec.) required to produce an audible sound, the abscissæ represent the values of frequency plotted logarithmically. There appears to be general agreement in the statement that the *maximum sensitivity of a normal ear occurs at a frequency near* 2000 *to* 2500 *p.p.s.* At this frequency, Fletcher and Wegel have found that *the ear can*

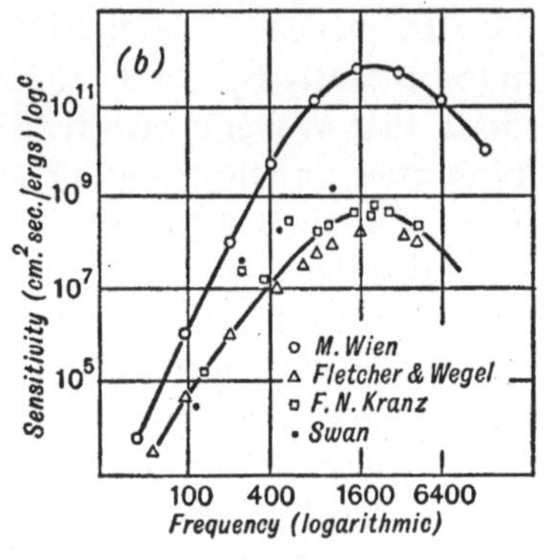

FIG. 110 (*b*)

* See H. Fletcher, *Bell System Techn. Journ.*, **4**, p. 375, 1925; and *Speech and Hearing* (Macmillan). † *Loc. cit.*

respond to a pressure-amplitude of the order of 10^{-3} *dyne/cm.*^{2} or 10^{-9} of an atmosphere. In air this corresponds to a displacement-amplitude of 10^{-9} cm., which is about 1/30th of the diameter of a molecule of oxygen, or about 10^{-4} of the mean free path of the molecules in air at N.T.P. It is interesting to note that Rayleigh * obtained a value of minimum condensation audible $s=4{\cdot}6 \times 10^{-9}$ by observing the period of time t that the vibrations of a tuning-fork (384 p.p.s.) can be heard. From a knowledge of the initial amplitude of the fork (at $t=0$) and the exponential decay constant (the 'log dec.') of its vibrations, the amplitude after a time t is readily calculated. Similar results were obtained by Wien † by an entirely different method.

Abnormal Hearing. Masking Effect of Other Sounds – The observations mentioned above refer only to 'normal' ears under ideal conditions of test. It is evident that the lower curve of fig. 110 (*a*) may have to be raised considerably if the observer is somewhat deaf, a greater pressure amplitude being required in this case to reach the threshold of audibility. Experiments by J. P. Minton ‡ not only demonstrate this effect, but also show that in certain cases of nerve deafness a person may be absolutely deaf to a certain limited range of frequency, whilst remaining sensitive to other frequencies (analogous to colour blindness). Minton has obtained audibility curves (1) for normal ears, (2) for ears diseased only in the middle ear, (3) for ears diseased only in the internal ear, and (4) combinations of (2) and (3). The results appear to have an important bearing on theories of audition. Abnormality in the conditions of tests may also modify the normal curve considerably. It is obvious that the presence of extraneous noises during the test will reduce the apparent sensitivity of the ear and raise the threshold curve of audibility. Measurements of this nature have been made by R. L. Wegel and C. E. Lane,§ a telephone receiver being supplied with current of two different frequencies of adjustable intensities. The amount of 'masking' by tones of frequency 200 to 3500 p.p.s. was determined for frequencies from 150 to 5000 p.p.s. The magnitude of a tone was taken as the logarithm of the ratio of its pressure to the threshold value, and masking was taken as the logarithm of its threshold value with masking to that without. The curves of masking as a function of magnitude were found to approximate to straight lines. Except when the frequencies were so close

* *Sound*, **2**, p. 439.

† *Wied. Ann.*, **36**, p. 834, 1889.

‡ *Phys. Rev.*, **19**, p. 80, 1922.

§ *Phys. Rev.*, **23**, p. 266, 1924.

together so as to produce beats, the masking was greatest for tones nearly alike. When the masking tone was loud it masked tones of higher frequency better than those of frequency lower than itself; there was little difference, however, if the masking tone was weak. At intensities considerably above minimum audibility they found that there was no longer a linear relation between the sound pressure and the response of the ear, this non-linearity accounting for the combination tones and 'subjective overtones' observed in certain cases. H. Fletcher * has also made experiments on the masking effect of one tone on another, and has found, contrary to statements by A. M. Mayer,† that higher tones may mask lower ones. This problem of masking is not only of importance in relation to theories of hearing, but has a very wide, practical significance in telephony and signalling in general, where it is often necessary to listen intently to a particular sound in the presence of other disturbing noises.

Perception of Quality and Pitch – A fundamental law in physiological acoustics was formulated by Ohm ‡ (the author of Ohm's law in electricity), in the statement that the ear recognises as pure tones only those due to simple harmonic vibrations, and that it resolves any other complex vibration into its harmonic components, perceiving them as a summation of pure tones. This amounts to saying that the ear is capable of analysing a complex tone into a Fourier harmonic series of simple tones. As everyone is aware, the ear is particularly sensitive to that characteristic of a sound which we call *quality*, which depends essentially on the number, intensity, and distribution of the harmonic components into which a sound can be analysed. Notes of the same pitch from a tuning-fork, a violin, a piano, a clarinet, and an organ are, to the ear, entirely different in quality and are instantly recognisable. Using such a wide selection of 'quality' in sounds, H. Fletcher § has recently made experiments on the effects of cutting out certain harmonic components. He used a telephone system to reproduce the sounds, and electrical filter circuits (see p. 457) to eliminate any desired frequency-range. As a result of his observations he claims to have proved that *only the quality and not the pitch* of such musical sounds is changed when a group of low- or high-frequency components is cut out.

* *Frank. Inst. Journ.*, **196**, p. 289, 1923.
† *Phil. Mag.*, **11**, p. 500, 1876.
‡ G. S. Ohm, *Pogg. Ann.*, **59**, p. 513, 1843 ; and **62**, p. 1, 1844.
§ *Phys. Rev.*, **23**, p. 427, 1924.

For example, he maintains that the pitch remains the same when the fundamental and the first seven overtones were eliminated from the vowel 'ah' sung by a baritone voice at an ordinary pitch.

Ohm's law says nothing about the *relative phases of the harmonic components* into which the ear analyses the sound, although in the Fourier analysis of a complex sound phase relationships are just as important as amplitudes and frequencies. To examine this point, König * made experiments with double sirens, constructed in such a manner that the phase of one component sound could be shifted relative to the other by displacing the 'fixed disc' of one siren relative to its normal position. His observations led to the conclusion that the phase effect, if it existed at all, must be due to secondary causes. Experimenting with a harmonic series of electrically-maintained tuning-forks, Helmholtz found that the quality depends solely on the number and relative strengths of the partial, simple tones, and in no respect on their differences of phase. Lloyd and Agnew,† by a more refined electrical method of producing the tones and varying the phases, have confirmed Helmholtz's conclusion.

Minimum Duration for Identification of Pitch. 'Persistence of Audition.' Aural Fatigue – Experiments by Kohlrausch to determine the lowest number of vibrations required to excite the sensation of pitch, indicated that a fair estimate (within 3 per cent.) may be formed if only two impulses reach the ear. Listening to musical sounds in a telephone circuit which could be completed for a known fraction of a second, G. Gianfranceschi ‡ found that the minimum duration required for identifying the pitch was the same for all frequencies, and amounts to 1/40 sec. A specially trained ear might reduce this period to 0·01 second. Similar experiments have been made by E. Lubcke § in the case of a sound of frequency 1000 p.p.s. In air or in water, with a feeble or strong signal, a sound lasting 9 or 10 periods (0·01 sec.) was found to be necessary for the production of a definite 'tone-impression.'

Experiments which have an important bearing on this question have been made, in an entirely different manner, by Weinberg and Allen.‖ A continuous sound issued through a hole in a felt-

* See Rayleigh, *Sound*, **2**, p. 469 ; and *Wied. Ann.*, **14**, p. 392, 1881.
† *Bureau of Standards Bull.*, **6**, 255, 1909.
‡ *Accad. Lincei. Atti*, May 1914.
§ *Zeits. Techn. Phys.*, **2**, 1921.
‖ *Phil. Mag.*, **47**, pp. 50, 126, and 141, 1924.

lined box, the hole being alternately closed and opened by a rotating disc with four symmetrical holes, corresponding in size with that in the box. The speed of the disc was gradually increased until, at a certain critical value, the pulsation or 'flutter' of the sound just disappeared, the sound then appearing continuous to the ear. The 'persistence' of the sound at this critical speed was found for various frequencies and intensities to lie between 0·0127 and 0·0215 seconds. This 'after-effect' in the ear, or 'persistence of audition,' is analogous to 'persistence of vision,' the eye being unable to distinguish the interruptions in a flickering beam of light if the frequency of flicker exceeds 40 per second. In a similar manner also the ear is fatigued after exposure to a loud sound, and takes time to recover its normal sensitivity after such exposure. In order to measure this fatigue, Weinberg and Allen (using the above method) determined the critical 'flicker' frequency after the ear had been subjected to different degrees of exposure to another sound. It was found when the 'fatiguing' tone agrees in frequency with the 'test' tone, the fatigue is a maximum. When the frequency is about 8 p.p.s. above or below that of the 'test' tone, the fatigue effect is negligible.

Audible Limits of Frequency – The intersection points of the two curves shown in fig. 110 (*a*) correspond to the upper and lower limits of frequency which the ear can perceive as sound. These limits vary greatly for different observers, and, in the case of the upper limit, the two ears of the same observer often differ appreciably. In many of the earlier observations of the upper limit of frequency insufficient attention was paid to the measurement of amplitude as well as frequency, and in some cases the limit appeared low on account of the small amplitude of the sound at higher frequencies. Other causes of differences of opinion regarding both the lower and the upper limits of frequency may have arisen through the simultaneous excitation of *other* frequencies, higher or lower than the one under test. For example, in determining the lower limit by Helmholtz's method of the loaded string, or the loaded tuning-fork of Edelmann, it is difficult to ensure that overtones are not excited. The vibrations of air in large organ pipes are also liable to contain low frequency overtones which may be mistaken for the fundamental. There appears to be general agreement, however, that the lowest audible frequency is not far from 20 p.p.s. Below this frequency the oscillations are detectable individually as direct pressures on the ear-drum—that is, they do not blend into a smooth note. At the other extreme of the frequency scale the divergence of opinion is still greater. Rayleigh * states that bird-calls cannot be heard above 10,000 p.p.s., although a sensitive flame can still detect the vibrations up to 50,000 p.p.s. There is no doubt, however, that the vibrations of short steel bars, electromagnetically excited, can be heard up to 20,000 p.p.s., and by some persons at frequencies even higher than this. A good average range of frequency for the human ear may be taken as 20 to 20,000 p.p.s. At very high frequencies, however, the ear loses its power to discriminate variations of pitch. Electrical wave-filter experiments on the reproduction of speech and music indicate that frequencies above 4000 or 5000 p.p.s. need not be considered, for above this frequency the ear is very insensitive and has lost its power of appreciating pitch. The frequency-range employed in music extends from about 40 to 4000 p.p.s.

Perception of Direction. Binaural Audition – The characteristics of hearing which we have hitherto considered apply equally well to one ear alone or to a pair of ears acting as independent receivers. One of the most important functions of the ears—namely, perception of direction—can only be fulfilled satisfactorily, however, when they are used *as a pair*. This instinctive 'sense of direction,' which is so familiar, presents a scientific problem which is still only partially solved. For the progress already made towards its solution we are indebted principally to Rayleigh.† Binaural hearing has been likened by some writers ‡ to stereoscopic vision, but the analogy is of doubtful application, for it is possible to estimate with considerable accuracy the *direction* of a source of light, using one eye only, but in the case of sound *both* ears are necessary. The two eyes help us to estimate *range* as well as direction, but the two ears are incapable of this additional function. An average person with closed eyes is usually capable of estimating the direction of a sound within a few degrees, whether the sound be continuous or of short duration. The directional accuracy is fairly good if the sound proceeds from the right or the left, but may be in error by 180° if the sound approaches from behind or in front. The estimation of direction is usually spontaneous and does not require a movement of the head.

Intensity and Phase Theories – For high-pitched sounds of short wave-length these effects might be explained by the difference

* *Sound*, **2**, p. 433. † *Sound*, **2**, p. 440.

‡ See, *e.g.*, Geiger and Scheele, *Handbuch der Physik*, **8**, p. 538.

of intensity of the sound reaching the two ears, the head acting to some extent as a screen to the ear farther away from the sound source. At lower frequencies, however, when the wave-length exceeds the circumference of the head, the intensity theory alone cannot explain the observations, for the directional effect is still good in spite of the fact that the intensity difference at the two ears must be very small. Regarding the average head as a rigid sphere of radius a, Rayleigh * has calculated the variation of sound intensity around it for sounds of wave-length λ proceeding from a distant source. Denoting the front, rear, and side positions by A, B, and C respectively, the relative intensities are given in the table, where $2\pi a$ is the circumference of the 'head.'

$2\pi a/\lambda$.	$=2{\cdot}0$.	$=1{\cdot}0$.	$=0{\cdot}5$.
A	0·69	0·50	0·29
B	0·32	0·28	0·26
C	0·36	0·24	0·23

For still smaller values of $2\pi a/\lambda$, the difference of intensities A and B is given approximately by $\frac{3}{4}\left(\frac{2\pi a}{\lambda}\right)^4$, that is, the difference varies inversely as the fourth power of the wave-length. Regarding the 'head' as a sphere of circumference, $2\pi a=2$ ft. at a frequency of 256 p.p.s., the ratio $2\pi a/\lambda=0{\cdot}5$, and the difference of intensities for positions A and B is only about 10 per cent. —*i.e.* would be just detectable under favourable conditions. When N=128 p.p.s. the difference would be less than 1 per cent., and at lower frequencies would be still less. Now the ears are capable of perceiving sound direction without difficulty at frequencies considerably lower than 128 p.p.s., so the intensity theory must fail. Rayleigh says, "When a pure tone of low pitch is recognised as being on the right or the left, the only alternative to the intensity theory is to suppose that the judgment is founded upon the *difference of phases* at the two ears. But even if we admit, as for many years I have been rather reluctant to do, that this difference of phase can be taken into account, we must, I think, limit our explanation upon these lines to the cases of not very high pitch. . . . Now it is certain that a

* *Nature*, **14**, p. 32, 1876; *Phil. Mag.*, **13**, p. 214, 1907; and *Scientific Papers*, **5**, p. 347.

phase retardation of half a period affords no material for a decision that the source is on the right rather than on the left, seeing that there is no difference between retardation and an acceleration of half a period. It is even more evident that a retardation of a whole period, or of any number of whole periods, would be of no avail. . . . It would seem that at high pitch (above 512) the judgment is based upon intensities, but that at low pitch, at any rate below 128, phase differences must be invoked." Rayleigh found if the same tone were led to the two ears by different paths, that the sound could be made to appear to come from 'right' or 'left' at will, by adjusting the relative path length, *i.e.* by varying the relative phase. The sound was always stated to be on that side on which the phase is in advance by less than half a period. This result has been confirmed by many other observers, but it remains unexplained. G. W. Stewart * has made careful observations of the apparent change of direction produced (*a*) by *varying the relative intensity* at the two ears of an observer, keeping the phases constant, and (*b*) by *varying the phase difference* and keeping the intensity constant. In the former case, writing θ as the apparent direction, and I_R and I_L the intensities at the right and left ear respectively, it was found that θ was approximately equal to $\kappa \log (I_R/I_L)$, the 'constant' κ decreasing somewhat with frequency for certain individuals. In series (*b*) the direction θ was given by $\kappa'\delta$, where δ is the phase difference and κ' is a 'constant' which is a linear function of frequency.

Stewart found that the upper limit of the phase-difference effect lies between 1000 to 1500 p.p.s., and concludes that this is the most important effect in the localisation of a pure tone of frequency lying between 100 and 1200 p.p.s. He points out also that there are more factors in *actual* localisation than the theory assumes, for example, reflection and changes of quality may have an important influence. Above the limit of the phase effect, 1200 p.p.s., the intensity must become an important factor. Rayleigh suggested, in explanation of the phase-difference phenomena, that "the nerve processes must themselves be vibratory, not in the gross mechanical sense, but with preservation of the period and retaining the characteristics of phase." This view of the matter raises physiological difficulties. It has been suggested that the phase effects may still resolve into a difference

* *Phys. Rev.*, **15**, pp. 425 and 432, 1920. See also Hartley and Fry, *Phys. Rev.*, **13**, p. 373, 1919, and **18**, p. 431, 1921.

of intensity due to interference between the direct and transmitted vibrations. The latter are supposed by Myers and Wilson * to pass from one ear to the other through the bones of the head, the resultant, depending on the phase difference at the ears, ultimately passing to the auditory or 8th nerve centre in the brain.

Time-difference Theory – Hornbostel and Wertheimer † have suggested that the directional properties of the ears might be due to the *time* difference of arrival of the sound at the two ears. Their measurements indicate a possible variation from 0 to $6{\cdot}3 \times 10^{-4}$ second, as the source passes from the median plane to a 90° position (directly right or left). The time difference $6{\cdot}3 \times 10^{-4}$ second corresponds to a path difference of 19 cms., which is not necessarily the distance apart of the ears, but a constant k in the relation $d = k \sin \phi$ where d is the path difference and ϕ the direction-angle which the sound makes with the median plane of the head. Hornbostel and Wertheimer attempt to reconcile this theory to explain the case of continuous pure tones, which the ears can readily locate even if they were 'stopped' at the instant when the sound commenced. W. Kunze ‡ has pointed out that if the head is immersed in water, when a source of sound is moved through 180° (from 90° right to 90° left), the impression is received that the source moves only from about 20° right to 20° left.

An interesting observation made by S. P. Thomson § in 1877 is the production of *binaural beats*. A sound led into one ear, through a telephone earpiece, is capable of giving beats with another sound, of slightly different frequency, led into the other ear through another earpiece. Combination tones are not produced under these conditions. This subject has been studied recently by C. E. Lane,‖ who considers the experimental evidence to indicate that the beats are not due to cross-conduction through the head, but are of central origin and result from the sense of binaural localisation of sound by phase. If the beats are slow they are generally recognised as an alternate right and left localisation, the sound appearing to 'revolve round the head.' Fast beats (more than one per second) are generally recognised as an intensity fluctuation.

* *Proc. Roy. Soc.*, **80**, p. 260, 1908. See also Bannister, *Phil. Mag.*, **2**, 144 and 402, 1926.

† *Preuss. Akad. Wiss. Berlin*, Ber. 20, p. 388, 1920.

‡ *Phys. Zeits.*, **22**, p. 649, 1921; and *Zeits. Techn. Phys.*, **3**, p. 40, 1922.

§ *Phil. Mag.*, **4**, p. 274, 1877. See also Stewart, *Phys. Rev.*, **9**, p. 502, 1917.

‖ *Phys. Rev.*, **26**, p. 401, 1925.

Rayleigh's phase theory of hearing involves a principle which has an important application in the directional transmission and reception of sounds (see p. 412).

Combination or Sum and Difference Tones - When the ears are assailed by two sounds of nearly equal frequency N_1 and N_2, the phenomenon of beats (see p. 120) is observed—that is, the sound appears to have a mean frequency $N=(N_1+N_2)/2$, fluctuating in intensity with a frequency (N_1-N_2). In 1745 Sorge, a German organist, and in 1754 Tartini, an Italian violinist, discovered independently that the union of two distinct musical tones sometimes produced a third or 'resultant tone,' which was quite distinct from either of the primary tones. For a long time it was supposed that this resultant tone was simply the 'beat tone' of frequency equal to the *difference* of the primary frequencies (N_1-N_2). The discovery by Helmholtz of a second resultant tone of frequency equal to the *sum* of the primary frequencies (N_1+N_2) reopened the question, and the 'beat tone' theory had to be abandoned. Helmholtz described the combination tones as the 'sum and difference tones.' It is essential for the satisfactory production of combination tones that the two primary tones should be loud and sustained, and that their frequencies be so chosen that the frequency of the combination tone should be about the middle of the audible range. Thus to hear a difference tone two high notes separated by a frequency interval of, say, 400 or 500 p.p.s. should be sounded loudly. The summation tone is more difficult to hear, and it is preferable that *low* notes should be used. The writer has observed the difference tone, 600 p.p.s., due to two intense but almost inaudible primary tones of frequencies 12,000 and 12,600 p.p.s. (obtained by exciting two small steel diaphragms electromagnetically). Barton * gives details of methods of observing sum and difference tones when the primary tones are produced by different instruments (double whistle, two fifes, two organ pipes, harmonium, or pianoforte). To make the combination tone more easily observed he suggests sounding first that primary which lies near in pitch to the combination tone to be produced. As we have seen, when two notes of the same intensity produce beats, the intensity varies from zero to four times that of one primary alone, even when the primaries are feeble. Actually, however, the difference tone is only heard when both primaries are loud, and even then it is of much smaller intensity than either of them. The question arises as to the

* *Sound*, p. 386.

objective existence of the combination frequencies of vibration in the medium itself, it being at least possible that the effect is entirely subjective, the 'combination' being performed in the ear. The objective reality of the tones has been demonstrated by many observers, by using a sensitive resonator tuned to the expected tone (N_1-N_2) or (N_1+N_2). For this purpose tuning-forks,* Helmholtz resonators,† and other devices for indicating the presence of a particular tone have been used. As we shall see, this does not exclude the possibility of subjective combination tones also.

Objective Combination Tones – As already stated, it is often necessary for the production of these tones that the primary sounds should be loud. In such cases the restoring forces called into play are not proportional to the displacements, as in small-amplitude vibrations, but require a term involving the square of the displacement. In other words, the relation between force and displacement is not linear, and the displacement due to the simultaneous operation of two forces is not the simple sum of the separate effects. Neglecting damping forces, the equation of motion of a particle of the medium (*cf.* p. 37) is consequently of the form

$$md^2x/dt^2+\alpha x+\beta x^2=F_1 \cos p_1t+F_2 \cos (p_2t-\delta),$$

the solution of which is expressed in terms of $\cos(2p_1t)$, $\cos(2p_2t)$, $\cos(p_1-p_2)t$, and $\cos(p_1+p_2)t$. That is, the resultant tone has component tones which are octaves of the primaries, and also the sum and difference tones which we are now considering. The amplitude factors of these tones require F_1 and F_2 to be large, before they are appreciable. This theory is due to Helmholtz, and is sufficient to explain the principal experimental observations on objective combination tones.

Subjective Combination Tones – It is not an uncommon circumstance to observe a difference tone when the primary tones are relatively feeble. The effect cannot be explained either on the 'beat tone' theory or the Helmholtz theory of large amplitudes. An explanation which appears generally satisfactory has been given by Waetzmann,‡ who attempts to reconcile Helmholtz's theory of combination tones and König's theory of beat tones. He lays stress on the fact that there is not only asymmetry in

* Rucker and Edser, *Proc. Phys. Soc.*, **13**, p. 412, 1895.

† Helmholtz, *Sensation of Tone*; C. V. Boys, *Phil. Mag.*, **14**, p. 186, 1882; and Rayleigh, *Sound*, **2**.

‡ *Zeits. f. Physik*, **1**, pp. 271 and 416, 1920.

the restoring force in the medium vibrating with a large amplitude, but also there is asymmetry in the ear itself, consisting as it does of a diaphragm (the drum) heavily and unsymmetrically loaded on one side by the bones of the ear. As an example of such effects he records the vibrations of a diaphragm with a central mass attached to one side only. The free vibrations are unsymmetrical about the position of zero displacement. When such a diaphragm is excited by means of two pure tones of different frequency, for example by means of two tuning-forks, the resultant record is similar to that of beats (see fig. 7, p. 20), but the curve lies mainly above the line of zero displacement. The effect is analogous to that of a poor quality rectifier in an electrical circuit passing current of two different frequencies. A carbon granular microphone is, in fact, a convenient means of demonstrating the asymmetry of motion of a diaphragm; oscillograph records obtained with such microphones always show this asymmetry, but the effect is partly due to another cause with which we are not concerned here. Waetzmann's view of the matter is therefore that the ear acts as an asymmetric vibrator and a partial rectifier of the vibrations incident upon the drum. On this view the various overtones, and sum and difference tones, are all readily explained.

Discord and Harmony – Before the Christian era the Greeks, and possibly others before them, studied musical harmony and discord. Pythagoras made the noteworthy observation that a string, divided into two vibrating segments, emits sounds which are more nearly in perfect harmony the simpler the ratio of the lengths of the two parts, *i.e.* the combination of two notes is more pleasing to the ear the smaller the two numbers which express the ratio of their vibrations. Sauveur in 1700 recognised that "octaves and other simple concords, whose vibrations coincide very often, are agreeable and pleasant because their beats are too quick to be distinguished." Helmholtz confirmed this observation that discords are due to 'unpleasant beats,' the unpleasantness arising through fatigue and irritation of the ear. As in the optical case, a very slow 'beat' or flicker is not irritating, the ear (or eye) having time to recover between successive stimuli. At a high rate of flicker the 'persistence' of audition (see p. 358) is such that the *sensation* is continuous, in spite of the fact that the stimulus is intermittent, hence there is no irritation in this case either. Weinberg and Allen * give values in the neighbourhood

* *Loc. cit.*

of 80 per second as the upper limit of the rate of sound flicker which is detectable, the frequency of the sound being 280 p.p.s. When the frequency of the flicker lies between these extreme cases, however, the effect becomes irritating to the auditory nerve, which is at one moment stimulated very strongly and at the next moment allowed to recover, but before recovery is complete is stimulated again, and so on. At frequencies in the neighbourhood of 500 p.p.s., Helmholtz found that the maximum irritation or discord occurred at a beat-frequency of 33 per second, the effect disappearing altogether when this frequency reached about 80 per second. The harshness or discord produced by two notes of different frequency depends upon (1) the beat-frequency, (2) the quality of the sounds, (3) the absolute pitch or sound-frequency of the notes. With regard to (2) it is not difficult to understand that beats of irritating frequency may be produced when two tones, rich in harmonics, are sounded together. Although the fundamental tones are separated by an interval too great to produce discord, it frequently happens that certain overtones of one sound may 'beat' with other overtones of the second sound. As an example of this we may refer to Tyndall's analysis * of the tones emitted from two strings, the ratios of whose fundamental frequencies are 1:2, 2:3, 3:4, 4:5, and 5:6, the fundamental tone of one string being always 264 p.p.s. The overtones form a complete harmonic series in each case. The lowest beat-frequencies between any pair of partials are for these combinations 264, 132, 88, 66, and 53 respectively, indicating that there is no serious tendency to discord until the ratio of the fundamental tones reaches 4:5. A tuning-fork vibrating with a small amplitude at a frequency of 250 p.p.s. will harmonise with another fork vibrating at any frequency above 350 p.p.s., the beat-frequency being sufficiently high and tuning-forks free from harmonics. Similar notes on a stringed instrument, a piano for example, would be discordant, for the third harmonic (750 p.p.s.) of the first tone would 'beat' at an irritating frequency (50 p.p.s) with the second harmonic (700 p.p.s.) of the second tone. Although the harmony or discord due to two notes depends mainly on their frequency-ratio and the harmonics present, it is also dependent on the absolute frequency. Thus 200 and 300 are concordant whatever harmonics are present; but 100 and 150, the same ratio, are discordant if the first four harmonics in each are present. In general, the lower the notes the simpler must be the ratio, or the

* *Sound*, p. 302.

fewer the harmonics, to avoid discord. At very high frequencies beats are rarely noticed. The choice of chords and musical intervals must recognise the foregoing principles. A consideration of music and harmony would, however, lead us too far astray. The reader is recommended to consult more specialised text-books on these subjects.

Structure of the Ear – It is beyond the scope of a book of this nature to deal at all adequately with a complex anatomical subject such as the detailed structure of the ear. For further information a standard textbook of anatomy or physiology should be consulted. The following remarks must be looked upon as a guide rather than a description. In human beings the 'organ of hearing' is divided anatomically into three principal parts: the outer, the middle, and the inner ear. The general arrangement is shown in fig. 111A. *The outer ear* consists of the relatively large external 'shell' or collector which leads into the auditory canal and thence to the drum ('membrana tympani'). In the lower animals the external shell is provided with muscles and is very movable, presumably to collect sounds from different directions; whereas in man this function is rudimentary, so that he can hear almost as well with his ear cut off. The auditory canal in the average adult has a length from 2·1 to 2·6 cm., a volume of about 1 c.c., and an opening of 0·3 to 0·5 cm.^{2}. *The middle ear* (or tympanum) is separated from the outer canal by the drum, a thin layer of fibrous tissue covered with skin externally and with mucous membrane internally. Its average dimensions are 1·00 cm. major axis (horizontal), 0·85 cm. minor axis (vertical), the area being approximately 0·65 cm.^{2}. Then comes the chain of three bones : the hammer (malleus, 23 mg.), the anvil (incus, 25 mg.), and the stirrup (stapes, 3 mg.), which stretch across the middle ear and connect the outer auditory canal to the inner ear. The 'handle' of the hammer is attached eccentrically to the inner side of the drum-skin, dragging the latter slightly inwards. The middle ear leads *via* the Eustachian tube to the throat, the tube being normally closed but opening in the act of swallowing. The object of this connection is clearly

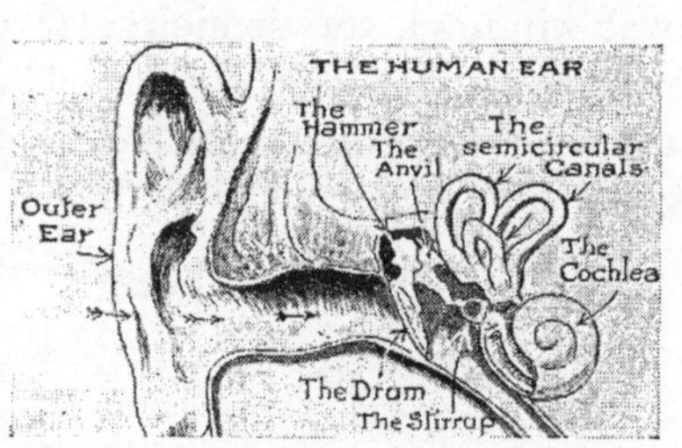

FIG. 111A
(From Bragg, *World of Sound*)

to provide a means of equalising the pressure on opposite sides of the drum-skin, thereby preventing damage due to possible large variations of pressure. The inner wall of the middle ear presents two openings known as the oval and the round windows. The former opens into the 'vestibule' of the inner ear and is normally closed by the base of the 'stirrup' bone and a marginal membrane. The round window leading into the 'scala tympani' of the cochlea is closed by an elastic membrane. *The inner ear* (the labyrinth) is highly complex, and contains the nerve-endings which respond to the vibrations of sound. It connects to the middle ear *via* the oval window (end of stirrup) and the round window. The labyrinth is a bony cavity (200 mm.^{3}) containing a membranous labyrinth of similar form filled and surrounded by lymph (*endo*lymph inside and *peri*lymph outside)—a fluid like water. The cavity consists of three parts: the vestibule (at the oval window), the semicircular canals (the organ of 'balance'), and the cochlea (the 'snail shell' containing the hearing processes and nerve-endings). *The cochlea* is a remarkable structure—a spiral tube of bone (20 to 30 mm. long), divided along its length

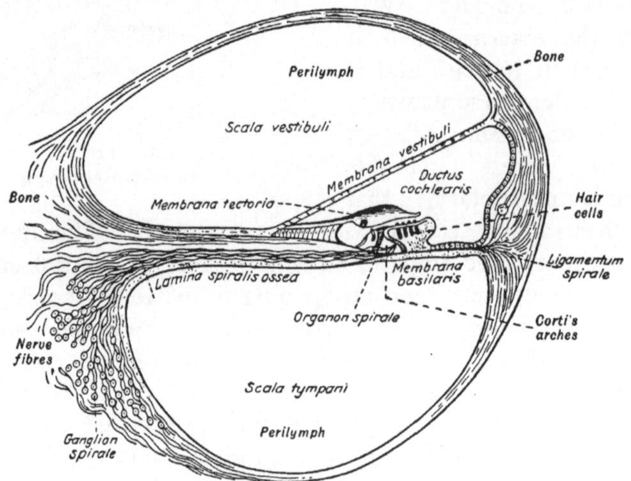

FIG. 111B—Section of one turn of Cochlea
(Geiger and Scheele, *Handbuch der Physik*)

by the 'lamina spiralis' and the *basilar membrane* into two cavities, the 'scala tympani' and the 'scala vestibuli,' which intercommunicate at the tip of the spiral at the end of the lamina spiralis. A section of one turn of the cochlea is shown in fig. 111B. The membranous labyrinth, almost triangular in section, is formed by the 'membrana vestibuli' and 'membrana basilaris.' The latter is kept in tension by the fibrous structure ending in the spiral ligament (see fig. 111c). Forming an arch, of triangular section, over a portion of the basilar membrane are the '*rods of Corti*' which support the hair-like cells and nerve-endings over which hangs the membrane tectoria. All these structures gradually diminish in size as the cochlea winds from its base near the oval window to its tip.

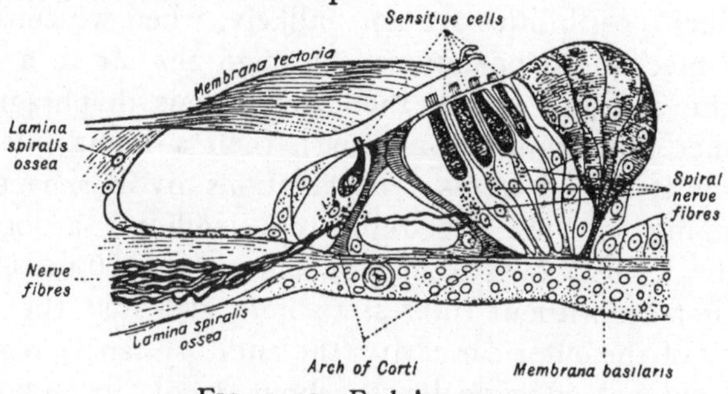

FIG. 111C—End Apparatus
(Geiger and Scheele, *Handbuch der Physik*)

Theories of Hearing - The progress of a sound wave down the auditory canal, affecting the drum-skin and the three bones and passing along the cochlea over the basilar membrane, can readily be imagined, but the actual conversion of sound energy into nervous energy is not so easy to follow. The gradation in size of the structures from the base to the tip of the cochlea forms a strong temptation to suggest that they were designed to respond sympathetically to vibrations of all frequencies within the range of audibility. It was thought at first that the rods of Corti served this purpose; but more recently the basilar membrane, with its fibres stretched radially like a harp throughout the spiral, has received special attention. This view of the matter is generally spoken of as *Helmholtz's Theory of Audition*,* but there are earlier references to a similar theory. According to this theory, the sound wave advancing along the cochlea excites into resonance that particular radial strip of the basilar membrane which is in tune with it. The vibration thus set up is communicated directly to the corresponding *hair cells* and the spiral nerve fibres which are supported between the rods of Corti and the basilar membrane (see fig. 111c). Any theory of the action of the ear must, of course, account for all the experimental observations to which we have referred above. The facts of Ohm's law, that the ear can resolve a complex tone into its harmonic

* *Sensations of Tone.*

components; the fact that a musician can name the notes forming a chord struck on the piano; the wide range of intensity and frequency over which the ear is sensitive; binaural perception of direction, and many other, as yet, obscure phenomena, must all be explained. The form of the curve of sensitiveness of the ear with variation of frequency (see fig. 110 (*b*)) suggests that the mechanism is either multi-resonant or considerably damped, or both. Such possibilities are not unlikely, when we consider the chain of mechanical operations; through the air in a resonant cavity (the auditory canal), the membranous diaphragm, three solid bones (not rigidly connected), then a liquid-filled cavity containing stretched fibres. Observations by Barkhausen * on the damping of the 'ear-resonators' indicate a logarithmic decrement $\Delta=0{\cdot}12$ over a range of frequencies 500 to 1500 p.p.s. It may not be without interest to note also that the resonant frequency of the outer ear cavity (the auditory canal), regarded as a closed pipe of effective length about 3 cm., approximates to 2700 p.p.s., which coincides approximately with the frequency at which the ear is found to be most sensitive (see fig. 110). The behaviour of the three bones is an interesting mechanical study in itself. There appears to be an attempt here to reduce the displacement and increase the pressure-amplitude in transferring the vibrations of the drum (in air) to the liquid-filled cochlea, very low- or very high-frequency vibrations would not pass well through such a system. Returning to the far more complicated problem of the function of the cochlea and basilar membrane, various theories have been advanced as alternatives to that of Helmholtz.† An objection which has been raised to the resonance theory is that there is not enough difference between the lengths of the fibres of the basilar membrane to account for the range of sounds which it is possible to perceive and analyse. It has therefore been suggested by Ewald ‡ that the whole basilar membrane vibrates in the form of stationary waves, so that a nodal pattern is impressed on the nerve-endings, which are stimulated, presumably, at the antinodes.

Another theory which avoids the difficulty altogether is due to

* *Phys. Zeits.*, **25**, p. 537, 1924.

† See Fletcher, *Journ. Frank. Inst.*, p. 312, 1923. H. E. Roef, *Phil. Mag.*, **43**, p. 349, 1922. A comprehensive account of the ear and Theories of Hearing, is given in Geiger and Scheele's *Handbuch der Physik*, **8**, 'Akustik,' p. 477, where numerous references to publications will be found. See also H. Fletcher, *Speech and Hearing*.

‡ Pfluger's *Archiv.*, **93**, p. 485, 1903.

Rinn (1865), and assumes that the basilar membrane vibrates as a whole. The impulses ascending the whole auditory nerve differ in quality, reproducing physiologically the character of the primary physical disturbances. On this view, the *whole* process of analysis is relegated to the auditory centre in the brain. The complicated structure of the cochlea would seem, however, to be quite unnecessary for such a purpose.

None of the theories of hearing hitherto proposed can be regarded as completely satisfactory, but the resonance theory certainly appears to hold most promise.

SENSITIVE FLAMES AND JETS

The upper limit of frequency at which the ear becomes useless as a means of detection is in the neighbourhood of 20,000 p.p.s. At frequencies much lower than this, however, it is insensitive, and alternative methods of detection of the vibrations are preferable, if not indeed necessary. Perhaps the oldest and best-known receiver or detector of high-frequency sounds is the sensitive flame which is generally associated with the name of Tyndall who, subsequently to its discovery by Leconte,* brought it to a high degree of perfection. Those interested in the subject should read Tyndall's detailed description † of sensitive flames and jets as used to detect high-frequency sounds. As the pressure of the gas supply to a flame is gradually increased, the flame increases in size, but if the pressure exceeds a certain critical value the flame shortens and flares. This effect is most marked if the gas issues through a pinhole orifice, in which case the flame is long and narrow until it begins to flare. If such a flame be adjusted to the critical pressure, just on the point of flaring, it will be found to be sensitive to external sounds, particularly to those of high pitch. Tyndall found that the taller the flame the more sensitive it becomes; he succeeded in obtaining a flame 24 in. high, 'the slightest tap on a distant anvil' reducing its height to 7 in. Such flames require rather greater gas pressures than the gas companies supply in the mains; but it is possible to operate shorter flames of somewhat less sensitivity at ordinary pressures, provided the diameter of the jet is suitably chosen.‡ Any high-frequency sound, the rattle of a bunch of keys or coins, a hiss or a sharp metallic sound are equally effective. Tyndall

* *Phil. Mag.*, **15**, p. 235, 1858. See also Barrett, *Phil. Mag.*, March 1867.

† *Sound*, p. 230.

‡ See, for example, S. R. Humby, *Proc. Phys. Soc.*, **39**, p. 435, 1927.

showed that the effects are practically the same if *un-ignited* jets of coal gas, carbon dioxide, hydrogen, or air are used, such jets being rendered visible by smoke. 'Smoke jets' of this character were found to exhibit a sensitiveness in some cases even greater than that of the flames. Another method of making the sensitiveness of an air-jet visible is to allow it to play on a flame which merely serves as an indicator. A somewhat different type of sensitive flame is one in which the coal gas is ignited on the upper side of a wire gauze held at a suitable distance from the jet. Rayleigh * describes a further modification which requires smaller gas pressures and is sensitive to sounds of lower frequency also. Ridout † found that fishtail flames, formed by the union at a small angle of two similar jets (as in an acetylene cycle-lamp), exhibited directional properties with respect to the source of high-frequency sound. The response was zero when the direction of the sound was at right angles to the line of the jets (*i.e.* in the plane of the flame). Screening a long sensitive flame from the incident sound waves from a high-frequency source, it is a simple matter to demonstrate that the tip of the jet is the seat of the sensitivity of the flame. This sensitiveness is undoubtedly due to the instability accompanying vortex motion at the jet. The phenomenon is very complicated and is imperfectly understood. That it is due in some way to the viscous drag on the moving jet of the fluid as it leaves the orifice cannot be doubted. Rayleigh's view of the matter ‡ is as follows: "At the root of the jet, just after it issues from the nozzle, there is a near approach to discontinuous motion and a high degree of instability. If a disturbance of sufficient intensity and of suitable period have access, the regular motion is lost and cannot afterwards be recovered. But the instability has a very short duration in which to produce its effect. Under the influence of viscosity the changes of velocity become more gradual, and the instability decreases rapidly if it does not disappear altogether. Thus, if the disturbance be insufficient to cause disintegration during the brief period of instability, the jet may behave very much as though it had not been disturbed at all, and may reach the full development observed in long flames and smoke-jets. This temporary character of the instability is a second feature differentiating strongly these jets from those of Savart, in which capillarity has an unlimited time of action." E. G. Richardson § has shown

* *Sound*, **2**, p. 402.

† *Nature*, **18**, p. 604, 1878.

‡ *Sound*, **2**, p. 408.

§ *Nature*, **116**, 171, 1925.

that the relation between the velocity V and the length l of a jet of water, in a tank of water, when turbulence sets in is approximately a hyperbola. In a jet issuing from a circular nozzle vortices are formed having a frequency N given by V/ND = constant, where D is a linear quantity depending on the diameter of the nozzle. On this view a jet should show a maximum response when the incident sound has a frequency N corresponding to the particular velocity V of the jet, and on account of the hyperbolic relation Vl = constant, the change of length over a certain small range of jet velocity might be very great. The frequency-velocity relation suggests that the higher the velocity of the jet the higher the frequency to which it gives maximum response. Humby * refers to sensitive flames responding to particular frequencies, but gives no data relative to the dimensions of the jets. He states, however, that the frequencies of response were unaffected by small changes of gas pressure, although the sensitiveness fell off rapidly as the pressure was lowered from the critical stage at which roaring begins.

Sensitive flames may be used as indicators of vibrations in air at practically all frequencies up to, and possibly beyond, 100,000 p.p.s. A quartz rod oscillating in a Pierce circuit (see p. 145) at this frequency provokes a vigorous response. With such an arrangement almost all the optical analogies between sound and light waves can be readily investigated. In metrical work, a suggestion made by W. E. Benton † is worth remembering, viz. that a small gas-pressure governor as used by gas companies would keep the jet-pressure constant within 2 per cent. for a 100 per cent. increase in the supply pressure. The chief difficulty in the practical use of sensitive flames is the maintenance of a *constant* gas pressure.

FACTORS GOVERNING THE EFFICIENCY OF SOUND RECEIVERS

Before proceeding to deal in detail with other forms of sound receiver it may be well to consider certain features which have an important bearing on the 'efficiency' of receivers of all types.‡ A sound receiver may, in general, be regarded as a device for converting the energy of wave-motion in the medium into some

* *Loc. cit.* † *Proc. Phys. Soc.*, **39**, p. 443, 1927.

‡ See F. Aigner, *Unterwasserschalltechnik*, p. 138 (Berlin, 1922); and Geiger and Scheele, *Handbuch der Physik*, **8**, 'Akustik,' pp. 117 and 544; and H. Lichte, *Phys. Zeits.*, **18**, p. 393, 1917, and **19**, p. 17, 1918.

other form of mechanical, thermal, or electrical energy. In some cases, to which we shall refer, the 'receiver' also acts as an amplifier, either of displacement or pressure, but for the moment we are not directly concerned with this aspect of the matter. In the first place, we shall regard the receiver as represented by a mass controlled by a spring and opposed by a certain resistance to motion (damping). This system is subjected to an alternating field of force. The vibrations are therefore represented by

$$m\ddot{x}+r\dot{x}+sx=\mathrm{F}\cos pt \quad . \quad . \quad . \quad (1)$$

The solution of this equation and its physical significance have already been discussed (see p. 37, etc.). In every sound receiver, of whatever type, the damping factor r in this equation is composite. In the simplest case, which we shall now consider, it consists of two parts: (*a*) the *external damping factor* r_e, which represents the loss of energy, from the vibrating element of the receiver, to the medium, *i.e.* the damping due to re-radiation; and (*b*) the *internal damping factor* r_i, representing the energy absorbed by the receiver. If therefore we write

$$r=r_e+r_i \quad \text{and} \quad \frac{r_e+r_i}{2m}=k, \quad \frac{s}{m}=n^2 \quad \text{and} \quad \frac{\mathrm{F}}{m}=f \quad . \quad (1a)$$

we find (as on pp. 37 and 39) that

$$x=\frac{f\sin\epsilon}{2kp}\cos(pt-\epsilon),$$

$$x_{\max}=f/[(n^2-p^2)^2+4k^2p^2]^{1/2} \quad . \quad . \quad (2)$$

and

$$\dot{x}_{\max}=px_{\max}=pf/[(n^2-p^2)^2+4k^2p^2]^{1/2} \quad . \quad . \quad (3)$$

Now the mean power E absorbed in internal damping is $\frac{1}{2}r_i\dot{x}^2_{\max}$, that is

$$\mathrm{E}=\frac{p^2f^2r_i}{2[(n^2-p^2)^2+4k^2p^2]} \quad . \quad . \quad . \quad (4)$$

A small receiver absorbs energy from an area of the wave-front greater than the actual area exposed to the wave. As we have already shown, in the case of resonant receivers, this effective area S_1 may exceed greatly the actual area S. In the present case, the energy passing per second through unit area normal to the direction of propagation (*i.e.* the *intensity* of the sound) is given by $\mathrm{I}=\frac{1}{2}\frac{\mathrm{P}^2}{\rho c}$ (see p. 54) where P is the pressure-amplitude. The

energy passing per second through an area S_1 is therefore $\frac{1}{2}\frac{P^2S_1}{\rho c}$.
But this is the power absorbed by the resonator of effective area S_1; that is

$$E=\frac{1}{2}\frac{P^2}{\rho c}S_1 \quad \text{whence} \quad S_1=\frac{2\rho cE}{P^2} \quad . \quad . \quad (5a)$$

or

$$S_1=\frac{\rho c p^2 f^2 r_i}{P^2[(n^2-p^2)^2+4k^2p^2]} \quad . \quad . \quad . \quad (5b)$$

Resonant Receivers – If we write $n=p$ in the above expressions, we obtain the conditions for tuned receivers. Thus the power E absorbed by the resonator is from (4)

$$E=\frac{f^2r_i}{8k^2}=\frac{F^2r_i}{2(r_e+r_i)^2} \quad . \quad . \quad . \quad (6)$$

It will be observed that this quantity E *has a maximum value* when

$$r_e=r_i\ (=r' \text{ say}) \quad . \quad . \quad . \quad (7)$$

whence

$$\left.\begin{aligned} E_{max}&=\frac{F^2}{8r'} \quad \text{or} \quad \frac{P^2S^2}{4r} \\ \text{Similarly} \qquad x_{max}&=\frac{F}{2nr'} \quad \text{or} \quad \frac{PS}{nr} \\ \text{and} \qquad \dot{x}_{max}&=\frac{F}{2r'} \quad \text{or} \quad \frac{PS}{r} \end{aligned}\right\} \quad . \quad . \quad . \quad (8)$$

where r the total damping $=2\times$(internal or external damping), and S is the actual area of the resonator.

These results are very important in the design of a good receiver. They indicate, for resonant receivers, that the rate of absorption of energy is a maximum when two conditions are fulfilled, viz.:—

(*a*) *when the internal damping r_i is as small as possible*, and
(*b*) *when the external damping r_e is equal to the internal damping r_i—that is, when the energy absorbed is equal to the energy radiated.*

These conclusions, expressed in (6), (7), and (8), may profitably be compared with the analogous electrical relations.

It appears necessary, therefore, in the design of a resonant receiver to ensure first of all that the internal damping is reduced to a minimum, and then to adjust the dimensions so that the damping condition (*b*) is satisfied.

Effective Area of a Resonant Receiver – At resonance the effective area S_1 of the receiver may be derived from equation (5*b*) by substitution of the values of f and k from equation (1*a*) and writing $n=p$, thus

$$S_{1max}=\frac{\rho c F^2}{4P^2 r'}, \text{ or } \frac{\rho c S^2}{4r'} \text{ (since } F=PS) \quad . \quad (9)$$

where S is the *actual* area of the receiver. It has been shown by Rayleigh * that the radiation resistance of a small surface vibrating with uniform amplitude and radiating in *all* directions is given by

$$r=\pi\rho c S^2/\lambda^2 \quad . \quad . \quad . \quad . \quad (10)$$

Substituting this value for r' we obtain

$$S_1=\lambda^2/4\pi \quad . \quad . \quad . \quad . \quad (11)$$

By a similar process, commencing at equation (4), we may show that a resonator with no internal damping ($r_i=0$), *i.e.* one which re-radiates *all* the energy which it receives, obtains this energy from an effective area

$$S_2=\lambda^2/\pi \quad . \quad . \quad . \quad . \quad (12)$$

a result to which we have already referred (see p. 190). In any practical form of resonator some internal damping is inevitable, in which case the most efficient receiver possible has an effective area given by equation (11). For example, if a resonator of frequency N=1000 p.p.s. (corresponding to a wave-length $\lambda=33$ cm. in air) and an actual area S of 0·6 cm.², the *maximum* effective area S_1 would be $\lambda^2/4\pi=87$ cm.² and *the displacement-amplification factor* $S_1/S=145$. The volume v of this resonator, calculated from $N=\frac{c}{2\pi}\sqrt{\frac{D}{v}}$ (equation 13, p. 186), would be 33 c.c. Let us now return to consider the practical importance of equations (6), (7), and (8), which require the external damping to be equal to the internal damping and *both* to be as small as possible for maximum response. Suppose we require to determine the best form and size of Helmholtz resonator for use at a particular frequency N, which we may therefore regard as fixed. After choosing the type which has least internal damping, *i.e.* making r_i as small as possible, we proceed to determine the best

* *Sound*, **2**, p. 165.

area S of neck, varying the volume, of course, in accordance with S to ensure the same frequency. From equation (8) we see that the maximum amplitude in the neck is given by $x_{max}=F/nr$ or, since F is proportional to S and (from equation (10)) r is proportional to S^2, x_{max} is inversely proportional to S. From this point of view, therefore, it is desirable to *reduce* the area of the neck, and consequently the volume of the resonator as much as possible. But the internal damping r_i, due to viscosity and eddies, increases rapidly as the size of the neck diminishes, so we require on this account as *large* a neck as possible. There is consequently a *best area of neck* which, as we have shown, must be determined by the equality of external and internal damping. The exact value of this area can be calculated in a particular case when the law of internal damping, which gives r_i in terms of S, is known.

Presence of a Detector – This brings us to a second point of importance. It is usually, but not always, necessary to introduce into the receiver some form of detector, which itself absorbs energy and thereby increases the damping. In such a case, which represents by far the greater proportion of resonant receivers used in practice, *the energy consumed by the detector may be regarded as useful*, whilst that lost in viscosity and eddies is wasteful. Consequently we must write $r_i=r_u+r_w$, where r_u is the useful damping in the detector and r_w is the wasteful damping. Equation (6) for a resonant receiver must consequently take the form

$$E=\frac{F^2 r_u}{2(r_u+r_w+r_e)^2} \quad . \quad . \quad . \quad (13)$$

Differentiating this with respect to r_u and equating to zero, we obtain

$$r_u=r_w+r_e \quad . \quad . \quad . \quad . \quad (14)$$

as *the condition for maximum overall mechanical efficiency.* The *internal mechanical efficiency* in this case is $r_u/(r_u+r_w)$, and the overall mechanical efficiency is $r_u/(r_u+r_w+r_e)$, which may reach a maximum value of 50 per cent. when $r_u=(r_w+r_e)$. Returning again to the Helmholtz resonator illustration, when the particular form of detector has been chosen, for example, a hot-wire microphone, and its damping r_u determined, the size of the neck should be such that the damping due to radiation r_e and viscosity r_w should together be equal to that of the microphone. To take another illustration, the damping of a small resonant diaphragm in water varies in proportion to the area, the amplitude being theoretically a maximum when the energy radiated is equal to the energy absorbed internally (friction and vibration of supports, etc.). If, in addition, some form of detector, such as a microphone or an electro-magnetic device, be attached to the diaphragm, the *best size of diaphragm* will be larger than before, in order that the radiation (and friction) damping may be increased to equality with the damping of the detector. In all cases it is desirable in the first instance to reduce the wasteful damping to a minimum.

Untuned Receivers of Small Area - The effective area S_u of an untuned receiver is given by equation (5*b*). Writing S_t as the effective area of a *tuned* receiver, we have therefore

$$\frac{S_u}{S_t}=\frac{E_u}{E_t}=\frac{4k^2p^2}{(n^2-p^2)^2+4k^2p^2},$$

which was obtained on p. 43, equation (15). This ratio falls to $\frac{1}{2}$ when $(n^2-p^2)^2=4k^2p^2$—that is, when $p^2/n=1\pm k/n$ (approx.). In the case where k is small, only a very small departure from exact tuning of the receiver will result in a drop of 50 per cent. in the energy received. If the frequency-ratio departs seriously from unity the sensitiveness of the untuned receiver will be almost negligible compared with that which is tuned.

The Receiver as an Obstacle to Sound Waves - A point of some importance which has hitherto been disregarded relates to the behaviour of the sound receiver as an obstacle which reflects or scatters the energy incident upon it. As we have seen on p. 282 the energy scattered from a small obstacle varies directly as the square of its volume and inversely as the fourth power of the length of the waves incident upon it. A small diaphragm telephone receiver, for instance, in air will reject a large proportion of the energy incident upon it, in spite of the fact that its diaphragm may be tuned. The effect on the sound field of the vibration of the solid surface of the diaphragm is relatively negligible (see also p. 423 on Sound Measurement).

Receivers of Large Area - If the dimensions of the receiver are large, *i.e.* of the order of several wave-lengths, the *effective area* S_1 approximates to the actual area S. The power available in this case is therefore

$$E=\frac{1}{2}\frac{P^2}{\rho c}\,.\,S=\frac{1}{2}r\dot{x}^2_{max};$$

and at resonance, when $\dot{x}^2_{max}=F^2/r^2$, we find $r=\rho c S$, which is equal to the radiation resistance of a corresponding area S of the medium. In this case, therefore, the resonator may absorb the *whole* of the energy incident upon it. Unlike the small resonator, it does not draw energy from a relatively large surrounding area.

Pressure and Displacement Receivers - The energy in the sound wave may be detected in as many ways as there are changing physical characteristics in the sound wave: particle displacement, velocity or acceleration, pressure and density variations, temperature changes, and so on. Of these, pressure and displacement variations are of most frequent application. Of necessity, a 'pressure' receiver must also require a little displacement for its satisfactory operation, or there could be no absorption of energy (for energy absorbed is given by $\frac{1}{2}P_{max}\dot{x}_{max}$). Similarly a so-called displacement or velocity receiver also requires a small pressure variation. Neither a 'pure' pressure receiver nor a pure displacement receiver could therefore be *efficient*, for both would serve as total reflectors of the incident energy. On this basis, therefore, they could have no practical value. A pressure receiver absorbing no energy implies a completely sound-resistant body (analogous to an electrical insulator), whilst a displacement receiver absorbing no energy would behave as a completely sound-conducting body (analogous to an electrical conductor of infinite conductivity in which no potential difference is possible). The distinction between pressure and displacement receivers is, however, a useful one, and in some respects is analogous to that between voltmeters and ammeters used in electrical measurements. A voltmeter is essentially an electrical pressure-measuring device although it requires a small current, whilst an ammeter is primarily a current-measuring instrument requiring a small potential difference across its coils. The same energy and damping conditions which we have just been considering apply, however, to both types. Outstanding examples of these types are readily found. As a typical displacement receiver we may instance the Helmholtz resonator, fitted with a sensitive device for detecting the motion of the air in the neck where the pressure-fluctuation is relatively small (but not zero). A good example of pressure receiver is the Langevin piezo-electric quartz disc (see p. 147) used for generating and receiving sounds of high frequency under water. The pressure-fluctuations at the surface of the receiver in the water compress the quartz slightly, and thereby produce corresponding fluctuations of electrical charge which can be detected by suitable apparatus. This receiver depends *essentially* on pressure fluctuation, but the elastic displacement of the quartz cannot be disregarded. Various types of diaphragm receivers are generally classed as pressure receivers, although the displacement amplitude may be considerable. If we employ the same receiver in air and in water it may serve as a pressure receiver in the former and a displacement receiver in the latter, because the acoustic resistance of the water is about 3500 times that of air. Similarly the same receiver in a medium of high acoustic resistance will offer a resistance which varies with the frequency of the incident sound wave—a diaphragm, for example, changing from a displacement receiver to a pressure receiver as the frequency is decreased. In air, with its low acoustic resistance ($\rho c=40$), all receivers having solid surfaces (diaphragms, piezo-electric crystals, etc.) may be regarded as pressure receivers, and the above considerations do not arise. In water, however, a medium of much higher acoustic resistance, a diaphragm may be regarded either as a pressure or a displacement receiver according to the frequency of the incident sound relative to that of the diaphragm.

Mechanical Transformers. The Lever Principle – Just as it is often necessary to transform electrical power from high to low voltage, with corresponding current changes, so it is frequently necessary also to transform mechanical power from high to low pressure and *vice versa*. For example, in transforming sound energy from water to air, or conversely, we are confronted with a serious difficulty, viz. that the radiation damping factor (proportional to $\rho c S^2$) is, for a given area of vibrating surface, far greater in water than in air ($\rho c=40$ for air, and $1{\cdot}5\times10^5$ for water). If, however, we can devise some method of transforming the relatively large pressure fluctuation (of small displacement amplitude) in water, to large displacement fluctuation (of small pressure amplitude) in air, the sound energy may become more readily available. Examples of mechanical vibration transformers are to be found in the gramophone sound-box and in various types of sound measuring and 'amplifying' devices. For such purposes, particularly for use in underwater sound generators and receivers, Hahnemann * has devised a *mechanical lever* of a novel type. An ordinary lever, with loose pivots and links, would be useless as a means of transmitting vibrations of even

* W. Hahnemann and H. Hecht, *Phys. Zeits.*, p. 187, 1920; and *Inst. Radio. Eng.*, **11**, p. 9, 1920. See also F. Aigner, *Unterwasserschalltechnik*, p. 133 (Berlin, 1920).

moderate frequencies. In a vibrating system, such as a diaphragm or a tuning-fork, the 'mass' and the 'spring' of the vibrator are inseparable, an alteration of one necessitating an alteration of the other. The case is analogous to an inductance coil having self-capacity and consequently its own free period of oscillation. Hahnemann prefers to separate 'mass' and 'spring' for tuning purposes, just as it is more convenient to separate inductance and capacity in tuning an electrical circuit. The principle of Hahnemann's 'lever'—the mechanical analogue of the electrical transformer—is illustrated in fig. 112 (*a*). Two rigid masses m_1 and m_2 are connected by a spring of length l and strength s. At resonance the system will vibrate about its centre of gravity O, this point being undisplaced and functioning as a node at which the system may be supported. When the system vibrates longitudinally it is evident that m_1 and m_2 move in opposite directions, so as to make $m_1\ddot{x}_1+m_2\ddot{x}_2=0$, where $\ddot{x}_1$ and $\ddot{x}_2$ are the accelerations of m_1 and m_2 respectively. Neglecting the sign of these accelerations we find

$$\frac{m_1}{m_2}=\frac{\ddot{x}_2}{\ddot{x}_1} \quad \text{similarly} \quad \frac{m_1}{m_2}=\frac{\dot{x}_2}{\dot{x}_1}=\frac{x_2}{x_1} \quad . \quad . \qquad (1)$$

that is, the *accelerations, velocities, and displacements are inversely proportional to the respective masses.* The relative amplitudes are consequently

$$\frac{a_1}{a_2}=\frac{m_2}{m_1}=\frac{l_1}{l_2} \quad . \quad . \qquad (2)$$

where l_1 and l_2 are the lengths into which the spring is divided by the point O, the centre of gravity of the system. The relative energies of the two masses ($\frac{1}{2} . mv^2$) are also inversely proportional to the masses. These are very important facts, applicable to all forms of resonant sound sources and receivers. The frequency of vibration of the system is obviously that of the mass m_1 on a spring of length l_1 or the mass m_2 on a spring of length l_2. Consequently

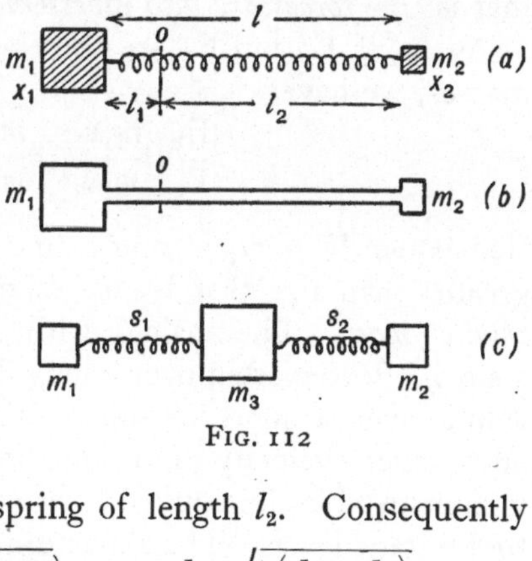

FIG. 112

$$N=\frac{1}{2\pi}\sqrt{\frac{sl/l_1}{m_1}}=\frac{1}{2\pi}\sqrt{\frac{s}{m_1}\left(1+\frac{m_1}{m_2}\right)} \quad \text{or} \quad \frac{1}{2\pi}\sqrt{s\left(\frac{1}{m_1}+\frac{1}{m_2}\right)} \qquad (3)$$

If one of the masses, m_1 say, is very great compared with the other, the frequency N becomes $\frac{1}{2\pi}\sqrt{\frac{s}{m_2}}$ and the point O practically coincides with m_1, so that $l_1=l$ and $l_2=0$, the motion of m_1 being extremely small. A practical attempt to construct such a system at 1000 p.p.s., for example, soon shows the necessity of replacing the spiral spring of fig. 112 (*a*) by an arrangement in which the elastic element combines the smallest possible mass with the greatest elastic strength, *e.g.* a longitudinally strained rod or tube as shown in fig. 112 (*b*). A stiff diaphragm with a central load serves a similar purpose. In the case of the rod of elasticity E, area of cross-section A and length l, the frequency of the system becomes

$$\mathrm{N}=\frac{1}{2\pi}\sqrt{\frac{\mathrm{EA}}{l}\left(\frac{1}{m_1}+\frac{1}{m_2}\right)} \quad . \quad . \quad . \quad (4)$$

neglecting the mass of the rod. A correction for the latter is easily applied. As we have seen, the system acts as a lever in amplifying or reducing the motion imparted to one of the masses. Similarly, at resonance, if a force $\mathrm{F}_1 \cos nt$ be applied to the mass m_1 we must have

$\mathrm{F}_1 \cos nt=r\dot{x}_1$ (since $m_1\ddot{x}_1+s_1x_1=0$ at resonance), and at m_2 we have

$$\mathrm{F}_2 \cos nt=r\dot{x}_2, \quad \text{whence} \quad \frac{\mathrm{F}_1}{\mathrm{F}_2}=\frac{\dot{x}_1}{\dot{x}_2}=\frac{m_2}{m_1}. \quad . \quad (5)$$

that is, *the forces are also inversely proportional to the masses.*

Again, if E_1 and E_2 are the respective kinetic energies of the masses, we have

$$\frac{\mathrm{E}_1}{\mathrm{E}_2}=\frac{m_1\dot{x}_1^{\,2}}{m_2\dot{x}_2^{\,2}}=\frac{m_2}{m_1} \text{ from (5)} \quad . \quad . \quad . \quad (6)$$

Consequently if m_1 is much smaller than m_2, E_1 will be much greater than E_2—that is, *the energy will reside principally in the smaller mass.* The analogy with the 'static lever' is very close; a small effort exerted over a long distance producing a large force acting over a short distance. The familiar terms 'mechanical advantage,' 'velocity ratio' and 'mechanical efficiency' may, with circumspection, be applied, therefore, to the vibratory as well as to the static lever. The above simple deductions have a very wide practical application in the design of sound sources and receivers. Almost all forms of vibrating system are reducible to this mechanical lever or transformer. An elastic diaphragm mounted on a heavy ring vibrates with a large amplitude and small force, whereas the ring (of large mass m_1) has a negligible amplitude, but may exert a considerable pressure. As we have just shown, the energy of the vibrating system resides principally in the smaller mass, *i.e.* in the diaphragm. A tuning-fork on a heavy base provides a similar illustration. In the above arrangement the 'mechanical coupling' between the two masses is perfect. Circumstances sometimes require, however, that the coupling should be relatively 'loose.' This result can be achieved by means of a mass m_3 which is common to both m_1 and m_2 as in fig. 112 (*c*). Here m_1 and m_3 are the masses of one system, s_1 the connecting elastic element, m_2 and m_3 the masses of the other system, and s_2 the elastic connection. The 'coupling coefficient' k of the two systems is expressed by

$$k^2=\frac{m_1m_2}{(m_1+m_3)(m_2+m_3)},$$

analogous to the coupling coefficient $k^2=\mathrm{M}^2/\mathrm{L}_1\mathrm{L}_2$ of two electrical circuits of mutual inductance M and self-inductances L_1 and L_2. When m_3 is zero we arrive at the simple case considered above, $k^2=1$ where the 'coupling' is 'perfect.' If m_3 is very large compared with m_1 and m_2 the value of k^2 is very small and the coupling said to be 'loose.'

By means of the above principles of the mechanical transformer and the electrical analogy it is possible to deal with somewhat complicated mechanical systems used in sound signalling. Hahnemann has designed on these principles very efficient sound generators and receivers for use under water. A diagram of his submarine electromagnetic transmitter (or receiver) with its accompanying mechanical 'circuit diagram' is shown in fig. 113(*a*). A steel diaphragm on a heavy mounting '*a*' is actuated by a powerful electromagnet, the armatures *b* and *c* of which are connected by elastic tubes or rods *ff*, the magnet alternately attracting and repelling *b* and *c* against the elastic-restoring forces of the tubes. As will be seen, there are two oscillating systems, one of which consists of the masses *b* and *c* with the spring *ff*, the other of the masses *a* and *b* with the spring *ee*. The coupling

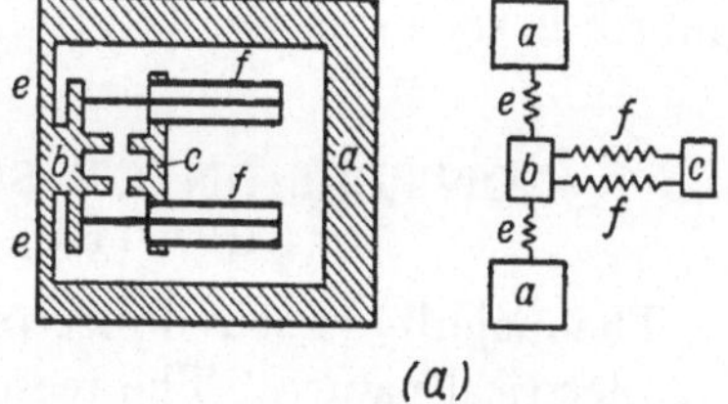

Fig. 113 (*a*)—Sound Transmitter (Hahnemann)

coefficient is determined by the relation between the common mass b and the two others a and c. To obtain the best coupling coefficient and the other desired results it is also necessary to tune the separate systems and to have suitable damping in each

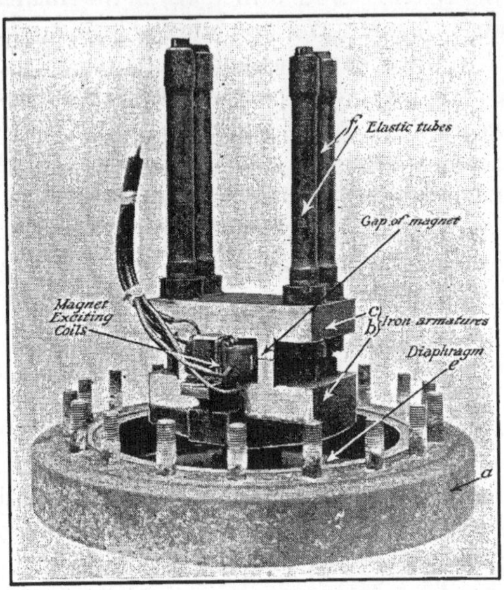

FIG. 113 (*b*)—Sound Transmitter (Hahnemann)

of them in relation to the coupling employed, just as is usually done in electrical circuits. A good electromagnetic transmitter, designed on such principles, forms a satisfactory receiver also, for the lever principle is reversible. A photograph of this transmitter is shown in fig. 113 (*b*).

CONVERSION OF SOUND ENERGY INTO ELECTRICAL ENERGY

The majority of sound receivers in use at the present day are of an electrical nature. The reason is not far to seek, for electrical methods offer a wide variety and flexibility in their application. By electrical methods sounds may be received in one place and reproduced or recorded in another. Electrical amplification offers a means for utilising special forms of sound receiver which would otherwise be too insensitive to be of service. By electrical methods, also, the effects of two or more receivers may be added vectorially, the necessary adjustments of phase being a comparatively simple matter in the electrical circuits. The frequency and intensity of sounds of feeble intensity (or outside the range of audibility) may be determined conveniently by electrical methods. These are sufficient to indicate the many advantages of the conversion of sound energy into electrical energy.

In many cases, with which we shall have to deal, an electrical sound generator is also used as a sound receiver. This reciprocal process is analogous to that in which an electrical machine may be driven mechanically as a dynamo to generate electrical power, or conversely the machine may be supplied with electrical power to drive it as a motor to produce mechanical power. Many forms of electrical sound-apparatus may be regarded either (*a*) as a sound receiver, converting vibratory mechanical motion into alternating electrical power (the dynamo principle) or (*b*) as a sound generator, converting electrical power into mechanical vibratory power (the motor principle). This analogy between sound apparatus and electrical machinery is of more than superficial significance, the principles underlying the one system being very closely applicable in the other.

Some writers * have gone so far as to classify sound receivers as—(*a*) those which convert the energy of sound vibrations *directly* into electrical energy (on the dynamo principle), and (*b*) those in which the sound vibrations control the electrical energy *indirectly*. In the former class, the electrical energy which the receivers generate bears a definite relation to the mechanical energy of vibration; whilst in the latter, the amount of electrical energy 'released' by the vibration may, at any rate in principle, be as large as desired. Outstanding examples of these two types are (*a*) the electromagnetic or electrodynamic receiver (a simple telephone earpiece), and (*b*) the granular carbon microphone. Although this method of classification is 'at first sight' very attractive, there are border-line cases which do not fall readily into either category. For example, certain forms of hot-wire microphone, the condenser microphone, piezo-electric and magnetostriction receivers could be regarded equally well in class (*a*) or class (*b*). We shall, therefore, avoid such a definite classification in what follows, although there is nothing better to offer. Other alternatives which may be suggested classify receivers as resonant and non-resonant, displacement and pressure, atmospheric and submarine, directional and non-directional, and so on. Little is

* See Geiger and Scheele, *Handbuch der Physik*, **8,** 'Akustik,' pp. 551 and 553; article by H. Sell.

gained, however, by such classification, and there is grave danger of repetition.

Electromagnetic Receivers – (Moving Iron Type.) The oldest, simplest, and most familiar of all electrical receivers is the electromagnetic telephone, invented in 1876 by Graham Bell. Numerous modifications are in existence, designed with the object of improving its efficiency or output, but the fundamental principle remains the same. In its most common form,* a thin diaphragm of magnetic material (steel or ferrotype) is clamped rigidly at the edge, whilst the central portion bridges across the pole pieces of a permanent magnet. The diaphragm is usually about 2 in. diameter and ·006 to ·012 in. thick, and is situated with respect to the magnet poles so as to be just not pulled into contact. The pull on the diaphragm is approximately proportional to the square of the magnetic flux B from pole to pole through the diaphragm. Small increases or decreases of flux, due to changes of magnetic reluctance as the diaphragm approaches or recedes from the poles, produce corresponding changes of current in the coils wound on the tips of the magnet poles. The vibrations of the diaphragm on which sound waves are incident therefore induce corresponding variations of current in the windings. The

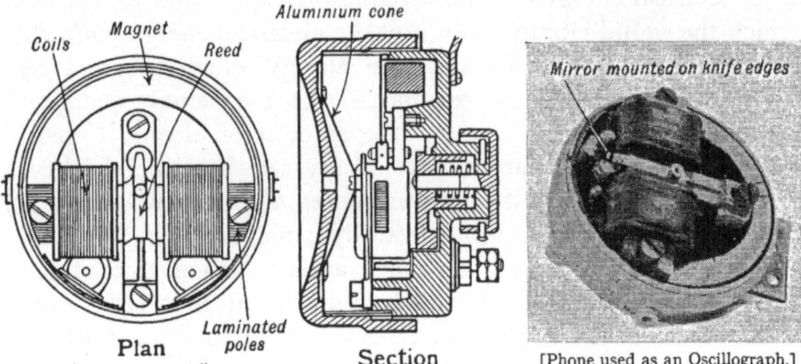

FIG. 114—Brown Reed Telephone
(By courtesy of S. G. Brown & Co.)

sensitiveness is proportional to the product BdB/dH, where B is the magnetic flux passing through the diaphragm. Provided B is not too small and does not approach too near the saturation value, where dB/dH is small, the sensitiveness may be fairly high. The vibrating diaphragm is in certain forms of receiver, notably

* See *Dictionary of Applied Physics*, **2**, p. 843, for description of construction and theory of Electromagnetic Telephone Receivers.

one designed by S. G. Brown,* replaced by a clamped-free reed (see fig. 114). The latter is a strip of soft iron or mild steel clamped firmly at one end, whilst the free end lies across the gap in the magnetic field. The poles of the magnet are adjusted to approach the reed until a critical position is reached, beyond which the stiffness of the reed is suddenly overpowered by the attraction of the magnet. Attached to the reed by its apex is a light aluminium cone which serves to collect the sound energy incident on the receiver. In other forms of receiver an electromagnet is used to provide the steady magnetic flux, independent windings on the pole-tips being employed as before to carry the alternating current developed by the vibration of the diaphragm or the reed. The efficiency of such a receiver may be determined by making a series of measurements of its electrical (motional) impedance at various frequencies near resonance by the method † described on p. 73. Such measurements indicate the amount of electrical power consumed by the receiver, when used as a transmitter, at these frequencies. The difference between two series of observations with the diaphragm or reed, (*a*) free to vibrate and (*b*) clamped, give the amount of mechanical power consumed by the vibrating diaphragm. A convenient approximate view of the behaviour of the diaphragm is to regard it as equivalent to a tuned 'L. C. circuit' shunting a part of the telephone winding.‡ In air, only a small proportion of the power consumed by the diaphragm is radiated as sound, the energy losses due to insufficient clamping of the edges, internal friction, etc., being usually very serious in comparison. The efficiency of an electromagnetic telephone receiver used in air rarely exceeds 1 per cent. at resonance. If, however, the receiver is used under water (the electromagnetic system being mounted on a suitable diaphragm enclosing a watertight cavity), the efficiency may be greatly increased, on account of the increased radiation damping of the water being more nearly comparable with the internal electromagnetic damping of the receiver. Designed on the lines of Hahnemann's electromagnetic transmitter, a receiver may reach a very high efficiency when used under water.

Generally speaking, diaphragm receivers exhibit marked resonance at certain frequencies. When used to receive signals of a definite frequency such resonance is desirable, but when it

* See Patent, Specification 29833/1910.

† See, for example, E. Mallett, *Journ. I.E.E.*, **62**, p. 517, 1924.

‡ See W. H. Eccles, *Handbook of W/T.*, p. 280.

is required that the sound shall be faithfully reproduced special care must be taken in the design to avoid resonance. In such cases it is usually the practice to arrange the diaphragm frequency to be very much higher or very much lower than the frequency-range it is desired to receive. Alternatively the diaphragm should be heavily damped, such damping unfortunately being wasteful. These methods of avoiding resonance necessitate a considerable loss of sensitiveness, but the use of a non-selective valve-amplifier (*e.g.* resistance-capacity type) will, to a large extent, compensate for such loss. The theory of electromagnetic receivers of the diaphragm type for use in air has been developed by R. L. Wegel.* On account of the relatively high iron losses (hysteresis and eddy currents) in such receivers, it is of course essential to laminate all iron sections which have to carry alternating flux. Hysteresis loss is proportional to frequency and eddy current loss to the square of the frequency, consequently it is necessary in a good receiver to use thin laminations of high resistivity iron (silicon-iron or stalloy). At frequencies much above 14,000 p.p.s. the losses become very serious in spite of such precautions—it is fortunate that the ranges of speech and music lie mainly below this frequency. The action of all forms of electromagnetic receivers is reversible, sound being produced by the diaphragm when the coil is supplied with alternating current, and current being generated when sound falls upon the diaphragm.

Electrodynamic Receivers - (Moving Coil Type). A conducting wire or strip of length l carrying a current i in a magnetic field of strength H experiences a force Hli, tending to displace the wire at right angles to its length and to the direction of the field. Conversely, if the wire is displaced in the magnetic field by the same force, a current i will be generated in it. These simple principles are fundamental to the design of electric motors and generators. The *same* principles apply to electrodynamic 'sound motors and generators.' If a coil of wire attached to a vibrating system, such as a diaphragm, lies in a strong magnetic field, a vibration to and fro in the field will generate corresponding alternating e.m.f.'s in the coil; or conversely, alternating currents in the coil will cause it to oscillate at the same frequency in the magnetic field and the diaphragm to which it is attached will generate sound waves. In the form shown in fig. 115 a coil having a large number of turns of fine wire is wound on a thin cylindrical former attached to the vibrating member (a diaphragm or a reed and cone). The coil

* *Journ. Amer. Inst. Elec. Eng.*, **40**, p. 791, 1921.

lies in a corresponding cylindrical gap (with radial field) of a 'pot' electromagnet. The more intense the field the greater the e.m.f. developed in the coil and the greater the *e*lectromagnetic (useful) damping. When used to receive sounds under water the area S of the diaphragm (of predetermined frequency) and the dimensions of the electromagnetic circuits are chosen as far as possible

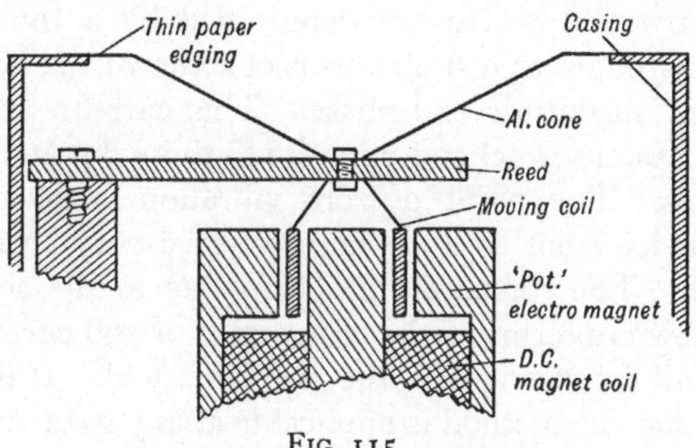

FIG. 115

to make the electromagnetic damping equal to the radiation damping ($\propto \rho c S^2/\lambda^2$), use being made of Hahnemann's lever principle, if required, to secure increased amplitude of vibration of the coil lying in the magnetic field. The efficiency of such a receiver may be determined from measurements of its motional impedance.

The Marconi-Sykes receiver * is an example of the electrodynamic type in which the elastic forces and moving masses are very small and the system is heavily damped (mechanically). It consists essentially of a 'pot' electromagnet with a thin annular coil of aluminium wire suspended in the gap. The coil in one method of mounting is supported on an annular pad of cotton-wool, whilst in another it is held in the gap by a trifilar suspension of silk thread. Such a receiver, although very insensitive, gives remarkably good reproduction of complex vibrations, such as those of ordinary speech. On this account it has been used extensively in 'broadcasting.'

Perhaps the simplest type of electrodynamic receiver or generator is *the vibrating strip receiver of Gerlach and Schottky*,† similar in principle to the Einthoven string galvanometer. A very thin strip (3 μ) of metal foil (aluminium or Al-alloy), a few millimetres wide, lies without appreciable tension in a magnetic field *parallel to the plane of the strip*. When it is set in vibration by the

* H. J. Round, *Wireless World*, **15**, p. 260, Nov. 1924.
† *Phys. Zeits.*, **25**, pp 672 and 675, 1924.

incident sound waves, electrical currents are induced in it. The magnetic field is so strong that the useful (electromagnetic) damping may reach the optimum value, *i.e.* equal to the radiation damping, when the receiver becomes very efficient. This device has been used by Gerlach * as a means of measuring absolute sound intensities (see p. 440). The forces on the strip due to the incident sound waves are determined by a 'null' method. The strip is brought to rest by means of a known auxiliary current, adjustable in magnitude and phase. This current sets up forces in the strip exactly equal and opposite to those due to the incident sound waves. The point of zero vibration is detected by an auxiliary device such as a stethoscope or a very small sensitive microphone. The forces on the strip due to the sound waves are then known in terms of the dimensions of and current through the strip, and the strength of the magnetic field. It is important to observe that the method is applicable at any frequency whether at resonance or not, the sensitiveness reaching a maximum at resonance when small forces produce large displacements.

Induction Type – A novel and extremely efficient form of electrodynamic sound generator and receiver is the *Fessenden oscillator* † for use in sound signalling under water. The apparatus (see fig. 116) consists essentially of an electrical transformer, the *secondary* of which is free to move in a very strong radial magnetic field of strength B. The primary winding is wound on the inner pole face of the electromagnet in close proximity to the secondary. The latter consists of a single turn of copper (a tube, in fact), which has a very low electrical resistance. The current i induced in it is consequently very great, and the force Bli is also very great ($B = 15{,}000$ lines/cm.^{2} approx.). The copper cylinder therefore experiences a large alternating force corresponding to the alternations of current. This alternating force is applied to a steel diaphragm, in contact with the water, the amplitude of which at resonance becomes very large. As a transmitter this apparatus has a sound-output of several kilowatts at a high efficiency. When used as a receiver of signals of the same frequency the process is reversed. The vibrations of the diaphragm with the attached copper cylinder induce e.m.f.'s in the latter which are 'stepped up' in the coil on the pole face. This coil is connected through a valve amplifier circuit to a pair of telephones or an

* *Wiss. Veröfft. d. Siemens-Konzern*, **3**, 139, 1923 ; and **3**, p. 67, 1924.

† See *Electrician*, **59**, p. 484, 1907 ; *Elect. Review*, **60**, p. 369, 1907 ; also W. Meissner, *Zeits. f. Phys.*, **3**, p. 111, 1920.

indicating device. A similar principle is involved in Hewlett's sound generator described on p. 161. In this case the diaphragm itself acts as a single-turn 'winding' of a transformer, inducing currents by its vibration, in adjacent 'pancake' coils.

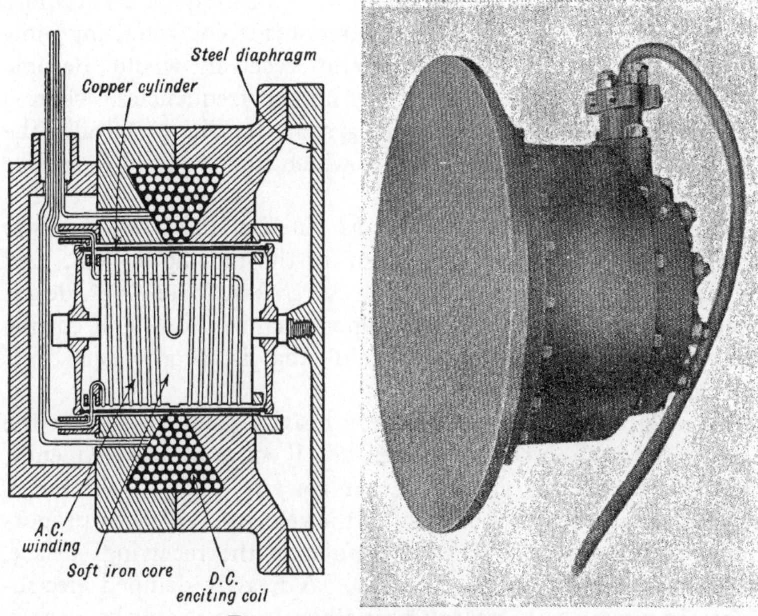

FIG. 116—Fessenden Oscillator

A great advantage of the electrodynamic receiver is the facility with which the mechanical and electrical quantities involved in its design can be adjusted. The frequency, size, and damping of the receiving surface (for example, a diaphragm) can be calculated, and the design of the electromagnetic system and its associated circuit adjusted to give the maximum sensitivity. From the electrical point of view it is superior to the electromagnetic (moving iron) type on account of the entire absence of troubles due to hysteresis and eddy current losses in iron. Care must be taken, of course, to ensure that eddy current losses in non-magnetic conductors are reduced to a minimum. The coil carrying the high-frequency alternating current must therefore be kept at a distance from solid masses of metal; or, as in the case of the Fessenden and Hewlett types of receiver, the vibrating element must form a closely coupled short-circuiting secondary to the primary winding, thus eliminating primary self-induction and eddy currents in neighbouring metallic masses (*e.g.* the iron of the electromagnet). Under such conditions it is possible to deal efficiently with very high frequencies which would be entirely suppressed in an electromagnetic (moving iron) system. On this account a moving coil or strip receiver is more suitable for the faithful reproduction of speech sounds, which require an accurate rendering of the high-frequency components characterising individual speech. An electromagnetic receiver would become decreasingly efficient towards the higher frequencies, whereas the losses in the electrodynamic system remain practically the same up to the highest frequencies which it is required to receive (of the order 10,000 p.p.s.).

Electrodynamic receivers may be divided into three classes according to the method of control of the moving system. All such receivers are essentially '*velocity receivers*,' the e.m.f. developed in the moving coil being proportional to the rate of cutting lines of force in the magnetic field—that is, proportional to $\dot{x}$ the velocity of the coil.

Type (*a*). If the receiver is very heavily damped, so that the equation of motion can be written $r\dot{x}=\mathrm{F}\cos pt$ at *any* frequency, then the velocity is given by $\dot{x}=\mathrm{F}\cos pt/r$—that is, the e.m.f. generated, which is proportional to the velocity, is directly proportional to the pressure $\mathrm{F}\cos pt$ applied to the receiving surface, and is *independent of the frequency*. A heavily damped electrodynamic system is therefore essentially a *pressure* receiver, equally sensitive at all frequencies, but of low sensitivity (since r is large in the expression for $\dot{x}$). Such a system, approximately achieved in the Marconi-Sykes receiver to which we have referred, is consequently of particular value for its faithfulness of reproduction.

Type (*b*). If the vibrating system may be regarded as a 'pure mass m'—that is, the frictional damping and the restoring force are both small and the system is of very low natural frequency, then the equation of motion is sufficiently represented by

$$m\ddot{x}=\mathrm{F}\cos pt \quad \text{or} \quad \dot{x}=\mathrm{F}\sin pt/mp.$$

The electrodynamic receiver will in this case develop an e.m.f. *inversely* proportional to the frequency of the incident sound.

Type (*c*). Similarly, if the damping and the mass may be neglected, we have a very high-frequency system, in which

$$sx=\mathrm{F}\cos pt, \quad \text{whence} \quad \dot{x}=p\mathrm{F}\sin pt/s,$$

that is, the e.m.f. in the moving coil is *directly* proportional to the frequency of the incident sound.

By suitable choice of the form of the receiving surface and the mechanical system it is therefore possible to arrange that the response of the electrical system is either dependent in a certain manner, or is entirely independent, of frequency. Such flexibility is the great asset of these receivers. The chief drawback is the relative insensitiveness (compared with other forms of receiver described later), but this may be easily overcome nowadays by the use of valve amplifiers. Here, again, the flexibility of the system may be extended to the design of the amplifier, which can be made to reinforce particular frequencies or frequency ranges, or to give a uniform amplification over a wide range. Electrodynamic receivers combined in this way with amplifiers are invaluable in all kinds of metrical observations on sound.

Piezo-Electric Receivers – The phenomena of piezo-electricity and the application of piezo-electric crystals in the production of high-frequency sounds have already been considered (see pp. 140 *et seq.*). The alternating mechanical stresses set up in such a crystal (quartz or Rochelle salt, for example), when alternating e.m.f.'s are applied to its faces, produce corresponding pressure fluctuations in the surrounding medium. Conversely, if alternating mechanical pressures are applied to the faces of the crystal, corresponding electrical fluctuations take place. As already stated (p. 146), the Langevin quartz 'transmitter' is also used as a receiver of sounds of the same frequency, viz. the echo of a transmitted beam, or signals from a similar transmitter in tune with it. As in the case of electrodynamic receivers, the 'dynamo and motor' principle is applicable to piezo-electric receivers also, the only difference being that we are now dealing with electrostatic effects. Rochelle salt, which is very sensitive in comparison with quartz, has been found by Nicholson (see p. 151) to respond readily to mechanical vibrations, in proof of which he has used a crystal as a gramophone 'pick-up.' A great increase of efficiency was obtained by thorough desiccation, static compression, application of torque, and a suitable choice of poles and the method of growth. The best effects are obtained when the applied forces, due to the incident sound waves, *twist* the crystal about its principal axis. Russell and Cotton* have also used Rochelle salt crystals in a piezo-electric gramophone reproducer. They have used such crystals, attached to metal diaphragms, as receivers of sounds under water (piezo-electric hydrophones). The sound waves falling on the diaphragm produce an alternating torque

* *Loc. cit.*, p. 151.

on the Rochelle salt crystal, the resulting electrical effects being amplified and detected in telephones. The crystal may be used directly—that is, without the intervening diaphragm, as a means of detecting sound waves. In this condition, however, it is very insensitive, but the electrical fluctuations are a very faithful reproduction of the alternating mechanical pressures. The lack of sensitivity may be compensated by valve-amplification.

Magnetostriction Receivers – Under 'magnetostriction' is classed a large group of magnetic phenomena * in which there are changes in dimensions accompanying magnetisation, and, conversely, changes in magnetic properties accompanying mechanical stresses. As in the case of piezo-electricity there is a reciprocal relation: † If a magnetic field increases the length of a ferro-magnetic rod, then a mechanical pull will increase its permeability beyond that of the unstretched rod. The phenomena of magnetostriction in iron, nickel, cobalt, and manganese and their various alloys have received the attention of a large number of investigators. The effects are best demonstrated with a bar or wire of nickel magnetised by means of a solenoid carrying direct current on which is superposed an alternating current of the desired frequency. The alternating magnetisation produces corresponding variations in length of the rod which emits a note of the same frequency. If this frequency coincides with the resonant frequency of the rod the vibrations may become very powerful. G. W. Pierce ‡ has applied this principle to the production of sounds having a wide range of frequency, from audible to supersonic, depending on the dimensions and mode of vibration of the magnetostrictive material. He has also made use of the converse principle in the design of sound receivers. Sound vibrations falling on a 'piston-diaphragm' (*i.e.* a disc having uniform amplitude all over its surface) caused the latter to compress and stretch alternately a rod or a series of rods of nickel (or other suitable material), to which it is rigidly attached. A magnetising coil surrounding each nickel rod serves to detect the variations of magnetisation due to the alternating stresses, the induced currents being detected, after amplification, by telephones or a measuring instrument. Pierce employs the 'lever principle' when

* A good summary and bibliography of magnetostriction phenomena is given by S. R. Williams, *Journ. Opt. Soc. Amer.*, **14**, 5, p. 383, 1927. See also numerous papers by Honda, *Phil. Mag.*, 1902, etc.

† J. J. Thomson, *Application of Dynamics to Physics and Chemistry*, p. 43, 1888.

‡ Patent No. 283116, Magnetostrictive Devices, 1928.

the apparatus is used to generate or receive sounds in air. The rod or strip of nickel which produces large *pressure* fluctuations with small displacement forms part of a lever system, the complementary part of which vibrates with a much larger displacement at reduced pressure. Magnetostriction sound receivers may be made non-resonant for faithful reproduction (when large valve-amplification is required), or resonant for maximum sensitivity at one particular frequency. In the latter case, the best effects are obtained by using a tuned lever system, the mechanical advantage of the lever being adjusted according to the medium from which the sound energy is to be abstracted. It may be of interest to observe in passing, that Pierce has used magnetostriction vibrators as frequency meters, frequency stabilisers, vibrators for transmission and reception of underwater sound signals, as a generator of alternating current, and in association with microphones as a wave-detector or audio-frequency amplifier.

The Condenser Microphone – (Electrostatic.) The history of this type of microphone commences in 1863 with the discovery of the 'speaking condenser' by William Thomson. Pollard and Garnier used a polarised condenser as a telephone receiver in 1874, two years before the invention of the electromagnetic receiver by Bell. Numerous other individuals have experimented since that time on the production of a condenser transmitter and receiver of sounds, the most successful being E. C. Wente * and H. Riegger.† The singing of a condenser carrying alternating current of audible frequency is well known, the effect being due to the varying force of attraction between the plates as the current rises and falls. Conversely, if a charged condenser, connected to an external circuit, be subjected to vibratory mechanical stresses producing variations in the separation of its plates, then an alternating e.m.f. will be set up in the circuit due to the changing capacity of the charged condenser. Wente's condenser microphone shown in section in fig. 117 is based on this principle. It consists essentially of a steel diaphragm (0·002 in. thick) stretched nearly to its elastic limit on a massive steel ring. This diaphragm forms one plate of an electrical condenser, the other plate being an insulated steel disc lying parallel to the diaphragm and separated from it by an air gap of about 0·001 in. This condenser of capacity

* *Phys. Rev.*, **10,** p. 39, 1917 ; **21,** p. 450, 1918 ; **19,** p. 498, 1922. See also W. H. Martin and H. Fletcher, *Journ. Amer. I.E.E.*, **43,** p. 230, 1924.

† *Wiss. Veröfft. a. d. Siemens-Konzern*, **3,** 2, p. 67, 1924 ; and *Elec. Techn. Zeits.*, p. 655, 1909.

C_0 is connected in series with a battery of voltage E and a resistance R. When the diaphragm is set in vibration by the incident sound waves the capacity of the condenser at any instant is given by $C=C_0+C_1 \sin pt$, and the voltage across the resistance R varies

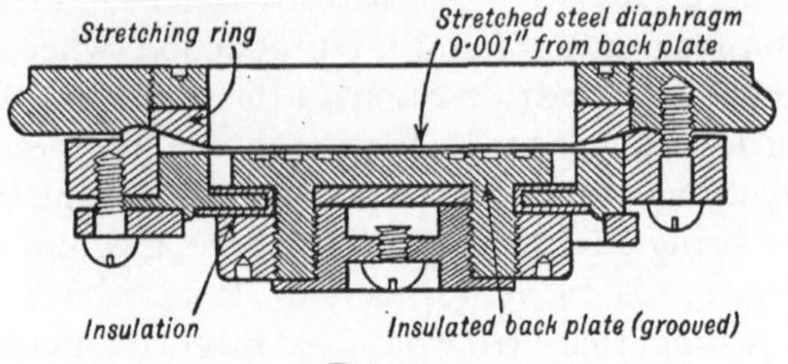

FIG. 117

accordingly. Wente shows that the alternating voltage e across R (when this resistance is large compared with $1/C_0p$ and C_1 is small compared with 2 C_0) is given by

$$e=Ri=\frac{EC_1R}{C_0\sqrt{(1/C_0^2p^2)+R^2}} \cdot \sin(pt+\phi),$$

a relation which indicates that the condenser microphone may be regarded as an alternating current generator giving an open circuit e.m.f. $(EC_1/C_0) \cdot \sin(pt+\phi)$ and having an internal impedance $1/C_0p$. The alternating e.m.f. is proportional to the applied direct voltage E, and in this respect the device may be regarded as a 'microphone.' Its action is reversible, however, an applied alternating voltage producing mechanical vibration of the diaphragm and emission of sound energy. In this respect it is more properly regarded in the same category as the electromagnetic and electrodynamic receivers. [It is important to note that a *receiver of sound* is often described, particularly in telephony, as a *transmitter* (of alternating electrical power)—thus Wente's condenser microphone, used as a receiver of sound, is described by some writers (including Wente) as the condenser transmitter. Throughout this book, however, a transmitter refers to a *sound* generator or source of sound.] The permissible voltage E which may be applied to the condenser microphone is about 300 volts; above this there is risk of breaking down the insulation, 0·001″ of air, between the plates (the minimum value of the sparking potential in air is about 400 volts). The steel backing plate forming part of the condenser is grooved radially so as to give the diaphragm the desired natural frequency and damping. Provision is also made for equalising the pressure inside and outside the microphone by means of a flexible rubber membrane, which serves also to exclude moisture. Wente determined the *sensitivity of the microphone* (*a*) at low frequencies by means of a small reciprocating piston device ('pistonphone') which applied measurable alternating pressures to the diaphragm arranged to form one wall of a small air-enclosure, and (*b*) at high frequencies, by means of a thermo-

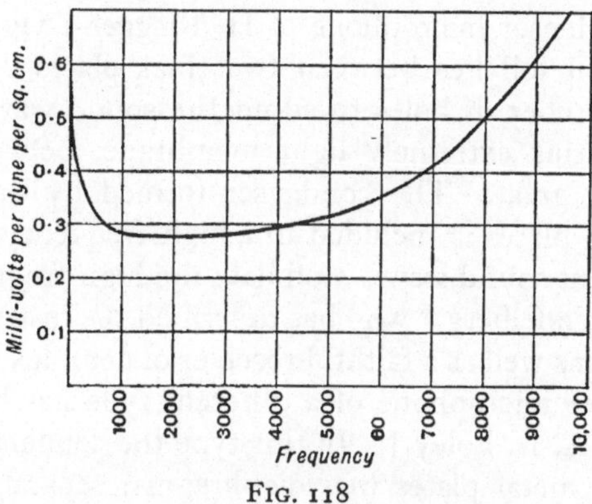

FIG. 118

phone method due to Arnold and Crandall.* Fig. 118 is a curve for a particular microphone having a fairly constant sensitivity of 0·35 millivolt/dyne over a range of frequency from 500 to 5000 p.p.s. Even outside this range, from 0 to 8000 p.p.s., the variation of sensitiveness is comparatively small. The natural frequency is over 10,000 p.p.s. and the damping constant 14,000 approx. A better idea of the sensitiveness may be obtained from the fact that the male voice in ordinary conversation exerts a pressure of about 10 dynes/cm.^{2} at a distance of 3 cm. from the mouth of the speaker. With a three-stage amplifier, thermocouple and galvanometer, pressures as low as 0·01 dyne/cm.^{2} may be measured, whilst it is possible to measure pressures barely perceptible to the ear if a pure tone from an electrical source is used and measurements are made with an alternating current potentiometer. Combined with an amplifier of ordinary design the device has an overall sensitivity which is practically uniform from 25 to 8000 p.p.s. It is therefore remarkably well adapted for the measurement of intensities of complex tones and for the distortionless conversion of speech sounds into the corresponding electrical oscillations. For this reason it has been used in a large number of investigations relating to speech analysis and recording.

* *Phys. Rev.*, **10,** p. 22, 1917.

On account of the heavy air-damping and the non-resonant character of the diaphragm the sensitivity of the condenser microphone is not very great. By means of electrical tuning and valve amplification, however, it becomes sensitive and selective over a wide range of frequencies. When properly calibrated, the instrument may be used in the converse process (the motor principle) as a precision source of sound.

In the condenser microphone of H. Riegger a very thin membrane of metal foil lies between two thick plates, one of which contains a number of holes to admit the sound waves. The air damping on this extremely light membrane, between two air-films, is very great. The condenser formed by the membrane and the metal plates is included in a high-frequency valve circuit so that incident sound waves modulate the high-frequency oscillations. F. Trendelberg,* who has described the instrument, states that it functions well as a faithful receiver of complex sound waves.

A condenser microphone of a different type has been recently described by A. L. Foley.† In this type the condenser is formed by two *thick* metal plates (*not* diaphragms) separated by an air space. The sound waves passing *between* the plates produce alternate condensations and rarefactions of the air. As the air serves as the dielectric of the condenser formed by the plates, the rapid changes of density produce corresponding fluctuations of electrical capacity, with the accompanying alternations of current in the circuit. Such a condenser microphone eliminates diaphragm resonance altogether, and, by suitable design, resonance in the air cavity formed between the plates may also be rendered unimportant within the range of speech frequencies. Foley's condenser microphone has a possible application in broadcasting and in the faithful recording of vibrations.

The Carbon Microphone – Almost simultaneously with Bell's invention of the electromagnetic telephone, Hughes (1878) discovered the action of the 'microphone.'‡ He found, what is now a very familiar observation, that a loose contact in a circuit containing a battery and a telephone may give rise to very loud sound in the telephone, due to vibrations of the loose contact. The sounds arise from the current fluctuations in the circuit as the

* *Wiss. Verofft. a. d. Siemens-Konzern*, **5**, 2, p. 120, 1926 ; and **3**, 2, p. 43, 1924.

† *Nat. Acad. Sci.*, April 1929. See also *Nature*, p. 733, May 11, 1929.

‡ The original microphone, a carbon rod held loosely between a pair of carbon blocks, is now in the Science Museum, South Kensington.

resistance of the contact varies. The precise behaviour of carbon or metal surfaces held in contact by a light variable pressure has formed the subject of numerous researches, but the information available is still of a general character only. R. Holm,* investigating the variation of resistance with pressure between spherical and plane surfaces of metal or carbon, has arrived at the conclusion that the *actual* area of contact is made up of a large number of very small areas (almost 'points'), the sum of which is usually far less than the theoretical area of contact determined from the applied pressure and elasticity of the surfaces. For very small pressures the effective contact area remained almost constant, but as the pressure increased the number of contact points increased until at last the actual area approximated to the calculated area of contact. Such observations explain the general behaviour of a microphonic contact. Single-contact microphones are now rarely used, having been replaced by the very common form of telephone transmitter microphone. The latter consists essentially of a diaphragm of metal or carbon covering a small cavity containing loosely packed carbon granules or pellets. The pressure between the pellets is gravitational, and the current is led in and out through two carbon plates, one of which may be the receiving diaphragm. The small pressure-variations produced in the microphone when sound waves excite the diaphragm produce changes of electrical resistance of the multiple contacts, which are indicated by current fluctuations when a battery and telephones are connected to the microphone. The process is essentially a *relay action*, since the alternating electrical power obtainable from the microphone circuit is drawn from a battery, and may greatly exceed the mechanical power absorbed by the microphone. By increasing the battery power in the primary circuit of the microphone (using a transformer to separate the alternating from the direct current) the electrical power output for a given amplitude of vibration is also increased. The limit is set by heating and arcing at the carbon contacts, these effects resulting in loud hissing and 'frying' sounds heard in the telephone. This burning of the carbon contacts, once started, tends to grow more serious, with eventual failure of the microphone. Commercial forms of microphone are described in various textbooks on telephony.† Two of these types are shown in fig. 119. Type (*a*) is similar to those in general use in telephony. A carbon or metal diaphragm about 2 in. diameter serves to

* *Zeits. f. Techn. Phys.*, **3**, pp. 290, 320, 349, 1922; **6**, p. 166, 1925, and **8**, p. 141, 1927.

† See *Dictionary of Applied Physics*, **2**, p. 585.

receive the sound waves and to compress the carbon granules by its vibrations. The diaphragm is suitably damped by means of flannel or cotton-wool washers, to avoid pronounced resonance effects which would tend to distort the voice of a speaker. The second type (*b*), the 'button' microphone, is smaller (about $\frac{1}{2}$ to $\frac{3}{4}$ in. diameter) and is designed for attachment as required

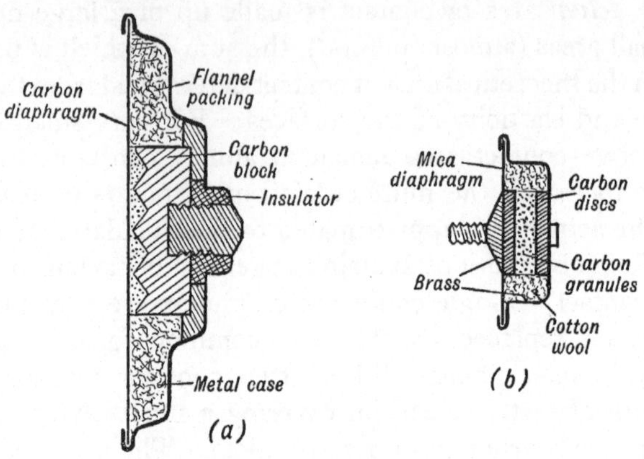

FIG. 119—Carbon Granule Microphones

to a diaphragm or any other form of vibrator. Two parallel carbon discs are supported in a little brass capsule, one on a mica disc, the other soldered to the brass body. A soft felt ring surrounds the carbons and presses lightly on the mica disc. Two-thirds of the cavity thus formed is filled with carbon granules or pellets. Such a microphone, screwed on the vibrating surface, may be used either with its back fixed ('*solid* back') or free ('free back' or 'inertia'). In the latter case the wire connection to it must be very light and flexible, exercising no appreciable restraint on the vibration of the microphone. Such microphones attached to diaphragms form very sensitive receivers of sound, but in spite of their great sensitivity they have many drawbacks. One trouble is known as 'packing,' the carbon granules clinging together, due to moisture or wedging in the lower part of the cavity. In such cases the resistance falls to an abnormally low value and the microphone becomes insensitive. The 'frying' noises to which we have referred not only set a limit to the maximum current which the microphone can carry safely, but also to the use of valve amplifiers which magnify both the signals it is desired to receive and also the 'parasitic' noises. For general purposes, as in telephones or in submarine hydrophones, where accurate reproduction is not essential, granular microphones are extremely useful.

They undoubtedly introduce distortion, however, (*a*) due to the free vibrations of the microphone itself, the mica disc with attached button forming a resonant system with only a moderate degree of damping, and (*b*) due to the 'push' and 'pull' action of the vibrations producing unequal changes of resistance, *e.g.* a microphone of 100 ohms normal resistance cannot possibly vary more than this amount in one direction, but may exceed it many times in the other direction; the curve connecting resistance and pressure (including tension) is not a straight line. The distortion due to (*a*) can be reduced by increased mechanical damping, whilst that due to (*b*) can be compensated by using a double 'push-and-pull' microphone, equivalent to two microphones arranged in opposite ways on the same vibrating surface. A novel type of granular microphone, sometimes used in 'broadcasting,' employs a large area (about 3 in. square) of granules covered by a non-resonant sheet of insulating material.* The current through the microphone flows from side to side, instead of from face to face as is customary, and the sound waves fall on the thin lamina covering the granules. It is claimed that these 'transverse current microphones' give a fairly uniform response to sound waves of all frequencies encountered in speech and music. The damping effect of a microphone attached to a diaphragm may be considerable, as shown in the oscillograph records of a rubber microphone on a steel diaphragm in fig. 51, p. 162. When such a combination is used to receive sounds of a particular frequency, therefore, it is desirable for maximum sensitiveness to choose the area of the diaphragm so that the radiation damping is equal to the microphone damping,† the thickness of the diaphragm being subsequently chosen to give the desired frequency. R. W. Boyle ‡ has constructed carbon granular microphones to respond to the very high-frequency sounds emitted from quartz oscillators and short steel rods. The granules are held loosely in a cavity formed between two small steel discs with attached carbon plates, electrically insulated from one another. These steel discs have a high natural frequency corresponding to that of the sounds they are to receive; they are relatively insensitive to sounds of audible frequency. In using the high-frequency microphone it is necessary to couple the secondary of its transformer to a 'heterodyne' circuit so that the beat note may, after amplification, be heard in

* The Marconi-Reisz and the Igranic microphones are examples of this type.

† See, for example, J. H. Powell, *Proc. Phys. Soc.*, **37**, p. 84, 1925.

‡ *Trans. Roy. Soc. Canada*, **3,** xvi, p. 293, 1922.

telephones. Alternatively the H.F. secondary currents may be measured on some sensitive form of electrostatic voltmeter (*e.g.* a Moullin voltmeter or a Lindemann electrometer). Boyle used microphones in this way up to a frequency of 42,000 p.p.s. for detecting sounds under water.

Unless precautions are taken, granular microphones are unsuitable for accurate metrical work. As detectors of feeble sounds they have a wide application in submarine signalling, and the detection of ship's noises and explosion impulses at long ranges. Whenever possible, however, they are being superseded by electromagnetic or electrodynamic receivers (magnetophones), coupled with valve amplifiers. The carbon granular microphone is a good example of an electrical sound receiver which does *not* operate on the dynamo principle; its action is irreversible.

Thermal Receivers. Hot-Wire Microphones. Resonators – We have hitherto considered only the direct mechanical conversion of sound energy into electrical energy, using diaphragms or similar vibrators as intermediaries. A method of sound reception which has attained considerable importance makes use of the thermal effects produced by an alternating current of air on an electrically heated wire or strip. In certain respects the method may be regarded as the converse of the thermophone principle (see p. 203), in which alternating electrical currents set up an alternating flow of air in the neighbourhood of a heated wire. The cooling effects produced on an electrically heated wire in streams of air of different steady velocities have been studied by L. V. King,* A. E. Kenelly,† J. S. G. Thomas,‡ and others, 'hot wire anemometers' having been devised on this principle. As a means of detecting and measuring alternating air-flow (as in a sound wave), the hot-wire principle has been developed mainly by W. S. Tucker and E. T. Paris.§

The *Tucker hot-wire microphone* was designed in the first instance for the detection of enemy guns during the war, but latterly it has been modified so that it can be used for the detection and measurement of continuous sounds. The instrument consists of an electrically heated 'grid' of fine platinum wire mounted in the neck of a Helmholtz resonator. The platinum

* *Phil. Trans.*, **214**, p. 373, 1914.

† *Amer. Phil. Soc. Proc.*, **53**, p. 55, 1914.

‡ *Phil. Mag.*, **39**, p. 518 ; and *Proc. Phys. Soc.*, **32**, p. 291, 1920.

§ *Trans. Roy. Soc.*, **221**, p. 389, 1921. See also R. S. Maxwell, *Phil. Mag.*, **6**, p. 945, Nov. 1928.

grid is held in a small rod of glass enamel mounted on a mica disc, the latter having a pair of annular silver electrodes attached to its faces (see fig. 120). The grids are made in the first place of Wollaston wire, the silver coating being removed by means of nitric acid after the wire has been mounted in position. The wire, about 0·0006 cm. diameter, carries a maximum current of about 30 milli-amperes. The average resistance of the grid at 10° C. is about 140 ohms, and about 350 ohms when carrying a safe working current of 25 to 28 milli-amperes, which heats the grid to just below a red heat. The natural frequency of the resonator, of course, depends on the volume v of the container and the form of the neck, $N=\frac{c}{2\pi}\sqrt{\frac{\kappa}{v}}$ (see p. 186). With a neck 2·2 cm. long, 0·75 cm. internal diameter, a volume of 290 c.c. gives the resonator a frequency of 116 p.p.s., whilst a volume of 68 c.c. gives it a frequency of 240 p.p.s. For frequencies below 200 p.p.s., brass tubing 2 to $2\frac{1}{2}$ in. diameter is used, whilst at higher frequencies tubing about 1 in. diameter is the most suitable. It was found that the material from which the resonator was made, and the thickness and rigidity of the walls have a marked effect on its resonating properties. For experimental purposes it is often desirable to use a microphone whose natural frequency can be varied continuously. This is easily arranged by fitting a plunger inside the brass container, or by making the container in two tubular parts, one sliding over the other. Another method of tuning is to alter the length of the neck, but on account of the effects of draughts on the hot grid it is inadvisable to make the neck shorter than 1 cm. When the air in the neck is set in vibration by a sound of suitable frequency, the hot grid is cooled

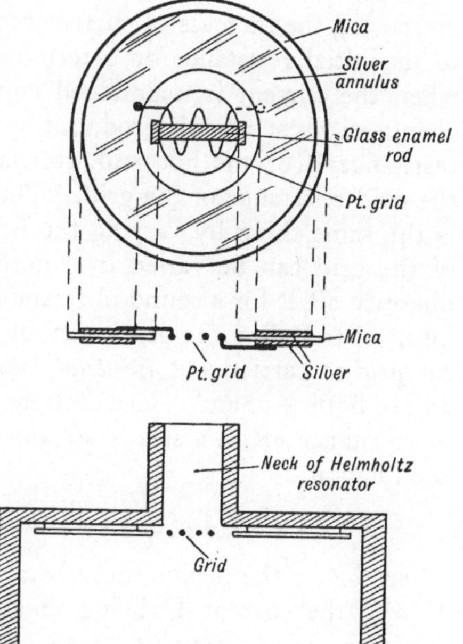

FIG. 120—Hot-Wire Microphone (Tucker)

and a change in resistance, which may be regarded as partly oscillatory and partly steady, takes place. There are thus two ways of using the microphone:

(a) *The Amplifier Method* – The oscillatory effect is introduced into a low-frequency valve-amplifier circuit, and the output current passed through telephones or a tuned vibration galvanometer. With a suitably designed amplifier the magnitude of the output current may be used to estimate the amplitude of a sound.

(b) *Wheatstone's Bridge Method* – This is preferable to method (a) on account of its greater simplicity and freedom from variations of sensitivity. The hot-wire grid forms one arm of a balanced bridge circuit, the bridge current being adjusted to give the requisite heating current through the grid (360 ohms when hot). In most cases it is sufficient to take the deflection of the galvanometer as a measure of the intensity of the sound affecting the microphone; but other methods may, of course, be used, such as measuring the increase of current required to bring the grid back to its initial resistance or determining the change of resistance when the current is maintained constant. It is found that the change of resistance δR produced by a sound varies with the total resistance R of the hot grid, so that $\delta R = 0{\cdot}2(R - R_0)$, R_0 being the cold resistance of the grid. Therefore by varying R, or what is the same thing by varying the heating current, the sensitivity of the grid can be varied in a perfectly definite manner. The quantity $\delta R/R$ for a sound of 'standard' intensity might therefore be used to define the sensitivity of the grid. *The calibration of the grid* is carried out in *steady air streams* of known velocity, and in both $+ve$ and $-ve$ directions. It is found that the change of resistance δR in a steady stream of velocity U is expressed by the relation

$$\delta R = \delta R_0 + a(U - V_0)^2 + b(U - V_0)^4 + \ . \ . \ . \text{ etc.,}$$

where δR_0 is the maximum *increase* in resistance occurring when $U = V_0$, the current U being taken as $+ve$ when flowing *into* the resonator. Applying these observations to *alternating air-currents* of small amplitude, the resultant effect is found to consist of three parts: (1) a steady drop in resistance $\delta R_1 \propto U^2$; (2) a periodic change of resistance $\delta R_2 \propto U \sin pt$; and (3) a periodic change $\delta R_3 \propto U^2 \cos 2pt$. The third effect is found to be quite negligible, the note of *double* frequency $2p$ being utterly obscured by the fundamental p in (2). It will be seen, therefore, (a) that *the steady drop in resistance*, proportional to U^2, *is a measure of sound intensity*, and (b) *the alternating effect* in (2), proportional to U, is *a measure of amplitude*. The former of these quantities is measured directly on the Wheatstone's bridge, and the latter by the amplifier method. Introducing the appropriate constants in δR_1 and δR_2 from the 'direct flow' calibration, the *absolute* value of U can be determined. It should be noted that this value of U is the particle-velocity in the neck of the resonator, *not* in the undisturbed sound field. Tucker and Paris checked these deductions by measuring the sound intensity, by the Wheatstone's bridge method, at different distances from an electrically maintained tuning-fork in the open air. It was found that the relation between the vibration galvanometer deflection δ and the distance r was $\delta \propto 1/r^2$, indicating that *the deflection is proportional to the sound intensity*. The instrument has a wide application in comparative measurements of sound intensity and distribution of sound—Tucker and Paris refer to its use, for example, in determining the magnifying and directional properties of trumpets, the diffraction effects with a large circular disc (see p. 310), etc. In addition to its use in metrical work, the microphone serves as a sensitive tuned detector of signals at long ranges whilst remaining relatively insensitive to neighbouring sounds of other frequencies. In this respect it is much superior to the granular carbon microphone, for example. The damping of the micro-

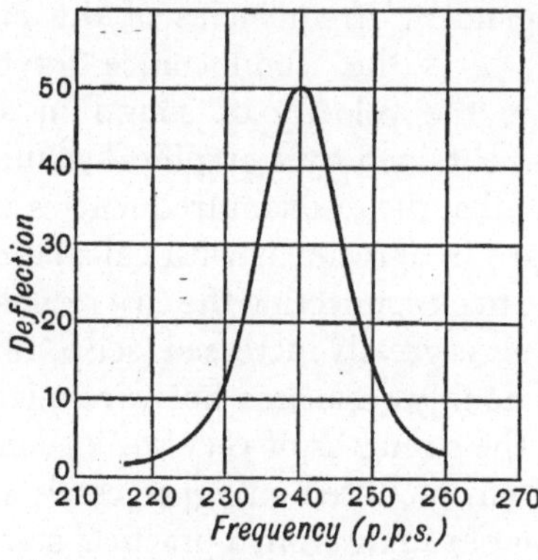

FIG. 121A—'Resonance Curve' of Hot-Wire Microphone—Single Resonator

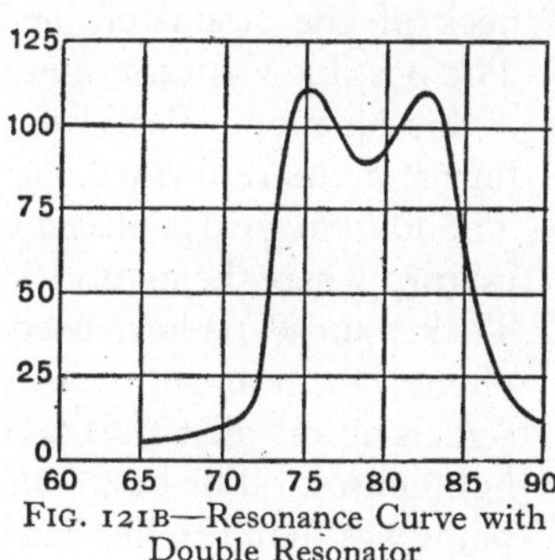

FIG. 121B—Resonance Curve with Double Resonator

phone is fairly small, in a particular case of frequency 240 p.p.s., shown in fig. 121A, the response fell to 0·1 of its maximum value by detuning 7 per cent.—*i.e.* the ratio $n/p = 1 \pm 0{\cdot}07$. The magnitude of the response depends, of course, on the degree of damping, and therefore to obtain great sensitivity and '*selectivity*' the damping must be small. For some purposes, however, sharp resonance may be a disadvantage, as, for example, when the source of sound is liable to small variations of pitch, or when allowance must be made for the Doppler effect—whilst at the same time it may be desired to retain a high degree of sensitivity. To meet such requirements a modified Boys' *Double Resonator* (see p. 189) has been used.* This may consist of two parts, (i) a resonator of the 'stopped pipe' variety, and (ii) a resonator of the Helmholtz pattern. These are combined by inserting the neck of the latter into a small hole in the stopped end of the pipe. Such an arrangement has two resonant tones, which are separated by a frequency-interval depending on the relative dimensions of the various parts of the compound resonator. The theory of the method is given by Rayleigh and has been extended in various papers by E. T. Paris,† who also deals with combinations of two or more resonators arranged in various ways.† In the simple case with which we are now dealing, the resonator shows the usual characteristics of a coupled system having two resultant tones, given by

$$\tan\frac{\pi n}{2n_0} = -\frac{2\pi\sigma}{c\kappa}n_1\left(\frac{n}{n_1} - \frac{n_1}{n}\right),$$

where n_0 and n_1 are the 'independent' frequencies of the pipe and the resonator respectively; κ is the 'conductance' of the neck of the resonator, and c is the velocity of sound in air. The possible values of n are best obtained by a graphical solution of this equation. Paris has found that the resonant frequencies and forms of the resonance curves are in agreement with calculation. The hot-wire grid is placed in the neck connecting the two resonant cavities, where the motion of the air is greatly increased (see p. 189). Tucker and Paris have used the doubly resonated hot-wire microphone as a means of measuring the strengths of very weak sound-signals at long distances from sirens mounted on light-vessels and lighthouses. The best condition of tuning from a practical standpoint was found to be that in which two resonance peaks of equal magnitude are obtained with a relatively small dip between them (see fig. 121B). In the midst of disturbing sounds this instrument has been used to pick up, at long distances, signals which

* See C. V. Boys, *Nature*, **42**, p. 604, 1890; Rayleigh, *Sound*, **2**, p. 45; and E. T. Paris, *Proc. Roy. Soc.*, **101**, p. 391, 1922; *Phil. Mag.*, **48**, p. 769, 1924, and **2**, p. 751, 1926.

† *Loc. cit.*

are quite inaudible.* When used out in the open or on board ships the aperture of the resonator is protected from disturbing draughts by a 'loofah' screen. As might be anticipated, a thermal microphone of this nature has a certain lag determined by the heat capacity of the wire. Consequently it becomes increasingly sensitive towards the lower frequencies. In fact, the hot-wire microphone was first employed to detect 'infra-sound'—that is, the very low-frequency sound emitted from the mouth of a gun (*onde de bouche*), the Helmholtz resonator in such a case having a large volume (10 to 20 litres). Using this type of microphone, vibrations of the order of 1 p.p.s. from a large explosion in Holland were recorded at Woolwich (262 miles away). The effect of such low-frequency 'sounds' may be simulated by the opening and closing of a door. The upper limit of usefulness of the microphone is reached at a frequency near 500 or 1000 p.p.s., when the thermal lag becomes more serious.

As we have stated above, the effective area of a resonant receiver which re-radiates the whole of its energy to the medium is equal to λ^2/π, where λ is the wave-length of the incident sound. In the most efficient receiver which utilises 50 per cent. of the total energy, the effective area is $\lambda^2/4\pi$. This implies that the energy density in the neck of a small resonator is greater than that in the undisturbed field, in the ratio $\lambda^2/\pi S$ or $\lambda^2/4\pi S$ respectively, where S is the actual area of the orifice. Since the driving force Sp is not amplified by the resonator, these ratios also represent *the velocity or displacement amplification coefficients*. The radiation resistance of an area S in the undisturbed plane wave is equal to $\rho c S$ (see pp. 55 and 376), the radiation resistance R of the orifice (diameter small compared with λ) is $\rho c S \times \pi S/\lambda^2$—that is

$$R = \pi \rho c S^2/\lambda^2 = \rho n^2 S^2/4\pi c \text{ (where } \lambda = 2\pi c/n\text{)},$$

which is much less than that of the same area S of wave-front. The pressure variation therefore produces a much larger velocity amplitude in the neck of the resonator than in the medium.† The expression for R must be multiplied by 4 if half the energy of the resonator is absorbed in a detecting device, *e.g.* the hot-wire microphone.

A. v. Hippel ‡ has examined, theoretically and experimentally, the possibilities of using thermal methods for the faithful conver-

* See W. S. Tucker, *Roy. Soc. Arts Journ.*, **71**, p. 121, 1923.

† See I. B. Crandall, *Sound, etc.*, p. 175.

‡ *Ann. der Physik*, **75**, p. 521, 1924; and **76**, p. 590, 1925.

sion of mechanical vibrations into alternating electrical currents. In the first place he has determined the 'node effect'—that is, the *direct* heating effect in a thin wire due to pressure fluctuations in the surrounding medium.* A wire of platinum 5 μ thick placed at a point in air where the pressure amplitude is 10^{-4} atmosphere (about 100 dynes/cm.^{2}) at a frequency of 100 p.p.s. will have a temperature-amplitude $3{\cdot}5 \times 10^{-6}$ degree C. The effect is therefore too small to have any practical significance. He also deals with the velocity (or antinode) effect in cooling a heated wire, and shows that a convection current of air flowing past a hot wire is necessary if the cooling-frequency is to be the same as the sound-frequency—the analogy being that of the permanent magnet in an electromagnetic telephone—otherwise the frequency would be doubled.

Underwater Receivers. Hydrophones† - The detection of sounds under water presents analogous problems to those already considered in the case of air. The *general principles* to be observed in the design of efficient underwater receivers are the same as in air, but the *methods* of applying these principles may be markedly different. The difference arises principally from the fact that the acoustic resistance (ρc) of water is of the order $1{\cdot}5 \times 10^5$, whilst that of air is only 40, the ratio of these quantities water/air being 3750 : 1. Now it has been shown, p. 54, that the intensity I in a sound wave is expressed by

$$I = \frac{1}{2}\frac{p^2_{\max}}{\rho c}, \quad \text{whence} \quad p_{\max} \propto \sqrt{\rho c}$$

for sounds of the same intensity in different media. Consequently the relative pressure amplitudes for waves of the same intensity and frequency in water and air would be 61 : 1 approximately. Similarly the expression for intensity I may be written in terms of velocity or displacement-amplitude a; since $p = c\rho\dot{\xi}$ (see p. 55) or $p_{\max} = c\rho\dot{\xi}_{\max} = c\rho n a$; we have

$$I = \tfrac{1}{2}\rho c n^2 a^2, \quad \text{whence} \quad a \propto 1/\rho c$$

for sounds of the same intensity and frequency in different media.

* See also Friese and Waetzmann, *Ann. der Physik*, **76**, 39, 1925.

† See C. V. Drysdale, *Journ. I.E.E.*, **58**, p. 575, 1920 ; and *Mechanical Properties of Fluids*, p. 290 ; also F. Aigner, *Unterwasserschalltechnik*, pp. 202–238 ; W. H. Bragg, *World of Sound*, p. 169 ; and *Engineering*, **107**, p. 776, 1919 ; *Nature*, July 1919 ; F. L. Hopwood, *Nature*, Aug. 1919 ; H. C. Hayes, *Amer. Phil. Soc. Proc.*, **59**, 1, pp. 1 and 371, 1920 ; and A. Troller, *La Nature*, **49**, p. 4, 1921.

The relative displacement amplitudes for sounds of the same intensity and frequency in water and in air is therefore 1 : 3750. It is not surprising to learn, therefore, that the most sensitive sound receivers in air are 'displacement' receivers (Helmholtz resonator type), whereas in water they are 'pressure' receivers. Of course, it is sometimes advantageous, when sensitiveness is not of primary importance, to use pressure receivers in air and displacement receivers in water. The majority of subaqueous receivers are, however, of the pressure type (see, however, p. 379).

One of the earliest forms of underwater receiver consists simply of a thin membrane stretched tightly across the end of a tube, the lower end of which dips under the water, whilst the upper end leads to the observer's ear. Such a device was used, for example, by Colladon and Sturm in Lake Geneva to receive the sound of a bell struck under water. This simple device has been elaborated in various ways, *e.g.* in *the Broca tube*, which consists of a small metal capsule (like an aneroid barometer) fixed to the tube. A similar, but less resonant, receiver of this nature employs a thick-walled rubber bulb or tube to replace the diaphragm. This modification, in addition to reproducing sounds more faithfully, possesses the advantage of more efficient transmission of the vibrations from water to air; this latter improvement being ascribed to the fact that the radiation resistance of rubber is intermediate between that of water and air (see p. 277). Such simple receivers, although fairly effective, are neither very sensitive nor very convenient to use. No provision is made for amplifying the sound, and the loss of intensity and change of quality in transmission through long and winding pipe-lines is very serious. At the present time, therefore, practically all underwater sound-receivers, or *hydrophones* as they are called, are of an electrical character. They consist usually of microphonic, electromagnetic, or electrodynamic detectors attached in various ways to diaphragms in contact with the water. Microphones are frequently used on account of their great simplicity and sensitivity, but with the development of valve-amplifiers electromagnetic and electrodynamic detectors are in many circumstances preferable. The simplest and most common form of hydrophone is shown in fig. 122. It consists of a massive lenticular metal disc on which is rigidly clamped a metal diaphragm with a carbon 'button' microphone screwed on a small boss at the centre. A cable passing through a watertight gland connects the microphone to a battery, transformer, and telephones at the listening-post. The frequency and damping of the diaphragm may be calculated in the manner indicated on p. 159. The loading effect of the water is an important consideration in underwater diaphragm receivers. This load is equivalent to a mass of water $\frac{2}{3}\pi r^3$ spread uniformly over, and moving with, the diaphragm of radius r (or it may be regarded alternatively as a mass of one-fifth this amount 'attached' to the centre). The vibrating mass in a receiver of appreciable diameter may therefore be fairly large. Consequently, in accordance with the lever principle (see p. 380), it is important that the mass of the metal ring on which the diaphragm is mounted should be many times greater than the effective mass of the diaphragm. Otherwise an appreciable proportion of the energy of vibration of the whole system would reside in the mounting. It is important also that the diaphragm should be *rigidly* attached to the mounting, otherwise excessive vibration of the edge of the diaphragm will be set up and energy will be lost in local flow and in frictional damping. In the case of electromagnetic receivers the damping due to radiation and friction losses must be made approximately equal to the useful damping. This can be done, of course, with microphone detectors, but it is not usually so important in this case. A rough idea of the sensitiveness of an ordinary hydrophone (granular microphone type) may be obtained from the observation that the sound of a Fessenden oscillator (emitting $\frac{1}{2}$ H.P. at 500 p.p.s.) is readily detected, without amplification, at 20 or 30 miles distance in the sea. It is of interest to compare this with the case of a siren of 200 H.P. used to transmit signals in air, the limiting range with the most sensitive detector available being in the neighbourhood of 10 miles. In air, as we have seen, the efficiency of power-sources of sound is very low and the attenuation in transmission is very great, whereas underwater sound sources are very efficient and attenuation losses are not so serious as in the atmosphere.

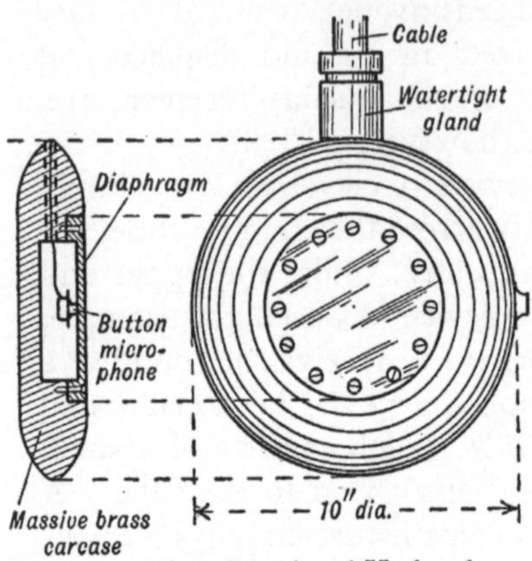

FIG. 122—Non-directional Hydrophone

For signalling purposes, the diaphragm and microphone of a receiver may both be tuned to the frequency of the signals, the resonant properties of a diaphragm being in such cases advantageous. If, however, it is desired to use a hydrophone as a means of detecting and *recognising* complex sounds under water, such as the propeller noise of a submarine, it is better to use a non-resonant receiver. For this purpose, hydrophones have been constructed of thick rubber to which a microphone with a rubber diaphragm (to replace the mica) is attached. With a heavily damped microphone of this nature the noise of a ship is more characteristic. Even with a resonant diaphragm hydrophone it is easily possible to distinguish between the reciprocating 'beat' of a trawler and the turbine 'rush' of a cruiser. Hydrophones carried by ships are of various kinds: (*a*) the portable type (used chiefly on small vessels, drifters, etc.), which is slung outboard when required; (*b*) inboard type built flush with the hull, the diaphragm being in direct contact with the water; (*c*) inboard type, in which the receiver hangs in a water-filled tank inside the hull—the sound reaching the receiver after penetrating the hull and the water in the tank; (*d*) towed hydrophones in which microphones are enclosed in rigid or flexible streamline bodies towed astern of the ship.

The principal trouble, common to all types of microphonic hydrophones, is that due to 'water-noise'—the movement of the ship through the water, breaking waves, eddies and bubbles carried down from the bows, etc., all contributing towards the general noisiness of the microphone, the granules of which are particularly sensitive to any form of shock excitation. Consequently electrodynamic and electromagnetic receivers (magnetophones) are now more generally used, their lack of sensitiveness being compensated by means of valve amplifiers. Fixed hydrophones, mounted on tripods resting on the sea-bed, are relatively free from disturbing noises, and are used as sentry hydrophones in particular localities, or on surveyed base-lines for sound-ranging purposes (see p. 483). In modern underwater signalling apparatus the sound generator, as we have seen, often serves as a receiver also. Examples of this are to be found in the Fessenden and Signalgelleschaft transmitters, and the Langevin piezo-electric high-frequency transmitter. We shall refer to other more important types of hydrophone in dealing with directional reception of sound.

DIRECTIONAL RECEPTION OF SOUND

The problem of directional reception is analogous in most respects to that of directional transmission. In the latter case we are concerned, however, mainly with questions of energy, the range obtainable with an approximately parallel beam being obviously greater than with a spherically divergent beam. In signalling, also, questions of secrecy may be involved. As regards reception, it is of little advantage in many cases to have an extremely sensitive receiver if it gives no indication of the direction of the source from which the sound emanates. The question of directional reception is, therefore, at least as important as that of sensitiveness. The receivers to which we have hitherto referred may be regarded as non-directional, the sensitivity being the same in all orientations. As in the case of transmitters, directional properties may be obtained by the use of surfaces of large dimensions, compared with a wave-length of the incident sound, *e.g.* the use of reflectors, trumpets, etc., with a small receiver at the focus, or by means of a large sensitive disc such as the quartz plate in a Langevin piezo-electric receiver. Alternatively we may obtain directional effects by means of two or more receivers suitably arranged as in the case of multiple sources of sound. The principles involved in such methods of directional reception are essentially the same as in directional transmission described on p. 215, etc. In what follows we shall therefore refer only to special types of receivers, most of which may be classified in the manner just indicated.

Binaural Methods - The perception of direction by means of a pair of ears has already been considered on p. 359. The principles involved in this binaural (two-ear) method have been widely extended for sound reception both in air and in water. The binaural effect, according to Rayleigh's view, is dependent on the phase difference of arrival of the sound wave at the two ears; but there can be little doubt that other factors, such as intensity differences, must also be invoked in order to explain all the relevant facts. The phase-difference theory, however, provides a useful basis and has at any rate been fruitful in its application to 'artificial' binaural reception. For example, if it is desired to locate a distant aeroplane at night when the unaided ears can only form a very rough estimate of its direction, a pair of large receiving trumpets mounted so as to rotate in azimuth and elevation may be used. In the first place, the amount of sound energy reaching the ears is increased by means of the trumpet-collectors ; and, secondly, the directional accuracy is increased by increasing the distance apart of the trumpets. Assuming the axes of the trumpets to be in the same plane as the source of sound, rotation of the platform supporting the trumpets will vary the path difference of the sound reaching the two ears (one connected to each trumpet). As this path difference will decrease on one side and increase on the other, the impression received will be that of the sound crossing over from one ear to the other. At the exact point where the 'cross over' occurs, *i.e.* of '*binaural balance*,' the common axis of the trumpets will be directed towards the source of sound. The accuracy obtainable in air is of the order of ± 2 or 3 degrees under good listening conditions. Errors are likely to arise from the motion of the sound source, and due to temperature and wind refraction. In practice, the necessary corrections for such effects are tabulated. Actually four trumpet receivers were mounted on a common platform, one pair in azimuth and the other pair in elevation (about 12 ft. apart). The former had an observer who operated the control (for *all* the receivers) in azimuth, whilst a second observer, listening on the other pair, made the elevation adjustment. The method was employed extensively during the war in the location of aeroplanes, a number of binaural receivers at triangulated positions serving to give a 3-dimensional instantaneous position of the rapidly moving source.

The same principle has been developed, mainly in America, for the directional location of sources of sound under water, more particularly in the location of submarines.* In this case it is necessary to increase the separation of the receivers to obtain the same directional accuracy, for the velocity of sound in the sea is about $4\frac{1}{2}$ times that in air, the phase differences being reduced in corresponding proportion. In order to avoid the practical difficulties of rotating a long arm carrying the receivers under water, the principle has been slightly modified.

Binaural Compensator – Instead of rotating the line of receivers in order to determine the direction of zero phase difference, it is sometimes more convenient to compensate for this difference by introducing artificially an increased path-length in the side which is nearer the source. The principle is similar to that employed to demonstrate interference between sound waves from the same source, travelling by different paths (two tubes) to the same receiver,† the phase difference of arrival being measured by the change of length necessary in one of the tubes to bring the two trains of waves directly 'in phase' or 'out of phase.' Let us suppose two receivers M_1 and M_2 a distance d apart, respectively distant x and $(x+\delta_w)$ from a source of sound S, to be connected

* See Hayes, Drysdale, and others, *loc. cit.*

† See Tyndall, *Sound*, p. 261.

through suitable circuits to two telephone earpieces T_1 and T_2, these in turn being connected to the ear through two air-tubes of adjustable lengths l_1 and l_2. When $\delta_w=0$, that is, the source is symmetrical with regard to M_1 and M_2, let $l_1=l_2=l$. If the receivers are in water in which the velocity of sound is c_w, we must have for binaural balance $\delta_w/c_w=\delta l/c_a$, where the suffixes w and a refer to water and air respectively, and δl is the 'compensation' necessary to make the total path-lengths from the source S to the two ears of the observer equal. Consequently δl, which is directly observed at the point of binaural balance, is a measure of δ_w, and therefore of the bearing θ of the source of sound S relative to the line M_1M_2 joining the receivers. For $\delta_w=d\cos\theta$ if d is small relative to the distance x from M_1 or M_2 to S, whence $\theta=\cos^{-1}(\delta_w/d)=\cos^{-1}(\delta l\,.\,c_w/dc_a)$, which, in the case of transmission through water $c_w/c_a=4{\cdot}5$, $\theta=\cos^{-1}(4{\cdot}5\delta l/d)$. In the application of the method a circular form of compensator (known as the American Binaural Compensator) is used, shown diagrammatically in fig. 123. It consists essentially of a fixed plate in which two circular grooves, not quite complete, are cut as shown. A rotatable coverplate, fitting closely over this fixed plate, carries two stoppers *ss*, which close the grooves. On both sides of these stoppers are cut in the upper plate two radial bypass grooves which connect the inner and outer groove of the lower plate. The sounds from the telephones T_1 and T_2 pass around a portion of the outer groove and return *via* the inner groove to the stethoscope which connects to the ear. Rotation of the upper plate, and consequently the stoppers *ss*, increases the length of one of these paths and diminishes the other, a pointer attached to the plate indicating directly the bearing θ of the source of sound. The zero position of the scale is obtained by adjustment on a source known to be symmetrical with respect to the receivers ($\delta_w=0$). The principle is applicable to any type of receiver in air or in water. The use of one pair of receivers in this way

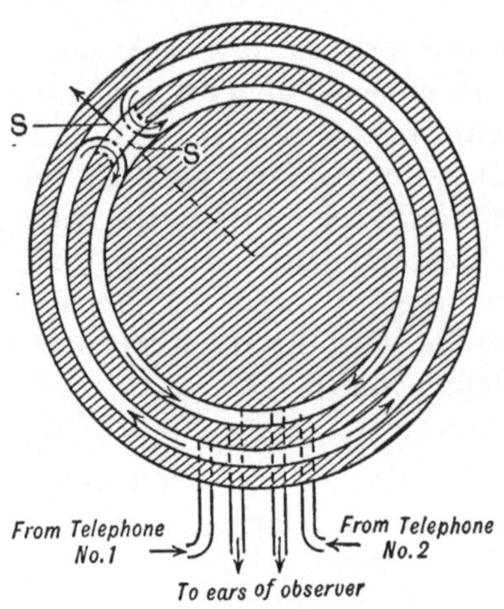

FIG. 123—Binaural Compensator (American)

sometimes leads to an ambiguity in direction, only the *line* of the source being determined. If, however, a third receiver is added, the three forming an equilateral triangle, 'binauralling' on each pair removes the ambiguity.

Binaural Geophones * – The binaural principle has been used not only to determine the direction of sound in the atmosphere and in the sea, but in the ground also. The requirement of a sound-direction apparatus arose during the war in connection with the location of enemy trenching and mining operations. Apparatus known as the geophone was devised to detect the sounds involved in such underground operations, and a pair of receivers employed to give the direction from which the sounds proceeded. A single

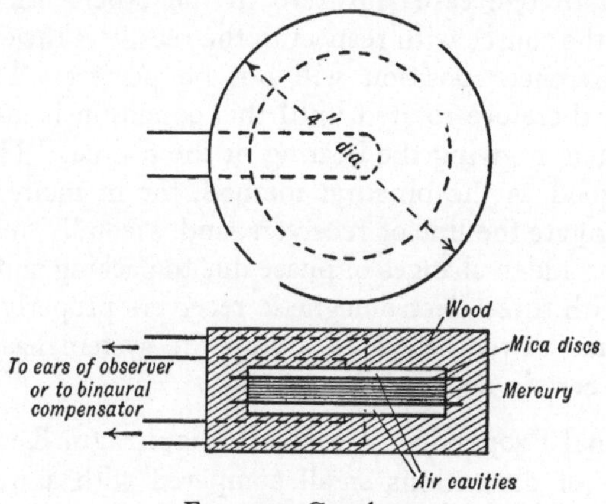

FIG. 124—Geophone

geophone, used simply as a detector of underground sounds, consists of a hollow wooden box, 3 in. in diameter by 2 in. deep, divided into three compartments by two mica discs (see fig. 124). The space between the discs is filled with mercury whilst the two outer air cavities connect to the ears by stethoscope tubes. On account of the large inertia of the mercury, vibrations affecting the box as it lies in contact with the ground produce fluctuations of volume of the air cavities and corresponding compressions in the ear of the observer. The device is very sensitive to 'thudding' sounds. When two geophones are used on the binaural principle only one cavity in each geophone is required. The observer adjusts the line of the geophones on the ground until the binaural 'cross over' is obtained. Assuming the direction of propagation

* See H. S. Ball, *Inst. of Mining and Metallurgy*, April 10, 1919; see also Bragg, *World of Sound*, p. 179.

in the ground is rectilinear (probably true for distances of 100 ft. or so), this gives the direction of the source of sound. Similar observations at a number of points are sufficient to give an approximate *position* of the source.

Sum and Difference Method – This is an alternative to the binaural method when electrical receivers are used, and it is also possible to swing the line of the receivers. In this method the secondary circuits of the two microphones are connected alternately to assist and to oppose each other's effects on the observer's telephones. Assuming the microphones are equally sensitive and equidistant from the source, the resultant secondary e.m.f. should be doubled in one case and zero in the other. In any other position of the source with respect to the receivers the cancellation in the 'difference' position will not be perfect. The line of receivers is therefore rotated until this condition is attained, the angle of rotation giving the bearing of the source. This method is not so good as the binaural method, for in many cases it is difficult to rotate the line of receivers, and, secondly, microphones are liable to sudden changes of phase due to packing and analogous effects. With three electromagnetic receivers properly spaced by spreaders and towed astern of a ship the system has, however, met with a certain degree of success.*

Directional Properties of a small Diaphragm Receiver – A diaphragm (of dimensions small compared with a wave-length) exposed to the free medium on one side only is a non-directional receiver. It responds to pressure variations in the medium, the response being the same in all orientations. If, however, the diaphragm, mounted in a heavy annular support, is exposed to the medium on *both* sides, the conditions are entirely changed. As we have seen on pp. 156 and 217, such a diaphragm exhibits marked directional properties as a *source* of sound, and the converse is also true—*i.e.* the diaphragm is a directional *receiver* of sound—if a suitable detector is attached to it. It is unnecessary to repeat the mode of action of such a receiver, but it is of some importance to describe a very successful directional hydrophone designed on this principle (which is equally applicable to such a receiver in air). The hydrophone shown in section in fig. 125A consists of a heavy brass streamline annulus about 10 in. in diameter, in which is mounted rigidly and symmetrically a metal diaphragm (about 4 in. in diameter). At the centre of this diaphragm is

* See Drysdale, *Proc. I.E.E.*, **58**, p. 577, 1920.

turned a small watertight cavity in which is fixed a carbon granular microphone (button type, free-back). Sound waves reach *both* sides of the diaphragm, either directly or by diffraction around the annulus, and the resultant vibration is indicated by the response

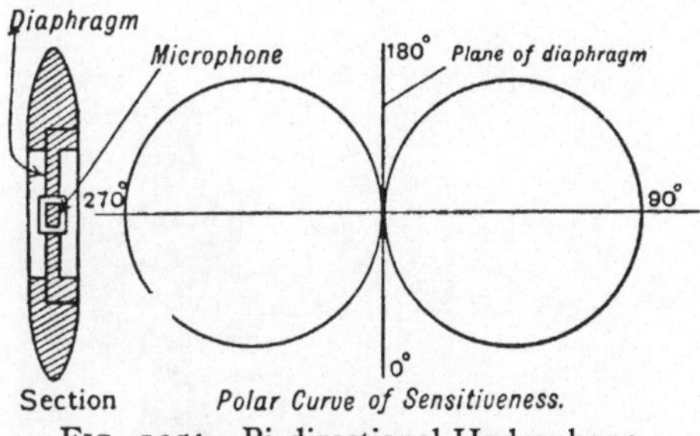

FIG. 125A—Bi-directional Hydrophone

of the microphone. When the plane of the diaphragm coincides with the direction of the sound waves, there is no pressure difference between the two sides of the diaphragm, and the microphone is 'silent.' If, however, the hydrophone is broadside-on to the incident sound waves, the pressure-amplitude on the front face of the diaphragm exceeds that at the back face, which is screened somewhat by the heavy annulus. There is probably a phase effect also, but we need not be concerned with it here. In the broadside position, therefore, the response is a maximum. It will be seen that a rotation of the hydrophone through 360° will result in the microphone response passing through two positions of maximum and two of minimum (zero) intensity. On account of the sensitivity of the ear to weak sounds the zero setting is most accurate in practice. This position gives the *line* of propagation of the sound, but is still ambiguous, for the source may lie in either of two diametrically opposite directions. The hydrophone is for this reason designated '*bi-directional.*' By means of a suitable screen or 'baffle plate' mounted at the correct distance on one side of the hydrophone as in fig. 125B, the directional properties are changed in such a manner that there is now only one maximum and one minimum, when the unscreened and screened faces of the diaphragm are respectively facing the source of sound The exact behaviour of the baffle, which must contain air and must be suitably damped (a hollow xylonite disc filled with lead shot), is not understood completely, but in practice the device is very successful. In average weather a directional accuracy of $\pm 3°$ is possible with the *uni-directional* hydrophone. The distribution of pressure and displacement-amplitude around a bi-directional and a uni-directional hydrophone in a sound field has been determined experimentally by F. B. Young and the writer,* who have also advanced a theory of the behaviour of the hydrophone and baffle. The theory indicates that the maximum

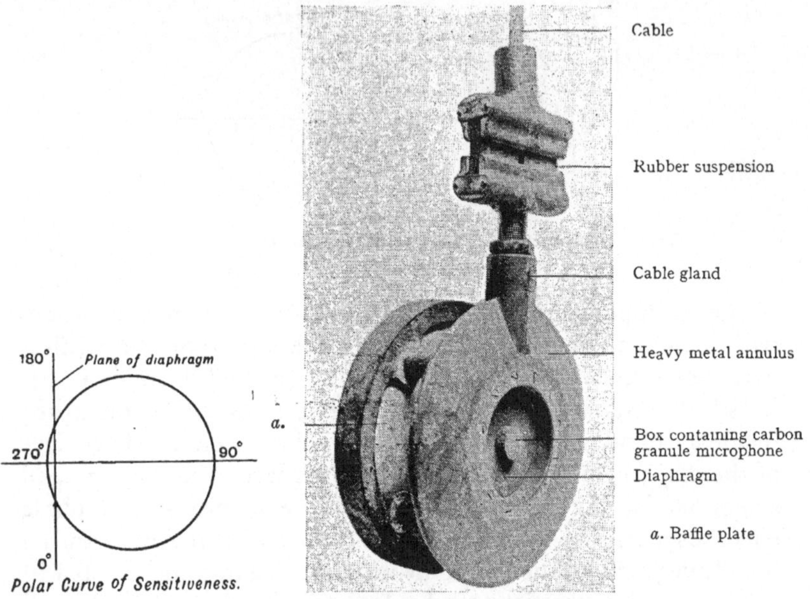

FIG. 125B—Uni-directional Hydrophone

sensitiveness (*i.e.* in the 'broadside' position) of the bi-directional hydrophone should be approximately a quarter of that of a similar hydrophone with one side only in contact with the water—a deduction which was confirmed by experiment.

Directional Displacement Receivers. Directional Properties of 'Button' Microphones – A special type of directional receiver, which is applicable more particularly to water as a medium of sound transmission, is due to W. H. Bragg (1916). He suggested a simple form of directional hydrophone, consisting of an elongated body of rigid but light construction, in which was mounted a carbon granule microphone of the button type. Such a body, lighter than the medium (water) which it displaces, was expected to have a displacement amplitude greater than that of the surrounding medium, the amplitude of the motion depending on the *shape* of the body. If, for example, the body is ellipsoidal, its motion will be greatest when the longest axis coincides with the direction

* *Proc. Roy. Soc.*, **100**, p. 261, 1921.

of the sound wave. The response of the enclosed microphone will vary therefore with the orientation of the ellipsoid, and the 'light-body' will constitute a directional hydrophone. Professor H. Lamb supplied the necessary data with regard to the loading effect of the medium (water), and F. B. Young and the writer * made a series of experiments to test the theory. Preliminary tests with a hollow glass ellipsoid indicated a direction-ratio of 20 or 40 : 1, whereas the theory indicated only 2 : 1. This discrepancy led to an examination of the button microphone, which was found to show marked directional properties. In some cases a ratio of 40 or 50 : 1 in current amplitude was observed, when the axis of the microphone was parallel and at right angles respectively to the direction of vibration. To test the light-body principle, therefore, a hollow ellipsoid of ebonite was constructed as in fig. 126, the microphone being mounted in a hollow, rotatable plug fitted tightly in the ellipsoid as shown. A series of

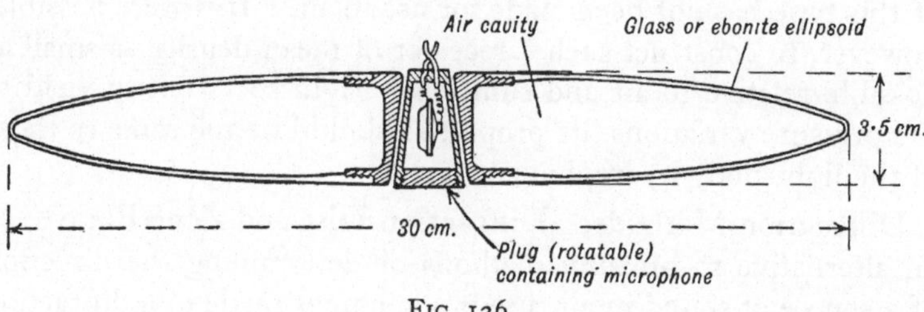

FIG. 126

observations of the apparent direction of the sound, for different orientations of the microphone relative to the major axis of the ellipsoid, yielded the information that the direction-ratio of the ellipsoid (apart from the microphone) was 1·56 as compared with the value 1·60 calculated from Lamb's data. The maximum possible direction-ratio of any ellipsoidal light body consistent with practical limits of rigidity and mean density was estimated to be about 2·3. This is negligible compared with the direction-ratio obtainable with a well-mounted microphone. Hence it appears that there is little to be gained by adhering to the ellipsoidal form, a microphone mounted in a light sphere being equally effective and possessing the advantages of greater compactness and rigidity. *Light body hydrophones* of this spherical type have been constructed with success of glass, ebonite, and paraffin-wax. F. B. Young and the writer * have used such hydrophones (1 in.

* *Proc. Roy. Soc.*, **100**, pp. 252 and 261, 1921. The experiments were made in 1916.

diameter (see fig. 96)) in exploring the sound distribution (direction and amplitude) around various bodies submerged in water. Larger hydrophones containing electromagnetic detectors have also been constructed on this principle.*

The 'displacement,' or velocity hydrophone, as it should be called, is less sensitive than a 'pressure' (diaphragm) hydrophone of comparable dimensions, but it serves not only as a detector but also as a direction indicator. The pressure variations in the medium have no direction, but the displacements take place in the direction of propagation. The small spherical 'light body' therefore vibrates with maximum amplitude in this direction and with zero amplitude parallel to the wave-front. Consequently the carbon granular microphone gives its maximum response when its axis coincides with the direction of the source and zero response in the 90° position.

So far as the writer is aware, a directional displacement receiver of this type has not been made for use in air. If it were possible, however, to construct such a receiver of mean density as small as possible relative to air and sufficiently rigid so as not to yield to the pressure variations, its properties should be the same as those of the light-body hydrophone.

Diffraction Methods. Diffraction Disc and Zone Plate – As an alternative to binaural methods of determining the direction of a source of sound in air, use is sometimes made of a diffraction effect, to which we have already referred (see p. 308). If plane waves of sound fall normally on a circular disc, which either reflects or absorbs sound, the 'shadow' cast behind it on any parallel surface will have a diffraction pattern surrounding its centre point. On the axis of the disc the intensity will be a maximum and equal to that in the undisturbed sound wave. At points slightly off the axis the intensity rapidly diminishes, eventually falling to zero. Still farther from the axis the intensity rises again to a secondary maximum, succeeded by a second zero position, and so on. If, now, the source of sound is *not* on the axis in front of the disc the primary maximum of intensity (the bright spot) will not fall on the axis behind the disc, but at some point displaced in accordance with the direction of the incident sound waves. Consequently, if a sensitive detector, such as a hot-wire microphone, be fixed at a point on the rear axis and the whole receiver (disc and microphone) rotated slowly, the response of the microphone will be a maximum when the axis of the disc

* See Drysdale, *loc. cit.*

points to the source of sound (see p. 310). Similar effects are observed when a zone-plate is used (p. 309). On account of the long wave-lengths of the low-frequency sounds emitted by an aeroplane, discs of large diameter (20 ft. or so) have been employed for accurate location by the diffraction method. Such large discs being *fixed* in a horizontal position, the relation between the direction of incidence of the sound and the position of maximum intensity may be calculated. This method provides one of the most reliable means of determining the direction of air-borne sounds.*

Receivers of Large Area - As in the case of directional transmission of sound waves, a surface of dimensions large compared with a wave-length has directional properties. Sound reflectors 20 ft. or more in diameter have been used to 'focus' low-frequency aeroplane sounds, but, as we have already seen (pp. 149 and 307), such a focus is in reality a diffraction pattern; just as a sound beam transmitted by a reflector with a small source at the focus, consists of a primary beam surrounded by a succession of secondary beams of diminishing intensity. In fact, it may be stated as a general principle that transmission and reception of wave energy are reciprocal problems—a directional receiver will act as a transmitter having similar directional properties. We may therefore refer directly to p. 148 and pp. 215 to 220, where the directional transmission of sound from sources of large area has already been considered. Piezo-electric discs (Langevin type), mirrors, trumpets, and such-like receivers, all exhibit directional properties characterised by the relation $\theta = \sin^{-1} 0{\cdot}61\lambda/r$, where θ is the semi-angle of the primary cone (or the semi-angle subtended by the inner diffraction circle at the 'pole'), λ is the wave-length of the sound, and r the radius of the disc or aperture. It is important to observe that the sharpness of direction increases as λ diminishes with respect to r—that is, every receiver of this type has a tendency to favour the sounds of higher frequency—apart altogether from resonance and similar phenomena. A trumpet, disc, or a mirror should therefore be used with caution if it is desired to reproduce faithfully a sound of complex waveform—*i.e.* a sound containing harmonics of widely different wave-lengths. For example, a small parabolic reflector which would focus the sharp 'crack' of a pistol or an electric spark very efficiently, would be quite useless as a means of concentrating a

* *Ency. Brit.*, 13th ed. (suppl. vols.), 1926; *Sound*, 'Detection and Perception of Direction,' A. O. Rankine.

continuous wave of frequency 200 p.p.s. (λ=5 ft. approx.). Another way of looking at the matter is to regard the concave reflector in terms of Fresnel's zones. If the 'focal length' is considerable and the wave-length λ of the incident sound is not small, the first Fresnel zone is fairly large; consequently there is little to be gained by making a reflector of such dimensions concave. If, however, the sounds are of very short wave-length, the number of Fresnel zones in a reflector of moderate dimensions may be large and the energy concentration at the focus appreciable, the advantages of the concavity are then more apparent.

Multiple Receivers – The treatment which has been given for the directional properties of multiple sources (p. 217) is also applicable to multiple receivers. A straight line, a circle, or a surface covered with equidistant 'point' receivers will have a maximum *resultant* response when the incident sound wave reaches all the receivers in the same phase. If, for example, six similar receivers are spaced along a straight line at intervals of half a wave-length, the resultant response for different orientations will be that indicated in fig. 67 (p. 219) for a line of six similar sources. In such a case the maximum amplitude occurs when the line of receivers is 'broadside' to the incident wave. It is of course important that all receivers in the 'line' should operate in the same phase, when the sound wave arrives simultaneously, *i.e.* in the 'broadside-on' position. This necessitates that they should either be all non-resonant or should all be tuned exactly to the same frequency. The principle employing lines of receivers has been applied in submarine detection by M. Mason * on behalf of the American Navy. The M.V. hydrophone, as the system was called, consisted of a row of twelve simple receivers (non-resonant rubber 'teats' fixed on the ends of metal tubes) spaced at equal intervals on a line 10 ft. long. Two such rows of receivers were fitted below water on the port and starboard sides of a ship. The binaural compensator principle (described on p. 413) was extended, by a 'progressive compensation' in a single instrument, to twelve receivers. This 'line' receiver had marked directional properties, and under suitable conditions could locate a moving vessel with considerable accuracy. The principle has since been applied in marine 'depth sounding.' † A device employing a large number of similar receivers covering a large area was invented by Walser (in the French navy). A 'blister'

* *Proc. Amer. Phil. Soc.*, **59**, No. 1, 1920; and *Marine Review*, Oct. 1921.

† See H. C. Hayes, *Journ. Frank. Inst.*, **197**, March 1924.

on the side of a ship was covered with small diaphragms, and an observer inside the ship located the position of the primary maximum by means of a small trumpet and stethoscope receiver. This and other directional devices of a similar nature have been described by C. V. Drysdale.*

Location of Impulsive Sources - The methods described above are applicable when the source emits continuous waves of sound, and fail completely when the 'sound' consists of a single impulse—as, for example, the detonation of an explosive charge. In such cases special methods of location have been devised, which we shall consider later in connection with Sound Ranging (see p. 478).

SOUND MEASUREMENT. ANALYSIS AND RECORDING

The characteristics of a sound wave at any point in a medium may be regarded as completely defined when the amplitude, frequency, and phase of its Fourier components are known. The term 'amplitude' to which we refer here may relate to displacement, pressure, density, or to some other physical quantity which represents the condition of the medium at the point in question. The measurement of frequency and phase is, in comparison with the measurement of intensity (or amplitude) a relatively simple matter. Intensity measurement is full of pitfalls, and great care and judgment is necessary in making 'absolute' determinations. Many instruments have been devised which are said to *measure* intensity, but in many cases it is necessary to make certain assumptions, often unjustifiable, to convert their indications into absolute measure. All known measuring devices are limited to certain ranges of frequency and intensity; they are best suited to a particular medium through which the sound waves pass, and their indications are dependent on the dimensions relative to the wave-length of the incident sound. A condenser microphone, for example, would have very different properties as a receiver in air and in water, not only in regard to the frequency range but also in regard to its behaviour as an 'obstacle,' the wave-length of sounds of the same frequency being *in water* four and a half times the corresponding frequency in air. It is necessary to remember, therefore, that the resulting indication of any type of receiver is dependent, amongst other influences, on the following factors:

* *Proc. I.E.E.*, **58**, p. 577, 1920; and *Mechanical Properties of Fluids*.

(*a*) the wave-length of the incident sound; (*b*) the law of pressure-volume variation assumed in the neighbourhood of the obstacle (the measuring device and its mounting) placed in the path of the sound wave; (*c*) the scattering and diffraction of sound-energy from the obstacle, 'the receiver' (involving its compressibility, density, and modes of vibration, the frequency of the incident sound, and the nature of the 'incident' medium).

It is often assumed in sound intensity measurements that the pressure fluctuation at the surface of the receiver is the same as that which would occur in the undisturbed sound wave. Sometimes the other extreme view is taken that the pressure is doubled by reflection. It is clear that the former assumption approximates to the truth when the sound field influenced by the receiver may be regarded as a '*point*'—*i.e.* very small compared with a wave-length; whilst the latter assumption is more nearly correct when the receiver approximates to an infinite rigid plane, *i.e.* its dimensions are large compared with the wave-length of any sound falling upon it. In practice neither of these ideals can be attained, but it is often possible to define the experimental conditions so that the results agree approximately with one of them. It is interesting to note, in a recent paper by S. Ballantine,* that this aspect of sound intensity measurement is now receiving more consideration. Using a Wente condenser microphone to measure sound intensity, Ballantine attempts to make the uncertainty less serious by mounting the receiver in a 'standard spherical body.' The whole is then regarded as a rigid sphere amenable to Rayleigh's mathematical diffraction treatment.† In fig. 127 the curves, due to Ballantine, show the pressure-ratio p/p_0 ‡ at different frequencies for spheres 6 in. and 12 in. in diameter. It will be seen that the spherical receiver 12 in. in diameter can only be regarded as small at

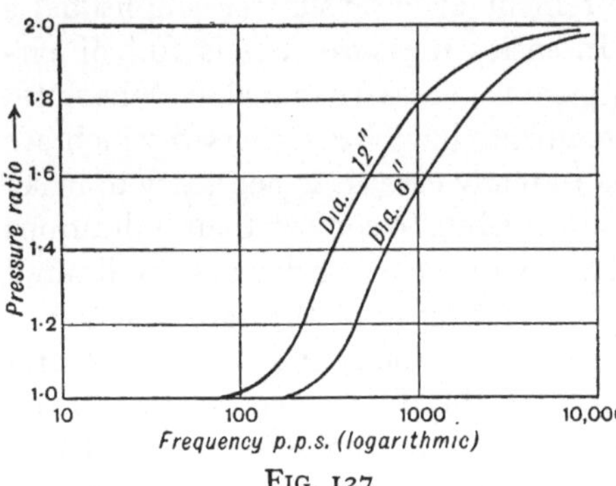

FIG. 127

* *Phys. Rev.*, **32**, p. 988, Dec. 1928.

† *Theory of Sound*, **2**, pp. 218 *et seq.*

‡ The ratio of pressure amplitude p on the receiver to that p_0 in the undisturbed medium.

frequencies below 80 p.p.s. Above this frequency the pressure-amplitude *measured* varies rapidly with frequency, ultimately reaching the double-pressure at very high frequencies (>10,000 p.p.s.). In this treatment it has been assumed that any form of pressure receiver may be used, and that the sphere is rigid. The latter assumption, which may easily be attained for air-borne sounds, is not so easy of attainment in a medium such as water. A body which may be regarded as *rigid* in water carrying sound waves is not known. In such a case, therefore, a further correction would be necessary to allow for the elasticity of the sphere. It must be remembered that the directional properties of a receiver depend on the ratio λ/D, where λ is the wave-length and D the diameter of the receiver (assumed circular). As the frequency increases therefore, λ diminishes, and the receiver becomes more and more sharply directional (see pp. 216 and 421). The orientation of the receiver with respect to the incident sound becomes increasingly important therefore the higher the frequency.

'Absolute' sound intensity measurement is very difficult. It must not be supposed, however, that sound intensity measurements are useless, for there are many forms of receiver which, when used with due caution, give results approximating to 'absolute' measure. Comparative sound-intensity measurements are possible with almost all forms of sound receiver, but some forms are decidedly better than others. Resonant receivers have proved of great value in comparing sound intensities at a particular frequency, but comparison of intensities of different frequencies with different tuned receivers is rendered uncertain on account of the difficulty of estimating the 'amplification factor' of such receivers. Certain forms of measuring device may be calibrated, so that their indications form an approximate measure of sound-intensity within prescribed limits of intensity and frequency.

To summarise: the term 'absolute' applied to a sound-intensity measuring device should always be used with great caution. Various types of sound receivers and recorders are known which give a fairly uniform response over a limited range of frequency and intensity; within such limits their indications may be correlated to the sound-intensity, and, subject to certain corrections, they may be used in sound analysis. Among the many methods used in intensity measurement and comparison, the following may be mentioned: (1) the Rayleigh disc (with or without resonator); (2) the sound-pressure radiometer; (3) hot-wire microphone (with or without resonator); (4) motion of a diaphragm observed by optical means (various types of 'phonometers'); (5) interferometer methods of observation of the variation of refractive index; (6) use of electromagnetic, electrodynamic, piezo-electric, or such receivers, in conjunction with auxiliary electrical circuits (valve-amplifiers) and recorders (oscillographs).

The frequency analysis of a complex sound by variable tuning devices (mechanical or electrical) or by oscillographic and stroboscopic methods is a relatively simple problem.

The Rayleigh Disc – It is shown in textbooks of hydrodynamics * that the lines of flow, or 'streamlines' around an obstacle of ellipsoidal form, are such that the obstacle tends to set in a stable position with its *least* axis in the direction of undisturbed flow. In the case of a plane lamina inclined at 45° to the stream, the lines of flow are as shown in fig. 128, these *being unaltered if the direction of flow is reversed.* The points A and B where the hyperbolic arcs meet the lamina are points of zero velocity, and therefore of maximum pressure. The fluid pressures on the lamina are therefore equivalent to a couple tending to set it broadside-on to the stream. In the case of a circular disc of radius a in a fluid of density ρ, the normal to the disc, making an angle θ with the direction of the undisturbed stream, the couple M tending to reduce the angle θ is given by

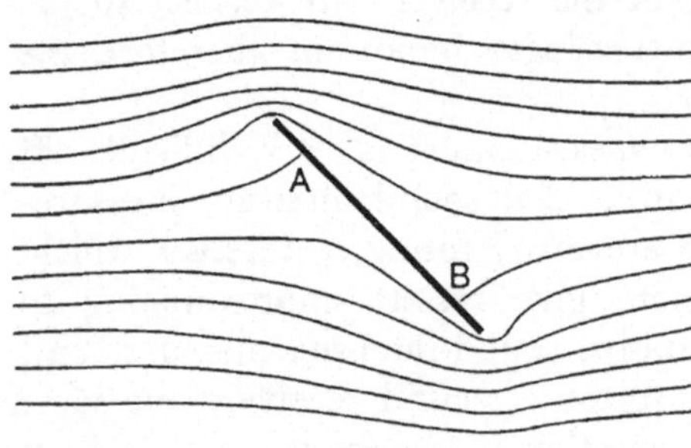

Fig. 128—Streamlines of Flow past a Disc

$$M=\tfrac{4}{3}\rho a^3 v^2 \sin 2\theta \qquad . \quad . \quad . \quad (1)$$

In this relation it will be seen that the torque is proportional to v^2—that is, independent of the *direction* of flow. The relation is therefore equally applicable to direct or alternating flow, provided the value of v^2 in the latter case represents the *mean* value over a cycle. The torque is a maximum when $\theta=45°$. The effect was demonstrated experimentally by Rayleigh,† who observed the rotation of a small paper disc, suspended on a fine fibre, at an angle of 45° to the plane of the opening of a Helmholtz resonator in

* Lamb, *Hydrodynamics*, pp. 94 and 178, 1895. † *Sound*, 2, p. 44.

vibration. Rayleigh subsequently designed a simple instrument, based on this principle, for comparing the intensities of sounds of definite pitch—the intensity being proportional to v^2, which is measured directly by the torque M. A tube $\frac{3}{4}$ λ in length is open at one end and closed with a glass plate at the other. At an anti-node, distant $\lambda/4$ from the closed end, a light circular mirror is hung by a fine fibre at an angle of 45° to the axis of the tube. A small glass window at the side allows the light reflected from the mirror to fall on a scale outside. When the tube is set in resonance the alternating air-flow round the mirror causes it to rotate. For small angles of deflection the rotation is proportional to the sound-intensity in the tube, and consequently to the intensity in the undisturbed field. The actual value of the torque may, if required, be determined by means of a torsion head which brings the mirror back to its initial position. A *double resonator* with such a mirror, supported by a fine quartz fibre in the connecting neck (see fig. 129), forms a much more sensitive arrangement (due to Boys,* see

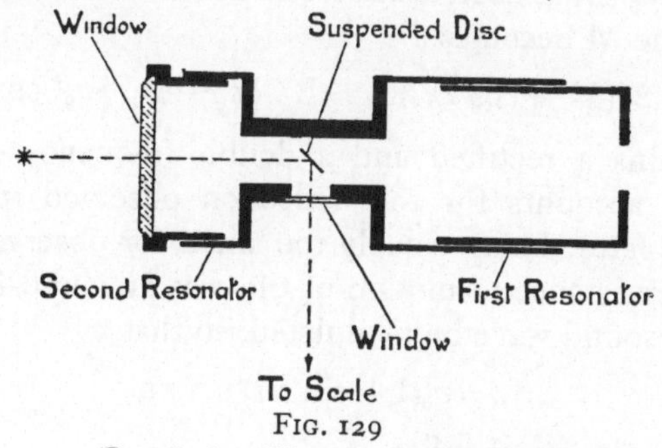

FIG. 129
(From Davis and Kaye, *Acoustics of Buildings*)

p. 189). The velocity of the air at the mirror is in this case greatly amplified by the second resonator. The sensitiveness is comparable with that of the ear. The system is, of course, very sharply tuned, and a number of similar instruments would, therefore, be required to cover a moderate range of frequency. The original form, suggested by Rayleigh, but fitted with a continuously variable tuning arrangement, is more generally useful, but is of course less sensitive. It is important to observe, in all forms, that the diameter of the disc should be small compared with a wave-length of the incident sound. The device has been used extensively as a means of comparing sound intensities, and has often been used to check other methods. Thus E. Meyer † has

* *Nature*, **42**, p. 604, 1890.

† *Elec. Nach.-Techn.*, **3**, p. 290, 1926 ; and **4**, p. 86, 1927.

shown that concordant results are obtained over a wide range of frequencies when it is compared with the thermophone (see p. 203) and the condenser-microphone (see p. 395). Mallet and Dutton * have used the Rayleigh disc to determine the efficiency of telephone receivers from the curves of amplitude-frequency near resonance, the results agreeing well with motional-impedance measurements. The device has a very wide application in explorations of sound fields.†

A modification in the method of use of a Rayleigh disc has recently been suggested by L. J. Sivian.‡ In this method the amplitude of the sound wave to be measured is modulated with a frequency equal to that of the free vibration of the suspended disc. The measurement consists in reading the amplitude of the oscillations corresponding to the modulating frequency rather than a steady deflection of the disc. The disturbances caused by spurious air currents are thus largely reduced. If in equation (1) above, we write $v=v_0 \sin nt$ as in a sinusoidal sound wave, the expression for the torque M becomes

$$M=\tfrac{4}{3}\rho a^3 \sin 2\theta(\tfrac{1}{2}-\tfrac{1}{2}\cos 2nt)v_0^2=B\,.\,\tfrac{1}{2}v_0^2-B\,.\,\tfrac{1}{2}v_0^2\cos 2nt \quad . \quad (2)$$

which contains a rectified and a double-frequency component. The former accounts for the deflection observed in the static method, the latter being usually too small for observation owing to the high resistance to motion of the disc at sound-frequencies. If, now, the sound wave be modulated, so that

$$v=v_0(1+\sin wt)\sin nt,$$

the torque, acting on the disc, is

$$M_t=B\,.\,\tfrac{1}{2}v_0^2[\tfrac{3}{2}+2\sin wt-\tfrac{1}{2}\cos 2wt](1-\cos 2nt).$$

The subsequent motion is therefore of the form $I\ddot{\phi}+R\dot{\phi}+S\phi=M_t$, the modulating frequency w being chosen so that $Iw^2-s=0$. In the steady state the deflections produced by the modulated torque are:

$$\left.\begin{array}{l}(1)\ D_1=\frac{3}{4}Bv_0^2\,.\,\frac{1}{S},\quad \text{that is,}\quad D_1\propto a^3/S\quad \text{since}\quad B\propto a^3\\[1ex](2)\ D_2=Bv_0^2\frac{I}{Rw},\quad \text{that is,}\quad D_2\propto 1/w\quad \text{since}\quad R\propto a^3\\\qquad\qquad\qquad\qquad\qquad (R=\frac{64}{5}\mu a^3\ \text{for a thin disc})\\(3)\ \text{A negligible effect of double frequency.}\end{array}\right\}\quad . \quad (3)$$

* *Journ. I.E.E.*, **63**, p. 502, 1925.

† Stewart and Stiles, *Phys. Rev.*, **1**, p. 309, 1913 ; **32**, p. 248, 1911 ; **33**, p. 467, 1911.

‡ *Phil. Mag.*, **5**, p. 615, 1928.

In the first of these expressions, D_1 is the deflection observed in the *static method*, proportional to a^3. The second expression giving D_2, the amplitude observed in the *dynamic method*, is independent of the disc radius, provided the period of the disc ($\propto 1/w$) remains constant. The distinction is very important in high-frequency measurements, since the diameter of the disc must always be small compared with the wave-length. The dynamic method, which measures intensity ($\propto v_0^2$) in terms of D_2, possesses the great advantage that it is not disturbed by slow drifts and sudden puffs of air, which are sufficient to vitiate entirely results obtained by the static method.

Pressure of Sound Waves

Sound waves, like light waves, exert a pressure on any surface, reflecting or absorbing, on which they fall, and radiometers for measuring sound energy have been constructed on this principle. The theoretical aspect of the matter has received considerable attention from Rayleigh, Larmor, Poynting, and others. In order to illustrate in as simple a manner as possible how this pressure arises, we cannot do better than quote Poynting:* "In sound waves there is at a reflecting surface a node—a point of no motion but of varying pressure. If the variations of pressure from the undisturbed value were exactly proportional to the displacements of a parallel layer near the surface, and if the displacements were exactly harmonic, then the average pressure would be equal to the normal undisturbed value. But consider a layer of air quite close to the surface. If it moves up a distance y towards the surface, the pressure is increased. If it moves an equal distance y away from the surface, the pressure is decreased, but by a slightly smaller quantity. To illustrate this, take an extreme case and, for simplicity, suppose that Boyle's law holds. If the layer advances half-way towards the reflecting surface the pressure is doubled. If it moves an equal distance outwards from its original position the pressure falls, but only by one-third of its original value; and if we could suppose the layer to be moving harmonically, it is obvious that the mean of the increased and diminished pressures would be largely in excess of the normal value. Though we are not entitled to assume the existence of the harmonic vibrations when we take into account the second order of small quantities, yet this illustration gives the right idea. The excess of pressure

* Address on Radiation Pressure, *Phys. Soc.*, 1905.

in the compressed half is greater than its defect in the extension half, and the net result is an average excess of pressure—a quantity itself of the second order—on the reflecting surface. This excess in the compression half of a wave-train is connected with the extra speed which exists in that half, and makes the crests of intense sound waves gain on the troughs." The following *simple treatment*, which proves the existence of radiation-pressure by an indirect method, is *due to Larmor*.* The case considered is that of plane waves incident normally on a perfectly reflecting wall. The latter is pushed with velocity v to meet the advancing train of waves. The energy density in the incident wave-train of velocity c is E, and in the case of a stationary reflector, the total energy density due to both incident and reflected wave-trains is 2E. In unit time the length of the wave-train incident on the wall is $(c+v)$. On account of the approach of the wall this is compressed into a space of length $(c-v)$. Consequently the energy density in the reflected wave is increased in the ratio

$$\frac{E+\delta E}{E}=\frac{c+v}{c-v}=1+\frac{2v}{c}$$

since v is very small, whence $\delta E=2vE/c$. In a region of length c in front of the wall, the increase in total energy is therefore $c \,.\, \delta E=2vE$. This must necessarily be the work done by the wall in compressing the radiation. Consequently, if P is the radiation-pressure, the work done by the wall in unit time is Pv, whence

$$Pv=2vE \quad \text{or} \quad P=2E \quad . \quad . \quad . \quad (1)$$

The radiation-pressure is therefore equal to the mean energy density in the medium in front of the reflector.

In the case of a perfectly 'absorbing' wall, there would be no reflected wave and P would be equal to E.

Rayleigh † has dealt with the problem by more rigorous mathematical methods. The gas is regarded as enclosed in a long cylinder of length l closed by a piston subjected to the additional pressure due to the sound waves reflected from it. The calculation is quite general and is applicable to a gas (or any fluid) in which the pressure is any arbitrary function of the density $p=f(\rho)$. If $p_1\rho_1$ denote the pressure and density at the piston and $p_0\rho_0$

* Article on 'Radiation,' *Ency. Brit.*, 11th ed., **22**, p. 786, 1911; see also Crandall, *Sound*, p. 179.

† *Phil. Mag.*, **10**, p. 364, 1905. See also *Phil. Mag.*, **3**, p. 338, 1902; and *Scientific Papers*, **5**, pp. 41 and 262.

the normal values in the absence of vibrations, then it may be proved that

$$\int(p_1-p_0)dt=\left\{\rho_0+\frac{\rho_0{}^2f''(\rho_0)}{2f'(\rho_0)}\right\}\iint\frac{\mathrm{U}^2dxdt}{l} \quad . \quad (2)$$

where U is the resultant velocity at any point and $f'(\rho_0)$ and $f''(\rho_0)$ are the first and second differentials of $f(\rho_0)$. The integral on the left of this equation is the mean additional pressure on the piston, *i.e.* the mean radiation-pressure P which is required. The equation expresses this pressure in terms of the mean kinetic energy. If the relation between pressure and density is adiabatic, that is, $(p/p_0)=(\rho/\rho_0)^\gamma$, equation (2) becomes

$$\mathrm{P}=\tfrac{1}{2}(\gamma+1)\rho_0\iint\frac{\mathrm{U}^2}{l}\,.\,dxdt \quad . \quad . \quad (3)$$

In this expression $\rho_0\iint\frac{\mathrm{U}^2}{l}\,.\,dxdt$ represents the volume density of the *total* energy—that is, double the volume density of the kinetic energy. Denoting the latter quantity by E, we have therefore

$$\mathrm{P}=(\gamma+1)\mathrm{E} \quad . \quad . \quad . \quad . \quad (4)$$

which, in the case of a gas obeying Boyle's law ($\gamma=1$), reduces to

$$\mathrm{P}=2\mathrm{E} \quad . \quad . \quad . \quad . \quad (5)$$

in agreement with the expression (1) obtained by Poynting and Larmor. As Poynting has explained, if the reflection of a train of waves exerts a pressure on the reflector, it must be due to the *momentum of the wave-train.* The value of the momentum M is obtained at once from the above expressions for P in accordance with the usual definition of this quantity M. Rayleigh shows that waves of finite condensation *propagated without change of type* ($p=$ const $-c^2\rho_0{}^2/\rho$) *exert no pressure* on a reflecting surface. "It would seem that pressure and momentum are here associated with the tendency of waves to alter their form as they proceed on their course."

R. W. Wood * demonstrated the existence of a radiation-pressure by means of a concave mirror which was used to focus intense sound pulses from powerful electric sparks, on a set of vanes like a radiometer, causing them to rotate when situated at the focus. In such a case the sound wave is distinctly unsymmetrical and a mechanical pressure would be anticipated on elementary grounds.

* *Phys. Zeits.*, **1**, p. 22, Jan. 1905.

It has been shown on p. 54 that the intensity I of sound waves is equal to the product of energy density E and wave-velocity c. Thus, if P is the radiation-pressure exerted on a reflecting surface, we may write

$$I = Ec = Pc/(\gamma+1) \quad . \quad . \quad . \quad (6)$$

a simple relation which forms the basis of an 'absolute' method of measuring sound intensity.

Sound Radiometers. (Torsion-Vane Pendulum) - The steady pressure exerted by sound waves on reflection from a plane solid wall was measured by Altberg.* A hole in the wall was closed by a loosely fitting piston attached to one end of the arm of a torsion pendulum, a balance weight (not exposed to the radiation) being fixed at the opposite end. The arm was supported, at the centre of gravity of the system, by a fine torsion wire or fibre, and a mirror (with lamp and scale) indicated any deflection of the piston. The latter was restored to its zero position by means of a torsion head. In the case of a piston of small area S, at an effective distance r from the point of suspension, the torque T is given by $T = SPr$. This is counterbalanced by the torsion of the suspension. If the torsion constant of the latter is k, and an angle of twist θ is required to restore the arm, against the pressure, to its initial position $SPr = k\theta$, whence $P = k\theta/Sr$, the value of the sound intensity being obtained at once from equation (6).

Altberg employed this method to measure the intensity of sound emitted from a Kundt's tube excited by the powerful longitudinal vibrations of a glass rod. Altberg's radiometer method was also used by N. Neklapajev † to measure the intensity of high-frequency waves, and to determine the law of absorption in air. T. P. Abello ‡ has employed the same method to measure sound absorption at high frequencies in various mixtures of gases. The same principle has also been applied by P. Langevin § and R. W. Boyle || to the measurement of high-frequency sound intensities under water. The torsion vane was a disc of metal of appropriate thickness (or

* *Ann. der Physik*, **11**, p. 405, 1903.

† *Ann. der Physik*, **35**, p. 175, 1911; see also P. Lebedew, *Ann. der Physik*, **35**, p. 171, 1911.

‡ *Phys. Rev.*, **31**, p. 1083, 1928; *Proc. Nat. Acad.*, **13**, p. 699, 1927; see also Altberg and Holtzmann, *Phys. Zeits.*, **26**, p. 149, 1925.

§ International Hydrographic Bureau, Monaco, Special Publication No. 3, p. 28, 1924, and No. 4, 1926.

|| *Proc. Roy. Soc. Canada*, **3**, p. 167, 1925.

alternatively an air-film enclosed between mica discs) supported from a fine phosphor bronze or steel wire. The deflection of the vane, under the action of the supersonic beam from a quartz transmitter, was observed by means of a telescope, and a torsion head was used to restore the vane to its zero position. In these measurements no 'guard plane' was used, so that it became necessary to correct the observations for diffraction effects at the edges of the vane. This troublesome correction is of course entirely avoided if the vane forms a part of a reflecting plane of large area.

Sound radiometers are very insensitive, and are consequently quite unsuitable for the detection of feeble sounds. They provide, however, a very convenient means of measuring the intensity of continuous sounds of large amplitude—particularly at high frequencies when the wave-length is small. In this respect they have a wide application in the measurement of the energy emission from sources of sound.

Forces of Reaction between Bodies in a Sound Field - It is convenient to refer here to a few interesting phenomena relating to the attraction or repulsion between various bodies in vibration, or under the influence of the vibrations of a medium carrying a train of sound waves.

(a) *Repulsion of Resonators* - Dvorak and Meyer * independently discovered that an air resonator of any kind, when exposed to a powerful source of sound, experiences a force of repulsion directed from the mouth inwards. Four light resonators mounted on a pivot, after the fashion of an anemometer, may be caused to rotate continuously under the action of this repulsion. Rayleigh (*Sound*, **2**, p. 42) has dealt with the mathematical side of the question, and has proved that the mean pressure inside a resonator is greater than that in the surrounding medium, *i.e.* there must be a force at the mouth acting *inwards*.

(b) *Attraction between a Vibrating Body and a Neighbouring Obstacle* - Kelvin has shown that the mean pressure at a place where there is motion is less than in the undisturbed medium:

$$\int(p_1-p_0)dt=-\tfrac{1}{2}\rho\int U_1^2dt,$$

where U_1 is the velocity and p_1 the pressure at any point in a medium of density ρ and undisturbed pressure p_0 (see also equations (2) and (3), p. 431). This theorem was used to explain

* *Pogg. Ann.*, **157**, p. 42, 1876; and *Phil. Mag.*, **6**, p. 225, 1878.

certain observations by Guthrie * relating to the attraction of a suspended disc of paper by the prong of a vibrating tuning-fork. The behaviour of the disc depends on the fact that the mean value of U^2 is greater on the face exposed to the fork than upon the back. The action depends essentially on the proximity of the source of disturbance. When the flow of liquid, whether steady or alternating, is uniform over a large region the effect on the obstacle is a question of shape (see, for example, the Rayleigh disc which *rotates* in a uniform sound field, whilst a sphere is unaffected).

(c) *Attraction or Repulsion between Spherical Particles in a Sound Field* – In a uniform sound field the force on a single sphere is zero. If, however, two spheres, at a moderate distance apart, lie in a line parallel to the alternating flow they will be repelled. If, on the other hand, the line joining their centres is at right angles to the flow they will be attracted. These statements are due to König,† who thereby explained the peculiar nature of the striations observed in the Kundt's dust-tube experiment. The forces between neighbouring particles are such as to cause the particles to collect in laminæ across the tube, the effect being a maximum at the antinodes. König showed that the force acting between two spheres of radii a_1 a_2, and situated a distance r apart, is given by (i) a repulsion $X=6\pi\rho a_1^3a_2^3v^2/r^4$ when the pair is 'end-on' to the stream of velocity v, and (ii) an attraction $Y=3\pi\rho a_1^3a_2^3v^2/r^4$ when 'broadside' to the stream. The force therefore increases extremely rapidly with diminishing distance of separation of the particles. R. W. Boyle and R. W. Wood have independently referred (*loc. cit.*) to the rapid coagulation of small particles of coke or other powder in a liquid through which high-frequency sound waves of great intensity are passing. It is not unlikely that the effects are, at any rate, partly due to this cause.

Interferometer Methods – The ideal method of measuring the intensity of a sound wave is one in which the 'receiver or indicator' does not modify in any way the sound distribution at the point of the medium under consideration. A solid obstacle, such as a microphone, electrodynamic receiver, or a resonator, does not comply with this condition, for it behaves as a secondary source and scatters some of the incident sound energy. A method

* *Proc. Roy. Soc.*, **19**, p. 35, 1870, and **19**, p. 271, 1871.
† *Wied. Ann.*, **42**, p. 353, 1891.

approximating to the ideal case is one to which we have already briefly referred (p. 184), viz. the interferometer method introduced by Töpler and Boltzmann,* and developed later by Raps.† In this method a beam of light traverses the medium through which the sound waves pass, and produces a set of interference fringes with a second beam of light outside the path of the sound waves. The changes of density in the medium traversed by the first beam result in corresponding changes of optical path, and displace the interference fringes accordingly. It is important to observe in such measurements that the interferometer mirrors and mountings are not set in vibration by the sound wave, otherwise large errors may occur. So far as the writer is aware, the method has only been applied to cases of large sound intensity—*e.g.* Raps studied the changes of density near the node of an organ pipe. As an approximation to an absolute method it is, however, worthy of more consideration than it has hitherto received.

Phonometers - *The Webster Phonometer* ‡ is a useful detecting and measuring device based on the principle of a tunable Helmholtz resonator (see fig. 130). The motion of the air in the mouth of the resonator is indicated by means of a small mica disc supported by three coplanar stretched wires, which serve also to tune the disc. The system is therefore doubly tuned. The movement

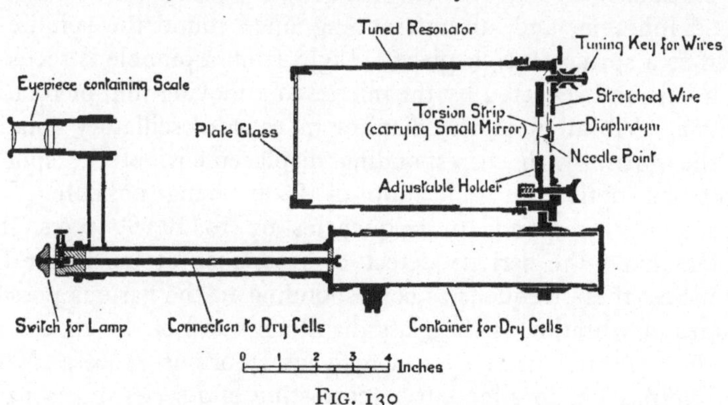

FIG. 130
(From Davis and Kaye, *Acoustics of Buildings*)

of the disc is indicated by means of a thin steel torsion-strip carrying a small concave mirror, which tilts when the disc vibrates. Light from a lamp (not shown in fig. 130) with a vertical, straight filament is reflected from the mirror into an eyepiece alongside it.

* *Pogg. Ann.*, **141**, p. 321, 1870. † *Ann. der Physik*, **50**, p. 193, 1891.
‡ *Nat. Acad. Sci. Proc.*, **5**, p. 173, 1919 ; and *Engineering*, **3**, p. 763, 1921 (R.I. discourse).

A calibrated scale in the eyepiece serves to measure the amplitude of the diaphragm. Using a total optical magnification of the order of 1500 times a deflection of 0·1 mm. in the eyepiece, corresponds to a displacement of 4×10^{-6} cm. of the disc. The instrument is calibrated, by an interferometer method, to give 'absolute' measurements of amplitude. L. V. King * used the Webster phonometer in measurements of the acoustic efficiency of fog-signalling machinery. The instrument was able to detect, and measure, the signals when out of range of the ear (see p. 211).

A somewhat similar device, in that it employs a Helmholtz resonator with a diaphragm-mirror combination, was used by M. Wien.† The usual opening of the resonator was, however, completely closed by a thin 'aneroid' diaphragm in tune with the resonator and the incident sound.

The Millar Phonodeik ‡ – The diaphragm sound receiver, devised by D. C. Millar to record musical sounds, is typical of many other forms of diaphragm receiver. A diaphragm of glass (0·003 in. thick) is held lightly between soft rubber rings and closes the apex of a conical horn. Behind the diaphragm is a minute steel spindle mounted on jewelled bearings, and attached to the spindle (of weight 0·002 grm.) is a tiny oscillograph mirror. A silk fibre, or platinum wire (0·0005 in. thick), is attached to the centre of the diaphragm, and, after wrapping once round the spindle, is fixed to a spring tension-piece. Light from a pinhole is focussed by a lens and reflected by the mirror to a moving film or rotating mirror. Vibration of the diaphragm causes oscillatory rotation of the mirror with corresponding displacements of the spot of light—an optical magnification of 2500 being possible. The arrangement responds to frequencies up to 10,000 p.p.s., but suffers from the serious defect of resonance at particular frequencies, these frequencies corresponding to the various possible modes of vibration of the glass diaphragm and of the air column in the conical horn. By a somewhat laborious process, Millar determines a calibration curve (consisting of a series of resonance peaks) for the instrument, and applies it to the correction of the record of a sound subject to analysis. S. H. Anderson § has

* *Frank. Inst. Journ.*, **183**, p. 259, 1917.

† *Wied. Ann.*, **36**, p. 834, 1889.

‡ Meaning 'to show, or exhibit, sound,' *Phys. Rev.*, **28**, 151, 1909; *Science of Musical Sounds*, p. 79.

§ *Opt. Soc. Amer. Journ.*, **2**, p. 31, 1925. See also C. J. Lapp, *ibid.*, **7**, p. 661, 1923; E. H. Barton and H. M. Browning, *Phil. Mag.*, **50**, p. 957, 1925, and **2**, p. 955, 1926.

recently described a modified form of phonodeik in which only the natural frequency of the diaphragm needs to be considered, the horn and tension spring being eliminated. The instrument is sensitive enough for the quantitative examination of the quality of musical tones within the range 128 to 8200 p.p.s. The device is calibrated, using electrically-maintained tuning-forks as sources of simple 'vibromotive forces.'

Various other types of diaphragm recorder have been introduced since the phonodeik—*e.g.* A. E. Kenelly used a tilting mirror arrangement described on p. 222. *The Hilger Audiometer** is a convenient form of sound recorder of the diaphragm type, but this, like the others, suffers from inconvenient resonances. In this device the diaphragm or membrane is an extremely thin film of collodion (about 1 μ thick) stretched on a metal ring. A mirror of silver or platinum is sputtered (cathodically) on a small area of the film half-way along a radius. The total mass of the film and its attached mirror is extremely small and the air damping fairly large. Sound waves falling on one side of the membrane set it in vibration and tilt the mirror, causing a spot of light to be deflected in the direction of the axis of a rotating drum carrying a 'stroboscopic observation strip' (see p. 226), or a photographic film. Various thicknesses of membrane have been used, having resonance-frequencies ranging from 80 to 3000 p.p.s. The most suitable frequency of membrane must be chosen according to the purpose in view. With this apparatus photographic records have been made of the sounds of speech and various musical instruments, and of the noises arising from machinery in motion. The actual interpretation of the records of noises which include the resonant frequencies of the instrument is necessarily a matter of considerable difficulty.

Nearly all the earlier methods of recording sound made use of a diaphragm as the sensitive receiver. In *the Scott-König Phonautograph* (1859) a membrane actuated by sound waves carried a stylus which traced the vibrations on a smoked paper carried on a rotating cylinder. The records were not only small in size but were considerably distorted by numerous resonances and by friction between the stylus and the surface of the paper.

König's Manometric Capsule (1862) employs a diaphragm to receive the sound and fluctuate the pressure of the gas which feeds a lighted jet. Nichols and Merritt,† using an acetylene flame of increased luminosity, photographed its vibrations when

* *Engineering*, **117**, p. 108, Jan. 1924. † *Phys. Rev.*, **7**, 93, 1908.

the exciting membrane was subjected to the sounds of ordinary speech. Alternatively, the vibrations of the flame are observed in a revolving mirror. A similar device has recently been demonstrated by E. W. Scripture.*

In the first *Edison Phonograph* (1877) the record of the movement of a diaphragm was made by indentations of a stylus in a sheet of tinfoil supported over a spiral groove in a metal cylinder. In more recent forms, the vibrations of the diaphragm cause a needle to cut grooves on the surface of a prepared cylinder or disc. Greatly enlarged copies of such records are obtained photographically by means of a tracing point carrying a small mirror, which reflects a spot of light. Not until recently, however, have such direct methods of sound recording, with diaphragm receivers, proved of value as a means of measuring or comparing sound intensities at different frequencies. All such methods are seriously restricted in application on account of resonance in the diaphragm or some other part of the apparatus.

Electrical Methods

Improvements in the design of non-resonant electrical receivers have led to a considerable development of electrical methods of measurement and recording of sound vibrations. Such methods have many advantages over the more direct mechanical methods. The electrical apparatus is often more sensitive and is more easily controlled. 'Distortionless valve amplifiers' may be used when the sounds are very feeble, and electrical oscillographs having a fairly constant sensitivity over a wide frequency range are available for observing or photographing the vibrations. Moreover, it is often possible in electrical circuits to compensate for defects in the receiver—*e.g.* a troublesome resonance peak or a region of reduced sensitiveness in the receiver may be reduced or increased respectively by suitable design of the amplifier circuits. It should be pointed out, to avoid misapprehension, that the design of mechanical sound-recording and reproducing apparatus (gramophones and loud-speakers) is nowadays proceeding on similar lines, with unmistakable improvement.

Non-resonant receivers such as the Wente condenser microphone, the Marconi-Sykes magnetophone (moving coil), the piezo-electric crystal and the like are now almost universally employed when it is desired to measure or record complex

* *Phys. Soc. Exhibition*, Jan. 8, 1929.

vibrations such as are encountered in music and speech. When the sound wave is simple in character, *i.e.* a pure tone, such receivers as the resonant hot-wire microphone, the Webster phonometer, the Rayleigh disc, and the Gerlach strip receiver are very suitable measuring devices.

Resonated Hot-Wire Microphone – As a means of comparing sound intensities at a particular frequency, this device, in the hands of W. S. Tucker and E. T. Paris, has developed into an extremely useful instrument of considerable precision. Resonators of continuously variable tuning are also very valuable in the frequency analysis of complex sounds. The instrument and its applications have already been described on p. 402, and further reference is unnecessary here.

Electromagnetic and Electrodynamic Receivers have a wide application in sound measurements. If the sound is a pure note the receiver may be resonant, *e.g.* of the simple clamped-diaphragm type. An example of the method of use of such a receiver (of the moving-coil type) is given by T. S. Littler.* When the receiver is placed in a field of sound the diaphragm is forced into vibration with an amplitude proportional to the pressure-amplitude of the sound wave. The coil, moving with the diaphragm, cuts the lines of the magnetic field at right angles, and an e.m.f. is generated in it. If the motion of the diaphragm is represented by

$$x = \mathrm{A}\cos wt,$$

the e.m.f. generated in the coil is

$$e = 2\pi rn\mathrm{NA}\sin wt,$$

where N is the flux-density, r the coil radius, and n the number of turns. The moving coil may be connected to the low-resistance primary of a step-up transformer, when the measured secondary e.m.f. is proportional to the sound-pressure amplitude for a constant frequency. An alternative method compares this secondary e.m.f. with another known e.m.f. of the same frequency obtained from the secondary of a variable mutual inductance. The e.m.f.'s to be compared are applied alternately to the input of a resistance-capacity amplifier, in the output of which is a rectifying device and direct-current measuring instrument. In another arrangement the transformer is replaced by additional inductance and a tuning condenser, which gives the required voltage-amplification without distortion.

* *Journ. Sci. Instrs.*, **4**, p. 337, Aug. 1927.

Null Method of Sound Measurement – An important principle to which we have already referred (p. 390) is due to Gerlach.* The vibrations set up by the incident sound waves in an electrodynamic strip receiver are neutralised in amplitude and phase by corresponding vibrations set up in the strip electrodynamically. This is achieved by passing a known alternating current, of adjustable phase, through the strip until some auxiliary detector (the ear or a small microphone which does not draw much energy from the vibrating strip) indicates that its motion has ceased. A knowledge of the alternating current, together with the electrical and magnetic constants of the apparatus, thus enables the observer to calculate the mechanical forces acting on the strip. The great advantage of the method lies in the fact that the motion of the measuring device (the receiver) is neutralised, and therefore it interferes as little as possible with the sound distribution in the medium. This null method has been called an 'absolute' method, but this description requires some qualification, for the pressure which is measured by the receiver is a function of the size of the receiver and the wave-length of the incident sound. The mounting of the strip interferes with the sound distribution, and it is not easy to determine whether the device measures the actual pressure of the sound wave or double that pressure (by reflection), or some intermediate value (see p. 424). Apart from such considerations, however, Gerlach's 'null' method is very valuable, and is applicable in principle to all forms of electrodynamic receiver. For example, in the Hewlett tone generator (see pp. 161 and 391), or the Fessenden oscillator, used in reception, the motion of the diaphragm can be stopped by a current of the correct strength, frequency, and phase injected into the coils—whence the force acting on the receiver is directly calculable. The method becomes somewhat involved, however, if the incident sound departs from the simplest case of a pure tone.

Another important method of sound intensity measurement, due to F. D. Smith,† in some respects resembles that of Gerlach, since it depends upon a balance between the effects produced on a sensitive amplifying circuit by the unknown sound wave and a known small electromotive force. A moving coil receiver is employed, and the received signal, after amplification, is compared with, or balanced against, the signal produced by a small known ‡

* *Wiss. Veröfft. d. Siemens-Konzern*, **3**, p. 139, 1923.
† *Proc. Phys. Soc.*, **41**, p. 487, Aug. 1929.
‡ See *ibid.*, **41**, p. 18, Dec. 1928.

electromotive force e injected in series with the receiver. When the two signals are equal in intensity it is shown that the following simple relation connects the total pressure P on the receiver with the electromotive force e,

$$P = eHl/z_m,$$

where H is the strength of the magnetic field in which the moving coil (of a length l of the wire) moves, and z_m is the motional impedance (see p. 74) of the receiver at the frequency of the sound. The choice of indicating instrument, to denote equality or balance between the two e.m.f.'s, depends on the frequency and intensity of the sound to be measured (a voltmeter, vibration galvanometer, or telephones may be used). Since the measurement is independent of the amplifying circuit, it is possible to use a high degree of amplification and very feeble sounds may therefore be measured. The method gives the total sound pressure P on the receiver when it is prevented from vibrating. In some special cases, the *absolute* value of the sound pressure p per unit area in the medium can be inferred. If S is the area of the vibrating surface of the receiver, then $P = kpS$, where k is a constant which may have any value between 1 and 2 as the linear dimensions of the receiver vary from a small to a large value compared with a wave-length of the incident sound.*

Sound Analysis

Non-resonant Receiver: Amplifier: Oscillographs–A recent investigation of speech sounds by I. B. Crandall† is typical of present-day methods of sound measurement and analysis, and it may be well to refer to it here in illustration. In this investigation Crandall employed a calibrated *Wente microphone* (see p. 395) coupled to a seven-stage valve amplifier. A special oscillograph of practically constant sensitivity up to 5000 p.p.s. was connected to the output terminals of the amplifier. The overall sensitivity characteristic of the system (microphone, amplifier, and oscillograph) showed that the amplitude of the oscillograph per unit of pressure-amplitude on the diaphragm of the condenser microphone remained practically constant up to 5000 p.p.s., a slight divergence being noticeable at frequencies below 200 p.p.s. The reader is referred to Crandall's original paper for details of the resistance-capacity distortionless amplifier,

* Ballantine, *loc. cit.*, p. 424.

† *Bell System Techn. Journ.*, **4**, p. 586, 1925. See also E. Lübcke, *Zeits. f. Techn. Physik*, **10**, p. 378, 1929.

or to numerous 'wireless' publications dealing with similar apparatus. A severe test of the accuracy of any recording system is to apply a 'square-topped wave,' or impulse, and compare it with the actual record. Crandall applied such a test to his apparatus in the following way. An electrode resembling the back plate of the condenser microphone was mounted in front of the diaphragm. Between the electrode and the diaphragm was applied a high potential which was made alternately positive and negative by a commutator. This arrangement produced the desired positive and negative displacements of the diaphragm. The calculated and observed wave-forms were in all cases found to be in close agreement, and indicated by harmonic analysis that accurate records could be expected over a range of 80 to 5000 p.p.s. The apparatus was sufficiently sensitive to record sounds spoken in the ordinary tone of voice, with the speaker's mouth about three inches from the microphone. A key was pressed by the speaker just before the sound was produced, this releasing a shutter placed before a rotating film drum on which the record from the oscillograph vibrator was traced. Alongside the complex speech record a separate 'timing-wave' was recorded on the film by means of a vibrator of known frequency. The speech records were analysed into Fourier series (see p. 25) in order to determine the relative amplitudes and frequencies of the component vibrations.

Somewhat similar methods of recording are described by Maxfield and Harrison,* who also deal with the mechanical problem of sound-reproduction based on the electrical analogy. S. Ballantine has used a Wente condenser microphone mounted in a rigid sphere to obtain an absolute measure of sound intensity (see p. 424). The microphone is first calibrated (see p. 397) by means of the 'pistonphone' and the 'thermophone,' and a correction for frequency applied in accordance with the curves shown in fig. 127, p. 424.

Crandall's work on speech-sound measurement and analysis is typical of modern practice in dealing with complex sounds. If desired, an independent frequency analysis may be carried out by the various resonance methods already described, or by means of electrical harmonic analysers.† Alternatively the oscillograph record is analysed by one of the many harmonic

* *Bell System Techn. Journ.*, **5**, p. 493, 1926.

† See Cockroft and Miles Walker, *Journ. I.E.E.*, **63**, p. 69. Manufactured by H. Tinsley and Co.

analysers mentioned by D. C. Millar (see p. 33). On account of the importance of wave-form recording and analysis of complex sounds by such methods, it may not be out of place to refer briefly to various types of electrical oscillograph in use at the present time and to indicate the range of possible application.

Electrical Oscillographs for Sound-wave Recording and Analysis - The name oscillograph was invented by Blondel * to denote an instrument for indicating the instantaneous value of an electrical current. The moving or indicating system of the oscillograph must be capable of following the current fluctuations even when these are very rapid. Progress in oscillograph design has consequently followed the lines of diminishing the mass of the moving system ; the heavy moving-iron system first used by Blondel having been succeeded by the relatively light bifilar strip-conductor of Duddell, the conducting quartz fibre used by Einthoven, and ultimately the beam of cathode rays, which is the ideal 'massless' moving element.

(a) *The Duddell Oscillograph* † has been described in many text-books,‡ particularly in reference to electrical engineering practice, but on account of its relatively high natural frequency it has proved of value in recording sound waves. The instrument consists of a powerful electromagnet NS which produces an intense magnetic field across a gap in which lies a bifilar strip of phosphor bronze carrying the alternating current from the electrical sound-receiver (or amplifier), see fig. 131. Between the strips *ss* is a narrow tongue of soft iron which reduces the magnetic reluctance of the gap. The alternating current passes up one strip and down the other, the plane of the strip tilting to and fro in the magnetic field, analogous to the rotation of a suspended coil of a galvanometer carrying current which alternately reverses in direction. The oscillation of the plane of the strips is indicated by a small mirror M (1 mm. ×0·3 mm.) attached to both at the midpoint and reflecting a spot of light from an arc to a moving film. By means of a tensioning device P the natural frequency of the system may be brought up to 10,000 p.p.s. in air, the tension in each strip being about 50 grm. In order to eliminate resonance in the strips, the clearances at the sides and the channels in which they lie are made as small as possible and the space around them is flooded with oil.

* *La Lumière Electrique*, **41**, p. 401, 1891.

† Duddell, British Assoc., 1897, *Electrician*, **39**, p. 636.

‡ The reader should consult Irwin's *Oscillographs* (Pitman) for a good general account of Oscillographic recording.

The latter produces some lowering of frequency (about 33 per cent.) and renders the system practically dead-beat. This oscillograph may be used without correction up to a frequency of about 1000 p.p.s. Above this frequency it requires careful calibration

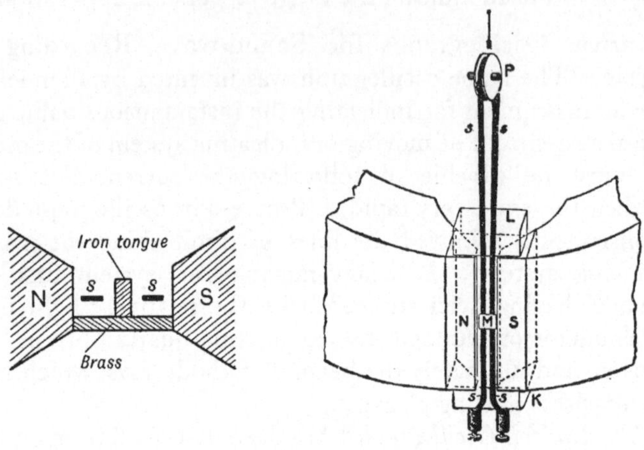

FIG. 131
(By courtesy of Cambridge Inst. Co.)

and must be used with caution. The instrument at its normal frequency is not very sensitive, requiring rather large alternating currents to produce measurable photographic deflections. This limits its use to recording sounds of fairly large intensity, or, alternatively, considerable valve amplification is required.

(b) *The Einthoven String Oscillograph* * consists essentially of a fine conducting wire or fibre stretched at right angles to a strong magnetic field B. Fig. 132A is a diagrammatic view of the arrangement and 132B is a photograph of an actual instrument. When a current i passes through the fibre of length l a force Bli tends to displace it in the field in a direction at right angles to the field and to the fibre. An alternating current produces corresponding oscillations of the fibre. The motion of the latter is magnified (about 500 times) by means of a microscope objective mounted in one pole-piece of the large electromagnet. In the opposite pole-piece is a second microscope objective serving as a condenser lens to illuminate the fibre for observation and projection on a photographic film. The light is obscured at regular intervals by the spokes of a wheel mounted on the spindle of

* See Irwin's *Oscillographs*, p. 23. Manufactured by Camb. Inst. Co. See also H. B. Williams, *Opt. Soc. Amer. Journ.*, **9**, p. 129, Aug. 1924, and **13**, p. 314, Sept. 1926.

a phonic motor controlled by a tuning-fork (see p. 126). The resulting record therefore appears with thin white lines, or 'time marks,' which divide the record into equal periods of time, for use in frequency analysis. The sensitivity of the oscillograph is governed by the strength of the magnetic field (of the order 20,000 lines/cm.^{2}) and the natural frequency of the fibre. The frequency

FIG. 132A FIG. 132B

(By courtesy of Cambridge Inst. Co.)

is controlled by means of a tensioning screw. The thickness of the fibre may vary from 1 or 2 μ quartz or glass, to 15 or 20 μ copper, aluminium, phosphor-bronze, or tungsten ($1\ \mu = 10^{-4}$ cm.). A silvered quartz fibre 2 μ diameter has a mass per cm. length of about 10^{-7} grm. and an area of cross-section $3{\cdot}1 \times 10^{-8}$ cm.^{2}. Taking the tensile strength of a quartz fibre as 10^{10} dynes/cm.^{2} (Kaye and Laby's tables, p. 28) and a safe working limit as 10^9 dynes/cm.^{2}, we can deduce the maximum safe tension T_{max} and frequency N_{max} of the fibre. Thus

$$T_{max} = 31 \text{ dynes} \quad \text{and} \quad N_{max} = \frac{2}{l}\sqrt{\frac{31}{10^{-7}}} = 3{\cdot}5 \times 10^4/l.$$

The upper safe limit of frequency of a quartz fibre of length $l = 10$ cm. would therefore be about 3500 p.p.s. With thin metal wires the maximum frequency is considerably lower than this on account of the increased mass per unit length and diminished tensile strength. The thinnest fibres of quartz or glass are almost critically damped (or dead-beat) in air at ordinary frequencies, but this is not the case with the heavier metal filaments. Irwin* has described a method of damping the natural vibrations of

* *Loc. cit.*

an Einthoven fibre by means of a 'resonant shunt.' The latter consists of a condenser C, inductance L, and resistance R in series. The best combination of C, L, and R for critical damping has been determined by S. Butterworth, A. B. Wood, and E. H. Lakey.* The shunt constants must be so chosen that the free electrical oscillation in the shunt when short-circuited is an exact copy of the free mechanical oscillation of the string when short circuited. This fixes the values of LC and R/L for the shunt. The absolute value of L depends upon the degree of damping required, small inductances giving large damping. For practically undamped strings, critical damping is obtained if the coil reactance at the resonant frequency of the string is equal to one half the resistance of the string. Subject to such conditions, records have been obtained with copper or phosphor-bronze fibres (14 μ thick) which showed no trace of the resonant oscillation. The damping of a high resistance, air-damped, silvered quartz fibre is relatively simple. The resonant shunt method of damping a fibre makes it possible to obtain the maximum degree of sensitiveness consistent with the condition of critical damping. If accurate records of sound vibrations are to be obtained the latter condition is essential, and a high, free frequency of vibration of the fibre is desirable. Records (*a*) and (*b*) in fig. 133 illustrate the effect of the resonant shunt, when a small voltage was suddenly applied (the 'square-wave test') to a fibre of frequency 1500 p.p.s. The advantages of the resonant shunt are obvious. The shunted record still shows traces of the

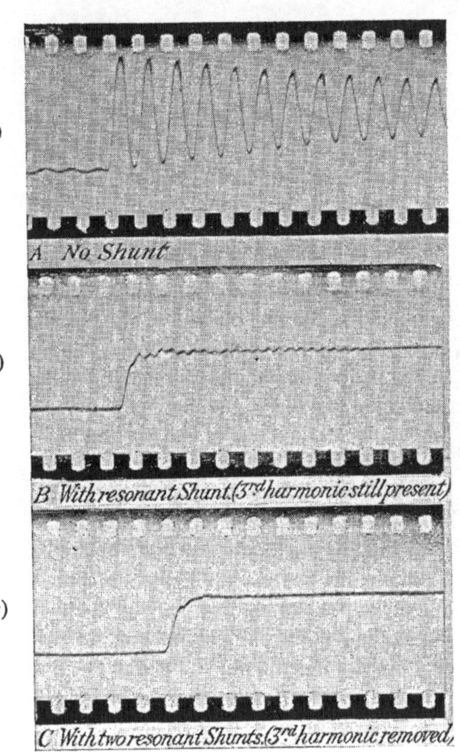

FIG. 133—Oscillograph Records illustrating damping of Einthoven fibre by 'Resonant-Shunt' Method

* *Journ. Sci. Instrs.*, **4**, p. 8, Oct. 1926.

third harmonic (4500 p.p.s. of the string; this is removed, if necessary, by a second shunt of the corresponding frequency—see fig. 133 (*c*)). The Einthoven oscillograph is frequently used in the analysis of sound waves. It is considerably more sensitive than the Duddell oscillograph, and consequently may be used to record sounds of reduced intensity, or, alternatively, less amplification of the currents from the sound receiver is required. On the other hand its upper limit of frequency, unless exceptionally short fibres are used, is somewhat low—as we have seen a very high value is 3500 p.p.s. for a 10 cm. fibre of quartz.

(c) The *Abraham Rheograph* * is an interesting example of a moving coil galvanometer used as an oscillograph. Abraham * succeeded in making the inertia and damping of the moving coil negligible by introducing new electrical forces to balance them—the coil being shunted by a special compensating circuit containing inductance, capacity, and resistance. The arrangement is particularly suitable for recording instantaneous or impulsive effects. It possesses the great advantage of simplicity, and the optical arrangements permit of a fairly large mirror attached to the moving coil.

Another simple form of oscillograph is a modified telephone receiver of the Brown reed-type (see fig. 97). This device, sometimes described as an *optiphone*, employs a small optical lever to magnify the motion of the tip of the vibrating reed when small alternating currents (*e.g.* from a microphone) are passed through the windings. A small oscillograph mirror is mounted by means of rubber solution across two parallel knife edges about 0·1 mm. apart. One of these knife edges is carried by the tip of the vibrating reed and the other by a fixed pillar screwed into the body of the telephone. Vibration of the reed tilts the mirror and produces corresponding displacements of a spot of light on a revolving mirror or moving photographic film. Although its indications are only faithful at comparatively low frequencies, this inexpensive and simple arrangement has many useful applications.

(d) *Electrostatic Oscillographs* – An E.S. string oscillograph for recording voltage oscillations has been designed by Ho and Kato.† A fine fibre carrying the alternating electric charge lies in a strong electrostatic field, which causes the fibre to oscillate with the frequency of the electrical supply. A *piezo-electric oscillograph* ‡ (see p. 146) has been designed by the writer, but this is generally

* See Irwin, *Oscillographs*, p. 51 ; and *Journ. de Phys.*, **4**, p. 265, 1909.
† Manufactured by Camb. Inst. Co.
‡ A. B. Wood, *Phil. Mag.*, p. 631, Sept. 1925.

too insensitive for use in recording sound vibrations of average intensity.

(e) *Cathode Ray Oscillographs* – For low frequencies (up to 1000 or 1500 p.p.s.) the Duddell and Einthoven oscillographs fulfil a very useful purpose, but they fail entirely at very high frequencies. Even at frequencies above 1000 these oscillographs require the use of troublesome correction factors involving the relation between the natural frequency of the moving element and the impressed frequency. The cathode ray oscillograph, on the other hand, *is equally sensitive at all frequencies* from zero to the highest frequency which is conceivable in an electrical circuit; for the cathode ray is, after all, an electron in motion, and can therefore execute any oscillation which can be performed by other electrons oscillating in electrical circuits. As we have seen, the mass per unit length of the thinnest quartz fibre obtainable is 10^{-7} grm., whereas the mass of a low-velocity electron is of the order 10^{-27} grm. (e/m $1{\cdot}772\times10^{7}$ E.M.U. and $e=1{\cdot}59\times10^{-20}$ E.M.U.). The mechanical inertia of a moving electron is therefore a quantity which is entirely negligible in comparison with that of a mechanical system such as a fine Einthoven fibre or Duddell strip. The first attempt to design an oscillograph employing a beam of cathode rays, an electron stream, is due to Braun. He made use of the facts that a pencil of cathode rays in a vacuum is easily deflected by magnetic and electrostatic fields, and that the rays produce phosphorescent and photographic effects under suitable conditions. The *Braun tube* consists, in principle, of a vacuum tube in which a stream of cathode rays is produced by means of a high tension supply of the order of 50,000 volts. Some of these cathode rays pass through a fine tube or pinhole in the anode and thence between the poles of an electromagnet and a pair of electrostatic deflecting plates. Variations of current in the magnet (or of voltage on the deflecting plates) produce corresponding deflections of the pencil of cathode rays, which ultimately falls on a screen of phosphorescent material (willemite, calcium tungstate, etc.) and is thus rendered visible. Attempts to photograph alternating current wave-forms with this oscillograph were unsuccessful until an automatic method was devised which caused the cathode rays to retraverse exactly the same path a large number of times in succession. Such a multiple-exposure method is in many cases impossible, *e.g.* if the electrical effect to be recorded is a single impulse or an oscillation having a complex wave-form. In such cases it is essential that the record should be obtained in a single traverse of the film. The first oscillograph achieving this condition was designed by M. Dufour.* The instrument is essentially an improved form of Braun tube in which the cathode stream, produced at a voltage of 60,000, is very intense. With this oscillograph Dufour has obtained beautiful records of electrical oscillations up to a frequency of 10^8 p.p.s. Now the electrostatic sensitivity of a cathode ray oscillograph is inversely proportional to the voltage V accelerating the cathode stream, consequently Dufour's instrument is somewhat insensitive. To overcome this difficulty Sir J. J. Thomson † employed an electrically heated, lime-coated filament as the source of the cathode stream, and accelerating voltages of the order of 3000 volts were found to give satisfactory results. With this apparatus, used in conjunction with a piezo-electric (tourmaline) receiver, D. A. Keys ‡ obtained records of the rapidly varying pressures of gaseous and under-water explosions. A more robust and reliable form of this type of oscillograph was designed by the writer.§ A section through one form of this instrument is shown in fig. 134, which, in view of the foregoing remarks, will be self-explanatory. On account of the relatively low velocity (1000 volts) of the cathode rays their penetrating power is very low and they are rapidly absorbed by the gelatine of ordinary plates. It is advisable, therefore, to use plates with a low gelatine, *i.e.* high silver bromide, content so that the maximum amount of energy of the rays may be expended in ionising the silver granules. The 'spot' is traversed across the photographic plate at a known speed by an electromagnetic method, depending on the finite time of rise of a current in an inductive circuit. Whilst it crosses the plate, the electrical oscillation applied to the E.S. deflecting plates produces the necessary deflection at right angles and the wave-form is recorded on the plate. The 'square-wave test' to which we have referred above in the case of other oscillographs, reveals the great superiority of this form over all others. A record of a voltage instantaneously applied to the deflecting plates is shown in fig. 135 (*a*), the spot traversing the plate in 0·001 second. As far as it is possible to judge, the change from one position to the other is absolutely instantaneous, and there is no overshoot. That is, the record is perfect from the point of view of recording high-frequency

* This instrument is described in a paper by the writer, *Journ. I.E.E.*, **63**, p. 1046, Nov. 1925.

† *Engineering*, **107**, p. 543, 1919.

‡ *Phil. Mag.*, **42**, p. 473, 1921.

§ A. B. Wood, *Phys. Soc. Proc.*, **35**, p. 109, 1923; and *Journ. I.E.E.*, **63**, p. 1046, 1925.

oscillations. The oscillograph is essentially an electrostatic instrument, and for that reason is best used in conjunction with electrostatic sound receivers, *e.g.* the piezo-electric (Rochelle salt)

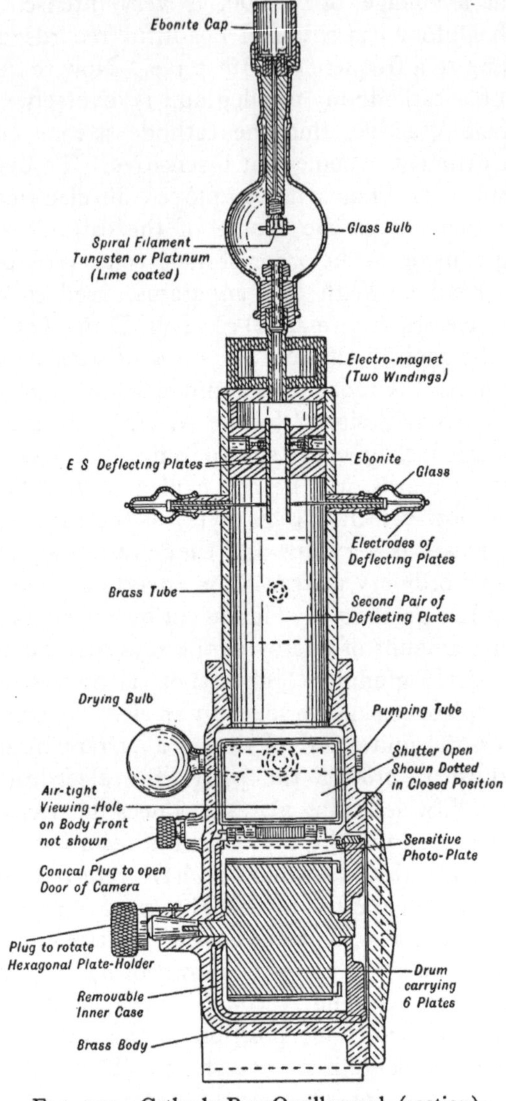

FIG. 134—Cathode Ray Oscillograph (section)
(By courtesy of the Physical Society)

type. A record of a complex, heavily damped sound-impulse under water is shown in fig. 135 (*d*). Fig. 135 (*c*) and (*b*) are similar records of a pure tone and one with superposed harmonics. The voltage-sensitivity of the instrument depends entirely on the potential applied to accelerate the cathode rays and on the dimensions of the deflecting plates (length, distance apart, and distance from photographic plate). In the records shown in fig. 135 the sensitivity was approximately 4 volts per cm. deflection.

With a condenser microphone, or a piezo-electric receiver, and a good distortionless amplifier (*e.g.* capacity-resistance type) the cathode ray oscillograph gives good results. The oscillograph

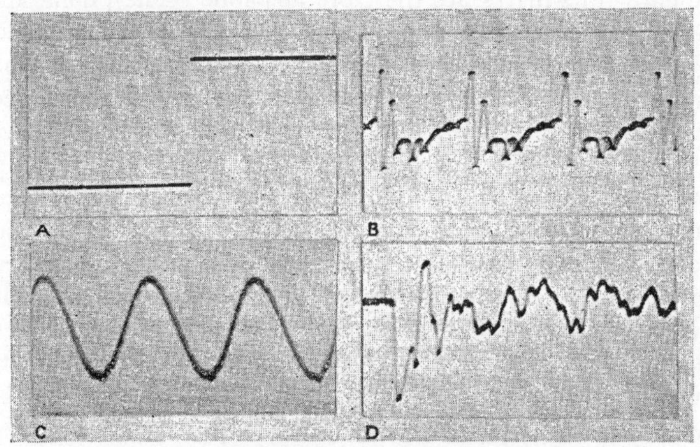

FIG. 135—Cathode Ray Oscillograph Records

itself is quite free from distortion, and the limits of accuracy are defined by the imperfections of the receiver and the amplifier. As we have seen, a piezo-electric receiver has a very high natural frequency, and may therefore be expected to give faithful reproduction at all frequencies within the audible range. A third type of cathode ray oscillograph, operating on 300 volts, has been designed by J. B. Johnson * for visual observations. It is very sensitive, but is generally unsuitable for observing transients or complex wave-forms. No provision is made for photography, and the 'soft' vacuum is liable to result in distortion of wave-form. For general laboratory purposes, particularly in the observation of stationary patterns and Lissajous' figures, it is very convenient.†

Measurement of Frequency

In dealing with typical sources of sound, reference has often been made to the determination of the frequency of vibration.

* *Opt. Soc. Amer.*, **6**, p. 701, 1922.

† See D. W. Dye, *Proc. Phys. Soc.*, **37**, p. 158, April 1925.

The various methods of frequency measurement may be classified broadly into 'absolute' and 'comparative.' It will be sufficient now to draw attention to some of the methods already described, and to add a few others.

Absolute Methods – In this class we include all methods in which the number of complete vibrations per unit time is determined by direct reference to a standard clock. All methods in which the wave-form is directly recorded on a film crossed by standard time-marks are therefore absolute. The phonic motor method of counting the vibrations of a tuning-fork (described on p. 129) is an absolute method, applicable to electrically maintained forks. As we have seen (p. 132), a tuning-fork standardised in this way, or driven directly from a standard clock, forms an electrical standard of frequency over a range extending up to 10^6 p.p.s. The multi-vibrator circuit controlled by such a standard fork then becomes a standard of radio-frequency, and as such is used to determine the frequencies of quartz piezo-electric vibrators, which may then be regarded as substandards (see p. 132).

A convenient laboratory method of determining absolute frequency is the electrically-driven siren fitted with revolution-counting mechanism. The siren note is matched, by the 'beat' method, to the note whose frequency is required, and the absolute frequency determined from a knowledge of the number of holes in the siren disc and the revolutions indicated by the counter in a known period of time.

Comparative Methods – The frequency of any pure tone may be determined by direct comparison with a standard—the latter having been checked by an absolute method. Tuning-forks serve as frequency standards over the audible range of frequency, and in addition to their use as sound sources for such purposes are also used to provide 'time scales' in comparative records (*e.g.* smoked drum, photographic, etc.). If the frequency to be determined is sufficiently near to that of the fork, the simplest method is that in which the number of beats per second is counted, care being taken to observe which has the higher pitch—the standard or the 'unknown.' This observation is easily made by adding a *small* load to the fork and observing the change in the number of beats per second. A number of standard forks may be used to calibrate a monochord (or sonometer), which may subsequently be used as a substandard of frequency having a continuous variation within the limits of calibration. Graphic methods are frequently employed to compare a standard frequency (*e.g.* that of a fork) with the frequency of an unknown, and possibly complex, sound. If the latter is of feeble intensity it may be received by a microphone, amplified and recorded by means of an oscillograph, the record being traversed by standard time-marks. Observation of Lissajous' figures is another method equivalent to the beat method (but of wider application) often used in frequency comparison. The frequency-analysis of a complex sound is conveniently made by means of a Helmholtz resonator of continuously variable tuning (using a piston or a rising water surface), and fitted with a sensitive device such as a hot-wire microphone (see, for example, Fage's analysis of the sounds from an airscrew, pp. 188 and 214). The response curve of the microphone will show a peak at each predominant frequency, the latter being determined from the known position of the piston in the resonator. *Stern's 'tonvariator'* * is a resonator of continuously variable tuning used as a source of sound for comparison of frequencies. The frequency of a vibrating body may in many cases be determined by stroboscopic methods (see p. 224). Sounds of very high (inaudible) frequency may be compared by 'beat' or heterodyne methods. The high-frequency sound is converted into an electrical oscillation, and superposed upon a second electrical oscillation of variable frequency. After rectification a beat or heterodyne note will be heard in the observer's telephone, the frequency of which can be determined by the usual methods. A knowledge of the superposed high-frequency oscillation and the beat-frequency is sufficient to determine the unknown frequency of the sound. The method of Lissajous' figures may also be applied in the case of high frequencies, using a cathode-ray oscillograph with two pairs of electrostatic deflecting plates (mutually at right angles) to indicate when the unknown and the known frequencies coincide. Frequency may be determined indirectly by measurement of wavelength, and assuming a value for the velocity of sound in the medium through which the waves are passing. For example, to determine the frequency of a quartz oscillator emitting high-frequency sound waves, Pierce's method (p. 258) is very convenient. Kundt's tube serves a similar purpose in the case of sound waves of high or low frequency.

Interrupter Method of Analysis – An interesting method of sound analysis has been developed by G. Barlow and H. B. Keene.† The receiver must be of an electrical type. The

* *Deut. Phys. Ges. Verh.*, p. 302, 1904.
† *Roy. Soc. Phil. Trans.*, **222**, p. 131, 1922.

alternating electrical currents generated by the sound waves pass through a *direct* current galvanometer and a motor-driven interrupter, of which the speed can be varied over the whole range of frequency to be investigated. Generally the intervals during which the circuit is open and closed are equal. When the interruptions synchronise with any component A sin *nt* of the current, the galvanometer responds by giving a steady deflection of magnitude depending on the phase difference. In making the analysis, the frequency of interruption is slowly increased over the whole range. Barlow and Keene used the method to analyse the sounds from a tuning-fork, a voice, an organ pipe, diaphragms, and resonators. Using hydrophone receivers (microphonic and electromagnetic), they also analysed sounds produced under water by ships' propellers.

Vibration Galvanometers are often employed in the frequency analysis of sounds received electrically. If the sound to be analysed has a frequency below 200 p.p.s. the Campbell * vibration galvanometer is a very suitable instrument. It consists essentially of a small coil of wire suspended in a strong magnetic field by a phosphor-bronze strip of adjustable tension which controls the tuning. The Duddell vibration galvanometer responds to higher frequencies (up to 1500 p.p.s.), and consists of a bifilar conducting phosphor-bronze strip (with mirror attached). Adjustments in tension and in length provide for a continuous frequency control. In both these instruments the principle involved is simply that of a moving coil galvanometer in which the coil and suspension are *tuned mechanically to the frequency of the current*. This results in a resonant vibration, and a spot of light reflected from the mirror is drawn out into a continuous band. At resonance the band has its maximum length on the scale. A similar instrument has been designed by P. Rothwell † with a number of elements of the Campbell type tuned in succession over a range of frequencies. In frequency analysis the alternating electrical currents set up in a microphone receiving sound waves are passed through the vibration galvanometer, the frequency of which is varied continuously, the width of the light-band being observed at the various frequency settings of the instrument. The galvanometers have a graduated scale of frequencies, calibrated by alternating currents of standard frequencies. They are not only useful in frequency analysis, but also serve a very useful purpose in comparing small

* A. Campbell, *Phil. Mag.*, **14**, p. 494, 1907; and *Proc. Phys. Soc.*, **26**, p. 120, 1913.

† *Journ. Sci. Instrs.*, **2**, p. 251, 1925.

alternating currents of any frequency within the range of the instrument. That is, they may be used in the *comparison of sound intensities* if the sound is pure and the receiver generates an electrical current. At a particular frequency the length of the band of light on the scale is proportional to the alternating current, the latter being a measure of some function of the amplitude of the sound (according to the type of receiver employed). F. B. Young and the writer * (in 1916) used a vibration galvanometer to calibrate a shunted-telephone system for comparing sound intensities with a microphone receiver under water.

Measurement of the Displacement-Amplitude of a Vibrating Solid (Sound Source) – Before leaving the question of measurement of sound waves, it may not be inappropriate to refer to one or two interesting methods of measuring the displacement-amplitude of a solid surface, such as a diaphragm in vibration and emitting sound waves. The *first method* employs a very simple and effective device due to Sir W. H. Bragg.† *Bragg's amplitudemeter*, as it may be called, is really an acceleration meter. The principle is as follows. A small mass *m* supported by a spring *s* is brought into contact with the vibrating surface, and the clamped end of the spring displaced a distance A towards the surface (see fig. 136), until there is no longer 'chattering' between the small mass and the vibrator (let us say a diaphragm). Up to this point the maximum acceleration of the diaphragm, tending to throw off the small mass, exceeded the acceleration of the mass (under the action of the spring) *towards* the diaphragm. At the point at which chattering just ceases the maximum acceleration of the diaphragm is just equal to that of the mass attached to the spring. Thus if *a* and *n* and A and N represent the maximum amplitude and the free frequency of the diaphragm and of the loaded spring respectively, we must have

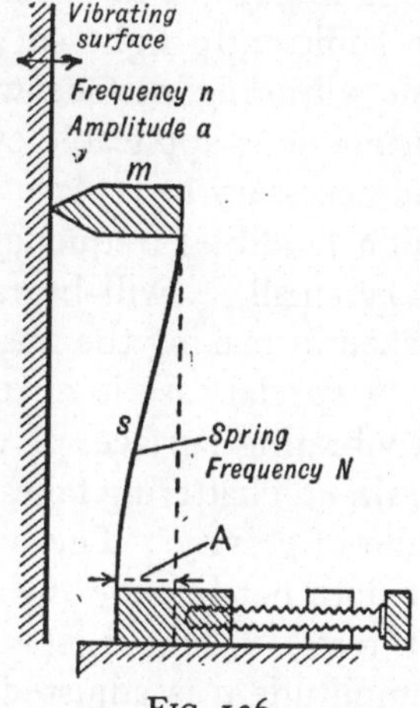

FIG. 136

$$\text{Maximum acceleration} = an^2 = AN^2 \quad \text{or} \quad a = AN^2/n^2,$$

* *Loc. cit.*

† *Engineering*, p. 777, June 13, 1919; and *Journ. Sci. Instrs.*, **6**, p. 196, June 1929.

a very simple relation giving the displacement-amplitude a of the diaphragm in terms of the displacement A of the spring and the ratio of the squares of the frequencies of the loaded spring and the diaphragm. If N is small compared with n, as will always be the case, the displacement A will be large compared with a, the 'magnification' being in the ratio n^2/N^2. In practice, the chattering is observed electrically, a pointer instrument or a pair of telephones being connected in series with a battery and the 'microphonic' contact formed between the loaded spring and the diaphragm (which must be electrically conducting). The frequency of the loaded spring, if sufficiently low, may be 'timed' directly, or determined by stroboscopic or photographic methods if too high for direct observation. For example, if N=1 p.p.s., $n=1000$ p.p.s., and A is found by experiment to be 1 cm., then $a=10^{-6}$ cm. The method permits of the measurement of very small amplitudes provided the surface is vibrating sinusoidally. It is important to observe, of course, that the measuring device is sufficiently light so as not to interfere with the motion of the vibrating surface whose amplitude is required. The same principle is applicable even at very high frequencies when it may be necessary to replace the loaded spring by a vibrator of fairly high (audible) frequency. Under such circumstances when a is *very* small, A will be rather small also, and great care must be taken in making the measurement.

A special case is of interest. When objects rest by gravity on a vibrating surface, of which the motion may be represented by $a \sin nt$, chattering takes place if an^2 is greater than g, the acceleration of gravity. The weight of the object is immaterial provided it does not impede the motion of the vibrating surface on which it rests. Thus if $n(=2\pi \times \text{frequency})$ is equal to 10^3, and the amplitude a is adjusted to the chattering-point, we have $an^2=g$, whence $a=10^{-3}$ cm. approximately. If a quartz crystal of frequency-number $n=10^5$ just produces chattering with a particle of sand on its surface, its amplitude is about 10^{-7} cm. As a simple means of determining the amplitude of sinusoidal vibrations of this nature Bragg's method is very valuable.

The *second method*, used by Webster,* makes a vibrating surface one of the reflectors of a Michelson interferometer, the second surface being fixed. The displacement of the interference fringes, photographed as a wavy line, gives a measure of the amplitude in terms of the wave-length of the monochromatic light which is

* *Nat. Acad. Sci.*, **5**, p. 179, 1919.

used. H. A. Thomas and G. W. Warren * have recently employed a modification of the method to measure the amplitude of small vibrations of reeds and diaphragms in 'loud-speakers.' The interference fringe pattern is observed by eye, and the alternating current in the loud-speaker is gradually increased from zero. When the vibration is very small the fringes become blurred and will ultimately disappear, but if the amplitude of vibration be increased till the peak value is $\lambda/4$ (λ being the wave-length of the light) they will reappear as a new set of fringes displaced half a fringe-width from the initial position. If the amplitude is still further increased, the fringes will again disappear and will reappear when the maximum amplitude is $2\lambda/4$, $3\lambda/4$, $4\lambda/4$, and so on (in general $m\lambda/4$). A curve may be plotted showing the values of current for various integral values of m in $m\lambda/4$. The method is applicable when the vibrating body is too light to permit of a glass reflector being attached, in this case a thin metal mirror being sputtered cathodically on the surface. It is important to note also that the accuracy of the method is in no way impaired if the surface is not optically flat, the observed effects being independent of the straightness or otherwise of the interference fringes. A diaphragm or other vibrator calibrated in this way may be employed as a standard source of sound—its absolute amplitude being known for various exciting currents.

Wave-Filters. Electrical and Mechanical

The passage of an electric current through a long telephone line or cable is, in the absence of special precautions, accompanied by distortion of wave-form. This effect is particularly noticeable in the transmission of complex wave-forms, as in the case of speech, through long lines. Heaviside showed that this distortion was due to the distributed electrical capacity of the line filtering out certain components of frequency from the wave during transmission, and he suggested the use of inductance or 'loading coils' to compensate for this distributed capacity. Heaviside's suggestions were first put into practicable form by Pupin and G. A. Campbell.† The latter extended Heaviside's theory and showed how it was possible in practice to distribute inductance and capacity in electrical transmission lines so that certain selected ranges of frequency are

* *Phil. Mag.*, **5**, p. 1125, 1928.

† *Phil. Mag.*, March 1903, and *Bell System Techn. Journ.*, Nov. 1922. See also various papers by Zobel, Carson, Johnson, Shea, and others, in *Bell System Techn. Journ.*, 1923, 1924; G. M. B. Shepherd, *Electrician*, **7**, p. 399, 1913; P. David, *Onde Électrique*, **5**, pp. 5 and 72, 1926.

transmitted with negligible attenuation, whilst other frequencies are suppressed almost entirely. In the simple form of the theory the transmission line is regarded as a repeated network or chain of impedances arranged in series and in shunt as in fig. 137. The

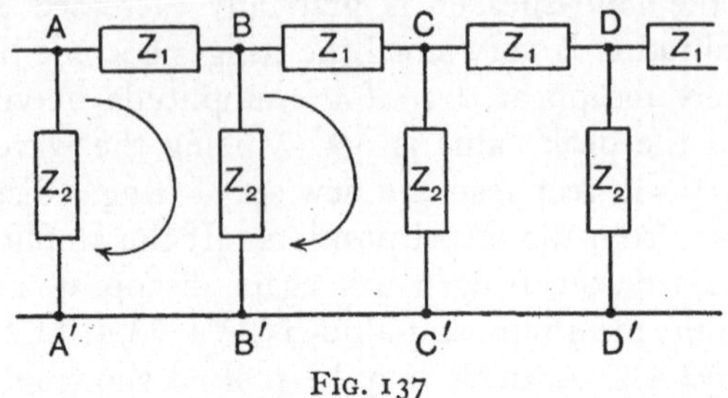

FIG. 137

current in its passage along the chain from left to right is attenuated, some frequencies more than others. Denoting each of the series impedances by Z_1 and the shunt impedances by Z_2, we can determine the attenuation factor. The line A′B′C′D′ in fig. 137 represents an 'impedanceless line,' for convenience all the series impedances being 'lumped' in the line ABCD. Denoting the currents in the parts AB, BC, and CD of the line by i_{m-1}, i_m, and i_{m+1} respectively, we must have: current through $BB' = i_{m-1} - i_m$ and through $CC' = i_m - i_{m+1}$, equating e.m.f.'s acting round the mesh BCC′B′ we obtain

$$Z_1 i_m + Z_2(i_m - i_{m+1}) = Z_2(i_{m-1} - i_m)$$

or

$$i_m(Z_1 + 2Z_2) = Z_2(i_{m-1} + i_{m+1}),$$

that is,

$$\frac{i_{m-1}}{i_m} + \frac{i_{m+1}}{i_m} = 2 + \frac{Z_1}{Z_2} \quad . \quad . \quad . \quad (1)$$

Now the ratio of the current in one section to that in the preceding one is a constant in such a chain system as this, consequently we may write

$$\frac{i_{m-1}}{i_m} = \frac{i_m}{i_{m+1}} = \frac{i_{m+1}}{i_{m+2}} = \ . \ . \ . \ = e^{\lambda} \quad . \quad . \quad (2)$$

where λ is defined as the 'propagation' constant. We may therefore write for equation (1)

$$\left. \begin{aligned} e^{\lambda} + e^{-\lambda} &= 2 + \frac{Z_1}{Z_2} \\ \text{or} \quad & \\ \cosh \lambda &= 1 + \frac{1}{2}\frac{Z_1}{Z_2} \end{aligned} \right\} \quad . \quad . \quad . \quad (3)$$

If λ is imaginary, the currents in adjacent sections differ only in phase. If, on the other hand, λ is real, the current is attenuated in transmission. In the former case, λ imaginary, $\cosh j\lambda = \cos \lambda$ $(j=\sqrt{-1})$ and

$$+1>\left(1+\frac{1}{2}\frac{Z_1}{Z_2}\right)>-1 \quad . \quad . \quad . \quad (4)$$

from which it follows that the *limiting values of no attenuation* are $Z_1/Z_2=0$ and -4. That is to say, there is no attenuation within a range of frequency corresponding to these limits, which therefore define the extent of the frequency transmission band of the network. (This simple theory refers to a resistanceless line, *i.e.* a line free from dissipation loss.) By suitably choosing the values of Z_1 and Z_2, therefore, it is possible to select a particular frequency range for transmission—the arrangement acting as a *wave-filter*. There are many types of filter according to the choice of Z_1 and Z_2. We shall now consider three of the most important types in which the line is made up of inductances and capacities only (resistance being regarded as negligible).

Type I - *Low-pass Filters.* $Z_1=nL$ and $Z_2=1/nC$, inductance L forming the series elements, with capacity C in the shunts ($n=2\pi \times$ frequency N) (see fig. 138). In this case $Z_1/Z_2=LCn^2$, when

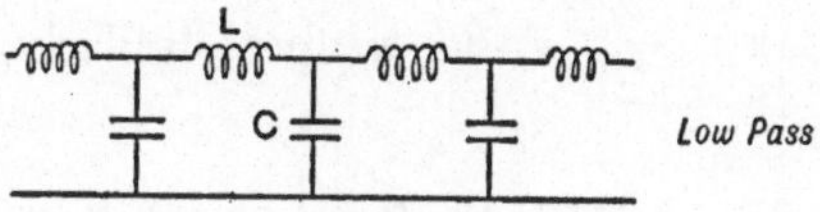

Low Pass

FIG. 138

$$\frac{Z_1}{Z_2}=0, \qquad n_1=0\,; \qquad \frac{Z_1}{Z_2}=-4, \qquad n_2=\frac{2}{\sqrt{LC}},$$

where $\frac{n_1}{2\pi}$ and $\frac{n_2}{2\pi}$ are the frequency limits of transmission. Consequently such a filter transmits all frequencies from zero to a frequency of $1/\pi\sqrt{LC}$. Above the latter frequency nothing is transmitted. The filter is consequently described as a 'low-pass' filter.

Type II - *High-pass Filters.* $Z_1=1/nC$, $Z_2=nL$, the series elements being equal capacities C and the shunts equal inductances L (see fig. 139). We have now $Z_1/Z_2=1/n^2LC$, when

$$\frac{Z_1}{Z_2}=0 \qquad n_1=\infty\,; \qquad \frac{Z_1}{Z_2}=-4 \qquad n_2=1/2\sqrt{LC}.$$

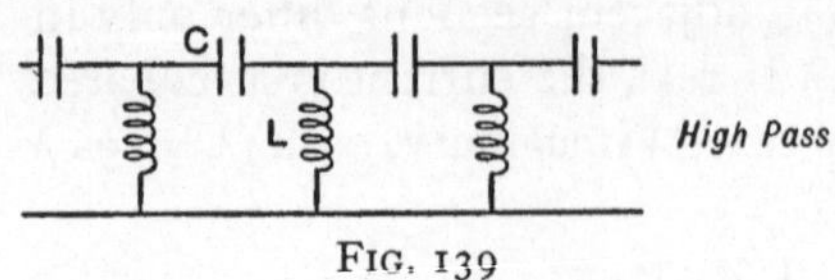

FIG. 139

The filter therefore transmits all frequencies from $1/4\pi\sqrt{LC}$ to infinity, and is consequently described as a 'high-pass' filter.

Type III (a) – *Band-pass Filters.* $Z_1 = 1/nC_1$, $Z_2 = (nL - 1/nC_2)$, capacities C_1 forming the series elements, with inductances L, and capacities C_2 *in series* forming the shunt elements (see fig. 140).

FIG. 140

Now

$$\frac{Z_1}{Z_2} = \frac{1}{nC_1\left(\frac{1}{nC_2} - nL\right)} = \frac{1}{C_1/C_2 - n^2LC_1},$$

when

$$\frac{Z_1}{Z_2} = 0 \quad \text{we have} \quad nL = 1/nC_2 \quad \text{and} \quad n_1 = 1/\sqrt{LC_2},$$

when

$$\frac{Z_1}{Z_2} = -4 \quad \text{we find, similarly,} \quad n_2 = \frac{1}{2}\sqrt{\frac{1}{LC_1} + \frac{4}{LC_2}}.$$

Type III (b) – *Band-pass Filters.* $Z_1 = 1/nC_1$, $\frac{1}{Z_2} = \left(nC_2 - \frac{1}{nL}\right)$; in this case the shunt elements are made up of inductance L and capacity C_2 in parallel, the series elements being capacities C_1 as before (see fig. 141). We now have

FIG. 141

$$\frac{Z_1}{Z_2} = \frac{C_2}{C_1} - \frac{1}{n^2LC_1}; \quad \text{when} \quad \frac{Z_1}{Z_2} = 0 \quad n_1 = \frac{1}{\sqrt{LC_2}}.$$

When

$$\frac{Z_1}{Z_2} = -4 \quad \text{we find} \quad n_2 = \frac{1}{\sqrt{L(C_2 + 4C_1)}}.$$

In Types III (*a*) and III (*b*) it will be seen that the filter transmits only between certain finite values of frequency (given by $n_1/2\pi$ and $n_2/2\pi$). Such an arrangement is consequently described as a 'band-pass' filter.

The foregoing simple examples will serve to illustrate the principles on which electric wave-filters are designed. The number of possible combinations is, of course, very large, and in actual cases ohmic resistance cannot always be neglected. The general behaviour of simple low-pass and high-pass filters is easily understood when it is remembered that the impedance nL of an inductance L *increases* with frequency, whereas the impedance $1/n$C of a capacity C *decreases* with frequency. Both the inductance and the capacity have therefore a selective influence in regard to currents of different frequencies, the inductance transmitting low and 'choking' high frequencies, whilst the condenser functions in the reverse manner. Electric wave-filters have been extensively used not only in practical telephony, but also in laboratories as a means of analysing complex sounds. The various sounds of speech, converted by a microphone receiver into the corresponding electrical currents have been analysed by wave-filters to determine the frequency range which is essential to intelligibility.* Other applications of a similar nature are self-evident.

We must now consider the question of *mechanical filters*, the theory of which is in all respects parallel to that of electrical filters. The development of mechanical or sound filters is due primarily to the work of G. W. Stewart.† A simple example of a filter is a Helmholtz resonator with a small ear-opening which transmits a sound wave of one particular frequency only. Cylindrical tubes such as those shown in fig. 142 also serve as filters, Type I *transmitting* the resonating frequency of the column a b, and Type II, a Quincke filter, *absorbing* that frequency. The filters with which we are now concerned, however, are somewhat different in principle, as they do not depend directly upon resonance effects but on the interaction between the successive links in the 'chain' of a mechanical transmission line (as in the electrical case we have just considered). The various sections or links in this chain have over-all dimensions which are small compared with a wave-length of the sound passing through them. There seems no reason for doubting that every electrical filter has its mechanical analogue,

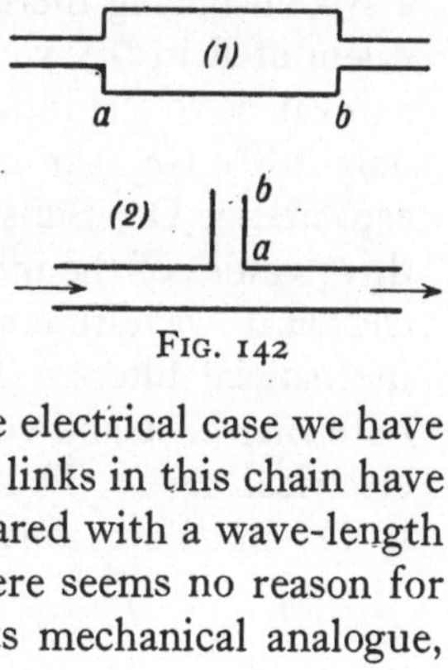

FIG. 142

* See, *e.g.*, E. K. Sandeman, *Proc. I.E.E.*, Nov. 1923.

† *Phys. Rev.*, **20**, p. 528, 1922. See also F. Canac, *Journ. de phys. et le radium*, **7**, p. 161, 1926; W. P. Mason, *Phys. Rev.*, **31**, p. 283, 1928.

although the latter may not always be practicable. Stewart deals with three kinds, analogous to those we have just considered in the electrical case. The following conditions are necessary to preserve the electro-mechanical analogy: (1) The length of any selected section of an acoustic 'line' or conductor must be regarded sufficiently small compared with a wave-length that no change of phase occurs within it. (2) *Acoustic impedance* is defined as the complex ratio of the pressure difference applied and the rate of change of volume displacement (analogous to the complex ratio of applied p.d. and current in the alternating electrical case). (3) The algebraic sum of the volume displacements at any junction of lines is zero (the equivalent of Kirchhoff's first law). Comparing electrical and mechanical equations (see p. 71) we have, in the usual notation

$$\left.\begin{array}{l} L\ddot{Q}+R\dot{Q}+Q/C=E\cos nt \text{ (electrical)} \\ \text{and} \\ m\ddot{x}+r\dot{x}+sx=P\cos nt \text{ (mechanical)} \end{array}\right\} \qquad (a)$$

The latter equation, when we are dealing with *volume* changes (Sx), must be written

$$\frac{m}{S^2}\ddot{x}+\frac{r}{S^2}\dot{x}+\frac{s}{S^2}x=p\cos nt \qquad (b)$$

where S is the area subjected to a pressure $P=pS$ and a linear displacement x. Comparing equations (a) and (b) in the case of a system having inertia only, and regarding p in (b) as the equivalent of E in (a), we find $m/S^2\equiv L$—that is, the *inertance* m/S^2 is equivalent to the inductance L; and in a system having stiffness only $s/S^2\equiv 1/c$ that is, the *compliance* S^2/s is equivalent to the capacitance C. Substituting for inductance L and capacitance C these values of the inertance and compliance, in our equations for electrical wave-filters, we obtain the analogous equations for mechanical filters. The mechanical values of Z_1 and Z_2 when both inertance and compliance are present, are obtained as in the electrical case. Thus, compliance and inertance in series (*e.g.* in a Helmholtz resonator) have a mechanical impedance Z given by

$$Z=\frac{nm}{S^2}-\frac{s}{nS^2}.$$

In a Helmholtz resonator the mass m is that of the air-piston in the neck ($m=\rho S_1 l$) and the stiffness s is that of the air in the cavity ($s = S_2 \delta p$). It is a simple exercise to prove that the frequency N is that obtained by other methods in equations (12), p. 186 (remembering that the corresponding electrical frequency $N = 1/2\pi\sqrt{LC}$).

Construction and Test of Mechanical Filters of the Three Types – Stewart * states: "*Low-pass filters* were made, for example, by two concentric cylinders joined by walls equally spaced and perpendicular to the axes. Each chamber thus formed had a row of apertures in the inner cylinder which served as the transmission tube. In one case the volume of each chamber was 6·5 c.cm., the radius of the inner tube 1·2 cm., and the length between the apertures 1·6 cm. A chamber and one such length of the inner tube is called a section. Four such sections were found to transmit 90 per cent. of the sound from zero to 3200 p.p.s. approx., where the attenuation became very high." Other similar filters of different dimensions showed transmissions having wider or narrower frequency ranges. The general construction of the low-pass filters and the experimental transmission curve are shown in fig. 143.

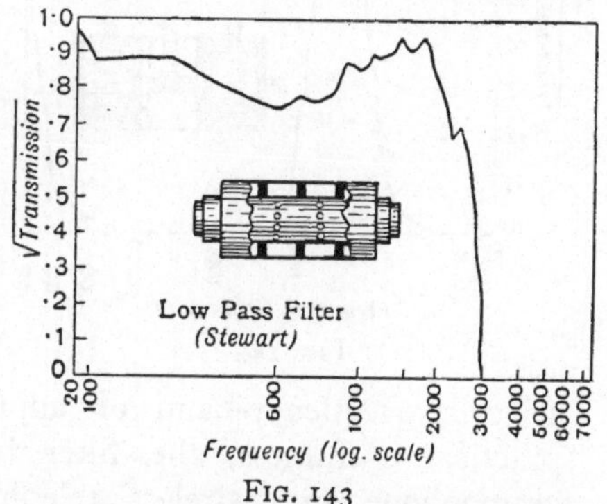

FIG. 143

"*High-pass filters* were made with a straight tube for transmission and short side tubes, for example, 0·5 cm. long and 0·28 cm. diameter, opening through a hole with conductivity 0·08 into a tube 10 cm. long and 1 cm. diameter. Six sections of such a filter would transmit about 90 per cent. of sounds above 800 p.p.s., but would refuse transmission to sounds of lower frequeney." The cross-section of a simple two-section filter of this type, with its transmission curve, is shown in fig. 144.

FIG. 144

* *Loc. cit.* See also V. L. Hartley, *Engineer*, **143**, p. 624, 1927.

Single band-pass filters are a combination of the other two types, having side tubes leading to chambers of considerable volume. Such a filter and its transmission characteristics are shown in fig. 145. The agreement with theory is fairly good, for it is possible to construct filters to meet certain specified conditions. In the laboratory, filters may be used in the elimination of undesirable components in a sound wave, and for certain types of sound analysis, *e.g.* in observations relating to the intelligibility of speech. A filter has the advantage over a resonating device in that it is able to select a frequency-band of adjustable width. Regarding its practical usefulness, the filter has direct application to the gramophone and wireless telephone in the removal of undesirable frequencies, *e.g.* needle-scratch or microphone and valve noises. Stewart suggests also that the device makes possible the simultaneous reception of an indefinite number of radio messages with the same antenna. Filters have also an application in the design of megaphones for use with loud-speaking devices. Maxfield and Harrison * have made use of the electrical analogy in the design of mechanical sound reproducers (the gramophone). The introduction of the filter phenomena may ultimately affect the design of musical instruments.

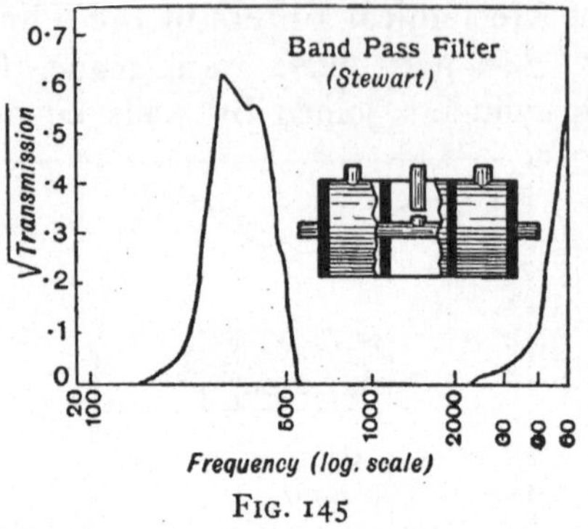

FIG. 145

A difficulty which is inherent in mechanical filters is due to the continuous structure of the medium within the transmission 'line.' The inertia and stiffness factors are not so definitely 'lumped' as in the analogous electrical case. Dissipation effects, due to the viscosity in the small holes and side tubes, cannot be ignored. In a later paper Stewart † takes these factors into account to explain certain experimental observations of multiple resonance effects in filters—accounting for certain cases of failure of the low-pass filter to suppress very high frequencies.

* *Bell System Techn. Journ.*, **5**, p. 493, 1926.

† *Phys. Rev.*, **25**, p. 90, 1925. See also Peacock, *Phys. Rev.*, April 1924.

SECTION V

TECHNICAL APPLICATIONS

In the present section it is proposed to deal with a few of the more important technical applications of the principles of sound, apart from those already mentioned. The last ten or fifteen years includes a period of marked progress in such applications of sound. The Great War stimulated the development of apparatus for detecting, identifying, and locating sounds at long ranges. Many forms of directional sound-receivers, sound-ranging, and signalling devices were realised and applied to urgent practical problems on land and sea. Since the war these methods have been applied for more peaceful purposes, *e.g.* echo depth-sounding, surveying, and navigation. Considerable advances have also been made in the study of the acoustics of buildings, initiated by the work of Sabine. The phenomena of piezo-electricity, discovered in 1880 by Curie, have developed into a means of standardising mechanical and electrical frequencies up to millions of vibrations per second. In this respect, also, improvements in tuning-forks and phonic motors have played an important part. Another important application of piezo-electricity, initiated by Langevin and developed by Boyle, employs the supersonic oscillations of quartz in depth-sounding at sea and in the echo detection of icebergs. At the other extreme of the frequency scale, Constantinesco has developed a system of transmission of low-frequency, alternating, mechanical power through water-filled pipes. For this purpose he employs generators, motors, transformers, and transmission 'lines' closely analogous to the corresponding electrical devices. The growth of radio-telephony and 'broadcasting' has stimulated the development of improved methods of reproducing sounds of audible frequency. The elimination of objectionable resonances from gramophones and 'loud-speakers' represents an important step in the reproduction of speech and music. The study of the characteristics of speech and hearing, notably the work of the Bell Telephone Laboratories, U.S.A., has not only increased our knowledge of these subjects but has proved of great value in the improvement of telephone apparatus, microphones, transmission lines, etc., and in the design of all forms of sound-reproducing mechanisms. In this respect also the introduction of the conception of acoustical impedance and the mechanical-electrical analogy of filter circuits, has proved of great value in dealing with complex mechanical systems. The general question of reduction of traffic noise is becoming increasingly insistent, and it is noteworthy that the problem is now receiving a certain amount of public attention. In a recent discourse before the Royal Aeronautical Society, W. S. Tucker has dealt with the similar problem of noise-reduction in aircraft.

Measurement of Distance by Sound

The principle involved in the most direct method of measurement of distance by means of sound waves is the converse of that employed in the determination of the velocity over a measured base line. Provided the velocity of propagation c is known, the unknown distance d between two points is determined by the time-interval t taken by the sound to travel from one point to the other, then $d=ct$. For example, if two microphone receivers, A and B, are situated a distance d apart, and are connected electrically to a recording device, *e.g.* an Einthoven oscillograph with time-marker, and a sound wave takes 5 seconds to travel from A to B, the distance apart of the microphones will be 5500 feet, assuming 1100 ft./sec. to be the value of the sound-velocity in air at the time of the experiment. A modified form of the method requires, under certain circumstances, only one receiver. In this case, which we shall describe as the *reflection method*, the sound wave after passing the receiver A subsequently falls on a reflecting surface and the returning wave passes A again after an interval $t=2d/c$, where d is the distance of the receiver from the reflector. The *direct method* of determining distance is not always possible, however, and another method, also very familiar, may be used instead. We shall describe this alternative as the *differential velocity method*—it is sometimes called the 'synchronous signalling' method.* Two sets of waves initiated at the same moment travel outwards with different velocities. An observer at a distant point receives the high-velocity v_1 wave first, and after a measurable time-interval t he receives the wave of lower velocity v_2. The distance d of the observer from the source is consequently derived from the simple relation—

$$t=\frac{d}{v_2}-\frac{d}{v_1}.$$

* See J. Joly, *Phil. Mag.*, **36**, p. 1, 1918.

If the velocity v_1 is very large compared with the velocity v_2, the relation simplifies still further to $t=d/v_2$. The waves need not necessarily be of the same type—*e.g.* those of high velocity v_1 may be electromagnetic waves (light or 'wireless') travelling with a velocity of 3×10^8 metres/sec. compared with the velocity of sound of the order 3×10^2 metres/sec. in air or $1{\cdot}5\times10^3$ metres/sec. in water. This case corresponds to that of all the earlier methods of measuring the velocity of sound in air (p. 234) or in water (p. 245), in which a flash of light signals to the distant observer the instant of firing a gun or a charge of explosive. Strictly speaking, the observer sees the flash slightly after the instant at which it occurs, but the error involved in the estimation of the distance is only that due to neglecting 10^{-8} in $(10^{-2}-10^{-8})$, an error which is quite insignificant. As we shall see, this principle has been extended to a very important method of locating a simultaneous source of sound and 'wireless' waves. The two sets of waves may alternatively be of the same nature but travelling in different media with different velocities. Thus sounds may be produced simultaneously in air and in water, the velocities v_1 (water) and v_2 (air) being of the order 333 and 1500 metres/sec. respectively. In such a case the distance d reduces to $d=430t$ metres, approximately, when t is measured in seconds. If the sound wave in air were replaced by a light flash we should have $d=1500t$ metres.

We shall now consider a few of the important technical applications of these simple principles.

(A) Echo Depth-Sounding

The process of measuring the depth of water beneath the keel of a ship has, from time immemorial, been described as 'sounding.' Until comparatively recent times this operation has been performed by the simple and direct method of the lead-line. Latterly this has been replaced on the larger ships by the well-known Kelvin sounding machine in which a weighted wire carrying a 'chemical tube,' closed at one end, is lowered to the bottom. The depth of water is estimated from the amount of compression of the air enclosed in the tube; the encroachment of the sea-water up the tube being indicated by the chemical action on the material coating the inside of the tube.

These methods, involving the use of a weighted line, require a ship to reduce speed to something below eight knots; this involves expense and loss of time. Again, in hydrographic surveys it is often necessary to take soundings in very deep water. Such operations are very slow and tedious; a single sounding in 3000 or 4000 fathoms might take several hours to carry out satisfactorily. The risk of long wires breaking under such conditions, and the errors arising from uncertain knowledge of currents at varying depths, make observations of this nature somewhat unreliable. It is not inappropriate therefore that the method of sounding by wire should be replaced by sounding by sound waves. This new method involves the simple principle of measuring the time interval for a sound wave to travel (with known velocity) from the surface to the bottom and back again. The velocity of sound in the sea under varying conditions of salinity and temperature is known with considerable accuracy (see p. 248). The various methods of echo sounding are distinguished primarily by the methods of determining the time interval. They differ in detail also as regards the type of sound source, receiver, and indicator. The advantages of the echo method are obvious. A sounding in 4000 fathoms can be made in about 10 seconds, as compared with several hours by the old 'wire' method. Again, the observations of depth can be made whilst the ship is in motion, up to speeds of 15 to 20 knots, resulting in a great saving of time and inconvenience. Echo-sounding systems have been devised in America, France, Germany, and Great Britain.*

(1) American Systems. (a) *Fessenden Rotating-Disc Method* – Fig. 146 illustrates diagrammatically a device invented by R. A.

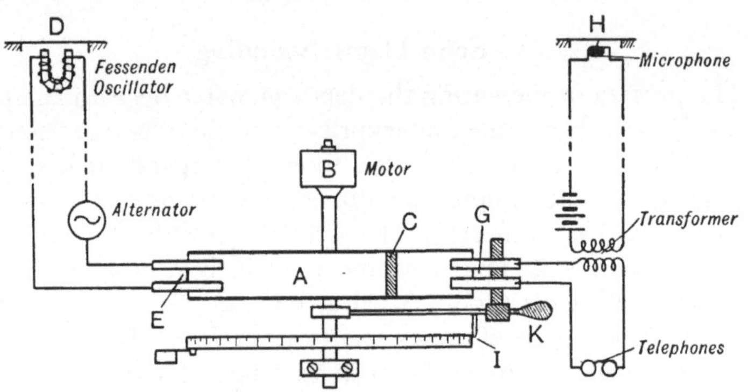

FIG. 146—Echo Sounding Apparatus (Fessenden)

Fessenden.† A is a disc of insulating material driven at constant speed by a motor B. The disc carries a conducting segment

* See H. C. Hayes, *Journ. Frank. Inst.*, **197**, March 3, 1924; and 'Echo Sounding,' Intern. Hydr. Bureau, Spec. Pubns., No. 3, 1924, and No. 14, 1926.

† U.S. Patent, No. 1217585.

C which closes the electrical circuit through a submarine sound transmitter (a Fessenden oscillator) D when it passes beneath the brushes E, thereby sending out a short sound signal. This segment subsequently closes a circuit through the telephones F when it passes across the brushes G. If the echo of the signal arrives at the microphone receiver H at the instant when the segment C short-circuits the brushes G it will be heard in the telephones. The time taken by the sound to travel from the transmitter to the sea-bed and back to the microphone H will then be equal to the time required for the segment C to travel the angular distance between the two pairs of brushes. This is indicated by a pointer I on a circular scale J. The adjustment is made by rotating the brushes G around the disc A by means of the handle K. This apparatus has a number of defects in actual use, *e.g.* the conducting segment C which closes the telephone circuit at each revolution of the disc produces a 'click' in the phone (due to contact e.m.f.) which might be mistaken for an echo. The Fessenden oscillator used to produce the sound signal is designed to give a continuous sinusoidal sound wave, whereas a sharp, heavily damped impulse is most suitable for echo purposes.

(b) *Angle of Reflection Method* – A system developed in the American navy measures the depth of water beneath the ship by observation of the angle of reflection of a continuous sound wave transmitted from one end of the ship and received at the other. In its simplest form the sound source may be the propellers of the ship, and the receiver any form of directional hydrophone fitted near the bows. If $2d$ is the distance separating the source and the receiver, and the received sound wave makes an angle θ with the horizontal, then the depth h of the water beneath the ship is given by $h = d \tan \theta$. In practice a Fessenden oscillator is used as the sound source and a line of Mason hydrophones (see p. 422), fitted with binaural compensator, gives the required direction. To eliminate errors due to a shelving bottom it is necessary to employ two sets of transmitter and receiver—transmitting first from one end of the base line (the length of the ship) and then from the other. The mean depth is then obtained from the 'mean angle' of reflection observed. This method of depth-sounding breaks down for depths so great that the angle θ approaches a right angle. It has been found in practice that the method gives reliable soundings to depths equal to three times the length of the base line—for most vessels this will cover depths up to 100 fathoms.

(c) *Stationary Wave Method* – A source of continuous sound

of variable frequency is employed (*e.g.* over a range 500 to 1500 p.p.s.), and an accurate frequency indicator included in the circuit which supplies current to the sounder. The operator's phones are connected one to the microphone receiver (the hydrophone), and the other to a coil coupled to the sound transmitter circuit. The two sounds heard in the phones will at all times have a definite phase relationship, which will depend on the phase of arrival of the sound waves at the receiver. If the operator adjusts the frequency correctly he can make the sound heard in the phones have the same phase and can recognise this condition by the fact that the resultant sound in the ears will be binaurally centred. If the sounds are in phase therefore, the sound path from transmitter to receiver, *via* the sea-bed, will be given by an integral number m_1 of wave-lengths at the frequency N_1. Calling this distance s, and the velocity of sound c, we have $s=m_1c/N_1$. The value of m_1 can be found by raising the frequency of the sound until a second condition of binaural centring is observed, then $s=(m_1+1)c/N_2=m_1c/N_1$, which, on eliminating m_1, gives

$$s=c/(N_2-N_1).$$

Knowing the separation of the sound source and the microphone, the depth h is readily deduced.

This method has proved accurate for measuring shallow depths —in which case a greater change of frequency ($N_2=N_1$) is required to change the number of waves in the standing-wave system. At greater depths it fails for similar reasons. A modification of this method consists in sending out a continuous series of short sound signals separated by equal time-intervals, and means for varying continuously the time-interval between successive signals from about 0·1 second to 10 seconds in such a manner that the time-interval between successive signals can be accurately determined.

(2) German Systems - A. Behm of Kiel has designed a depth-sounding apparatus based on a different principle, viz. measurement of the *intensity* of the echo. A source of sound on one side of the ship sends out continuous waves which are reflected from the bottom and received by a resonant chamber which causes a tuning-fork to vibrate. The intensity of reflection is measured by the amplitude of the fork, observed by means of a microscope— the arrangement being first calibrated in known depths of water. This method fails for a number of reasons. It is extremely difficult to keep a sound source operating continuously at a constant intensity and frequency, and to ensure also that the receiver shall

always be accurately tuned and of constant sensitivity. A more serious difficulty, however, is that of making allowance for variations of the coefficient of reflection from the sea-bed—a quantity which may vary rapidly (from rock to mud or sand) in relatively short distances. Realising the weakness of this method Behm developed an alternative method, now in use as the *Behm-Echolot System* in which the time between signal and echo is measured. The sound source is an electrically fired detonator which gives a single impulse to the water. The receiver on the opposite side of the ship, screened from the direct blow, receives the reflected impulse from the bottom. An electromagnetic device at the instant of firing the detonator releases a disc which commences to rotate. The rotation is arrested suddenly, however, when the echo reaches the receiver. The edge of the disc is graduated in metres to indicate depths directly. This apparatus gives reliable soundings up to 80 fathoms. It has also been used on airships as a means of determining heights above the ground.

(3) French System. *Langevin Piezo-Electric Oscillator* * – The Langevin system is based on a suggestion made, at the time of the *Titanic* disaster (April 1912), by L. F. Richardson, who proposed the use of a beam of supersonic waves as a means of detecting submerged objects, such as icebergs or wrecks, by 'echo.' The proposal was made possible by the development of Langevin's high-frequency piezo-electric oscillator to which we have already referred (see fig. 48, p. 147). The high-frequency vibrations of a slab of quartz, of diameter equal to several wave-lengths, set up corresponding vibrations in the water. The sound wave is distributed in the form of a primary beam with more or less insignificant secondaries. The angle of spread of this primary beam is $\sin^{-1}(1{\cdot}22\lambda/D)$; with a large ratio D/λ, therefore, the beam may be regarded as almost 'parallel.' To adjust the frequency of the quartz disc to a suitable value, *e.g.* 50,000 p.p.s., steel discs are rigidly attached to its faces as shown in fig. 48. The alternating e.m.f., from a high-tension spark oscillatory circuit, is applied to these discs and tuned to the resonant frequency of the 'steel-quartz-steel' oscillator. Under these conditions the supersonic transmitter, as it is called, works at its maximum efficiency. A signal of short duration, consisting of a single train of damped high-frequency waves, is emitted from the transmitter and, after travelling to the sea-bed, where it is reflected, falls upon the quartz

* See International Hydrographic Bureau, Special Publications on 'Echo Sounding,' No. 3, 1924, and No. 14, 1926.

oscillator again. During the short interval in which this takes place a switch connects the oscillator to the receiving circuit. The time-interval between the emission of the signal and the arrival of the echo may either be observed visually on a scale or may be directly recorded. The emission or reception of a sound pulse is indicated by means of a small oscillograph which reflects a spot of light on a scale graduated in fathoms or metres. This spot of light is caused to traverse the scale at a uniform speed, and the switches are so arranged that the supersonic impulse is transmitted and a slight deflection of the spot occurs at the instant when it is passing the zero position on the scale. On the arrival of the echo, the spot undergoes another slight deflection at a point farther along the scale. The graduations of the latter are so arranged that the reading at which the echo deflection occurs represents the depth of water beneath the transmitter. This is easily arranged when the velocity of sound in sea-water and the rate of traverse of the spot along the scale are known—remembering, of course, that the distance travelled by the sound wave corresponds to twice the depth of the sea. When it is required, signals are transmitted once per second and a continuous record of the depth is obtained. The advantages of such a procedure in navigation are obvious. The Langevin-Florisson system, as it is called, has been employed not only in echo depth-sounding but in the location of wrecks and icebergs. In the latter connection, the work of R. W. Boyle is noteworthy (see p. 150).

(4) British System * - The British Admiralty system employs a sonic, or audible-frequency, source of sound with a device for the indirect measurement of the time-interval between signal and echo. In some respects this method resembles that of Fessenden already described, but it differs in certain important particulars. The source of sound is a steel diaphragm (about 5 in. diameter) which is caused to emit, three times per second, a heavily damped train of waves (frequency about 2000 p.p.s.). The diaphragm is excited into vibration by a sudden blow from an electromagnetically operated hammer; current through an electromagnet causes the hammer to compress a spring which, at the instant the current is cut off, drives the hammer into sudden and violent contact with a boss on the diaphragm. A small hydrophone receiver (a button microphone mounted on a diaphragm on a water-tight container) receives the echo from the sea-bed and

* See F. E. Smith, *Proc. Roy. Inst.*, **24**, p. 342, 1924; also Admiral Douglas, *Roy. Geog. Soc. Journal*, **74**, No. 1, p. 47, July 1929.

a small amount of the direct sound. The transmitter and receiver are mounted in water-filled tanks on opposite sides of the ship, the hull forming a partial screen to the direct sound. The direct and reflected sound waves consequently pass through the hull of the ship to or from the sea. A constant-speed motor at the observing-post drives two commutators through suitable gearing. One of these, which consists of a metal disc with an insulated segment, serves to excite the electromagnetic transmitter three times per second, the hammer being released and striking the diaphragm when the insulated segment passes the brushes. The second commutator, running on the same shaft, short-circuits the telephones in the receiving circuit, except for a brief period (about 1/300 second) provided by an insulating segment; during this short period it is possible for the telephones to 'listen.' There will consequently be nothing heard in the telephones unless the insulating segment in the telephone-commutator happens to coincide with the position of the brushes at the instant when the transmitter is actuated or at the instant of arrival of the echo from the sea-bed. The position of the brushes in the telephone circuit can be displaced by hand relative to the brushes in the transmitter circuit so that an interval of time, proportional to the angular displacement of the brushes, elapses between the initial impulse and the instant at which the telephones 'listen.' When the transmitter gives three impulses per second, one completed revolution of the brushes is equivalent to a time-interval of 1/3 second, *i.e.* $360^\circ \equiv 1/3$ second. Assuming a velocity of sound in sea-water of 4920 ft./sec. at a particular temperature and salinity, a displacement of the telephone brushes of 60° to reach the echo position indicates a time-interval of $1/3 \times 60/360 = 1/18$ sec. and a depth of 136 ft. approximately. In the actual apparatus, calculations are unnecessary, the depths being indicated directly on a dial carrying the telephone brushes. In very shallow water allowance must be made for the inclination of the path of the reflected wave from the transmitter to the receiver—the correction is automatically applied by means of a simple cam device which causes the dial to advance at a somewhat different rate from that of the brushes. The general diagram of the apparatus is shown in fig. 147. The form just described is used to measure depths up to about 200 fathoms. Another type, employing a more powerful transmitter, pneumatically operated, and a different timing device, has been used successfully in oceanic depths up to 4000 fathoms. The shallow-water, echo-sounding apparatus is valuable in navigation as a quick and simple means of measuring depths without reducing the speed of the ship. A line of echo soundings, compared with those marked on a chart, is sufficient to give a navigator the ship's position, and may in this way assist and safeguard navigation

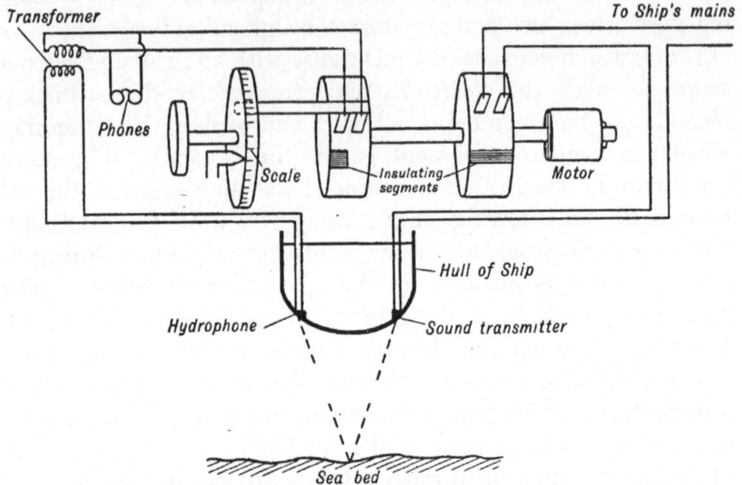

FIG. 147—Echo-Sounding Apparatus (British Admiralty system)

under conditions of bad visibility. The apparatus is extensively used on H.M. ships and survey vessels, in the mercantile marine, and in fishing-vessels. The latter have found the echo sounder valuable as a means of locating 'fishing-holes' in the North Sea and in the vicinity of Iceland.

Geological Prospecting by Echo and Seismic Methods—Various attempts * have been made to apply analogous methods to determine the depths of geological strata, containing valuable minerals or oil, below the earth's surface. Complications arise due to the heterogeneous nature of the earth's crust and the presence of 'faults' and inclusions of air or gas between the strata. In certain cases, however, where the general structure is known from independent geological data, seismic methods have been found sufficiently practicable. As a rule, the source employed is an explosive charge detonated on, or below, the surface, the receiver being a sensitive seismograph or some form of shock-microphone connected to an oscillograph (such as the Einthoven string type fitted with a time-marking device). An examination

* See, for example, Berger, *Schalltechnik*, **87**, 1926; and A. S. Eve, *Nature*, **121**, p. 359, March 1928.

of the records obtained for various positions of the explosive charge and the receivers sometimes reveals the required information.

Dealing with the application of physical principles to the problem of oil-finding, A. O. Rankine * has recently described in some detail a seismographic method of locating the depth and contour of oil-bearing strata. This method, which has been applied with some success in the oilfields of Texas and Persia, involves novel principles due to Professor Mintrop. Preliminary tests showed that ordinary depth-sounding methods, *e.g.* the British Admiralty system described on p. 472, were unsatisfactory on account of the heavy damping of sound waves or impulses in the surface layers of the ground. Consequently an entirely new principle † was invoked, depending on the fact that the velocity of sound or earthquake waves in the lower-lying strata is greater than that near the surface. For example, the following velocities have been measured:—

Locality.	Stratum.	Velocity, Metres/Sec.
Texas	Salt dome (oil-bearing) . . .	5300
,,	Clay and sand overlying salt domes .	2000
Persia	Limestone strata (oil-bearing) . .	4700
,,	Overlying soil	3700

In order to penetrate to the deeper layers in which oil is usually found, it is necessary to employ a very powerful initial impulse—in fact, an artificial earthquake wave is created by means of a large explosion on the surface of the ground. The explosion wave after travelling through the ground is received, by means of sensitive recording seismographs, at a number of points at varying distances away. The seismograph, which will record vibrations of low, inaudible frequency, is preferable to the microphone on account of the smaller damping of such vibrations. The instant of firing the charge is indicated by means of a wireless signal controlled from the firing circuit. Reference to fig. 148 indicates three possible paths for the explosion impulse at O to reach the seismograph at A, viz.:—

(1) The direct path OA, through the surface layer;

* *Inst. of Mining and Metallurgy Bull.*, **295**, p. 28, April 1929; and *Nature*, pp. 684 and 718, May 1929. Report of R.I. Lectures, Feb. 21 and 28, 1929.

† See H. Jeffreys, 'On Compressional Waves in Two Superposed Layers,' *Proc. Camb. Phil. Soc.*, **23**, p. 472, 1926.

(2) The path ODA reflected from the interface, travelling entirely in the surface layer; and

(3) The 'diffraction-path' OBCA, passing through both the surface and the underlying layers.

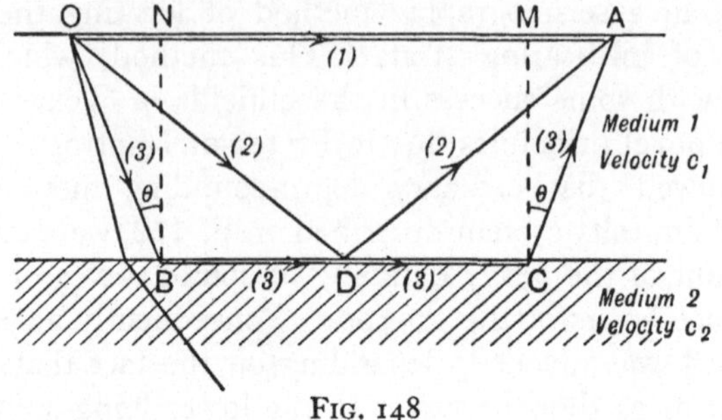

Fig. 148

If c_1 and c_2 are the velocities of sound in the upper and lower layers respectively, then it is seen that these disturbances arrive at A at times t_1, t_2, and t_3, given respectively by $t_1=OA/c_1$, $t_2=(OD+DA)/c_1$, and $t_3=2OB/c_1+(OA-2ON)/c_2$. The seismograph record should therefore indicate *three* arrival times. In practice, however, the reflected disturbance travelling by the path like ODA is obscured by the violent initial disturbance preceding it and continuing for a relatively long period. In the case of the third disturbance, however, *via* OBCA, above a certain critical value of the distance OA this impulse arrives *before* the direct impulse, since c_2 is greater than c_1. It is easily shown, in the critical case when the times t_1 and t_3 are equal and $OA=d$, that the distance d is related to the depth h of the interface in the following manner:—

$$\frac{h}{d}=\frac{1-\sin\theta}{2\cos\theta} \qquad \text{(where } \theta=\sin^{-1}(c_1/c_2)\text{).}$$

If a time-graph is plotted with the time-intervals t between the instant of the explosion and that of the *initial* disturbance of the seismograph as ordinates and the distances x between explosion and seismograph as abscissæ (see fig. 149), it will display a break at P where the times of arrival of direct and diffracted disturbances are equal. Moreover, the slope of OP, which corresponds to the direct disturbance, is proportional to $1/c_1$, whilst the slope of

Fig. 149

PQ, which relates to the indirect disturbance, is proportional to $1/c_2$, thus

$$\sin\theta = c_1/c_2 = (\text{slope of PQ})/(\text{slope of OP}),$$

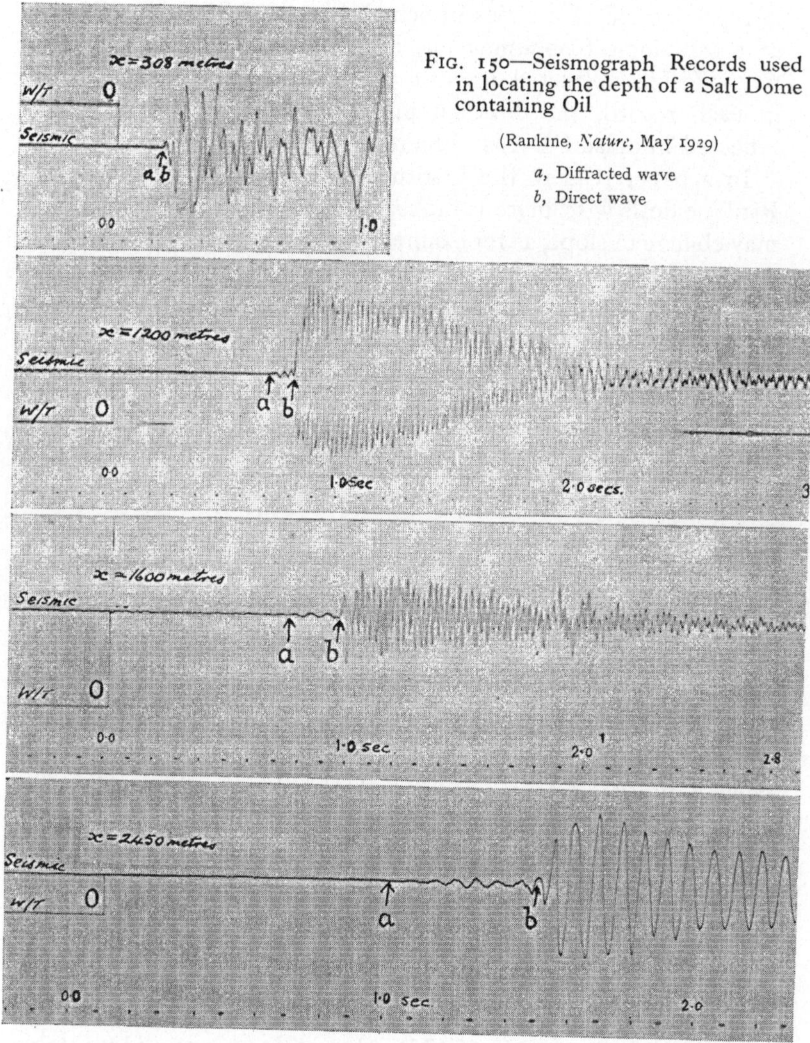

FIG. 150—Seismograph Records used in locating the depth of a Salt Dome containing Oil

(Rankine, *Nature*, May 1929)

a, Diffracted wave

b, Direct wave

whence the critical angle θ can be determined. Using this value of θ in the above equation, together with the value of the critical distance d (the value of x at the point P), the value of h, the depth of the interface may be at once calculated. For example, if c_1 and c_2 are 2000 and 5300 metres/sec. respectively (as in the Texas oilfields), and the discontinuity at P occurs at a distance $x=d=1000$ metres, we find

$$\sin\theta=c_1/c_2=0{\cdot}38, \quad \text{whence} \quad \theta=22{\cdot}5^\circ \quad \text{and} \quad \cos\theta=0{\cdot}924.$$

Consequently $h=d(1-\sin\theta)/2\cos\theta=335$ metres approximately. A reproduction of a series of actual records used in the location of a salt dome (containing oil) is given in fig. 150. The relatively feeble diffracted impulse (path OBCA) is shown at '*a*' in each record, the direct impulse OA arriving at '*b*.' The reflected impulse ODA is obscured by the after-effects of '*b*.'

In a paper read to the Institute of Mining and Metallurgy,* Rankine deals with more complicated cases in which the interface may change in slope, as for example at the edge of a salt dome. A survey of the Anglo-Persian oilfields has recently been made by the Geophysical Co., Ltd., employing very successfully the seismic method outlined above. A description of the various forms of portable recording seismographs used in such surveys is given in the translation of a book by R. Ambron.†

(B) Sound Ranging

The methods which we have hitherto described for determining the direction of a source of sound (see p. 411) are inapplicable when the sound consists of an isolated impulse or shock, as distinguished from a continuous or frequently repeated sound. Special methods have therefore been devised for locating sources of sound, such as a gun firing or a charge exploding. In fact, the location of hostile guns on land and of explosions under the sea was one of the most important practical applications of sound during the war. The method universally employed in such circumstances was known as the 'multiple-point method' of sound ranging. Since the war another more simple and direct method known as the 'radio-acoustic method' has been developed, but in this case the co-operation of the 'source' of the impulse is required, a condition which is clearly impossible in war. We shall consider the second method first, on account of its simplicity.

(a) *Radio-Acoustic Method* – This is a direct application of the 'differential-velocity,' or 'synchronous signalling' system, to which we have already referred (p. 466). It was first proposed by J. Joly, and has been developed recently by H. E. Browne and the writer for the British Admiralty ‡ as a means of locating explosive charges

* *Loc. cit.*
† *Elements of Geophysics* (M'Graw-Hill Publishing Co., 1928).
‡ See A. B. Wood and H. E. Browne, *Proc. Phys. Soc.*, **35**, p. 183, 1923.

fired under water. The following outline of the method refers to this particular application, but it will, of course, be understood that the same principles are applicable also to impulsive sounds in air. It consists essentially of the simultaneous emission of a 'wireless' signal and an explosion impulse, the former travelling with the velocity of light (186,000 miles/sec.), whilst the latter travels in the sea with a velocity approximately 1 mile/sec. (an accurate expression for this velocity is given on p. 249). At a shore station a multi-stringed Einthoven oscillograph records on one string the arrival of the radio-signal, *i.e.* the instant of detonation of the charge, and on other strings the arrival of the sound wave at the successive hydrophones. The record is crossed by time-marks every hundredth and tenth of a second. It is therefore a simple matter to determine the various time-intervals t_1, t_2, t_3, etc., between the instant of firing the charge and the arrival of the sound wave. The distances of the explosion from the various hydrophones are consequently given by ct_1, ct_2, ct_3, etc. Now the positions of the hydrophones are accurately known from an independent survey, so that the position of the explosion is accurately defined by the common intersection of circles of radii ct_1, ct_2, etc., described with the respective hydrophone positions as centre. In practice four (tripod) hydrophones were laid on the sea-bed in surveyed positions on a 12-mile base-line outside the Goodwin Sands, and connected by cables to a recording station at St. Margaret's Bay, near Dover. The six-stringed Einthoven galvanometer to which they were connected was fitted with an automatic developing and fixing arrangement, and the resulting record, on bromide paper strip, showed six parallel lines broken suddenly at the respective instants of arrival of the radio-signal and the sound impulse. Time-intervals could be read off quickly to $\pm 0{\cdot}01$ second ($\equiv \pm 50$ ft. in range), or to $\pm 0{\cdot}001$ second by estimation. The 'simultaneous' radio-acoustic signal was produced in various ways according to the degree of accuracy required. In the simplest case where only an approximate estimate of position was wanted, the W/T operator on the ship pressed his transmitting key at the instant of feeling the shock of the explosion in the water. This introduces the personal error of the operator. For more accurate work a double key is used. One half of the key sends the radio signal whilst the other fires the charge electrically. The error in time introduced by this method is negligible ($\pm 0{\cdot}001$ second$\equiv \pm 5$ ft. range). A 9-oz. charge of guncotton can be located accurately in this way at

40 miles, whilst larger charges have been located at 80 miles or more. At 80 miles the sound wave through the sea reaches the hydrophones $1\frac{1}{2}$ *minutes* after the charge is fired—the use of head phones and an ordinary stopwatch (indicating fifths of a second) instead of an elaborate recording system would therefore indicate the distance with a probable accuracy of $\pm 1/5$ mile. With the more accurate method of recording it will be realised therefore that the radio-acoustic method permits of very great accuracy, and has important applications in navigation and hydrographical survey. For navigational purposes, a ship calls up the recording station and asks for her position. The record is made and the position determined, within a radius of half a mile, in about seven minutes after receiving the request. In hydrographical surveys greater care is taken in reading the record; the method has been used successfully in fixing accurately the positions of certain buoys and light-vessels out of sight of land.

As an alternative to the Einthoven photographic recorder system of measuring time-intervals, J. M. Ford and the writer * in 1919 devised an instrument known as the *Phonic Chronometer*. This may be described as a stopwatch operated electromagnetically and fitted with three sets of dials for measuring three independent time-intervals. The speed of the chronometer is governed by a tuning-fork and a phonic motor, giving an accuracy of 1 in 10,000. Time-intervals are indicated directly on the dials to 0·001 second. Special explosion shock-receivers were designed for use with the chronometer. The radio signal, through a relay, started all the dials, these being stopped in turn as the sound wave passed the respective shock receivers. The three time-intervals were thus indicated directly on the dials. A single-dial phonic chronometer (see fig. 151) has, of course, many other applications involving the accurate measurement of a time-interval.

The Geodetic Survey of U.S.A. is at the present time using a modified form of the R/A method in surveying coastline and banks and in locating marking buoys.† The method has been thoroughly tested under working conditions. It has proved accurate and reliable by day or night, in rough or foggy weather, and at all seasons of the year. Joly ‡ has made a number of

* A. B. Wood and J. M. Ford, *Journ. Sci. Instrs.*, March 1924. See also Drysdale, *Mechanical Properties of Fluids*.

† U.S. Coast and Geodetic Survey, Spec. Pubn., No. 107, 1924 and 1928.

‡ *Loc. cit.*

ingenious proposals to safeguard ships in fog, by the application of radio-acoustic methods. For example, a light-vessel should send a simultaneous radio-acoustic signal followed by twenty radio 'dots' at intervals of half a second. A ship hearing the acoustic signal between the seventh and eighth dot in the W/T series would at once know it was $3\frac{1}{2}$ to 4 miles from the lightship (assuming the velocity of sound in water to be approximately 1 mile per second).

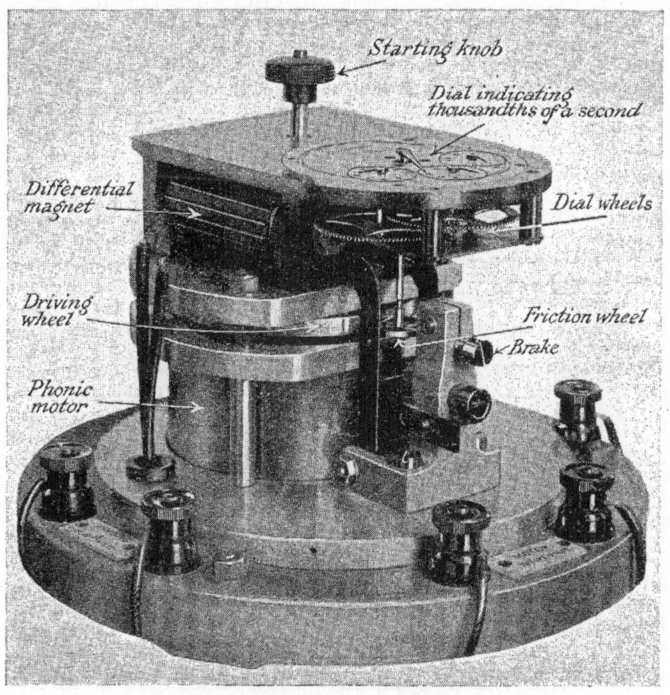

FIG. 151—Phonic Chronometer
(*Journal of Scientific Instruments*)

(b) *Multiple-point Sound Ranging* – This is primarily a 'war' method, for the source of sound, a gun or a mine, has no desire to co-operate in the 'location.' The observer must find the source of sound by means of the sound wave only. For this purpose three or more receivers are mounted at known positions on surveyed base-lines. Suppose the spherical wave-front WW (see fig. 152), which originated at the explosion source, passes in succession over the receivers 0, 1, 2, 3 at the instants o, t_1, t_2, and t_3 respectively. The simple construction shown in the diagram indicates how the origin of the wave-front may be determined.

If c is the velocity of sound appropriate to the medium, describe circles of radii o, ct_1, ct_2, and ct_3, with centres at 0, 1, 2, and 3 respectively. The wave-front WW must be tangential to all these circles; the origin of the sound wave, the explosion, must therefore be at the centre of a circle of which the circumference touches all the small circles. The explosion may be located more accurately by calculation. Taking any pair of receivers, it is evident that the explosion must lie on a curve such that the difference of the radius vectors from any point P on it to the respective receivers must be a constant (ct)—that is, the explosion must lie on the hyperbola, with foci at 0 and 1, and the difference $\pm(P1-P0)=ct_1$. Similarly, it must also lie on the hyperbolæ, in which the differences are ct_2 and ct_3—that is, the explosion lies at the intersection of these hyperbolæ. For approximate determinations of distant explosions it is sufficient to find the intersection of the asymptotes of the hyperbolæ, a special 'asymptote correction' being applied in accurate determinations.

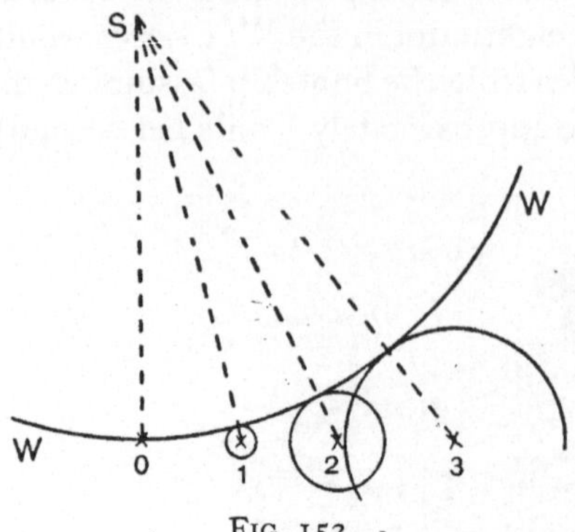

FIG. 152

(i) *Army Sound Ranging* – In the location of enemy guns a base-line 900 yards long with six receivers was generally used, the three additional receivers being employed in case of failures and to provide check 'locations.' A special type of microphone, the Tucker hot-wire microphone (described on p. 402), was used. This was mounted at the mouth of a large Helmholtz resonator of about 16 litres capacity, the system responding only to sounds of very low frequency such as the gun-wave (the '*onde de bouche*' or '*infra son*'), whilst ignoring sounds of higher, audible frequency. A six-stringed Einthoven oscillograph with timing device and automatic photographic arrangements was used to record the arrival of the wave at the various microphones. A typical record is shown in fig. 153 (*a*). The record was started when an advanced observer, or a 'sentry' microphone, heard the gun—a few seconds before the arrival of the pulse at the recording microphones. In all cases it was necessary to make corrections for wind and temperature. The corrections become somewhat involved when these quantities vary with height. Under certain conditions of wind-gradient, sound ranging was not possible. The corrections were in some cases avoided entirely by recording not only the enemy gun but also the 'answering' shell-burst, and 'adjusting' the latter to coincide with the former.

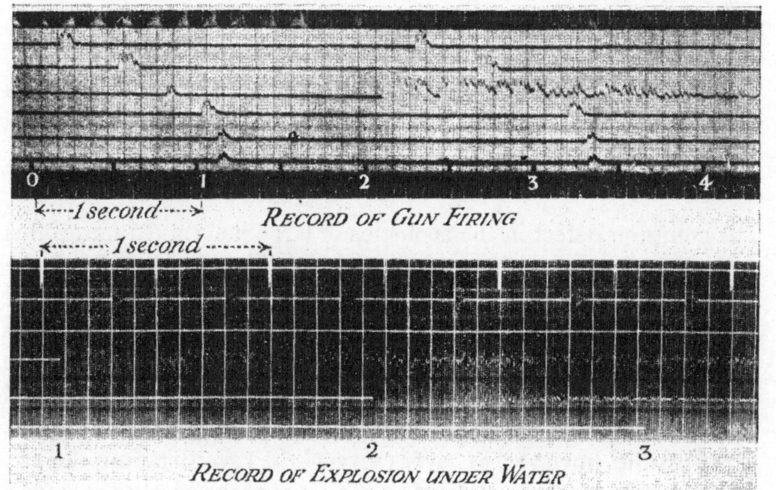

FIG. 153—Sound-ranging Records

(ii) *Navy Sound Ranging* – The same principles were applied to locate explosions of mines, depth-charges, and torpedoes in the sea. The only differences were in detail—hydrophones (steel diaphragms with carbon granular microphones) were used instead of hot-wire microphones, four of these being spaced over a base-line 12 miles long and about 20 miles from the recording station. As before, an Einthoven oscillograph with six strings recorded the sound impulses and measured the time-intervals. A specimen record is shown in fig. 153 (*b*). The hydrophones were carefully surveyed from Ordnance Survey points on shore and their positions and distances apart were known within a few feet. Explosions taking place up to 100 miles or so away from the base-line were located with great accuracy. The sea is an excellent medium for sound transmission, it is very homogeneous and there are no 'meteorological' corrections as in transmission through the atmosphere—it is only in very accurate work that tidal corrections (a few feet per second) have to be applied to the sound velocities. Values of the latter are tabulated for different temperatures of the sea (see p. 249) and an electrical thermometer, with recorder on shore, is laid on the base-line of hydrophones. The temperature fluctuations in the sea are, however, usually very slow and very small—covering a range from 6° to 17° C. annually in the North Sea between England and Holland. The multiple-point sound-ranging method has been used very successfully also in hydrographical survey. The positions located are extremely accurate, and may be obtained much more quickly than the alternative astronomical methods hitherto employed. As previously stated, a 9-oz. guncotton charge is sufficient to cover a range of 40 miles in the sea.

Acoustics of Buildings *

A common defect in the acoustics of large halls and auditoriums is that due to excessive reverberation. The slow decay of the sounds of speech or music in such a hall is due to the repeated reflections occurring at the hard, smooth surfaces of its walls before the energy is dissipated. The obvious cure is to introduce rougher and more porous surfaces—that is, to cover the walls and floor with sound-absorbent materials, such as hangings and carpets. It is possible, however, to 'over correct' the reverberation, thereby producing a 'deadening' effect on the voice of the speaker or singer. Sabine,† who first placed the subject on a scientific footing, laid down three simple rules which must be followed if satisfactory results are to be obtained, viz. (1) the sound heard must be loud enough; (2) the quality of the sound must be unaltered, *i.e.* the relative intensities of the components of a complex sound must be preserved; and (3) the successive sounds of speech or music must remain distinct, *i.e.* there must be no confusion due to overlapping of syllables. In certain cases distinct echoes may be heard in a hall. These may be reduced in intensity or removed entirely by changing the angle of the wall or by covering the existing surface with absorbent material. Curved surfaces, such as domed ceilings, sometimes focus sounds in an objectionable manner. From a scientific point of view the design of an acoustically perfect hall involves three principal factors—reverberation, interference, and resonance. The means of controlling these factors are the materials used in construction, the dimensions, and the shape of the hall.

* The following account of the subject is mainly derived from *Acoustics of Buildings*, by Davis and Kaye (Bell, Publishers), to which the reader is referred for further details.

† Collected papers on Acoustics, 1922.

Reverberation – The theoretical study of reverberation commenced in 1900 with the work of Sabine, and was extended later by Jäger,* Buckingham,† Watson,‡ and others. All theories are based on the fact that sound in an ordinary room undergoes two or three hundred reflections before it becomes inaudible. It is consequently assumed that the sound energy rapidly becomes uniformly distributed throughout the room—interference and such like effects being ignored. If I is the mean intensity (energy per unit volume) in the room at any given instant, then the change of intensity δI by absorption in a time δt is $\delta I = -Ian\delta t$, where a is the mean absorption coefficient and n the number of reflections per second. By a statistical method Jäger proved that $n = cS/4V$, where V is the volume of the room, S the total surface area, and c the velocity of sound. Denoting the rate of emission of sound energy from the source by E, the rate of energy change per unit volume of the room is given by

$$\frac{dI}{dt} = \frac{E}{V} - \frac{caSI}{4V} \quad . \quad . \quad . \quad . \quad (1)$$

a fundamental equation. The intensity at a time t after *stopping* the source is therefore

$$I = \frac{4E}{caS} e^{-caSt/4V} \quad . \quad . \quad . \quad . \quad (2)$$

The intensity at a time t after *starting* the source is

$$I = \frac{4E}{caS}(1 - e^{-caSt/4V}) \quad . \quad . \quad . \quad (3)$$

The intensity in the room finally attains a steady value given by

$$I_{max} = \frac{4E}{caS} \quad . \quad . \quad . \quad . \quad (4)$$

Equation (2) may therefore be written $I/I_{max} = e^{-caSt/4V}$, which is a very convenient form. Expressing the *standard initial condition* as an intensity I_{max} equal to 10^6 times the 'threshold' value (minimum intensity audible) we have $10^6 = e^{-caST/4V}$, whence $T = 55{\cdot}2V/caS$, where T may be regarded as the standard duration

* *Acad. Wiss. Wien*, **120**, 2*a*, p. 613, 1911.
† *Bureau of Standards Sci. Paper*, No. 506, 1925.
‡ *Acoustics of Buildings*, 1923.

of audibility. Assuming a value 1100 ft./sec. for the velocity of sound c, we have

$$T = 0{\cdot}05V/aS \quad . \quad . \quad . \quad . \quad (5)$$

as the *theoretical standard reverberation period*. The duration of reverberation therefore varies directly as the volume V of the room and inversely as the surface area S and mean absorption coefficient a of its walls. This relation has been accurately verified experimentally by Sabine. He first used automatic recording methods to measure the exponential decay of sound intensity in a room, but later preferred to rely on direct aural observation. The source of sound was an organ pipe blown at a definite pressure and emitting a note of definite pitch and uniform intensity. The instant of cutting off the sound and the instant at which the observer considered it inaudible were recorded on a chronograph drum. Sabine considered that an average accuracy of $\pm 0{\cdot}05$ second in the measurement of T was possible by this simple method, *e.g.* he obtained consistently for a particular lecture room values lying between 5·45 and 5·62 seconds, with a mean of 5·57 seconds, which was repeated after an interval of three years. It was found (1) that the duration of reverberation was almost independent of the positions of the source and of the observer in the room; (2) the effect of a given amount of absorbent was also practically independent of its position in the room. Sabine used an open window as his *standard unit of absorption*, expressing other results in '*open-window*' *units*. In this manner he determined the mean absorption coefficients of various absorbent materials, obtaining results varying from $a = 1{\cdot}0$ for the open window to 0·01 for hard surfaces, such as glass or plaster. He found that each member of an audience was equivalent in absorbing power to about 4½ square ft. of open window.

Optimum Reverberation – With a committee of musical experts Sabine determined the degree of reverberation most suitable for speech and music in rooms of different dimensions. The reverberation period was adjusted in each room until all were agreed that the conditions were neither too reverberant nor too 'dead.' For piano music a mean of all observations, ranging from 0·95 to 1·16, gave T a value of 1·08 seconds. Eckhardt,* using the reverberation equations given above, has discussed the optimum reverberation conditions for syllabic speech, indicating graphically how the syllables overlap seriously in rooms of low

* *Frank. Inst. Journ.*, **195**, p. 799, 1923.

absorbing power, whilst they may be entirely separated in a room of high absorbing properties. Estimates by other observers * appear to indicate that the optimum period increases with the volume of the room or hall and varies with the nature of the sound. In a theatre of volume 60,000 cub. ft. Sabine regarded 1·1 to 1·5 seconds as satisfactory, whereas 2·3 seconds was prescribed for theatres of volume 250,000 cub. ft. and 650,000 cub. ft. Broadly speaking, the optimum period appears to lie between 1·0 and 2·5 seconds; for speech and 'light' music towards the lower value, and for orchestral or 'grand' music it may approach the upper limit.

Determination of Absorption Coefficients – In order to correct a room for excess or deficiency of reverberation it is necessary to know the absorption coefficients of the various materials present in the room. The methods of measuring such absorption coefficients may be divided broadly into two classes: (*a*) full-scale measurements—reverberation methods, and (*b*) small-scale measurements—direct laboratory measurements using small test specimens.

(a) *The full-scale method* is exemplified by Sabine's experiments. The duration of reverberation was first measured in an empty room for various initial intensities of the source. To determine the rate of decay of sound intensity in a room observations are made of the durations of audibility t_1 when a certain source is used, and the duration t_2 with a source of m times the intensity. Since the intensity decays to the same limit in each case, it follows from equation (2), p. 485, that

$$\log_e m = (t_2 - t_1) ca\mathrm{S}/4\mathrm{V} \quad . \quad . \quad . \quad (6)$$

whence the mean absorption coefficient a can be determined, c, S, and V being known. The absorbing power of different objects or materials may similarly be determined by measuring the reverberation period for two strengths of source both with and without the objects in the room. As already mentioned, the procedure usually adopted by Sabine was a substitution method, the absorbing power of the object being expressed in terms of that of a known area of open window.

(b) *The stationary wave method* of measuring absorption co-

* Lifshitz, *Phys. Rev.*, **25**, p. 391, 1925; **27**, p. 618, 1926; Watson, *Frank. Inst. Journ.*, **198**, p. 73, 1924; Knudsen, *Phys. Rev.*, **22**, p. 206, 1923.

efficients described on p. 324 is an extremely convenient and accurate laboratory method in which only small specimens of absorbent material are required. The specimen, in the form of a flat sheet, closes one end of the tube, in which stationary sound waves are set up by means of a source at the opposite end (see fig. 99). The degree of 'imperfection' of the maxima and minima serves to measure the ratio of incident and reflected amplitudes, whence the absorption coefficient may be calculated in the manner indicated on p. 324. This method measures only the reflection at *normal* incidence, whereas the sound reflected to and fro in a room meets the walls at *all* angles of incidence. E. T. Paris * has examined this question theoretically, and has shown that as the angle of incidence θ increases from 0° at normal incidence to about 80°, the absorption coefficient increases slowly at first then more rapidly. In a particular case, that of a porous plaster, the absorption coefficient increased from 0·28 at normal incidence ($\theta=0$) to 0·76 at 83°. Thereafter the absorption dropped rapidly to zero at grazing incidence ($\theta=90°$). These deductions explain why the absorption coefficients obtained in large scale experiments (all angles of incidence) are always higher than those determined by the stationary wave method (normal incidence). Paris has obtained theoretically the correction necessary to the 'normal incidence' absorption coefficients to make them applicable to the case of 'all angles of incidence.' This correction brings the experimental values obtained by the stationary-wave method into good agreement with those obtained from Sabine's full-scale reverberation experiments.

Tables of absorption coefficients obtained by various methods are given by Davis and Kaye (*loc. cit.*). These values vary over wide limits according to the porosity of the material—*e.g.* 0·01 for various hard surfaces such as wood, glass, or metal; 0·5 for thick carpets, audiences, 'acoustic tiles and plasters,' and hair-felt; 1·0 for an open window. As the theory of absorption in porous bodies indicates, the absorption increases with the pitch of the incident sound; but other factors also enter into the question. For example, it was found experimentally that a certain hair-felt had values 0·09, 0·25, 0·43, 0·63, 0·33, 0·35 for the absorption coefficient at frequencies 128, 256, 512, 1024, 2048, and 4096 p.p.s. respectively, the maximum absorption occurring at a frequency near 1000 p.p.s.

Two factors are involved in the selection of a good anti-

* *Proc. Roy. Soc.*, **115**, p. 407, 1927.

reverberation material; it must *reflect* as little as possible and must *absorb* a maximum of the incident sound energy. The former condition implies that the acoustic resistance of the absorber must be as nearly as possible equal to that of the incident medium, and the latter condition requires a maximum amount of viscous damping to suppress the transmitted sound in a relatively thin layer of absorbent material. The methods of correction of defective acoustical conditions in a room or hall by the introduction or removal of absorbent surfaces will be sufficiently obvious. Materials, such as porous tiles and plasters, containing asbestos or cork constituents are specially manufactured as a means of adjusting the reverberation periods of large auditoriums to the optimum value. Such materials, their properties and uses, are fully described by Davis and Kaye (*loc. cit.*).

Study of the Shape of an Auditorium – The reflecting characteristics of various shapes of auditorium are conveniently studied by the *spark-pulse and the ripple-tank methods* described on p. 340. The spark method was first applied to this purpose by Sabine in 1913. A sectional model of the auditorium is mounted between a 'light'-spark gap and a screen (or a photo-

(*a*)

FIG. 154—Ripple Tank (with model section of auditorium)

(From Davis and Kaye, *Acoustics of Buildings*)

graphic plate), a 'sound-spark' being produced at the required point inside the model. The shadow cast on the screen or plate reveals the position of the pulse after reflection from the walls of the model. A succession of photographs at varying time-intervals after the production of the sound-spark indicates the progress of the sound-pulse and its reflection from the walls of the model auditorium. Results of a similar nature may be obtained more easily by means of the ripple tank (see p. 344). This method is based on the fact that the wave-length of the ripples on the surface of water is usually comparable with the size of any models concerned. The general arrangement is shown in fig. 154 (*a*). Fig. 154 (*b*) and (*c*) are comparative photographs showing the marked resemblance between sound-pulse and

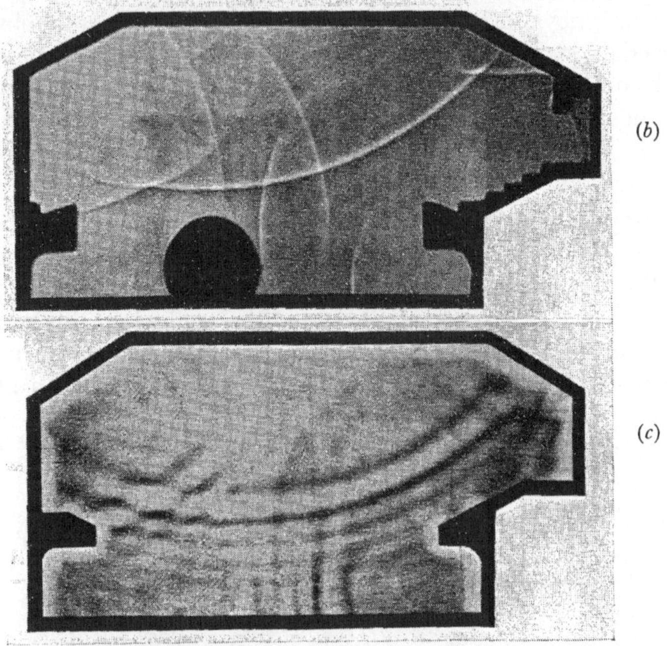

Fig. 154—Comparative Spark and Ripple Photographs (A. H. Davis)

(From Davis and Kaye, *Acoustics of Buildings*)

ripple photographs for similar model sections (Davis and Kaye). Cinema records of this nature are valuable as a means of following the progress of the multiply-reflected waves as they spread from a source, a speaker, or singer to an audience. The *three-dimensional* study of echoes has not yet been undertaken on a model scale, but Watson * has described a ' ray method ' of tracing the reflections of a high-pitched sound in a large auditorium. In his experiments Watson used as the source of sound the hiss from an alternating current arc at the focus of a parabolic mirror. The light from the arc assisted in the location of the sound

* Univ. of Illinois Eng. Expt. Station, Bull. No. 3.

echoes. Such model and full-scale experiments indicate that markedly smooth curved surfaces should preferably be avoided from an acoustic standpoint, as they result in the focussing of sound in particular directions. It is considered by those who have wide experience of such matters that reflected sounds arriving within an interval of $\frac{1}{15}$ second (equivalent to a path-difference of 75 ft.) will contribute usefully in raising the level of loudness of speech, whilst reflections arriving later are undesirable and should be weakened as much as possible by absorption or scattering. Troublesome echoes are not likely to arise, therefore, except in large halls, more especially those with domed or curved ceilings. It is important that such surfaces are covered with absorbent material or broken up by coffering to scatter the sound. Resonant surfaces, panelling, etc., should be suitably damped, otherwise sounds of appropriate pitch will be reinforced with unpleasant results. Such resonant vibration is most noticeable when the source of sound is very powerful, *e.g.* a church organ will frequently set in vibration resonant panelling or window panes of low frequency. The phenomenon of interference gives rise to maxima and minima of sound, but the effect in a room 'lagged' with sound-absorbent materials is usually unimportant.

The Reproduction of Sound *

(*a*) **Gramophones and Loud-Speakers** – The history of the mechanical recording of the sounds of speech and music may be said to date from 1864, when the Phonautograph of Scott and of König was invented. In 1877 Edison obtained sound records by a somewhat similar device, but instead of using a smoked surface the recording stylus engraved its vibrations on a tinfoil or waxed cylinder. A re-traverse of the 'record' by the stylus resulted in the reproduction of sounds by the diaphragm somewhat similar to the sounds which produced the record. From these early efforts have developed the excellent recorders and reproducers of sound which may be heard to-day. The great progress made in the development of such devices is largely due to the

* The reader is recommended to consult (*a*) *Proc. Inst. Elec. Eng.*, 'Discussion on Loud Speakers, etc.,' Nov. 1923; (*b*) Maxfield and Harrison, *Bell System Techn. Journ.*, **5**, p. 493, 1926; (*c*) S. T. Williams, *Journ. Frank. Inst.*, **202**, p. 413, Oct. 1926; and (*d*) A. Whitaker, *Journ. Sci. Instrs.*, **5**, p. 35, Feb. 1928, from which much of the following information is derived. See also 'Wireless Loud-Speakers,' N. W. McLachlan (*Wireless World*, Pub.).

application of electrical principles to mechanical operations. The striking electro-mechanical analogy which has been found so convenient in studying complex mechanical problems is nowhere more clearly illustrated than in the design of sound-reproducing apparatus such as the gramophone and the loud-speaker. Before dealing with this aspect of the question, however, it may be well to refer to some of the requirements of good sound-reproduction. First of all, it is desired to produce sounds at one place which are a faithful copy of those originating at another place. Secondly, it is desirable that the reproduced sounds should be of considerable intensity—approximating at least to the intensity of the original sounds. In other words, the ideal 'reproducer' would be a secondary source equivalent in every respect to the primary source of whatever nature. Mechanical transformation of complex wave-forms, *e.g.* from one *type* of vibration to another, or from one *amplitude* of vibration to another; or conversion of mechanical into another form of energy (*e.g.* electrical or thermal) involves *distortion*. Only the simple harmonic type of vibration, the pure tone, is transmitted without distortion; but in speech and music we are almost invariably concerned with the transmission of complex types. The chief danger is that due to resonance. In aiming at loudness there is a temptation to employ resonance—the majority of diaphragm sound-reproducers (telephones, loud-speakers, etc.) have resonance peaks more or less flattened by damping. Such resonances, unless great precautions are taken to counteract them, inevitably lead to distortion, those components of the sound to be reproduced which coincide most closely with the resonant frequencies being disproportionately emphasised. It is perhaps fortunate that the ear is sufficiently accommodating to ignore fairly large defects in sound reproduction. For example, a 10 per cent. error in intensity, or even more than this, would generally pass unnoticed. The accuracy of *intensity reproduction* need not be very great to conform with such a standard. With regard to *frequency reproduction*, however, the ear is much more exacting, slight errors in pitch (or to be more precise, slight errors in the relative pitch of component sounds) being readily observed. It is fortunate that the practical difficulties involved in frequency reproduction are far less serious than in intensity reproduction. Reproduction of sound involves the preliminary processes of reception, and possibly of recording, transmission, and amplification. In dealing with sound reception and recording we have already considered various types of mechanical and electrical receivers and the important factors influencing the choice of a receiver for distortionless recording or transmission. The process of sound recording and reproduction is best illustrated in the case of the gramophone. Recent development * has followed the lines of regarding the mechanical system as analogous to an electrical band-pass filter circuit, which may be so designed that it transmits a certain range of frequencies without loss whilst disregarding all other frequencies. Both the recording and reproducing systems are good examples of the use of this analogy. It would require too lengthy a description, however, to deal with this analogy in detail, but the diagram shown in fig. 155, taken from Whitaker's paper, may serve as an illustration of the point. In recording it is necessary to have a mechanism such that for equal input energies at any frequency the maximum velocity of

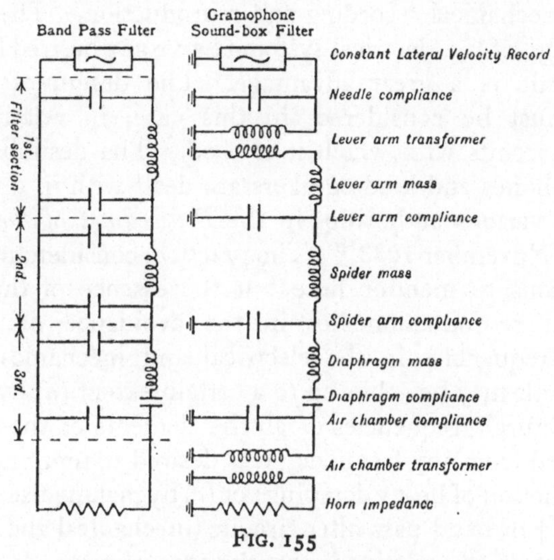

FIG. 155
(After A. Whitaker, *Journal of Scientific Instruments*)

the cutting stylus is constant, the amplitude of vibration varying inversely as the frequency. Velocity in the mechanical system is equivalent to current in the electrical network. Stiffness represents capacity, or compliance the reciprocal of capacity. Masses represent inductances, and frictional forces or non-reactive loads are equivalent to resistances. In deriving the equivalent circuit of any mechanical device a stiffness between two consecutive

* See, for example, Maxfield and Harrison or Whitaker, *loc. cit.*, also Bell Laboratories Record, Nov. 1928.

moving members is represented by a shunt capacity, and between a moving member and a rigid support a series capacity. A great difficulty in designing such mechanical systems is the lack of satisfactory non-reactive mechanical resistance. Maxfield and Harrison describe a recording system, designed on the lines of a filter system, which has no prominent resonances within the range of speech frequencies. The frequency characteristic of such a recording system is approximately a horizontal straight line (constant amplitude) from 250 to 5000 p.p.s. A gramophone reproducer is designed on similar lines. Recently the electrical gramophone reproducer has developed; the needle is attached to a 'pick-up' which converts the mechanical vibrations into electrical oscillations, these being amplified and used to operate loud-speakers. This system and the corresponding electrical recording system have many advantages over the older method of direct mechanical recording and reproduction. The flexibility in the design of the electrical system to give any desired frequency-characteristic is a great advantage. The design of the loud-speaker must be considered, in this case, in relation to the electrical circuits with which it is used. The desirable features in gramophones and loud-speakers are dealt with in a number of papers by various authorities in the *Proceedings of the Electrical Engineers*, November 1923.* The practical considerations are far too numerous to mention here, but the essence of the problem of faithful reproduction lies in the avoidance of prominent resonant frequencies in the electrical and mechanical systems. Such a result may be achieved to a certain extent (*a*) by arranging that the natural frequencies of all the elements of the system are far removed from any frequency it is desired to reproduce; (*b*) by the introduction of heavy damping; or (*c*) by making use of multiple resonance † or band-pass filter circuits (mechanical and electrical). On account of the relative insensitiveness of methods (*a*) and (*b*) the third method is often preferred. Nicholson has employed piezo-electric crystals of Rochelle salt, having a very high natural frequency, as a means of reproducing sounds electrically (see p. 393).

(*b*) **Optical Methods of Recording and Reproducing Sound.** *The Photophone. Phonofilms* – Hitherto it has been customary to record sounds optically by transverse vibrations such as, for example, the trace made by a vibrating spot of light on a moving

* See also *Proc. Phys. Soc.*, **36**, p. 114, Feb., and p. 211, April 1924.

† See Rothwell, *Nature*, Feb. 24, 1923; and Porter, *Phys. Soc. Proc.*, Feb. 1924.

photographic film. The record thus obtained does not lend itself readily to sound reproduction. If, however, the sounds are recorded photographically by means of changes of *intensity* of a beam of light, reproduction is a relatively simple matter. This process was tested by Rühmer * in 1900, but his methods were not very satisfactory. A simple and effective method is described by A. O. Rankine,† making use of a device known as the '*photophone*.' In this device, designed for the transmission of speech by light,‡ the light from a point source is collected by a lens of about a metre focal length and an image formed on a small concave mirror which is attached to the diaphragm of a gramophone sound-box or to the reed of a telephone earpiece. The light diverges and passes through a second similar lens, which projects it to the distant station. Two similar grids are mounted—one in front of each lens. An image of the first grid is superposed on the second by reflection in the small concave mirror. When the latter oscillates under the vibrations of speech the dark spaces of the image-grid move over the openings of the second grid, thus producing fluctuations of intensity in the beam. In the photophone this fluctuating light is received by a collecting lens or mirror and focussed on a selenium cell in circuit with a battery and telephones. The resistance changes of the selenium result in the reproduction of the original sounds in the telephones. If, however, the sounds are to be recorded, the beam of fluctuating intensity is focussed on a slit of which an image is formed by a lens on a moving photographic film. Examples of the resulting record are shown in fig. 156, which represent the fluctuations of light intensity due to a tuning-fork and various spoken words. The sound picture actually represents an 'instantaneous' view of the succession of compressions and rarefactions in the associated sound wave. For reproduction of the sounds light from an illuminated slit is focussed on the film, behind which is placed a selenium cell connected to a telephone or loud-speaker circuit. As the film moves, the intensity of the light reaching the cell varies in accordance with the density of the film, and the corresponding sound is reproduced. A somewhat similar process is used in the production of *phonofilms* or 'speaking pictures.' §

* Ernst Rühmer, *Wireless Telephony*, 1907.

† *Proc. Phys. Soc.*, **32**, p. 78, 1920.

‡ *Proc. Phys. Soc.*, **31**, p. 242, 1919. See also *Proc. Opt. Convention*, pt. ii, 1926.

§ See C. F. Elwell, *Wireless World and Radio Review*, **15**, p. 466, Jan. 7, 1925.

The sounds from the speaker are received by a sensitive microphone, amplified, and the current variations used to fluctuate the light emitted from a special form of lamp (invented by T. W. Case). The light is focussed on a fine slit placed close to the cinema film and occupying a width of 0·1 in. of the film. The variations of intensity are thus recorded on the continuously moving film in

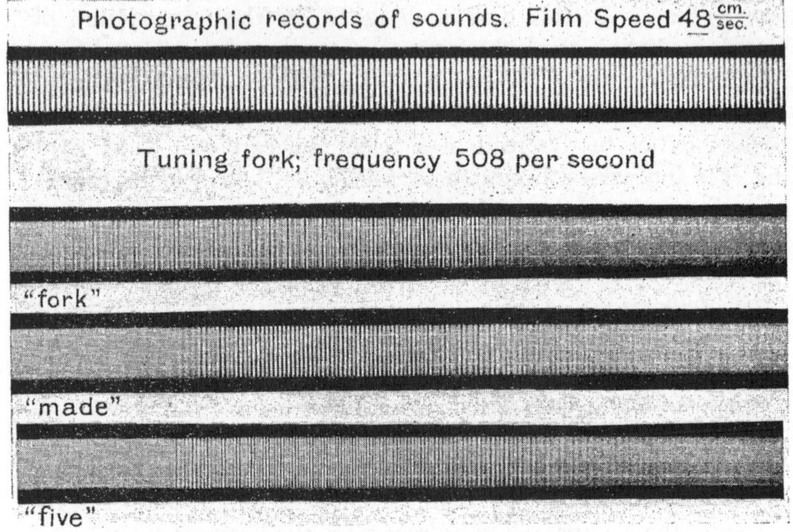

FIG. 156—Sound-film Records

(By courtesy of Prof. A. O. Rankine and The Optical Convention)

the form of a band of varying blackness. The reproduction of the sound takes place by passing light through a similar slit in contact with the continuously moving film, beyond which is a sensitive photo-electric cell—the Case Thalofide cell.* The current fluctuations on the cell, as before, are amplified and passed through a loud-speaker, which is mounted near the screen on which the moving pictures are projected. It is important to observe that the sound is recorded and reproduced on a *continuously* moving film, *i.e.* on the film before it enters the 'gate' where the motion is made intermittent (15 to 20 p.p.s.) for cinematography. This method of producing phonofilms guarantees perfect synchronisation between the sound and the picture, for the sound-record is made near the edge of the film simultaneously with the cinema picture.

The sound-recording system for talking films as developed by

* Thallium oxysulphide on a quartz disc with graphite electrodes.

the Bell Laboratories * utilises the variable density photographic method and the Wente 'light valve.' The latter comprises a loop of aluminium tape centred under tension in the gap of an electromagnet to form a window 2×256 mils. Light through this slit is accurately focussed as a line 1×100 mils. perpendicular to and near the edge of the standard photographic film. The resonant frequency of the aluminium strip is very high (about 7000 p.p.s.). As it vibrates it varies the amount of light passing through the slit—hence the description 'light valve.'

Numerous other optical methods have been examined, with varying degrees of success, to modulate a beam of light as a means of sound reproduction. Use has been made of the Kerr electro-optical effect, and the Faraday magneto-optical effect, and of the luminous electric discharge in gases. Kerr Grant,† in co-operation with Messrs. Hilger & Co., has recently succeeded in interrupting light at very high frequency by means of a quartz piezo-electric resonator placed between crossed Nicol prisms, but attempts to photograph the flashes at 144,000 p.p.s. were not successful. The intensity of the flashes could be modulated by varying the oscillating voltage applied to the crystal.

Alternating Power Transmission

The close analogy which exists between mechanical wave transmission and alternating current electric power transmission is demonstrated in a most impressive manner in a system devised by M. Constantinesco.‡ The method of transmitting hydraulic power with which we are most familiar utilises water at a pressure of, say, 1000 lb./in.^{2} supplied from a pump and forced through pipe-lines to a 'D.C. motor.' This may be described as the 'Direct Current' system, as it requires a continuous or direct flow of water. It is also possible to supply energy to a column of water enclosed in a pipe by applying *alternating* pressures at one end. By virtue of the elastic properties of the fluid in the pipe-line this energy is transmitted in the form of longitudinal vibrations through the column, and may be used to drive an 'Alternating Current' (mechanical) motor. A number of applications of this method have been patented by M. Constantinesco,

* See *Bell System Techn. Journ.*, **8,** Jan. 1929.

† *Loc. cit.*, p. 138.

‡ See *The Theory of Sonics* (Proprietors of Patents Controlling Wave Transmission, 132 Salisbury Sq., E.C., 1920); also *Mechanical Properties of Fluids*, pp. 221–228 (Blackie, 1923).

who has employed it to transmit large amounts of mechanical power by alternating pressures in water, the frequency being sufficiently high, *e.g.* 50 p.p.s., to be described as 'sonic.' The theory of the wave-transmission of power in pipes follows exactly the same lines as that of propagation of ordinary sound waves. The source A is a reciprocating plunger, which sends waves of compression and rarefaction along the water column; the 'receiver' B is the motor at the opposite end of the pipe-line. As in the case of ordinary sound also, stationary waves are set up by reflection at the end, or at any point of discontinuity of the pipe. If the system is used efficiently the 'receiver' piston B will absorb all the energy and there will be no reflection; if more energy is introduced by the piston A than is absorbed by B, reflected waves will be formed, and energy will accumulate in the pipe, which would ultimately burst. This is avoided by introducing near the piston a closed vessel filled with liquid, and having a volume large compared with the volume-variation produced by the piston. With a motor at the end B of the line, absorbing only a fraction of the energy emitted from the generator A, the stationary wave system is similar to that which we have already considered for a pipe closed by an imperfect sound reflector (see p. 324), *i.e.* the effect may be regarded as being due to two superimposed stationary waves of amplitudes $(a+r)$ and $(a-r)$, the nodes and antinodes of one being $\lambda/4$ distant from those of the other. (Alternatively the wave-motion may be regarded as being compounded of a simple stationary wave with a travelling wave conveying energy.) Under these conditions, therefore, a motor connected at *any* point of the pipe will be able to absorb energy to perform useful work. Constantinesco has devised a 'three-pipe system,' which is analogous in all respects to the 'three-phase' electrical system. The 'generator' in this case is a high-pressure reciprocating pump with three pistons and cranks 120° apart, the sonic power being transmitted through three pipe-lines to three-phase motors of similar construction. The three-phase system is usually employed as giving a more uniform torque and ease of starting. The mean pressure in the system is maintained by a pump, which returns any water leaking past the pistons. The principle of the transformer has also been introduced into the system. The single-phase system has been applied commercially to reciprocating rock drills and riveters, the alternating pressures being applied through flexible pipes comparable with electric cable. As in the corresponding electrical case there are energy losses in the transmission line due to friction (equivalent to electrical resistance), the maximum amplitude diminishing exponentially along the pipe. The theory of wave-transmission in liquid-filled pipes has been given by H. Moss.*

Other Technical Applications

(1) **The Problem of Noise Reduction** – *Vibration of Machines and Complex Structures.* We live in noisy times, and the problem of noise reduction is daily becoming more acute. In all parts of the civilised world there has been a steady increase in machinery and traffic noise during the past few years. Noise in excess contributes to fatigue and lowers the general efficiency. It has been proved, if proof were required, that excessive noise in an office results in a considerable lowering of output. Medical men have repeatedly expressed the opinion that noise is the cause of much suffering,† and also of 'educational waste' ('brain-wasting' noise is the term employed). Noises to some are pleasant, whilst to others they are the reverse; Milton liked to hear a cock crowing near his quiet retreat at Chalfont St. Giles; Carlyle was annoyed by it. A certain background of 'indefinite' noise is sometimes soothing, but when it exceeds a critical intensity and irregularity it may have the opposite effect. A world of utter silence, as in a desert or a mausoleum, would be unbearable, but the other extreme is undesirable also. The noises of machinery in workshops or in road vehicles are produced in an almost infinite number of ways. In any particular machine, however, it is often possible to locate the principal sources of noise and, in many cases, to provide a remedy. A considerable amount of research has recently been made, for example, to reduce the noise in electric motors and high-speed gears. Noise implies wear and is objectionable in other ways. By electrical recording of the noise and subsequent Fourier analysis of the wave-form, Metropolitan Vickers ‡ have succeeded in tracking down the various sources of the noise, and have as a consequence effected considerable improvements in the design of A.C. motors, and faults have been located in the master-wheels of gear-cutting machines. Vibration of turbine blades has also been studied with a view to rendering them more

* *Proc. Inst. Mech. Engs.*, 1923.

† See H. J. Spooner in *The Times*, May 2, 1928. See also April 7, 1928, Educ. Suppl. to *The Times*, and Aug. 31, 1928, p. 13.

‡ *Engineering*, p. 766, Dec. 21, 1928. See also T. Spooner and J. P Foltz, *Amer. Inst. Elec. Eng. Journ.*, **48**, p. 199, March 1929.

silent when running at very high speeds. It is manifestly impossible to deal, in a limited space, with more than a small fraction of the possible *causes of noise in machinery*,* but we may mention three of the more important types:

(a) *Lack of Balance in Rotating Parts* – A wheel or a cam which is dynamically out of balance and is rotating at a high speed produces an unbalanced reaction on its bearings once per revolution. This force of reaction is communicated to the base of the machine, or the body of the vehicle, and sets it in vibration. At certain speeds the frequency of the force may coincide with one or more of the natural frequencies of some part of the machine or its mounting. Large amplitude vibrations are then set up, and the sounds emitted may reach serious proportions. The remedy in such cases is obvious.

(b) *Shock Excitation* – Sudden accelerations or retardations of high-speed parts of a machine produce corresponding reactions on the framework. The impulsive forces thus generated are ideal for setting in vibration resonant structures, such as rods, plates, or tubes. Unless all such structures are carefully damped the noise may become excessive. A badly sprung vehicle travelling over an imperfect road is subjected to such sudden accelerations, which 'jolt' all loose parts, setting them in vibration and transmitting the impulses to other parts of the structure which may resonate at audible frequencies. The noise from a tramcar † or an improperly tended motor lorry with solid tyres is an extreme example of this. In a good engine or machine the only noises of this character are the high-frequency sounds resulting from the impact of valves on their seatings. Such noises can be reduced by screening.

A problem involving very low frequencies arises in the case of vibrations set up in railway bridges when heavy engines pass over them at various speeds. The unbalanced forces due to revolving and reciprocating masses in the engine impart shocks to the bridge. If these shocks are repeated at the appropriate time-intervals, *i.e.* at a certain speed of the engine, serious resonant vibration may be set up in the bridge, which may result in serious overstrain or damage to the structure.‡ In order to investigate the phenomena and to determine the fundamental

* See J. H. Hyde and H. P. Lintern, *Proc. Inst. Civil Engs.*, Jan. 1929.

† See 'Reduction of Noise in Tramcars,' G. H. Fletcher (*Elect. Review*, **105**, p. 90, July 12 and 19, 1929).

‡ Sir A. Ewing, 'Vibration of Railway Bridges,' *Roy. Soc. Arts*, **77**, May 1929. See also *Engineering*, pp. 143 and 187, Feb. 1929.

frequency of vibration of the bridge a special 'bridge-oscillator' was designed. This consisted of a heavy truck on which was mounted motors driving a pair of geared axles running in opposite directions and carrying heavy weights which could be varied according to the impulse it was desired to impart to the bridge at each revolution of the shaft. By varying the speed of the motor the frequency of the impulses could be adjusted to produce 'resonance' in the bridge. The oscillations of the bridge were recorded by means of the Collins' recording accelerometer (Camb. Sci. Inst. Co.). The resonant frequencies of numerous bridges were determined in this way, values lying between 2 and 12 p.p.s. being found for bridges varying in span from 40 to 350 feet.

The low-frequency vibrations of the hulls of ships have been studied by A. D. Browne, E. B. Moullin, and G. M'Leod Paterson* in conjunction with the Cunard Steamship Co. Damped harmonic vibrations are set up in the ship's plating due to the impact of a wave on the hull. The relation between the amplitude of vibration of the deck and the rate of revolution of the engines was also determined, recording seismographs being used to measure the amplitude and frequency of the vibrations. Resonance effects were observed at certain engine speeds. Fluctuations of propeller-torque, causing vibration of the hull, were recorded when the ship was pitching in a heavy sea.†

(c) *Explosive Noises* – The exhaust from the cylinders of oil-driven engines is a frequent cause of irritating noise. The gas from the cylinders is expelled at high pressure into a medium at low pressure. As it leaves the ejection pipe large-amplitude oscillations follow the main explosion pulse. The only remedy for this is an efficient 'silencer' or 'pressure reducer.' A large proportion of the noise from an aeroplane engine is due to this cause, and, in the case of civil aircraft, efforts are being made to reduce it. Motor bicycles are the most frequent offenders on the roads.

On a railway, noise proclaims defects in the track and looseness in the vehicles; on the track it tells the civil engineer that the road bed is uneven, and the mechanical engineer that the gears are not accurate and that the body and chassis are unsatisfactory. As an example of the research that is being undertaken to reduce such noises may be quoted the case of the underground tube railways in London.‡ These railways have a much more difficult noise-problem to solve than the surface railways, for the noise

* See *Engineering*, p. 650, May 24, 1929.

† *Loc. cit.*

‡ See *The Engineer*, Oct. 12, 1928.

made by the wheels and coaches as they roll on the steel rails is confined by the steel walls of the tube, and the sound reverberates in the tube with very little diminution of intensity. A coach which would be relatively silent on a surface railway might therefore be extremely noisy in a tube. Analysis of the noises on the underground railways * has led to further experiments to make the coaches sound-proof and to 'silence' the tube track. Rapid-motion photographs show that the wheels of a train are not in continuous contact with the rails—that is, they do not roll along the rails smoothly, but proceed in rapid jumps. The impact of the wheels as they jump on the steel rails produces noise of great intensity. Attempts to reduce this noise by shrouding the wheels proved unsatisfactory, however, but lagging the walls and roof with sound-absorbent material, *e.g.* asbestos quilting, led to a little improvement. The noise appears to increase at a rate considerably greater than the increase in speed of the train. Experiments have also been made in connection with the track, the space between the sleepers being filled with ballast to absorb and damp out the vibrations. A section of the Piccadilly Tube treated in this way showed some improvement.

The noise problem is also a very important one in connection with the development of civil aircraft. In fact, the noise at present experienced in the cabins of aeroplanes is sufficiently serious to be regarded as a hindrance to commercial development. The question has been studied by W. S. Tucker,† who has analysed the sources and principal characteristics of such noises. One remedy is to line the cabins with sound-absorbent material (*e.g.* balsam wool); this reduces the sound to about $\frac{1}{20}$th of that in a non-absorbent cabin. Hitherto little or no improvement has been possible in regard to the reduction of the propeller and exhaust noise. It is important to note, however, that such questions are now being regarded seriously, and that public interest has been aroused in the general problem of reduction of traffic noise.

The New Motor Car (Excessive Noise) Regulations, dated June 3, 1929, make it an offence for any person to use a motor vehicle which causes excessive noise as a result of any defect in design or construction, or lack of repair or faulty adjustment. It is also an offence under the regulations to use a vehicle which makes an excessive noise due to faulty packing or adjustment of the load. 'Excessive noise' is, however, a term which is very difficult to define.

* See *The Engineer*, Oct. 12, 1928. † *Proc. Roy. Aero. Soc.*, Jan. 19, 1928.

(2) Sound Signalling – Frequent reference has been made in the text to sound-signalling devices for use in air and in the sea. In the section on sound sources we have dealt with various forms of siren and the diaphone for use in air, and have mentioned the Fessenden oscillator and the Signal Gesellschaft electromagnetic transmitter for use in water. Various directional sound transmitters have also been described. In the section on the transmission of sound the important questions of attenuation and refraction have received particular attention in relation to sound signalling.

If the reader requires further information the following bibliography may prove useful:

Signalling in Air – Various published papers by L. V. King, W. S. Tucker, and E. T. Paris; *The Principles of Sound-Signalling*, by Hart and Smith.

Signalling in Water – Various published papers by R. A. Fessenden, Hahnemann, Hecht, and Lichte; *Unterwasserschall-technik*, by F. Aigner (Berlin, 1922); publications by the Submarine Signalling Co. (Friar's House, London) and Hydrographic Bureau, Monaco.

(3) Supersonics – The piezo-electric quartz oscillator as a source of high-frequency inaudible vibrations has been described (see pp. 140 *et seq.*). The technical applications of such oscillators (*a*) in the Langevin depth-sounding apparatus, (*b*) as a means of standardising the frequency of electrical circuits, and (*c*) as a tool in physical and biological investigations have been mentioned. Further reference should be made to the original papers.

(4) Musical Instruments – An adequate treatment of the technical applications of sound to the design and construction of musical instruments would require a volume in itself. On the grounds that anything short of such a detailed treatment would fail in its purpose, the writer has refrained in the text from entering into particular descriptions of musical instruments. Reference has been made to them only in so far as they serve to illustrate points of definite scientific interest raised in the text. It must not be supposed, on this account, that the writer has neglected so important a branch of the subject, but rather that he considers that the reader should consult specialised treatises. For example, D. C. Millar's *Science of Musical Sounds*, E. H. Barton's article in *The Dictionary of Applied Physics*, and E. G. Richardson *The Acoustics of Orchestral Instruments and of the Organ*, contain much valuable information.

APPENDICES

I.—TABLES OF VELOCITY (c) AND ACOUSTIC RESISTANCE (ϱc) (Calculated)

(a) *Solids*

Solid	Density	Bulk Modulus	Young's Modulus	Rigidity	Poisson's Ratio	$(k+\frac{4}{3}.\mu)$	Velocity (calculated) cm./sec. Longitudinal bar			Acoustic Resistance
							Longitudinal bar	Bulk	Torsion	
	ϱ	k	E	μ	σ		c_1	c_2	c_3	ϱc_1
		$\times 10^{11}$	$\times 10^{11}$	$\times 10^{11}$		$\times 10^{11}$	$\times 10^{5}$	$\times 10^{5}$	$\times 10^{5}$	$\times 10^{6}$
Alm	2·65	7·46	7·05	2·63	0·32	10·9	5·25	6·4	3·16	1·39
Copper	8·93	13·1	12·3	4·55	·35	19·1	3·58	4·6	2·26	3·20
Iron (cast)	7·7	9·6	11·5	4·4	·27	15·6	3·85	4·5	2·4	2·96
Steel	7·8	18·1	20·0	8·4	·28	29·3	5·05	6·1	3·3	3·93
Lead	11·4	5·0	1·62	0·56	·45	5·7	1·2	2·2	0·7	1·37
Nickel	8·9	17·6	20·2	7·7	·31	27·9	4·76	5·6	2·94	4·23
Sier	10·5	10·9	7·9	2·9	·38	14·7	2·74	3·7	1·66	2·88
Tin	7·3	5·29	5·4	2·0	·33	8·0	2·72	3·3	1·66	1·96
Brass	8·5	10·6	10	3·5	·37	15·2	3·42	4·25	2·0	2·90
Glass	2·5 to 5·9	3·5 to 6·0	5 to 8	2 to 3	0·2 to 0·3	6 to 8·5	4·5 to 5·6	4·9 to 5·8	3·1	1·12 to 1·4
Quartz	2·65	..	10·3	..	..	..	6·25	..	..	1·65
	2·65	..	7·85	..	..	..	5·45	..	..	1·44
Ivory	1·85	..	0·90	..	..	..	2·2	..	..	0·41
Pitch	1·1	..	0·051	..	..	..	0·68	..	..	0·075
Paraffin-wax	0·9	..	0·192	..	..	..	1·46	..	..	0·13
Rubber	0·95	..	0·0001	..	0·5 app.	..	0·03	..	..	0·003
Oak	0·8	..	1·3	..	..	..	4·05	..	..	0·32
Ice	0·916	..	0·936	..	..	..	3·2	..	..	..

(b) *Liquids* (at 290° abs. temperature T)

Liquid	Density ϱ	Volume of Unit Mass ν	Elasticity (iso-thermal) κ_θ	Specific Heat C_p. Calo-ries	Specific Heat C_p. Ergs	Coeffi-cient of Cubic Expan-sion α	Ratio of Specific Heats $\gamma = \frac{1}{1-\alpha^2\kappa_\theta\nu T/c_p}$.	Velocity c (cal-culated) cm./sec.	Acoustic Resist-ance ϱc
			$\times 10^{11}$		$\times 10^{7}$	$\times 10^{-3}$		$\times 10^{5}$	$\times 10^{5}$
Alcohol	0·79	1·27	0·132	0·59	2·46	1·10	1·31	1·44	1·14
Ether	0·74	1·35	0·069	0·56	2·34	1·63	1·46	1·17	0·86
Turpentine . . .	0·87	1·15	0·128	0·42	1·75	0·94	1·27	1·36	1·18
Pentane	0·7	1·4	0·032	0·45	1·9	1·59	1·21	0·75	0·525
Water (fresh) . . .	0·999	1·001	0·204	1·001	4·18	0·17	1·004	1·43	1·43
Water (sea, 3·5 per cent. salinity) . . .	1·0255	0·975	0·232	0·934	3·91	0·23	1·0090	1·51	1·55
Mercury . . .	13·6	0·0735	2·56	0·033	0·138	0·18	1·13	1·46	19·8

(c) *Gases* [at 0° C., $1{\cdot}013 \times 10^6$ dynes/cm.^{2} (760 mm. mercury when g=980·62)]

Gas	Density ϱ	Ratio of Specific Heats γ	Velocity c (calculated) cm./sec.	Acoustic Resistance ϱc
	$\times 10^{-3}$		$\times 10^{5}$	
Air . .	1·293	1·402	0·331	43
Oxygen .	1·429	1·400	0·315	45
Hydrogen .	0·090	1·419	1·263	11
CO .	1·250	1·401	0·336	42
CO_2 . .	1·977	1·300	0·258	51

II.—TABLES OF VELOCITY (Observed)

(a) *Solids* (Longitudinal Vibration of Rods)

	$\times 10^5$ cm./sec.		$\times 10^5$ cm./sec.
Aluminium	5·1	Brass	3·6
Copper	3·9	Glass	4 to 5·3
Iron (cast)	4·7	Quartz	5·50
Steel	5·1	Oak	4·2
Lead	1·2 to 1·35	Cork	0·43 to 0·53
Nickel	4·97	Ivory	3·01
Magnesium	4·60	Paraffin-wax	1·4
Silver	2·6	Rubber	0·03
Tin	2·5	Pitch	0·68
Ice	3·2		

(b) *Liquids*

Liquid	Temperature ° C.	Velocity cm./sec.	Method	Observer
Water (fresh)	8·1	$1·435 \times 10^5$	Bulk (Lake Geneva)	Colladon and Sturm, 1827.
Water (distilled)	13	1·441	..	T. Martini, 1908.
" "	19	1·461	..	
" "	31	1·505	..	
" "	5	1·439	High frequency (3×10^5 p.p.s.) quartz oscillator in tube	R. W. Wood, Hubbard and Loomis, 1927.
" "	15	1·477		
" "	25	1·509		
" "	35	1·539		
Water (2·5% salt solution).	15	1·510	..	
Sea water ($\Delta = 1·026$)	15	1·5041	Open sea	Marti, 1919.
Sea water (3·5% salinity).	6·0	1·4740 ± ·0006	Open sea	A. B. Wood, Browne and Cochrane, 1920–23.
	7·0	1·4773		
	16·95	1·5104 ± ·0003		
Mercury	25	1·469	Quartz oscillator (3×10^5 p.p.s.) in tube	R. W. Wood, Hubbard and Loomis, 1927.
"	35	1·468		
Chloroform	5	1·066		
"	15	1·027		
"	15	0·983	..	Dorsing, 1908 (see Geiger and Scheele, *Handbuch der Physik*, 8).
Alcohol, Ethyl	12·5	1·241	..	
Ether, Ethyl	0	1·145	..	
Turpentine	15	1·326	..	
"	3·5	1·371	..	Martini, 1888.
Petrol	7·4	1·395	..	Martini, 1896.
Benzene	17	1·170	..	Schmidt, 1905.

(c) *Gases*

Gas (at 15° C.)	Velocity cm./sec.	Observer
Air (dry)	$0·3398 \pm 1 \times 10^5$	Esclangon, 1917–1919.
" (in the open)	$0·3308 \pm 1$	Angerer and Ladenberg, 1922.
" "	0·3313	Hebb, 1919.
" (in tube)	0·3319	G. W. Pierce, 1925.
CO_2	0·2585	G. W. Pierce, 1925.
Hydrogen	1·261	
CO	0·337	
Oxygen	0·317	Dulong, 1829.
Water-vapour	0·401	Jaeger.
Argon	0·308	
Helium	0·971	
Coal gas	0·49 to 0·51	

INDEX

The numbers refer to pages. Names of persons are printed in italics.